# ICTs for Modern Educational and Instructional Advancement:
## New Approaches to Teaching

Lawrence Tomei
*Robert Morris University, USA*

Information Science REFERENCE

**INFORMATION SCIENCE REFERENCE**
Hershey · New York

| | |
|---|---|
| Director of Editorial Content: | Kristin Klinger |
| Senior Managing Editor: | Jamie Snavely |
| Assistant Managing Editor: | Michael Brehm |
| Publishing Assistant: | Sean Woznicki |
| Typesetter: | Michael Brehm, Michael Killian |
| Cover Design: | Lisa Tosheff |
| Printed at: | Yurchak Printing Inc. |

Published in the United States of America by
Information Science Reference (an imprint of IGI Global)
701 E. Chocolate Avenue
Hershey PA 17033
Tel: 717-533-8845
Fax: 717-533-8661
E-mail: cust@igi-global.com
Web site: http://www.igi-global.com/reference



Library of Congress Cataloging-in-Publication Data

ICTs for modern educational and instructional advancement : new approaches to teaching / Lawrence Tomei, editor.
p. cm.
Includes bibliographical references and index.
Summary: "This book explores current models and issues involved with online course development, assessment, and blended learning"--Provided by publisher.
ISBN 978-1-60566-936-6 (hardcover) -- ISBN 978-1-60566-937-3 (ebook) 1. Blended learning. 2. Education, Higher--Computer-assisted instruction. 3. Employees--Training of--Computer-assisted instruction. I. Tomei, Lawrence A.
LB1028.5.I1168 2010
378.1'7344678--dc22

2009035436

British Cataloguing in Publication Data
A Cataloguing in Publication record for this book is available from the British Library.

## Editorial Advisory Board

# Table of Contents

# Detailed Table of Contents



In a recent dissertation study, research was conducted to evaluate online instructors' characteristics and preferences concerning the use of a telementor or online instructor's assistant as a part on an online course. Those who participated in the anonymous survey came from a sample of two thousand online instructors from colleges and universities located across the United States. Of those contacted, 323 online instructors responded to the survey. Results presented in this chapter were produced using data from nine of the questions included in the survey. These Likert Scale questions specifically asked the instructors about their use of Theory of Multiple Representation, Gagne's Conditions of Learning, Instructional Transaction Theory, Cognitive Flexibility Theory, Three Form Theory, Dual-Coding Theory, Elaboration Theory, Theory of Transactional Distance, and Theory of Immediacy and Social Presence. Outcomes showed that a larger number of online instructors applied design theory when creating a course compared to the instructors who indicated that they did not apply design theory. Descriptive results presented illustrate how often the participants said that they utilized each of the different theories.



Differing methods of course development can lead to widely varying results. The University of Phoenix develops courses for both on-campus and on-line (e-learning) delivery, using electronic collaboration as well as in-person teamings. Course developers at the University rigorously measure feedback about course materials, and revise courses based on learners' input. This chapter describes a model for developing and delivering e-learning doctoral-level curricula based on current research and a learner needs analyses. Suggestions for further improvements and surprising results about the most effective method for deriving e-learning materials are explored.

This chapter proposes a multi-disciplinary strategy for identifying affective usability design aspects in educational geosimulation systems. It is based on the association of these aspects with an architecture that defines the basic components of a geosimulation system as well as the learning strategies used in this context. The authors' goal is to provide design strategies that might elicit positive emotional responses from the students in learning experiences. The chapter illustrates how these strategies have been used in a learning system by evaluating the students' emotional responses evoked during their interaction with the system.



When introducing the metadata standard LOM, objectives such as the ability to find or to reuse learning objects were followed. These objectives are actually achieved in LOM to a limited degree only, despite the designation as de-facto standard for description of electronic learning content. Based on the complexity of the standard, a high theoretical potential faces rejection in practice. One reason for this is that the process of metadata generation—for example, who creates which metadata attributes—is not defined in detail yet. This chapter illustrates an approach which guarantees a high quantity as well as a high quality of learning object metadata records, bringing together known ways of metadata creation and the new paradigm of users describing content as implemented in recent Web 2.0 applications. In the context of a concrete e-learning platform, the authors exemplarily illustrate who creates which metadata records of LOM in which way at what time. Finally, the authors show why this approach of creating metadata matters as they measure their metadata quality and compare it with other's findings.



This chapter includes practical and accessible overviews of some of the most commonly used and most useful technologies. The chapter serves as an idea generator, especially for teachers looking for ways to update their courses or to explore new concepts in learning. Technologies once only imagined are now opportunities to be implemented in the classroom. Audio and video conferencing, blogs, podcasts, RSS feeds, social bookmarking, and wikispaces are popular means of communicating in today's society. However, Web technology is developing at such an exponential rate that even the newest of these technologies, Web 2.0, may one day soon be a footnote in computer history. Once these newer technologies are better understood and appreciated, educators can evolve their teaching strategies to help their students remain competitive in the global society.



Advances in communication technologies, such as widespread use of the Internet, have opened new avenues for continuing higher education. These advances have allowed educators to provide for and satisfy individual variations in learning. Generally, online courses are adaptations of traditional courses; some courses are more suitable than others for such online instruction. As the trend to offer online courses continues, understanding the factors that lead to students' preference can be useful. Online courses can be costly to develop and to implement, and inappropriately designating courses for online participation can lead to lower student retention rates. This study focuses on students' perceptions of online courses. The results identify issues that affect students' perceptions, and this study concludes by suggesting ways for universities to design online programs that better suit the desires of students.



This chapter reports on action research (AR) that implements online learning community (OLC) software to foster conversation and community at a specific graduate school. Informed by theories of conversation, online learning, and social networking, the authors incorporate Web 2.0 technologies in the creation of a user-centric OLC. A distinguishing feature of the authors' software is that, rather than being centered on courses like traditional course management software (CMS), the authors' software is oriented towards and controlled by individuals. Results indicate that stakeholders—graduate students and faculty—appreciate and find value in the OLC that was implemented.



This chapter reports on two follow-up studies to "A Comparative Analysis of Online and Traditional Undergraduate Business Law Classes" (Shelley, Swartz and Cole, 2007) designed to further examine two critical areas of e-learning, that is, student satisfaction with, and student learning in, an online environment as compared with an onland, or traditional classroom environment. While the initial study found no significant difference between the two, the second study did find statistically significant differences

technical terms referred to in the research literature. The chapter then goes on to provide a detailed description of the study set-up and presents the main findings obtained from this study. The results are then discussed in relation to previous findings in the research literature



This research examines the relative impact of student's in-class behaviors, (i.e. attendance and participation) by assessing student perceptions on the value instructional technologies, such as eCollege management systems and instructors' PowerPoint presentations. The results of the study through exploratory factor analyses revealed that 13 items were divided into three factors (electronic presentations, online-course management, and effective classroom behavior) with 53% explained variance in instructional technologies' impact on student learning. ANOVA results indicated significant differences in online-course management and perceived impact of electronic presentations on students' classroom behavior among respondents who used the online-course management system. Respondents who used multiple online-course management features viewed it more favorably and did not believe that it had a negative impact on classroom behaviors, such as attendance or class participation compared to those who used fewer features. Implications for construct refinement and future research are discussed.



A platform-independent Java Web application named TSI (Teacher-Student Interaction) that supports communication between an instructor, teaching assistants and students in a traditional on-campus course is presented in this chapter. Using the TSI, the instructor and teaching assistants can handle most of the routine work: upload student personal information, send students personal emails, etc. The system can easily be installed and administered individually by an instructor inexperienced in computers. It is as simple as a pen for students. Students can check their personal data (scores and comments), download educational materials, etc. As part of the TSI, a VBA application is used to analyze the course log files. This tool is helpful in understanding individual and group students' behaviors. The TSI was successfully tested during six years at the University of Aizu (Japan) in an environment where English is one of the working languages and both students and professors are non-native speakers of English.

The impact of examination software on student attitudes was investigated. The Technology Acceptance Model (TAM) provides the theoretical foundations for studying the completion of examination on student laptop computers. The model applies TAM to link both faculty and technical support for the examination software to student attitudes towards the software, while it is mediated by the perceived usefulness and ease of use of the software. The model is empirically tested using questionnaire responses from 107 students enrolled in sections of a business core course using the examination software. The statistical technique used is structural equations modeling. Empirical results show that perceived usefulness and ease of use of the software have direct, meaningful impacts on developing positive attitudes toward the software. Furthermore, faculty support and ease of system use impact student attitudes in a meaningful fashion indirectly through perceived usefulness. These empirical results are discussed and implications for instructors are offered.



This chapter describes the development of an interactive PowerPoint module for use in an introductory accounting course in a business school. This use of PowerPoint gives students the ability to access additional information and provides students with immediate, appropriate feedback with explanatory details. This tool is designed to be used outside of the classroom at the student's own pace and can be used in disciplines other than accounting.



This chapter seeks to address three points. First, the authors explain the significant increase in school use of collaborative software resulting from four key drivers: the speed at which social software has been embraced by young people outside school and its adoption by educators in more formal school settings; the push to develop "knowledge construction skills" that are relevant to a knowledge-based economy; exposure of more pupils to access a wider curriculum; and, the promotion of intercultural education through citizenship. Second, the chapter considers the research that sustains the importance of basing inter-school work on theoretical models of learning and contact; the most frequently used learning models derive from the application of "communities of practice" and "knowledge-building networks." Third, and finally, the chapter discusses the implications of collaborative software and the theoretical models of learning presented in terms of pedagogy and learning outcomes and offers comments on the potentially disruptive impact of this approach on learning.

Pennsylvania is at the forefront of the public cyber charter schooling movement in America. As more and more students elect to transfer from traditional public schools into cyber charter schools – and their districts of origin are forced to forfeit their tuition allocations – a need for a public school alternative to cyber charter schools has emerged. Using current practices in Pennsylvania's public schools as a backdrop, this chapter presents a new model for district-level cyber schooling called the holistic model for blended learning that public schools in Pennsylvania (and elsewhere) can use to compete with cyber charter schools and meet the growing demand for K-12 online learning.



This chapter identifies the potential risks associated with primary school children using the Internet. This chapter evaluates the level of understanding of Internet safety that children have. This chapter identifies what children use the Internet for and what information that they are willing to reveal while online. The findings of this chapter are based on analysis of a survey carried out on 645 Irish primary school children in February 2007.



This research examines the impact of education on the ethical decision-making outcomes of adult learners in the area of information technology (IT). This study sheds light on the research question "Does IT ethics education make a difference?" and more specifically, "Do ethics courses influence decisions regarding IT ethical issues in adult learners?" In a field study of 78 pre and post-test surveys, the authors found that graduate students who took a course in IT ethics made different decisions than those made at the start of the term, for two of six ethical issues. The ethical issues described in this paper are particularly relevant in today's knowledge economy. Implications for IT ethics education and future research in the area are discussed.



It is difficult to understand students' social practices from artifacts of anonymous online postings. The analysis of text genres and discursive types of online postings has potential for enhancing teaching and learning experiences of students. This chapter focuses on analysis of students' anonymous online postings using Critical Discourse Analysis (CDA). The chapter argues that social practices reproduce during online interaction and artifacts embody such reproduction. A study involving over 300 commerce students at a higher education institution (HEI) used a special purpose anonymous online consultation tool, the Dynamic Frequently Asked Questions (DFAQ) and social practices embodied in the artifacts is analyzed using CDA. The analysis used the three dimensions of CDA i.e. description (text genres), interpretation (discursive type) and explanation (social practice) and insights into students' social practices were inferred. The chapter concludes that CDA of anonymous postings provided insight into social practices of students, in particular highlighted the tension between perceptions of inflexibility of traditional teaching practices and student demands for flexible learning. Finally, CDA as described in this chapter could be useful in analyzing email communication, short message service (SMS) interaction, webblog and podcasts.



Traditional forms of assessment used in face-to-face and distance learning education are insufficient to ascertain the increase of the knowledge acquired and the learners progress, therefore do not provide enough information to detect their learning gaps necessary to improve their competencies . Another point is that traditional assessment ways rarely involve the student in monitoring his own learning through his metacognitive abilities. Nowadays, professional skills to obtain a working position changes at the same velocity than the increase of knowledge and have to be considered by any professional and/or student to be qualified for a new job. This chapter presents a model for formative assessment and certification in Lifelong Learning based on cognitive and metacognitive measurements that will make possible the identification of the professional learning gaps showing a roadmap to obtain educational and conceptual certification for his/her competence. Moreover, it presents the architecture of a computational environment for student knowledge mapping that will allow identifying more specifically the learning gaps in order to supply the educational system with qualitative information.



The author of this chapter argues that although university lecturers delve into the 'shallow waters' of e-learning they do not do so in sufficient depth and resign themselves to the perpetuation of cognitivist, behaviourist and objectivist forms of knowledge without discovering more about the medium that could possibly liberate their restricted epistemologies. This chapter explores possible reasons for varying engagement with e-learning, assuming that these reasons are located within the dimensions of the unit of analysis of the study, namely, lecturers' changing theories of knowledge and teaching in first encounters with e-learning. Using Lee Shulman's table of learning (Shulman, 2002) as a heuristic, the author uses excerpts from personal narratives to highlight the epistemological and pedagogical transformation of nine lecturers as they engage with educational technologies in their work.

**Special Section**
**A Focused Discussion on Educational Technologies**

*This section revisits a 2007 special issue of the International Journal of Information and Communication Technology Education, which was edited by Dr. Bruce Howard. Following an introduction by Dr. Howard and Dr. Lawrence Tomei, editor of this collection, the next six chapters explore six key elements of the authors' examination of the classroom of the future. Particular emphasis on implications for science education is provided.*



This submission provides an overview of the final six chapters of the publication. Within these chapters, the authors describe their research on choosing and applying emerging educational technologies in the light of what they know about best practice teaching methods. Whereas many well-respected experts have addressed the need for new methodologies, the aim of this research is to focus on the process of choosing the technologies themselves. The authors set out to determine how to evaluate the individual promise an educational technology may hold and to provide guidelines to those who choose and use the technologies for teaching and learning.



The authors of this chapter reviewed published results from the last 15 years to compile a list of the characteristics of effective educational uses of technology. All the studies considered technical, administrative, and learning features, while more recent investigations emphasized administrative characteristics necessary to support No Child Left Behind reporting. Recommended characteristics have evolved over time as expectations for technology integration have shifted from a focus on technology skill development to integrated use of technology as part of effective teaching and learning practices. Technology literacy is now considered as an integrated component of curriculum support and professional development. A timeline of relevant historical milestones in the evaluation of educational technologies illustrates how the understanding of and expectations for effective use of educational technologies has progressed to keep pace with advances in technological affordances.

# Preface

## BLENDED LEARNING: THE OVERARCHING LEARNING STRATEGY

Blended learning is an instructional strategy that combines classroom instruction with technology-rich resources to enhance learning. It is realized by a team of instructional technology experts devoted to studying the effects that technology-based learning delivery environments have upon learning and educational institutions. Blended learning takes place in a traditional classroom directed by conventional teachers who interact with students face-to-face while controlling the content and pace of the lesson – they simply do it with technology tools.

If an observer were to look into a blended learning classroom from a secondary school hallway window for example, she would most likely observe students who were being taught by direct instruction (i.e., lecture or lecture-discussion) under the direction of a teacher using visual resources, an LCD projector and screen. During the same lesson, students would be noted participating in a web-based virtual tour or working together in small groups to enhance their social skills. Online instruction would be employed to appeal to those students who can work independently either to gain valuable remedial teaching opportunities for a learning objective missed or explore additional enhanced content while the teacher addresses other instructional tasks with other students.

Blended learning gives students the opportunity to receive personal attention while retaining the much-needed (for some) control provided by the face to face classroom environment. Simultaneously, students are building independence through learning with technology.

This text, ***ICTs for Modern Educational and Instructional Advancement: New Approaches to Teaching***, examines some of the instructional challenges that have inspired faculty improvements in lesson delivery using both new and familiar technologies. In some cases, the articles presented herein discuss classroom applications that encourage new kinds of learning experiences that would not, otherwise be possible. Some of the common themes addressed in this text include:

- Anonymous online postings to increase student engagement
- Internet Usage and the associated safety implications
- Exemplary educational technologies
- Design principles for 21st-century educational technology
- Evaluating educational technologies
    - Bringing a large and varied quantity of multimedia into classroom presentations
    - Creating a content collection (multimedia) that can be searched and expanded
- Use Learning Management Systems in relation to students' attitude and performance to enable students to share their analyses and reflections
- Online learning communities: connecting students with experts to encourage critical thinking

- Emerging educational technologies
- Classroom of the future

## INTRODUCTION

Despite common misconceptions and the time-honored views of some, teacher-directed learning can successfully incorporate technology. Self-directed learning can occur using the Internet. A blended learning approach seeks technologies to make face-to-face learning more effective by removing the mundane aspects of basic instruction and supplanting them with a technology-rich learning environment that is more engaging and interactive.

When students meet in the classroom, face-to-face instruction focuses on higher-level skills since technology has delivered and evaluated learner competency during presentation of most of the required basics. Instructor-led sessions can focus on knowledge transfer and application and not simply the memorization of facts and jargon.

Blended learning represents the best of both worlds: traditional in-class presentations coupled with effective technology-based learning. The multiple modalities work well together to complement the lesson and course content so that students are able to cover the necessary subject matter even with limited classroom time. For example, one lesson plan offered by a cyber charter school calls for students taking a blended learning biology class to attend a face-to-face instructional presentation to learn the basics of Earth's soil. Following the in-class presentation, students complete the majority of their class work by accessing course materials hosted by a popular learning management system. Each of the three component lessons includes part of a comprehensive online video provided by the renowned Annenburg Media (www.learner.org). The video entitled, Session 1. Earth's Solid Membrane: Soil, discusses how soil began to appear on the Earth, how it is formed, its role in certain Earth processes, its composition and structure, and its place in the structure of the Earth.

Learners are asked to instant message or email their instructor with any questions. As the lesson progresses, instruction is enhanced by in-class labs, providing the opportunity to do hands-on experiments under teacher supervision as well as web-based virtual tours of Earth's top three levels (core, mantle, and crust). Finally, students submit an electronic portfolio of images and text that they gleaned from the resources provided or found on the Internet.

Blended learning exposes students to a host of valuable tools not afforded the traditional in-class learner. Working online encourages students to improve their technology skills including computers, electronic mail, preparing electronic documents, and more. With a blended learning curriculum, students use technology to participate in lessons, take quizzes and exams, and communicate with teachers and other students. By doing so, they learn not only course material but also valuable real-world skills necessary in the knowledge-based world they will encounter in higher grades, college, and the work environment.

At the same time, schools save substantial financial investments often at a much low per-pupil cost. Greatly advanced from just 15 years ago, blended learning lessons require schools to access shared technology (e.g., desktop and laptop computers and Internet connections) that were once very expensive and have now been relegated to the status of commodity items for most schools and many parents.

Greater effectiveness, multiple modalities of learning, and advancement of individual technology skills are just a few of the ways in which blended learning contributes to the bank of instructional teaching strategies in today's classroom.

Let's look at some of the most popular blended learning technologies for the classroom – many of which are addressed in one or more of the chapters that follow. The technologies will be divided into asynchronous and synchronous categories to better discuss their advantages and limitations with respect to blended learning applications.

## BLENDED LEARNING TECHNOLOGIES

### Asynchronous Learning

Asynchronous teaching and learning involves the separation of instructor and student – in both physical and chronological perspectives. Often, asynchronous teaching involves lessons that are delivered online and do not require live attendance. Collaboration and engagement happen over time with the direct application of specific technologies. Online course environments, in particular, lend themselves to teachers and students communicating with one another anytime and anywhere.

There are several advantages to asynchronous delivery systems. For example, they are typically more flexible, allowing continuous access to teaching material without interruption or delay. Asynchronous materials give students time to reflect on content and formulate responses rather than the typical classroom expectation that requires immediate reaction to instructor questions or peer-initiated comments. They allow the learner to ponder new ideas and concepts, locate additional references, and prepare a more scholarly reply. If the online session is captured and archived, students have the additional capability to review the lesson either for content or to prepare for an assessment.

Asynchronous technologies allow access from home or work (aka situated learning). Since a growing number of students fit the characteristics of the "non-traditional learner," it seems reasonable that the delivery methods employed to teach these courses offer more options. Finally, asynchronous teaching is oftentimes more cost effective requiring less bandwidth and lower-end technologies to operate than was the case a mere 15 years ago. Issues of accessibility once restricted blended learning opportunities to only the wealthiest of schools and student populations; thankfully, that is no longer the case for many communities.

The primary drawback of asynchronous technologies is that they require some discipline to use in a community of practice. For example, participants in an online discussion board must take the initiative to join the session periodically to pick up the newly posted messages and respond in a timely manner. To many learners, asynchronous learning may feel "impersonal" to those who prefer the more personal synchronous technologies.

### Asynchronous Technologies

Asynchronous tools enable communication and collaboration over a period of time through a "different time-different place" mode. These tools allow learners to share the educational experience at their own convenience and according to their own schedule. Asynchronous tools are useful for sustaining dialogue and collaboration over a period of time and providing people with resources and information that are instantly accessible regardless of time. Asynchronous tools can engage learners from multiple time zones while chronicling the exchanges of a group for later examination either by other students or by the instructor.

Here are some of the most popular asynchronous technologies suitable for infusion in a blended learning curriculum.

***Email and Instant Messaging***. The two most successful asynchronous tools, by far, would be email and instant messaging. Email, e-mail or electronic mail is the transmission of messages (emails or email messages) over a network (most commonly these days, that would be the Internet). Instant messaging, abbreviated IM, is a communications service that offers an asynchronous version of a "chat room" (more on chat rooms in the synchronous portion of this discussion), Typically, IM is conducted with one other individual at a time and can be readily compared to a text-based telephone conversation using text rather than voice.

Instant messaging is more interactive than e-mail because messages are sent immediately. However, e-mail messages can be more extensive and sent to more recipients. Email options are more elaborate (e.g., page layout, attachments, etc.) while the IM message is short (often limited to 128 characters or less).

Since both email and IM are among the most ubiquitous technologies on the market and most educators are intimately familiar with both, this discussion will move quickly to the practical applications of these technologies for blended learning.

***Discussion boards***. A discussion board (aka electronic discussion group, digital discussion forum, online message board, and online forum) is a general term for any online bulletin board that offers the user the opportunity to post messages, track read/unread message, and reply to messages. Discussion board messages are usually sorted within categories, topics, or themes chosen by the host or monitor in a threaded or straight-line format.

The flat format displays messages in a strictly chronological order. Someone joining the conversation will see new messages appear at the end of the discussion thread regardless of which message is receiving the reply. The conversation may be more difficult to follow and the context of the discussion may require the user to scroll up and down the postings to find those directly related to the topic at hand, but the flat format provides a more suitable chronology of posts and responses when dealing with a single-focus topic.

The threaded format displays posts in a logical, conversational order similar to an outline with bullets and sub-bullets. A response (reply) is indented under the initial message making it clear which posting pertains to which message. The discussion flows clearly from one message to the next. For shorter messages, the entire message can be shown in the subject line, making conversations even easier to follow and saving on bandwidth since every message does not have to be opened to follow the conversation.

Blended learning uses discussion boards to help their learners master the complex skills of asynchronous collaboration. Bruck (2005) has identified *A Five-Step Model for High Impact Learning* that illustrates how this model fosters the communication of content, questions and answers, skills practice, apprenticing/ coaching, and teaching.

***Web logs (Blogs).*** Defining a web log or blog often takes the form of an explanation of its various purposes rather than any distinguishing characteristics. For example, there are personal blogs (i.e., diaries of not-so-private activities over time), corporate blogs for business purposes, and FAQ (frequently asked questions) blogs that focus on a particular problem or situation. Blogs are also defined by the media they employ; for example, a blog comprising videos is called a vlog. Blogs shared via cell phone are called moblogs (mobile blogs). Finally, some blogs are defined by their type: a teaching blog for educators, a techie blogs for technologists, and the like.

Generally speaking, blogs are online journals created by linking individual postings in reverse chronological order so that the most recent postings appear at the top of the web page. In other words, a blog is basically a journal made available on the Internet. The function of entering or updating a blog is called "blogging" and those who maintain a blog are called "bloggers." Blogs are characteristically updated on a daily basis using content management software specifically designed for creating and maintaining large amounts of frequently updated web pages.

A blog post has three basic elements: a title, links to related sites, and a narrative. Some blogs only have the narrative section, others always have all three. Most blogs require a title for purposes of tracking the posts and to serve as a permalink for the item (a URL that links to a specific posting that allows blog entries to be bookmarked by visitors).

Most blogs site permit short posts – a paragraph or two at most. Others provide for longer articles or stories along with an option to provide the browser a summary of the complete blog post.

There are many advantages of blogs in an academic setting over a web site. First, using CMS is easier than HTML, saving the teacher and student time and encouraging even novice web users to post their ideas. Second, blogs use templates, many of which are available for download from the web and present a professional image. Third, bloggers may post as often as they like; the same cannot be said for personal web pages. Fourth, spam filters do not block blogs because they are not considered an email communication. Finally, blogs are available on just about any subject; it may be impossible to determine exactly how many bog sites there are, but Blog Flux, for example, is currently featuring 137,198 blogs in their directory.

Providing a technological basis for blended learning lessons, blogs offer numerous benefits including student motivation, especially for those who otherwise might not become participants in classrooms. Blogs provide excellent opportunities to practice reading and writing. They are effective forums for collaboration and discussion. And, they are powerful tools to promote cognitive development. As an educational tool, blogs may be used to accommodate all style of learners. They can serve as a vehicle to foster a community of learners or serve as the host for student demonstration of learning. Regardless of the application, the imagination of teachers and their students define the only limitations to the creative use of such web-based technology in the classroom.

***eBooks.*** Electronic books, or e-books, are mobile devices that resemble an ultra-portable computer. They were meant to approximate the size of a standard paperback book with the convenience of storage and display of a hard-copy text. E-books offer a range of features that make them suitable for inclusion in a blended learning lesson. For example, they are intuitive to the learner; with less than 15 minutes of familiarization, most learners grasp the operation of an eBook. They store large amounts of material and high-quality backlit screens make for comfortable viewing in most lighting situations.

Compared to other standard storage media such as CDROMs, e-books are easier to use and transport. eBooks offer additional features including, hyperlinks, adjustable fonts for the visually impaired, text search capabilities, and a customized table of content. eBooks are expected to offer advanced multimedia capabilities as animation, video, audio, translations, and pronunciation guides in the future.

Eric J. Simon (2001) conducted a pilot study in which e-books were loaned to college students. During the semester, 22 participants in an introductory biology course volunteered to use e-books as their sole source of reading material for the course. His survey uncovered several trends that make using eBooks well suitable for some blended learning lessons.

***Streaming Audio and Streaming Video***. "Streaming " refers to a file format and software that permits extended audio or video files to be played simultaneously during the download of the entire file. Today's audio (and certainly video) files encompass megabytes of space and, depending on the target computer, may take from several minutes to hours to download. Typically, streaming technologies allow for 15 to 30-second downloads of the file to be captured to a buffer before the media begins to play. As the play continues, the buffer continues to accept downloaded content remaining just ahead of the materials being played and giving the appearance of continuous, uninterrupted play in the most seamless of virtual conditions.

Using application software such as Real Audio, for example, students are able to play most audio and video almost immediately. The software is also able to capture the entire stream for later should the network connection be unable to provide uninterrupted play (i.e., the buffer available is too small or the network connection too slow). A well thought out audio or video stream can store files up to 40 times smaller that other formats, making them quicker to download and requiring less disk space for storage.

Streaming audio and video is popular in a blended learning environment because of the breadth of possible applications. For example, streams are useful as samples; short sound snippets or video clips can be used to help pronunciations in a language class or a view of a complex step in a scientific biology experiment. Many schools already provide classroom lectures in streaming format; usually audio, but video is becoming more popular. Streamed lectures are valuable tools for learners who need additional time to absorb the content of the presentation, missed a class for whatever reason, or prefer to listen to a captured lecture in preparation for an upcoming examination.

***Slideshows (narrated).*** Somewhat akin to streaming audio and video in terms of its contributions to blended learning is the narrated slideshow. The major advantage of this technology is the simplicity with regards technical competencies needed to produce such a resource. Considered very useful in face-to-face settings, slide presentations are often less effective when used in a blended learning environment – particularly one that is primarily online. Either the simple slideshow contains insufficient information or it is too cumbersome (i.e., slow) to access.

When creating effective online slideshow presentations, teachers have a powerful tool to assist them in their efforts: narrated slides. This technology is not merely about an instructor reading directly from the slides. Rather, a narrated slideshow makes the learner feel that the instructor is talking directly to them by speaking in an informal manner; for example, many such presentations are narrated without a script. The goal is to make students feel as though they were actually sitting in the classroom with the instructor listening to a face-to-face lecture.

There are several options for creating a narrated presentation – some more practical than others. First, teachers can actually record a narration onto their PowerPoint presentation by using the "slide show" menu and "record narration" to each slide. Students are able to hear the narration as they move through the presentation at their own speed. However, PowerPoint files can be large when narration is included. Teachers must consider the delivery format before deciding on this option for producing a narrated slideshow.

Other software packages have been expressly developed to convert PowerPoint presentations into Flash-based slideshows. *Articulate Presenter* is one example of a package that provides learners with high-quality presentations even when beginning with a simply PowerPoint lesson. A file converted using this package is considerably smaller in size and therefore much faster to download.

*RealPresenter*, by Real Networks, allows an instructor to enhance his or her PowerPoint presentations by being able to add audio to the slide shows. By incorporating voice-over narration into their online presentations, instructors may now add an audio component to complement their slides' visuals and the text. When completed, the file format of such presentations may then be compressed for delivery across an online setting.

*Impatica* for PowerPoint also converts PowerPoint presentations; however, this package converts the slides into a web-based presentation optimized for viewing on any web browser (Internet Explorer, Netscape, Firefox, etc.). These slideshows play equally well over any Internet connection and any connection speed. Perhaps most importantly, presentations using Impatica do not require any special plug-ins before viewing – a common stumbling block for many novice online learners.

***Virtual Tours.*** Finally, under the general heading of asynchronous tools for blended learning is the web-based virtual tour. A virtual tour is a web-based teaching strategy which presents multi-sensory, multimedia instructional appropriate for student exploration and group learning experiences (Tomei, 2001).

A virtual tour consists of several important components including: introduction, lesson objectives, timelines for the lesson, and instructions. In addition to infusing text hyperlinks, the virtual tour will include image hyperlinks, animated graphics, sound and video files, and a self-assessment tool to measure progress toward achieving mastery of the assigned learning goals. The incorporation of all these features into asynchronous learning is an ideal match for blended learning lessons.

## Synchronous Learning

Historically, traditional educational learning environments (K-12, higher education, corporate, etc.) have been based on the transfer of knowledge from expert (teacher or trainer) to learner (student or employee) by means of lectures or training sessions. Mostly, this transfer was advanced by text books and face-to-face presentations. The classroom has been the location for learning from time immemorial; that is, until the virtual classroom became a reality.

In both traditional and online classrooms, synchronous learning describes the various forms of communications that occur at the same time between individuals while, at the same time, accessing information instantly. Teachers communicate with learners in real time using technologies that were once only imagined in the movies and comic books. Computers host discussions with two participants on separate continents. Presentations integrate electronic whiteboards and electronic slides under the control of the instructor who might physically be located miles from the classroom.

There are several advantages to synchronous delivery. For example, synchronous tools focus the group on the tasks at hand. They create a community of learners and classroom cohesion. Rapid feedback inherent in synchronous communication fosters consensus-building in classroom exercises. Finally, many of the tools discussed in this section share the common characteristic of lesson control; using these tools allows the instructor stronger influence over the sequence and velocity of the instruction.

For the instructor, it remains imperative that every effort is expended to overcome the few (but critically important) disadvantages of asynchronous learning. Research has found that students often feel isolated or less motivated without the face-to-face time human interaction of the traditional classroom (Woodfine, Baptista-Nunes, and Wright, 2006; Park & Bonk, 2007). In addition, asynchronous e-learning does not provide immediate feedback on a student's performance, leaving adjustments to training until after a subsequent evaluation is completed (Slack, Beer, Armitt, and Green 2003).

## Synchronous Technologies

Teachers who include synchronous, computer-mediated communication in their instruction increase student participation, motivate their charges with respect to learning (especially self-directed learning), and give their learners a higher level of comfort and confidence both inside and outside the classroom (Wang, 2008 and Kadirire, 2007). The following synchronous technologies have the best chance for application in a blended learning environment.

***Tele-Presence Technologies***. Under the general classification of synchronous tele-presence teaching comes audio, video, and web conferencing. Audio conferencing solutions provide an easy and cost-effective way to deliver instruction via telephone with a geographically separated group of learners. Audio conferences, whether traditional or voice-over internet protocol (VoIP) can be a cost-effective complement to face-to-face classroom meetings. Teachers can quickly set up and manage virtual meetings from any telephone with better communication and collaboration along with significant cost savings advantages.

When properly implemented, video conferencing provides valuable delivery alternatives to educators. There are two key factors when considering video conferencing for teaching. The number of students in the classroom, number of locations participating in the conference simultaneously, facility configuration, and the experience of the instructor with regards teaching with technology all play into the learning environment in which the video conferencing will take place. From a technical perspective, users of video must consider bandwidth speeds, compatibility and availability of equipment, and network reliability. In the early to mid-1990's, technical issues (e.g., the use of ISDN lines versus the Internet) made the use of video conferencing temperamental at best. Today's technology makes video one of the best delivery modalities for physically separated learners.

Web conferencing is perhaps the best of the best with regards to tele-presence instruction, combining equipment, software, and networking to reach new levels of delivery excellence for educators. Internet-based instruction gives educational institutions of all sizes state-of-the-art tools to conduct synchronous sessions in an environment that is affordable, easy to manage, and if needed, secure.

Offering and delivering effective audio, video, and web-based conferencing classes require a good deal of preparation on the part of the instructor. Visual materials should be organized and distributed to students prior to the session to allow the student to preview the material and address any confusing agenda items. Materials are distributed electronically using the document-sharing features of a learning management system, attachments from an instructor's email, or electronic media such as a CDROM. Instructors should provide clear instructions for the session and discuss protocols for interaction in preliminary communications before (or, if necessary, as an initial introduction during) the first online session.

Seasoned tele-presence instructors suggest that the use of synchronous conferencing forced them to re-evaluate (and thereby improve) their teaching practice (Hinger, Date Unknown). Conference-based classes allowed them to draw on many of the teaching skills they would use in a normal classroom while offering their students all the advantages of a distance education. Some of the most affected changes in teaching styles included the need to limit lecture time, incorporate interactive learning experiences, advance engagement learning opportunities, prior preparation using asynchronous tools, additional lead time to prepare conference class, and others. Honing one's course preparation techniques were mandatory and included such skills as ensuring the technology is ready; identifying a qualified educational technologist in case of emergencies; planning learning strategies and outcomes well in advance; ensuring appropriate design and integration of online tools based on accepted models of instruction, and more.

***Online chat rooms***. Chat rooms provide an online forum for synchronous discussions of pertinent lesson content. Most chat rooms are fully integrated into existing learning management systems, affording users a secure login with built-in tracking and assessment. Channels are provided to control the topics under discussion and archive participant comments. Other common features include text formatting options, private chats, emoticons, sounds, avatars, pop-up windows, and online help as well as enhanced security features such as banning visitors, ignoring users, and profanity filters.

Online chat rooms have perhaps more advantages and disadvantages than most other synchronous tools. On the plus side, students are more inclined to participate since the setting is less formal and most students are already familiar with the protocols and etiquette of a chat session. Chat rooms typically flex a learner's writing prowess (always a plus in today's audio and video-rich world) giving the student ample opportunity to craft their ideas in a more studious product often not possible during face-to-face classroom discussions.

In a properly monitored chat room, students are able to share their ideas with peers in their own class, other classes studying the same content, and even other students from around the world who share the same interests – all in a safe and non-harassing environment.

On the down side, chat rooms have certain limitations that must be considered and overcome. For example, chats represent poor pedagogy when it involves more than a handful of participants or when the instructor functions as a didactic lecturer delivering content and avoiding interactive conversation. And, of course, the chat room is notorious for online predators, identity thief's, spammers, and viruses.

***Interactive Whiteboards***. The interactive whiteboard is an often misunderstood technology. Some vendors, in particular, advertise their whiteboard as a peripheral device connected to an instructor's computer that offers electronically all the familiar features of a traditional classroom blackboard or ordinary whiteboard. Other suppliers refer to a whiteboard that is internal to a learning management system (commercially, this feature is often called an eBoard). In the online version of the whiteboard, an image window provides the online instructor with a drawing palette, file import, screen save, file-sharing, web page projections, and more.

Let's talk the peripheral whiteboard first. A whiteboard is touch sensitive, connected to a projector, computer, DVD, VCR, and other devices. Instructors control the images from a podium usually located at the front of the classroom. Using this device, the instructor projects digital as well as physical presentations, interact with the computer using a digital pen that displays virtual "ink," and captures the notes and images to a digital lesson file.

The eBoard has all the same features, more or less. The major exception is the use of the interactive technology as part of an online course. The software available with the whiteboard can take any virtual writing and transform it to printed text. The entire lesson, as it is unfolded before the eyes of the online learner, can be printed, saved, loaded online, e-mailed, or shared via a host of other ways to share with students who seek to use the material eitehr for remedial work, preparation ofr an assessment, or as a primary source of content for a lesson they might have missed.

There are numerous advantages to using interactive whiteboards, regardless of the type. Whiteboards work well with large groups; rather than crowding around a single computer screen, students can comfortably view the instructor's presentation. Any presentation or lesson is easily enhanced with integrated video, animation, graphics, text and audio. Resources from a number of venues such as CD-ROMs, websites, DVDs, VHS tapes and television are possible. The instructor is able to manage information directly during lessons and save changes or additions to a digital file. Students benefit from the interaction with the whiteboard; collaboration is enhanced, engagement is fostered, and peer-to-peer interaction within groups is promoted.

***Application sharing***. Application sharing allows instructors and students to share software or any part of the instructor's desktop. The host of the application (usually the teacher) grants remote control of the application(s) for purposes of simulated experiences, practical training, real-world demonstrations, or, in some cases, hands-on technical support. Participants see the screen-view of the shared application. Their coordinated mouse and keyboard movements enable collaborative work, software tutoring and e-synchronous learning. There is only one copy of the shared application running from a main server. As a result, institutions save money by avoiding purchases of multiple copies of the software and prevail over privacy and security issues. The main challenges of application and desktop sharing are scalability, reliability, true application sharing, operating system independence, and performance.

There are two models for application sharing: application-specific and generic. The application-specific model requires the developers to add this feature to their applications; for example, the latest version of Microsoft Office includes application sharing as a key feature. In the generic model, applications include word processors, Internet browsers, Power Point presentations, and others.

Application sharing refers to simultaneous access by two or more to a common application, document file, or video screen from different locations. Most application sharing programs offer the instructor a software interface that opens the program or document using a username and password for security. Once the program is available, the instructor takes control of the session (remember, this is a synchronous technology) and invites students into the application sharing session. Features allow the instructor to relegate control of the program to the learner who can modify, manipulate, edit, or apply the program.

Some application sharing software permits two or more users to edit a document or application simultaneously, encouraging true synchronous activity. Instructors are able to demonstrate complicated applications in real time and coach their students through otherwise complicated applications. Using application sharing activities in a blended learning environment helps to create a true sense of community among students and instructors.

## CONCLUSION

Blended learning is a mix of traditional classroom environment with distance learning components; in effect, it is the systematic combination of delivery procedures for the classroom. Blended learning is most successful when teachers are trained in face-to-face didactic instruction, group processes, and the technology. All three components of blended learning must interact to produce effective instruction.

For many schools and corporate training organizations, blended learning is replacing distance learning. Research by Bersin & Associates (2003) found that blended learning programs are perhaps the highest impact, lowest cost way to drive major corporate initiatives. Blended learning addresses the issues of speed, scale, and impact while offering the instructor an alternative to distance learning when distance learning has been shown not to be the most effective instructional delivery media for some learners.

Each delivery format has its own particular strengths and weaknesses. Technology offers online course management systems; audio, video, and web-based tele-presence; CDROM-based courses, etc. Face-to-face instruction submits text books, group discussion, Power Point presentations, and more. Collectively, blended learning capitalizes on the strengths and minimizes the weaknesses of each modality.

Blended learning forces you to think about the instructional goals. Together with instructional design models such as the ADDIE Model, Kemp Model, or the Dick & Carrey Model, teachers can use blended learning to more closely analyze the needs and constraints of the learner; design the best recipe of learning activities, assessment tools, and media; develop the instructional resources; implement the lesson; and, evaluate the results from all perspectives.

Finally, as the reader of this text will witness is the remaining chapters of the book, blended learning has less impact on the organization's budget. By combining the technologies addressed above with traditional face-to-face instruction, the tradeoff between development cost and delivery cost is minimized. Most teachers can build their own instructional content following a few in-service training sessions on the principles of blended learning and the application of technology.

Use ***ICTs for Modern Educational and Instructional Advancement: New Approaches to Teaching*** to set the stage for an evolution in teaching by introducing teachers to the theories, principles, and applications of blended learning in your institution. The scholars who have contributed to this book have already found success in using the technologies discussed to enhance their own teaching. Take advantage of their lessons learned.

*Lawrence A. Tomei*
*Robert Morris University, USA*

## REFERENCES

Bersin & Associates (2003), Blended Learning: What Works? e-LearningGuru.com. Downloaded: January 2009. URL: http://www.e-learningguru.com/wpapers /blended_bersin.doc.

Bruck, Bill (Winter 2005), Building Interactivity Into E-Learning. ELearning! Magazine. Downloaded January 2009. URL: http://www.elearning.b2bmediaco.com/issues/winter05/feature_0205.html

Hinger, David (Date Unknown). Blended Learning Through Videoconferencing. Association for the Advancement of Computing in Education. Downloaded: January 2009. URL: www.editlib.org/toc/index.cfm/files/paper_26746.pdf?fuseaction=Reader.DownloadFullText&paper_id=26746

Kadirire, J. (2007). Instant messaging for creating interactive and collaborative m-learning environments. International Review of Research in Open and Distance Learning, 8(2), 1-14.

Park, Y. J.; Bonk, C. J. (2007). Synchronous learning experiences: Distance and residential learners perspectives in a blended graduate course. The Journal of Interactive Online Learning. 6(3). http://www.ncolr.org/jiol/issues/PDF/6.3.6.pdf

Slack, Frances; Beer, Martin; Armitt, Gillian; and, Green, Sharon (2003). Assessment and Learning Outcomes: The Evaluation of Deep Learning in an On-line Course. Journal of Information Technology Education. Downloaded: January 2009. URL: jite.org/documents/Vol2/v2p305-317-29.pdf.

Simon, Eric J. (2001). Are e-Books Ready for the Classroom? Campus Technology, August 2001. Downloaded: January 2009. URL: http://campustechnology.com/Articles/2001/08/Are-eBooks-Ready-for-the-Classroom.aspx?Page=1

Tomei, Lawrence A. (2001). Teaching Digitally: A Guide for Integrating Technology into the Classroom. Christopher-Gordon Publishers, Inc. Norwood MA.

Wang, Shiang-Kwei (2008). The effects of a synchronous communication tool on online learners' sense of community and their multimedia authoring skills. Journal of Interactive Online Learning, 7(1), 59-74.

Woodfine, B.P; Baptista Nunes, M.; and, Wright, D.J. (2006). Text-based synchronous e-learning and dyslexia: Not necessarily the perfect match! Computers & Education Volume 50, Issue 3. Publisher: Elsevier Ltd. Downloaded: January 2009, URL: http://www.sciencedirect.com/science?_ob=ArticleURL&_udi=B6VCJ-4M93P3N-1&_user=2144460&_rdoc=1&_fmt=&_orig=search&_ sort=d&view=c&_acct=C000056294&_version=1&_urlVersion=0&_userid=2144460&md5=4d883261ae4ce4701164cd0e25ff3665

# Chapter 1
# A Description of Online Instructors Use of Design Theory

**MarySue Cicciarelli**
*Duquesne University, USA*

## ABSTRACT

*In a recent dissertation study, research was conducted to evaluate online instructors' characteristics and preferences concerning the use of a telementor, or online instructor's assistant, as a part on an online course. Those who participated in the anonymous survey came from a sample of two thousand online instructors from colleges and universities located across the United States. Of those contacted, 323 online instructors responded to the survey. Results presented in this article were produced using data from nine of the questions included in the survey. These Likert Scale questions specifically asked the instructors about their use of theory of multiple representation, Gagne's conditions of learning, instructional transaction theory, cognitive flexibility theory, three form theory, dual-coding theory, elaboration theory, theory of transactional distance, and theory of immediacy and social presence. Outcomes showed that a larger number of online instructors applied design theory when creating a course compared to the instructors who indicated that they did not apply design theory. Descriptive results presented illustrate how often the participants said that they utilized each of the different theories.*

## INTRODUCTION

Distance education has become an alternative when taking a course or earning a degree (Chu & Hinton, 2001). Researchers stated that taking an online course is one example of distance education through which students participate at different times from different locations (Simonson, Smaldino, Albright, & Zvacek, 2000). Colleges and universities that offer online courses have chosen to use course management systems because of the alternatives and flexibility options that they provide (Course-Management Systems, 2005).

One feature of a course management system often used by students and instructors is the asynchronous board feature, or venue for written discussion. Studies on the use of the asynchronous board tool showed that there are advantages and disadvantages to using the tool (Makrakis, 1998; Collins & Berge, 1996; Prestera & Moller, 2001). Providing instructors and students with a telementor, or online instructor's assistant, was one possible solution to reducing the disadvantages experienced during asynchronous discussions. A dissertation study was conducted to identify online instructors' characteristics and preferences concerning the utilization of a telementor. To help identify the instructors' characteristics, participants were asked how often they utilized nine specific design theories when developing an online course (Cicciarelli, 2006, 2007).

## REVIEW OF LITERATURE

Research showed that when online instructors design a course, they can use theory to guide the development process. Theories used by instructors tended to come from the three schools of psychology known as Behaviorism, Cognitivism, and Humanism. Behavior theories have made use of the environment to influence actions. Theories that are cognitive-based have focused on meaningful ways of learning that included authentic learning experiences, and tasks that are declarative and procedural. Humanistic theories, which attended to students' affective needs, concentrated on students' feelings, emotions, values, and attitudes. The nine theories presented in this article are a part of the three schools of psychology (Cicciarelli, 2006, 2007).

### Theory of Multiple Representations

The theory of multiple representations, a cognitive-based theory, held that the learner can take information and make it more meaningful by connecting multiple representations to the content. There are researchers who supported the use of this theory, and there are those who cautioned against its use during instruction (Gfeller, Niess, & Lederman, 1999; Huang & Liaw, 2004). Gfeller, et al. (1999) studied the perceptions of preservice teachers. They looked at their understanding of mathematical concepts and their ability to develop different representations of concepts that they would eventually teach in the classroom. The results showed that the teachers who had a mathematical background were better able to develop a number of representations which would make it easier for them to understand their future students' different views of the subject matter when compared to the preservice teachers with a scientific background.

### Cognitive Flexibility Theory

Researchers studied the process of thinking and learning as children developed. Cognitive theory has been used to guide the interaction between students, the instructor, and the content. When this theory has been applied, students take their conceptual knowledge about a situation and relate it to new situations. This helps their understanding of a concept so they can move from a more basic understanding to one that is more complex (Huang & Liaw, 2004). Jonassen (2003) explained that when students have been presented with a problem, the problems have tended to be presented in a structured way. He indicated that real life problems are not structured, and since it has been recognized that transferring problem-solving skills to real life situations was not always done readily, it was vital for instructors to help their students externalize what they knew. In order to externalize knowledge and understanding, he suggested the development of mental representations—making internal maps of problems and using tools to externalize problem representations.

## Bruner's Three Form Theory

Bruner (1990) stated that individuals see the world through three different ways; action, icons, and symbols. He said they used action when performing or demonstrating what it was they saw from their perspective. Icons or mental images, according to Bruner, were used to present a path, summary, or pattern. Symbolism, an abstract way of seeing reality, was used by individuals through words and numbers. Vacca and Vacca (1998) discussed Bruner's work on scaffolding and categories. They said that when instructors help students recognize what they know and what is new, they can then help them build new categories of information and learning.

## Dual-Coding Theory

When dual-coding theory is applied, a system of verbal and imagery processing is used. The verbal aspect helps while information is presented and processed. Aspects of imagery help the learner create images, sounds, actions, and emotional responses when nonverbal cues are not available during the learning situation (Huang & Liaw, 2004). Research indicated that individuals use aural and visual paths when processing information and making meaning. The theory indicated that the ways in which people make use of their aural and visual abilities differs from one person to the next, because one way is stronger for some and the other way is stronger for others. Whichever modality is used by an individual does not matter, because both modalities influence the way individuals perceive information. Instructors are encouraged to utilize aural and visual stimuli when presenting information to students to keep the learner from becoming confused and misunderstanding the content that is presented (Paivio, 1979, 1986; Simpson, 1997).

## Gagne's Conditions of Learning

Another form of instructional learning is Gagne's conditions of learning. This theory has been credited for the successful incorporation of instructional psychology into the instructional technology and design field. Gagne's Conditions of Learning is a descriptive theory made up of five outcome categories. The five categories are labeled as intellectual skills, verbal information, cognitive strategies, motor skills, and attitudes. Intellectual skills require individuals to have the ability and knowledge to categorize and use materials. Having the ability to show what something is or what something means refer to verbal information abilities. Individuals' own learning skills are connected with cognitive strategies, and the simple and complex movements that people make are connected to motor skills. Finally, attitudes are the feelings that individuals develop after interacting in situations that are either constructive or unconstructive. Nine practices have grown out of Gagne's work. The nine conditions are known as gaining attention, informing learners of the objective, stimulating recall of previous learning, presenting the content, providing learning guidance, educing performance, providing feedback, assessing performance, and enhancing retention and ability to transfer learning (Gagne, 1985; Smith & Ragan, 1996; Molenda, 2002; Gagne, Wager, Golas, & Keller, 2005).

## Merrill's Instructional Transaction Theory

Merrill's instructional transaction theory has posed a belief that motivation can be gained through processes of transactions that individuals use to make connections during the learning process. During this process, conventions are used while objects of knowledge are selected and sequenced (Huang & Liaw, 2004). When this theory is implemented, relationships between educational and technical factors can be made.

There are two facets to the thought of Instructional Transaction Theory: schemes of knowledge and procedures for applying knowledge. According to this theory, in order for learning to take place, more than one knowledge structure has to be in place for the information to make sense. When instructional transactions are taking place, different parts of knowledge can be grouped into one structure of knowledge. The theory has made use of transactions so that the content that students are supposed to learn is categorized into ways that are more meaningful for the students (Buendia, Diaz, & Benlloch; 2002).

## Elaboration Theory

Reigeluth developed elaboration theory. According to this theory, the materials in a course should be organized so that new learning is presented in a simple way for initial understanding, and then the instructor should gradually present the information and content in more complex ways. When utilizing this strategy, instructors have tended to begin the learning experience by presenting information that the students are already familiar with. Then, they carefully move on to more complex content. This makes it easier for students to make connections and retain new knowledge. Elaboration Theory relies on the cognitive structure of the individual learners. Learners will move from having an understanding of the simple content to the more complex content at a different pace because their cognitive abilities differ. When students learn, they go through a process of selecting, sequencing, synthesizing, and finally, experiencing the summarizing phase of the learning experience. Student capacity to move through each phase varies depending on student ability. When instructors have implemented strategies to help the students move through each phase, the ability for students to understand the information has improved (Ludwig, 2000; Huang & Liaw, 2004).

## Moore's Theory of Transactional Distance

Moore's theory of transactional distance is a distance theory. The affective influence on teaching procedures has brought many instructors to implement this form of learning into course design. The theory is made up of three dimensions. They are referred to as interaction, course structure, and learner autonomy (Huang & Liaw, 2004). This theory holds that when a course is highly structured, the understanding and connection between the student and the instructor is more connected because the interaction between the two is stronger (Moore, 1973; Moore & Kearsley, 1996; Laly & Barrett, 1999; Chen, 2001; Jung, 2001; Kanuka, Collett, & Caswell, 2002).

## Theory of Immediacy and Social Presence

Researchers have held that learning takes place through the interaction of three core components: cognitive presence, teaching presence, and social presence. When a deeper look has been taken at these three components, the researcher refers to the responses as affective, interactive, and cohesive. When analyzing the responses made by students, the researchers found that affective behaviors impacted student perception of the learning experience and the course. As a result, the researchers have encouraged instructors to meet the affective needs of their students. Two behaviors that appeared to influence student perception were quick response and presence. Students who received a quick response from an instructor after a post and students who believed that the instructor was a present part of the learning situation perceived the learning experience in a better way (Rourke, Anderson, Garrison, & Archer, 2001; Martyn, 2004).

## Summary

Research has shown that online instructors use design theory to accomplish such goals as improve feedback, help students develop problem-solving skills, and reduce isolation (McAlpine & Ashcroft, 2002; Huang & Liaw, 2004). Other research has shown that threaded discussions have become more accessible and beneficial as course management systems have improved (Levin & Ben-Jacob, 1998; Hathorn & Ingram, 2002; McAlpine & Ashcroft, 2002; Fauske & Wade, 2003-2004; Greenlaw & DeLoach, 2003; Im & Lee, 2003-2004). Greater focus on different forms of interaction in the form of learner-to-content interaction, learner-to-learner interaction, learner-to-instructor interaction, and learner-to-interface interaction has been advised by researchers as well (Moore, 1989; Moore, 1999; Chen, 2001; Huang, 2002). Williams (2001) indicated that online instructors play a significant role in the success of computer-mediated discussions. How often online instructors utilize the different design theories to help them reduce the number of disadvantages and guide interaction has not been researched.

# METHODOLOGY

## Purpose

The purpose of this quantitative exploratory study was to examine online instructors' teaching characteristics and perspectives of telementor support. Questions that asked instructors to share how often they utilized particular design theories were included in the survey because they helped illustrate the varying characteristics of the online instructors. Of the 29 survey questions posed in the survey, there were nine questions that targeted use of design theory. A descriptive research design was used because the researcher wanted to identify the participants' basic characteristics.

## Procedures

A convenience sample was used in this study because it was an expedient and accessible alternative for identifying possible study participants. The only restriction in the study was that the instructors had to have taught an online course. Discipline, years of teaching, and other existing conditions did not prohibit participation.

A book titled *Distance Degrees* was used to identify schools that offered online courses (Wilson, 2001). Once the appropriate schools were identified, online instructors' e-mail addresses were identified and recorded in a database. In all, two thousand online instructors' e-mail addresses were recorded. Since the participants were found through a convenience sample, the researcher did not intend to generalize the results to the population, thus the researcher obtained a 95 percent confidence level by keeping access to the survey open until at least 322 participants had responded t to the survey.

## Instrumentation

A survey, the Online Instructor Characteristics and Preference for Telementor Support (OIC and PTS) Survey-1, was developed as the instrument to be used. Questions were formed in a contingency arrangement, which meant that the participants were asked to continue or stop taking the survey depending on how they responded to a contingent question. For example, if an instructor indicated having never taught an online course, the participant was asked to stop answering questions and submit the survey.

## Validity and Reliability

For the purpose of providing validity and reliability; the survey questions were professionally reviewed. After the first draft of the cross-sectional survey had been developed, three university professors and one individual from the Teaching, Learning, and Technology Group (TLT) reviewed the questions. Adjustments were made to the questions based on the suggestions.

## Procedure for Data Collection

University and college online instructors from across the United States were sent a request to participate by an e-mail which included a link to the anonymous survey. Incorporated in the message was an explanation that identified the purpose of the study, a description of the researcher and the researcher's institution, and a polite participation request. In addition, the participants were provided with assurance of confidentiality. A reminder e-mail that contained the same information was also sent. Once 323 completed surveys had been submitted, access through the link to the survey was disabled.

## Procedure for Data Analysis

First, data cleaning steps were taken to identify any outliers. Then, individual frequency distribution tests were run on each variable to find odd data in the output, and the original data were corrected if any anomalies appeared. A univariate, descriptive level analysis of frequency distributions was run for each variable during the analysis. Each survey question was created in Likert Scale format, thus all responses were at the ordinal level. These questions were used to operationalize the independent variables in the study. Finally, the researcher examined the data results and looked for any patterns that had developed. Cross-tabulations were utilized to determine additional patterns.

*Table 1. Online instructor use of specific design theories*

| Design Theories | Always | MOTO | Occasionally | LOTO | Never |
|---|---|---|---|---|---|
| Theory of Multiple Representation | 32.8% | 32.8% | 20.7% | 7.3% | 6.4% |
| Cognitive Flexibility Theory | 30.4% | 36.7% | 22.4% | 5.4% | 5.1% |
| Three Form Theory | 14.6% | 27.5% | 25.9% | 17.8% | 14.2% |
| Dual-Coding Theory | 23.4% | 23.7% | 18.3% | 17.6% | 17.0% |
| Gagne's Nine Conditions of Learning | 29.2% | 42.6% | 13.8% | 9.0% | 5.4% |
| Merrill's Instructional Theory | 22.7% | 39.1% | 21.7% | 8.6% | 7.9% |
| Reigeluth's Elaboration Theory | 31.4% | 41.4% | 16.2% | 6.8% | 4.2% |
| Theory of Transactional Distance | 39.7% | 37.4% | 14.8% | 5.5% | 2.6% |
| Theory of Immediacy and Social Presence | 37.0% | 36.0% | 15.1% | 8.0% | 3.9% |

*Note: Rows may not add up to 100% due to rounding*
*MOTO=More Often Than Occasionally*
*LOTO=Less Often Than Occasionally*

## RESULTS

In all, 323 online instructors responded to the survey. There were instructors who chose not to respond to every survey question, and there were only six participants who indicated that they had never taught an online course. Presented in Table 1 below are the results which indicate how often the online instructors said that they used each of the nine design theories included in the study.

## CONCLUSION

In conclusion, the results from this part of the study provide educators with a general picture of how often instructors apply these different theories to course design. One can recognize that a greater number of instructors do incorporate theory into course design, but how the instructors blend the theories is not apparent. The greatest outcome of this study is the realization that since more instructors incorporate theory compared to those who do not, it would be of benefit to do further research on this topic. Learning exactly how instructors incorporate the theories, what strategies they use, and how they blend the theories when designing a course would add to the knowledge base on instructor use of theory. Exploring to find out whether or not instructors change the use of theory depending on the characteristics of the students would also be interesting. In addition, asking instructors who teach face-to-face and hybrid courses about how they make use of the theories when designing a course would be useful. Finally, investigating the use of other theories connected to the three schools of psychology would produce interesting results. The results from these possible studies could provide those connected to the field with a look at how pedagogy has shifted.

## REFERENCES


Bruner, J. (1990). *Acts of meaning.* Cambridge, MA: Harvard University Press.

Buchanan, E. A., Myers, S. E., & Hardin, S. L. (2005). Holding your hand from a distance: Online mentoring and the graduate library and information science student. *The Journal of Educators Online, 2*(2), 1-18.

Buendia, F., Diaz, P., & Benlloch, J. V. (2002). A framework for the instructional design of multi-structured educational applications. In P. Kommers & G. Richards (Eds.), *Proceedings of world conference on educational multimedia, hypermedia & telecommunications 2002,*(pp. 210-215). Chesapeake, VA: AACE.

Chen, Y. J. (2001). Dimensions of transactional distance in the world wide Web learning environment: A factor analysis. *British Journal of Educational Technology, 32*, 459-470.

Chu, H. C., & Hinton, B. E. (2001). *Factors affecting student completion in a distance*

*learning mediated HRD baccalaureate program.* Paper presented at the Academy of Human Resource Development 2001 Conference, Fayetteville, AR. (AHRD Reference No. 021).

Cicciarelli, M. (2007). Behavioral, cognitive, and humanistic theories that guide online course design. *The International Journal of Information and Communication Technology Education, 3*(4).

Cicciarelli, M. (2006). *Telementoring and computer mediated discussions: A description of online instructors' support.* Doctoral Dissertation, Duquesne University of Pittsburgh, Pennsylvania, 2004. (UMI N0. 3238573)

Collins, M., & Berge, Z. (1996). *Facilitating interaction in computer mediated online courses.* Paper presented at the FSU/AECT Distance Education Conference, June. Retrieved June 29, 2005 from, http://www.emoderators.com/moderators/flcc.html

Course-management systems. (2005). *Library Technology Reports, 41*(3), 7-11.

Fauske, J., & Wade, S. E. (2003-2004). Research to practice online: Conditions that foster democracy, community, and critical thinking in computer-mediated discussions. *Journal of Research on Technology n Education, 36*(2), 137-153.

Gagne, R. M., Wager, W. W., Golas, K. C., & Keller, J. M. (2005). *Principles of instructional design* (5th ed.). Belmont, CA: Wadsworth/Thomson Learning.

Gfeller, M. K., Niess, M. L., & Lederman, N. G. (1999). Preservice teachers' use of multiple representations in solving arithmetic mean problems. *School Science and Mathematics, 99*, 250-257.

Greenlaw, S. A., & DeLoach, S. B. (2003). Teaching critical thinking with electronic discussion. *Journal of Economic Education, 34*(1), Winter, 36-52.

Hathorn, L. G., & Ingram, A. L. (2002). Cooperation and collaboration using computer-mediated communication. *Journal of Educational Computing Research, 26,* 325-347.

Huang, H. M. (2002). Student perceptions in an online mediated environment. *International Journal of Instructional Media, 29,* 405-422.

Huang, H. M., & Liaw, S. S. (2004). Guiding distance educators in building Web-based instructions. *International Journal of Instructional Media, 31*(2), 125-137.

Im, Y., & Lee, O. (2003-2004). Pedagogical implications of online discussion for preservice teacher training. *Journal of Instructional Media, 31*(2), 125-137.

Jonassen, D. (2003). Using cognitive tools to represent problems. *Journal of Research on Technology in Education, 35*, 362-381.

Jung, I. (2001). Building a theoretical framework of web-based instruction in the context of distance education. *British Journal of Educational Technology, 32*, 525-534.

Kanuka, H., Collett, D., & Caswell, C. (2002). University instructor perceptions of the use of asynchronous text-based discussion in distance courses. *The American Journal of Distance Education, 16*(3), 151-167.

Lally, V., & Barrett, E. (1999). Building a learning community on-line: Towards socio-academic interaction. *Research Papers in Education, 14*(2), 147-163.

Levin, D. S., & Ben-Jacob, M. G. (1998). *Collaborative learning: A critical success factor in distance education.* Paper presented at the Annual Conference on Distance Teaching & Learning, August, Madison, Wi.

Ludwig, B. (2000). *Web-based instruction: Theoretical differences in treatment of subject matter.* Paper presented at the Annual Meeting of the Psychological Association, August Washington, DC. (ERIC Document Reproduction Service No. ED453708).

Makrakis, V. (1998). *Guidelines for the design and development of computer-mediated collaborative open distance learning courseware.* Paper presented at the 1998 World Conference on Educational Multimedia and Hypermedia & World Conference on Educational Telecommunications, June, Freiburg, Germany. (ERIC Document Reproduction Service No. ED428694).

Martyn, M. A. (2004). *The effect of online threaded discussion on student perceptions and learning outcomes in both face-to-face and online courses.* Doctoral dissertation, University of Akron. (UMI No. 3123389)

McAlpine, I., & Ashcroft, B. (2002). Turning points: Learning from online discussions in an off-campus course. In P. Kommers, & G. Richards (Eds.), *Proceedings of world conference on educational multimedia, hypermedia & telecommunications* pp. 1251-1257, Chesapeake, VA: AACE.

Molenda, M. (2002). *A new framework for teaching in the cognitive domain.* Syracuse, NY: ERIC Clearinghouse on Information and Technology (ERIC Document Reproduction Service No. ED470983)

Moore, M. G. (1973). Towards a theory of independent learning and teaching. *Journal of Higher Education, 44*, 661-679.

Moore, M. G. (1989). Three types of interaction [Editorial]. *The American Journal of Distance Education, 3*(2) 1-6.

Moore, M. G. (1991). Distance education theory [Editorial]. *The American Journal of Distance Education, 5*(3) 1-6.

Moore, M. G., & Kearsley, G. (1996). *Distance education: A systems view.* New York: Wadsworth.

Paivio, A. (1979). *Imagery and verbal processes.* Hillsdale, NJ: Lawrence Erlbaum Associates.

Paivio, A. (1986). *Mental representations.* New York: Oxford University Press.

Prestera, G. E., & Moller, L. A. (2001). *Facilitating asynchronous distance learning: Exploiting opportunities for knowledge building in asynchronous distance learning environments.* Paper presented at the Annual Mid-South Instructional Technology Conference, April, Murfreesboro, TN. (ERIC Document Reproduction Service No. ED463723.

Rourke, L., Anderson, T., Garrison, D. R., & Archer, W. (2001). Assessing social presentation asynchronous text-based computer conferencing. *Journal of Distance Education, 14*(2), 50-71.

Simonson, M., Smaldino, S., Albright, M., & Zvacek, S. (2000). *Teaching and learning at a distance: Foundation of distance education.* Upper Saddle River, NJ: Merrill.

Simpson, T. J. (1995). *Message into medium: An extension of the dual coding hypothesis.* Paper presented at the Annual Conference of the International Visual Literacy Association, October, Tempe, AZ. (ERIC Document Reproduction Service No. ED380084).

Smith, P. L., & Ragan, T. J. (1996). *Impact of R. M. Gagne's work on instructional theory.* Paper presented at the 1996 National Convention of the Association for Educational Communications and Technology, Indianapolis, IN. (ERIC Document Reproduction Service No. ED397841).

Williams, S. W. (2001). *Experiences of web-based instruction among African-American students enrolled in training and development graduate courses.* Paper presented at the Academy of Human Resource Development Conference, Raleigh, NC. (AHRD Reference No. 059).

Wilson, Mark. (2001). *Distance degrees.* Kearney, NE: Morris Publishing Company.

Vacca, R. T., & Vacca, J. L. (1998). *Content area reading: Literacy and learning across the curriculum* (6th ed.). New York: Addison Wesley Longman, Inc.

*This work was previously published in International Journal of Information and Communication Technology Education, Vol. 4, Issue 1, edited by L. Tomei, pp. 25-32, copyright 2008 by IGI Publishing (an imprint of IGI Global).*

# Chapter 2
# A Model for Online Instruction and Faculty Assessment

**Michael Thomas Shaw**
*SilkWeb Consulting & Development, USA*

**Thomas M. Schmidt**
*University of Phoenix, USA*

## ABSTRACT

*Differing methods of course development can lead to widely varying results. The University of Phoenix develops courses for both on-campus and on-line (e-learning) delivery, using electronic collaboration as well as in-person teamings. Course developers at the University rigorously measure feedback about course materials, and revise courses based on learners' input. This paper describes a model for developing and delivering e-learning doctoral-level curricula based on current research and a learner needs analyses. Suggestions for further improvements and surprising results about the most effective method for deriving E-learning materials are explored.*

## A MODEL FOR ONLINE INSTRUCTION AND FACULTY ASSESSMENT

University of Phoenix (UoPhx) is the largest for-profit university in the U.S. The institution is accredited by the Higher Learning Commission and offers degree programs at the associates, bachelors, masters, and doctoral levels. The University's central office is located in Phoenix, Arizona. Its mission as stated on its student home page is "to provide access to higher education opportunities that enable students to develop the knowledge and skills necessary to achieve their professional goals, improve the productivity of their organizations, and provide leadership and service to their communities" (University of Phoenix, 2006).

DOI: 10.4018/978-1-60566-936-6.ch002

A pioneer in adult education, the institution has broken new ground in electronic course delivery, originally using a fax system and now delivering course content and library materials over the World Wide Web (the Web). With over half of its 200,000 students enrolled in online programs, its materials must be geared to work effectively in an e-learning context, and must remain current and topical.

The University created the School of Advanced Studies (SAS) in 1998 to address the needs of students desiring to pursue studies beyond the level of a Master's degree. As part of this effort, several doctoral-level degree programs were created, including the Doctor of Business Administration, the Doctor of Health Administration, Doctor of Management, Doctor of Education, each with several specializations including Curriculum and Instruction and Information Systems and Technology.

The following paper will document the model around which online courses have been developed at UoPhx and discuss supporting literature This study will conclude with a revised model for online instruction and a faculty assessment tool to support the new model.

## THE ORIGINAL MODEL

Moskal (2006) investigated online student satisfaction. The authors found that 83% of learners indicated that they were satisfied with their program of study, citing convenience and flexibility as the major reasons. Eighty percent of students polled attributed their ability to complete their respective course of study to the online modality. Universities around the world are increasingly turning to this mode of education, making it necessary to review common approaches and practices.

The online environment possesses several advantages and some disadvantages over a classroom environment. Tallent-Runnels (2005) discussed best practices related to online instruction. They focused on five enabling factors for successful online courses, including organization of the platform, pace of learning, support for learning, resources available to students, and maintaining a welcoming environment.

The platform is the substructure that supports the online course. The methods used to organize the platform significantly affect learning outcomes. Ideally, the platform is logically organized into folders containing resources that meet the needs of both faculty and students. According to a review conducted by Tallent-Runnels, online courses should organize electronic resources and materials in weekly folders.

Pacing is another significant element of online instruction. Findings by Tallent-Runnels indicate that students appreciate the ability to move at their own pace. Not surprisingly, their evidence strongly supports asynchronous discussions and faculty feedback. Asynchronous discussions allow students to research and debate ideas and create a virtual community. To further this fostering of community, faculty should provide timely feedback. Without feedback, students tend to withdraw from discussion.

### Platform

The School of Advanced Studies programs are essentially hybrid in nature. Courses are delivered partially on location in Phoenix, Arizona in the form of residencies, and partially through a customized Learning Management System (LMS). Shifting away from reliance upon locally-installed, computer based software (such as Outlook Express), the LMS allows learners and faculty to use browser-based classroom software to engage in classroom discussions, submit assignments, and meet with team members from any Web-connected computer (i.e. not just their personal computer), allowing access from a range of locations.

Class discussions are held asynchronously, allowing learners and faculty the convenience of participating at times that best match their schedules. Althaus (1997) found asynchronous learning environments allowed students more time to read and respond to messages. In addition, the author found a positive relationship between discussion involvement and student grades. Rovai (2001) noted that asynchronous learning environments promote synthesis of knowledge and contribute to better-informed critical discussion. Furthermore, Heckman and Annabi (2003) found that asynchronous environments allow for more formal and careful responses.

While it provides a platform that encourages student involvement, the LMS used by Uophx lacks the immediate interpersonal contact of the classroom setting. According to Pappas and Jerman (2004), future online courses will use a blended approach that incorporates both face-to-face and online instruction. Lim (2006) conducted a comparative analysis of online and blended courses in undergraduate education to identify differences in learner satisfaction, learning, and application of learning. The authors measured the perceived degree of learning, learning application, and instructional quality of 125 undergraduate students through a combined close-ended and open-ended online questionnaire collected at the beginning and end of a course. Lim (2006) found that both online and blended groups had a significant increase in their actual and perceived learning. Online learners displayed a higher mean score for difficulty with the instruction. Learners in the blended delivery format felt that they had had significantly more learning support than online students. Both groups cited instructional effectiveness as the most influential factor for learning.

## Student Resources

A cursory review of research into student facilities that influence online learning success is needed before we can continue the discussion. Bee and Usip (1998) found a relationship between improved course performance and the use of supplementary materials, tutorials, and general course information. Cooper (1999) found that weekly folders enhance learning. Koszalka and Ganeson (2004) recommended that course organization in itself is not an independently sufficient predictor of student success; rather, all of the course features including student expectations, course philosophy, and the system management tools must be aligned. Erstad (2003) discussed the importance of using modern technologies as cultural tools to consider when seeking creation of knowledge and meaning for learners. The author argued that digital artifacts give students the ability to support their own learning without complete reliance on an instructor.

Hara and Kling (2000) investigated online-student frustrations. The authors collected data on six students using interviews and observation. Findings suggested that the most prominent frustrations faced by online students included a lack of non-verbal communication cues, instructor feedback, and technological support. Motteram and Forrester (2005) investigated the needs of new online students. The authors found that students' perceived advantages of learning online included the opportunity to work and learn with others in geographically dispersed locations. Students also found that electronic access to libraries and databases increased their ability to find and use sources. Disadvantages included technological limitations or infrastructure failures.

The Uophx branded LMS (rEsource) hosts course readings, topics, objectives, assignments, tutorials, and other related resources. Course materials are organized by lessons using folders. The rEsource platform also provides learners with links to an electronic library, electronic writing lab, financial services, grades, scheduling services, and many other education related services.

The electronic library portal hosts an extensive digital library that provides research and reference services to students. An inter-library loan and document delivery service supplements the online subscription. The inter-library loan, facilitated by the Michigan Information Transfer Source, provides a book loan from the University of Michigan Library.

While the LMS used by Uophx provides little in the way of non-verbal communication cues, instructors participate regularly in discussion forums, providing guidance and feedback to learners. This constant interaction enhances the learning experience for facilitators and students alike.

## Faculty Resources

In a traditional academic setting, individual faculty members are generally responsible for the design of their own course outlines and syllabi. It is assumed that the disciplinary grounding of full-time faculty members qualifies them to determine the content for a course based on the current body of knowledge. With a teaching faculty comprised largely of professional practitioners who are not full-time academicians, the University must ensure that course content reflects the current state of theoretical knowledge in a particular field. For this reason, teams selected from both full-time and practitioner members of the faculty are recruited and contracted to create expanded syllabi for each University of Phoenix course. The teaching faculty member is then able to marry the sound theoretical foundation contained in the course module to their knowledge of current practice in the profession. Gibson and Herrera (1999) found that faculty teaching online courses need assistance in the development of curricula.

Faculty members are assisted in the process of syllabus development by full-time faculty and University staff members who help ensure compliance with University standards, as well as suggest additional avenues beyond the standard syllabus and materials that facilitators might wish to include with the course materials. According to Ally (2004), preliminary planning is a crucial determinant of online instructor success.

Faculty members have access to all the features of rEsource that are available to students, including the University library and related search functions. In addition, faculty members are given the opportunity to revise and improve course materials through Faculty End-of-Course Surveys (FEOCS) and course revisions.

## Classroom Community

Student interaction was found by Garrison (1992) to be an essential element of learning. In addition, Fung (2004), Webb (1982), and Garrison (1993) found collaborative learning environments encourage dialogue, and enhance understanding and construction of knowledge. Doctoral classes are capped at 15 students but are generally kept at 10. Learners work at the same pace and often tackle assignments with the help of learning teams.

Learning teams are used in UoPhx classrooms to create new learning through discussion and collaboration in assignments that require integration and building of knowledge and theories generally set in real-world issues. Stacey (2002) found that online students working in groups of three to five students were able to more effectively relate to one another than students working individually.

Wiesenberg and Stacey (2005) argued that the online instructor plays an important role in creating a community of learners through modeling communication and fostering group cohesion. According to Stacey (2002) it is important for faculty to model communication patterns that build a community that fosters trust and support.

Learning Teams also effectively contribute to the development of community in the online environment. Knupfer (1997), and Wegerif (1998) suggested that the development of a sense of community in online courses could increase student engagement and communication. The authors further suggest that groups should be established early, and interaction in discussions should be kept high.

The online medium is inherently devoid of audible and visual cues. Hall (1999) found that an absence of body language and facial expressions inhibits trust, creating a unique and serious problem for online learners. Without trust, it may be difficult to create the open and free forum needed for debate and collaboration.

Hughes (2002) suggested that faculty increase collaboration by sharing positive teamwork stories with learners, develop icebreaking collaborative projects that hold a high probability for success, and give recognition for individual contributions. They also recommend that faculty become aware of student technical abilities and provide technical contact points. Knight (2006) suggested that university courses move toward a student-centered curriculum approach by utilizing collaborative and active learning activities, promoting student-faculty contact, and improving student engagement. The authors also saw an emphasis on competency over content and collaboration over competition as ways to improve student assessment. Furthermore, Hughes (2002) found that instant messaging systems facilitate the development of an online community by creating a social presence and adding the ability for learners to collaborate synchronously.

Faculty members are the managers of the resources discussed above, and create the links that connect tools to implementation. They also play a central role in the development of community within the online classroom. The following section will discuss this role in more detail and link it to the larger model.

## Faculty Members and Classroom Community

Motteram and Forrester (2005) investigated the needs of new online students. Creating a virtual community with a developed social environment was found to facilitate student progress. A sense of community decreased feelings of isolation and increased participation in forums.

Shih (2000) investigated factors that influenced learning in web-based courses. Responses to a questionnaire indicate that instructor communication is a significant issue for online learners, including social norms and protocols that should be communicated to students. In addition, an emphasis needs to be placed on participation. However, it is easy for the online learning environment to become clogged with excessive amounts of information. Motteram and Forrester (2005) suggest that faculty carefully organize information delivery timelines.

Anderson (2004) argued that online faculty should be learner-centered, knowledge-centered, assessment-centered, and community-centered. Salmon (2000) developed a model for online facilitation in which faculty move through a set of five progressive stages. In the first stage, faculty play a welcoming and motivating role, followed by the development of online social skills with the hope of developing an online community. In the third and fourth steps, faculty facilitate the process and assist in knowledge construction. In the last step, learner independence from the instructor and interdependence on the cohort is encouraged.

Berge (1995) identified four essential roles for online instruction. First, online instructors should be content experts. Instructors also perform the role of manager. In this capacity, they design and implement course foundational elements and procedures. Instructors also perform a social role in which they engage learners in critically reflective dialog. The final role is assisting in the delivery and navigation of electronic resources.

The chief role of UoPhx faculty is facilitation. Faculty shape discussions, create weekly lectures, and, most importantly, deliver weekly feedback based on cognitive evaluation. Faculty members also develop classroom discussion through feedback, follow-up questions, and developing a classroom community that values opposing viewpoints and debate. Peer agreement was found by Makitalo (2002) to impede interaction, while peer debates improved the quality of online interaction. Debates allow students to share alternative theories and ideas while also improving critical thinking skills.

Brookfield (1999) found that student engagement in discussions led to better understanding of the topic and the creation of new knowledge. The use of learning teams by UoPhx further enhances interaction by improving individual comfort levels. Faculty members encourage a safe, fair, and respectful learning environment, while also emphasizing student accountability.

Tallent-Runnels (2005) note that online instructors should teach in smaller rather than larger blocks of time. Palloff and Pratt (1999) add that faculty should invest up to 2 hours per day participating in an online class. This provides greater opportunities for regular and timely interaction and feedback.

Other opportunities for student feedback are provided by Uophx faculty members through reflective end-of-lesson activities called Implications and Conclusions. Kolodner and Guzdial (2000) found activities like reflection and meaningful feedback to enhance online learning outcomes.

## FINDINGS

The Uophx course model was analyzed using End of Course Survey Results, faculty feedback, and related literature. Findings were used to construct a revised instructional model. The analysis was also used to develop a faculty quality assessment model and worksheet.

For purposes of this analysis, 615 responses were obtained from learners enrolled in a Doctor of Management program at UoPhx using the instructional model discussed above. Statements included yes/no scaled response items. Some statement responses were considered proprietary or unrelated to the current study and as a result, were omitted from our analysis.

We first examined doctoral learners' opinions of learning teams. Results show that 80% (n=588) of participants indicated that their learning teams were a valuable part of the course. When curriculum factors were analyzed, only 69% (n=610) of respondents indicated that sufficient time was allotted to master content. Student-resource-related statements indicate general satisfaction with the existing platform: 81% (n=555) of respondents indicated that the student website was a useful resource in doing coursework. Respondents showed even more favorable reaction with 83% (n=615) acceding to the statement that "the online collection provided adequate library resources" and 82% (n=609) agreeing that "the presentation of course materials contributed to course objectives." When asked to respond to the statement "size of class contributed to an effective learning environment," 83% (n=573) of learners were satisfied. Although these scores are generally high, there remains room for improvement. These findings are summarized in Table 1.

Qualitative faculty feedback revealed additional information. Faculty members indicated that the time investment required for online teaching is far greater than that needed for traditional face-to-face instruction, confirming the results reported by Cavanaugh (2006) and Stacey and Rice (2002). In general, faculty found 8 to 10 students per class to be an optimal number. This size provides ample peer interaction and support, while also keeping the class time investment manageable. Faculty also overwhelmingly found extensive feedback to be key in the online learning environment. Faculty support and materials were cited by many instructors as critical for success. Furthermore, they indicated that having a venue for interfaculty communication provides excellent opportunities to share best practices and course materials.

*Table 1. Summary of student end of course survey response totals*

| Statement | Response total | Percentage yes |
|---|---|---|
| Learning team value | 588 | 80% |
| Time adequate to master content | 610 | 69% |
| Satisfaction with platform | 555 | 81% |
| Adequate library resources | 615 | 83% |
| Presentation of learning materials | 609 | 82% |
| Size of class | 573 | 83% |

## THE NEW MODEL

Teams were cited by both faculty members and students as an important element of online education. As such, these will continue to be a central element of the revised model. Team effectiveness is somewhat compromised by the distributed nature of online students, and the lack of in-person interaction. To assist with the question of immediacy in communication, synchronous communication options like Yahoo Messenger may improve the educational effectiveness of learning teams by developing a more collaborative social environment, and will be considered for addition to the platform. This is in accord with the research in this area. Hughes et al. (2002) found that messaging systems facilitate the development of a social presence in the online community. Fung (2004), Webb (1982), and Garrison (1983) found collaborative learning environments encourage dialog and enhance understanding and construction of knowledge.

### Formal Course Size Restrictions

Faculty feedback suggests that online doctoral courses operate best with eight to ten students. Fewer than eight students in a course inhibits interaction, while more than ten is difficult for faculty to effectively manage. This also allows for an effective learning team size, defined by Stacy (2002) as three to five students. The revised model incorporates a minimum class size of eight and maximum size of twelve.

### Work Organization Restructuring

According to student feedback, the current model, delivering a single lesson per week, is insufficient. Price (2005) recommends that 2 weeks be reserved for each lesson. According to the author, each lesson should contain an introduction, objectives, directions to proceed, class discussion, exercises, and assignments.

Student feedback and course materials were considered by many faculty members to be a crucial component of success. Providing rubrics to faculty will enhance feedback, standardize evaluation, and free additional faculty time for class discussion. Gibson and Herrera (1999) found that the extended demands of online instruction make the development of curriculum and other course materials necessary for success. These extended demands are highlighted by faculty comments that the time investment for online teaching is far greater than in face-to-face instruction. Although it has a high cost in time, online communication is an essential component of online learning. Brookfield (1999) found online course discussion to facilitate the development of critical thinking skills and helps connect learners to topics. To ensure that faculty members can devote the greatest amount of time to online interaction, we have added rubric development as a necessary element of the revised model.

## FACULTY QUALITY ASSESSMENT

Even the best instructional model is worthless without quality faculty to support it. The findings discussed in this chapter were used to develop a tool for assessing faculty quality in the online environment. The quality of online faculty can be a difficult dimension to assess, given that many work in settings far removed from the host school. This separation makes it more difficult to communicate culture, values, preferred practices, expectations, and curriculum development and delivery models and tools to faculty. This problem increases the disparity between on-site and online education models. Even the best curriculum model can be made ineffective by faculty using poor online instruction techniques.

Many schools have focused on the use of student and faculty surveys as a measure of educational effectiveness; however, it is less common for these schools to measure effectiveness using a robust model designed specifically for the online environment. The literature analysis presented above was used to construct a tool for measuring faculty quality in the online environment. This tool serves two central purposes: one to measure faculty from an individual and group perspective, the other being identifying training needs from both an individual and group perspective.

The assessment tool included below measures 4 primary focal points (community-centered, assessment-centered, learner-centered, and knowledge-centered) found by previous researchers to relate to online classroom quality. The importance of community was argued by Wegerif (1998), Wiesenberg and Stacey (2005), and Motteram and Forrester (2005), knowledge creation was highlighted by Kolodner and Guzdial (2000), the importance of holding a learner focus was discussed by Hara and Kling (2000), Shih (2000), while Hara and Kling (2000), Kolodner and Guzdial (2000), discussed the importance of assessment. These four pillars also match the four roles of online faculty discussed by Berge (1995).

This assessment tool measures each of these classroom elements equally through established and new metrics. Established metrics commonly used in the academic arena include student and faculty surveys and GPAs. The total Z scores in each of the columns (community centered, assessment centered, learner centered, and knowledge centered) are added up to generate a final faculty Z score. This final faculty Z-score will help administrators visualize individual faculty on the bell curve in comparison to the total faculty population. Cut off scores can be used to place faculty into quality quadrants.

The faculty quality model illustrated below is designed to be a living assessment tool. Additional measurement areas can be added to each of the four focal points; however, no focal point should exceed 25% of the total weight of the final score since each area holds equal importance. As the population mean changes, so to will the Z score. As Z scores change, cut-off scores for each faculty group can be modified.

Since each focal point represents a significant area of the online classroom, problem areas will be easy to identify and in most cases correct through training and further assessment. As such, faculty training programs should be structured around these pillars. Although this tool will likely be most affective if used to model all other aspects of curriculum delivery and development as well as faculty training, it is also possible for an institution to integrate this assessment piece into existing structures.

## CONCLUSION

Our model for online instruction stands on four pillars: a platform for instructional delivery, availability of student and faculty resources, creation of an online community, and faculty assessment. An asynchronous platform with constant availability and technical support will be a necessary component for a successful online doctoral course, but we suggest that this is not sufficient: adding synchronous communication abilities to the platform strengthens the platform as a facility, and also contributes to the better formation of an online community. The fostering of community likewise requires that group sizes be kept within strict limits, neither too large nor too small. Online doctoral courses should be kept at sizes of eight to twelve students, and formal organizational structures should support this goal.

Finally, we keep foremost in mind that the chief determinant of success in a course for motivated students remains the effectiveness of the faculty member. A course development model must focus on providing effective tools for faculty members to concentrate on their most critical functions in an online doctoral course- the fostering of community and a focus on students, knowledge, and assessment. We would argue that rubric creation thus is an essential component of online doctoral

*Table 2. Faculty assessment tool*

| Area | Raw Score | Population Mean | Z- Score | Score |
|---|---|---|---|---|
| ***Knowledge Centered (25%)*** | | | | |
| Fostering Knowledge Construction | | | | |
| Content Expertise | | | | |
| Design and Implement course foundational elements and procedures | | | | |
| Assist in the delivery and navigation of electronic resources | | | | |
| Organize information delivery timelines | | | | |
| *Total* | | | | |
| ***Learner Centered (25%)*** | | | | |
| Welcoming | | | | |
| Motivating | | | | |
| Support Learner Independence on Instructor | | | | |
| Engage learners in critically reflective dialog | | | | |
| Withdrawals | | | | |
| *Total* | | | | |
| ***Assessment Centered (25%)*** | | | | |
| Grade Variance | | | | |
| GPA Average | | | | |
| Sets attainable Goals | | | | |
| Feedback Timely | | | | |
| Feedback Regular | | | | |
| Feedback Meaningful | | | | |
| *Total* | | | | |
| ***Community Centered (25%)*** | | | | |
| Establish groups early | | | | |
| Maintain high level of interaction in discussions | | | | |
| Develop icebreakers | | | | |
| Develop collaborative projects | | | | |
| Recognize individual and team contributions | | | | |
| Foster online social skills | | | | |
| Foster peer debates and student engagement | | | | |
| Value student engagement | | | | |
| Support Learner Interdependence on Cohort | | | | |
| Participation | | | | |
| *Total* | | | | |
| **Total** | | | | |

course creation. In addition, faculty should be regularly assessed using either the model included in this discussion or another tool that considers quality from the areas discussed throughout this article.

The advantages of following this model for online doctoral course development are legion. Chief among them is the improvement of the overall environment for both students and faculty members, as indicated in formal surveys of satisfaction. The increased satisfaction with the classroom environment leads to greater retention of doctoral students, and to a larger number of students who complete their programs of study.

The limitations of our model lie chiefly in the increased exertion of course developers and administration. Adding more required components, and restructuring the organization to guarantee optimal course sizes involves an immediate cost with no immediate benefit, and it might be difficult to justify. Further limitations are of a technical nature: an online platform currently consumes a great deal of network bandwidth, and adding a synchronous platform will further tax both the university and the students' information technology resources. This model applies well to any University seeking to develop online doctoral-level instruction. It might also apply well in other distance-learning applications, especially with focus on platform and faculty support and assessment.

## REFERENCES

Ally, M. (2004). Foundations of educational theory for online learning. In T. Anderson & F. Elloumi (Eds.), *Theory and practice of online learning* (p.5). Athabasca University, Alberta: Creative Commons.

Althaus, S. L. (1997). Computer-mediated communication in the university classroom: An experiment with online discussions. *Communication Education, 46*, 158–174. doi:10.1080/03634529709379088

Anderson, T. (2004). Toward a theory of online learning. In T. Anderson & F. Elloumi (Eds.), *Theory and practice of online learning*. Athabasca University, Alberta: Creative Commons.

Bee, R., & Usip, E. (1998). Differing attitudes of economic students about web-based instruction. *College Student Journal, 32*, 258–269.

Berge, Z. (1995). Facilitating computer conferencing: Recommendations for the field. *Educational Technology, 35*(1), 22–30.

Brookfield, S. D. (1999). *Discussing a way of teaching*. Jossey-Bass: San Francisco.

Cavanaugh, J. (2006). Comparing online time to offline time: The shocking truth. *Distance Education Report, 10*(9), 1–6.

Clark, J. (2000). Collaboration tools in online learning environments. *ALN Magazine, 4*, 1.

Cooper, L. (1999). Anatomy of an online course. *T.H.E. Journal, 26*(6), 49–51.

Erstad, O. (2003). Electracy as empowerment: Student activities in learning environments using technology. *Young: Nordic Journal of Youth Research, 11*(1), 11–28.

Fung, Y. Y. H. (2004). Collaborative online learning: Interaction patterns and limiting factors. *Open Learning, 19*(2), 135–149. doi:10.1080/0268051042000224743

Garrison, D. R. (1992). Critical thinking and self-directed learning in adult education: An analysis of responsibility and control issues. *Adult Education Quarterly, 42*(3), 136–148.

Garrison, D. R. (1993). A cognitive constructivist view of distance education: An analysis of teaching-learning assumptions. *Distance Education, 14*(2), 199–211. doi:10.1080/0158791930140204

Gibson, J., & Herrera, J. (1999). How to go from classroom based to online delivery in eighteen months or less: A case study in online program development. *T.H.E. Journal, 26*(6), 57–60.

Hall, T. (1999). Intelligence community collaboration baseline study, final report. Retrieved January, 2006 from http://collaboration.mitre.org.

Hara, N., & Kling, R. (2000). Students' frustrations with a web-based distance education course. *First Monday*, *4*(12), 6–24.

Heckman, R., & Annabi, H. (2003). A content analytic comparison of FTF and ALN case-study discussions. In P*roceedings of the 36th Hawaii International Conference on System Sciences (HICSS'03)*. Honolulu, Hawaii, USA.

Hughes, S. C. (2002). Overcoming social and psychological barriers to effective online collaboration. *Educational Technology*, *5*(1), 86–92.

Kanuka, H., & Anderson, T. (1998). Online interchange, discord, and knowledge construction. *Journal of Distance Education*, *13*, 57–74.

Knight, W. E. (2006). Paper presented at the Annual Forum of the Association for Institutional Research (AIR). Chicago, IL.

Knupfer, N. N. (1997). Participant analysis of a multi-class, multi-state, online, discussion list. In *Proceedings of Selected Research and Development Presentations at the 1997 National Convention of the Association for Educational Communications and Technology.* Albuquerque, USA, 133-137.

Kolodner, J. L. (2000). Theory and practice of case-based learning aids (pp. 215-242). In D. Jonassen (Ed.), *Theoretical foundations of learning environments*. Mahwah, New Jersey: Lawrence Erlbaum and Associates.

Koszalka, T. A., & Ganeson, R. (2004). Designing online courses: A taxonomy to guide strategic use of features available in course management systems (CMS) in distance education. *Distance Education*, *25*(2), 245–256. doi:10.1080/0158791042000262111

Lim, D. H. (2006). *Online vs. blended learning: Differences in instructional outcomes and learner satisfactions.* Paper presented at the Academy of Human Resource Development International Conference (AHRD). Columbus, OH, USA, 809-816.

Makitalo, K. (2002). Mechanisms of common ground in case-based web discussions in teacher education. *The Internet and Higher Education*, *5*(3), 247–265. doi:10.1016/S1096-7516(02)00112-4

Moskal, P. (2006). Assessing online learning: What one university learned about student success, persistence, and satisfaction. *Peer Review*, *8*(4), 26–30.

Motteram, G., & Forrester, G. (2005). Becoming an online distance learner: What can be learned from students' experiences of induction to distance programmes? *Distance Education*, *26*(3), 281–298. doi:10.1080/01587910500291330

Palloff, R. M., & Pratt, K. (2003). *The virtual student: A profile and guide to working with online learners.* Wiley: San Francisco.

Pappas, J. P., & Jerman, J. (2004). *Developing and delivering adult degree programs: New directions for adult and continuing education*. Jossey-Bass: San Francisco.

Price, R. V. (2005). A model for designing web-based units of instruction. *Distance Learning*, *2*(1), 5–8.

Roach, V., & Lemastus, L. (2006). Satisfaction with online learning: A comparative descriptive study. *Journal of Interactive Online Learning*, *5*(3), 317–332.

Rovai, A. P. (2001). Building a sense of community at a distance. *International Review of Research in Open and Distance Learning*, *3*(1), 1–16.

Salmon, G. (2000). *E-moderating: The key to teaching and learning online*. London: Kogan.

Shih, C., & Gamon, J. (2002). Relationships among Learning Strategies, Patterns, Styles, and Achievement in Web-based Courses. *Journal of Agricultural Education*, *43*(4), 1–11.

Stacey, E. (2002). Social presence online: Networking learners at a distance. *Education and Information Technologies*, *7*(4), 287–294. doi:10.1023/A:1020901202588

Tallent-Runnels, M. K. (2005). How to teach online: What the research says. *Distance Learning*, *2*(1), 21–27.

University of Phoenix. (2006). *Student rEsource Page*. Retrieved on 2/22/06 from https://mycampus.phoenix.edu.

University of Phoenix (2007). *Internal SEOCS report*. Retrieved on 2/11/07 from internal data source.

Webb, N. M. (1982). Student interaction and learning in small groups. *Review of Educational Research*, *52*(3), 421–445.

Wegerif, R. (1998). The social dimension of asynchronous learning networks. *Journal of Asynchronous Learning Networks*, *2*, 34–49.

Wiesenberg, F. P., & Stacey, E. (2005). Reflections on teaching and learning online. Quality program design, delivery and support issues from cross-global perspective. *Distance Education*, *26*(3), 385–404. doi:10.1080/01587910500291496

# Chapter 3
# A Multi-Disciplinary Strategy for Identifying Affective Usability Aspects in Educational Geosimulation Systems

**Elizabeth S. Furtado**
*University of Fortaleza, Brazil*

**Vasco Furtado**
*University of Fortaleza, Brazil*

## ABSTRACT

*In this article we propose a multi-disciplinary strategy for identifying affective usability design aspects in educational geosimulation systems. It is based on the association of these aspects with an architecture that defines the basic components of a geosimulation system as well as the learning strategies used in this context. Our goal is to provide design strategies that might elicit positive emotional responses from the students in learning experiences. We illustrate how these strategies have been used in a learning system by evaluating the students' emotional responses evoked during their interaction with the system.*

## INTRODUCTION

The human computer interface (HCI) field has often attracted considerable attention from academia and industry, and particularly the use of concepts such as usability and adaptation during software design as a salient factor for obtaining more usable systems. Traditionally, the usability of a developed system has been evaluated to assure both its effectiveness (such as the number of successful task completions) and efficiency (such as the time required to complete an interactive task). Recently, these assumptions have been revisited and broadened to embed the concepts of the affective quality theories. Affective quality is related to the users' emotional responses (such as the affect, activity, and attitude) in regard to the system that they are experiencing (Chorianopoulos & Spinellis, 2006).

We claim that the affective aspect (such as users' feeling states and their involvement with the content) is particularly relevant in the context of educational systems, since learning strongly depends on how synergistic the relationship between teacher and student is. So, the affective dimension of the user interfaces of an educational system is an issue to be considered. In other words, it is important to identify the ways in which the interactive objects of an educational system's user interfaces can be perceived by the students when manipulating, creating, visualizing, or controlling these objects in their learning experiences. In this text, these perceived ways (how the student was persuaded to do something) are evoked by affective usability design aspects that are techniques (such as persuasive techniques and personalization) and characteristics (about the look, sound, and feel) applied to the user interfaces.

Despite the aforementioned advance in HCI, affective usability aspects are still not taken into account in the interaction design of educational systems. The reasons can be the following: There is not yet a strategy that shows how the integration of usability and affective quality concepts can be done with learning strategies and how such integration can be useful to evaluate the users' satisfaction in learning experiences.

This situation can be even worse if the educational system involves complex phenomena in urban centers, whereby the interaction with geographical information is intense. In many cases, these systems are based on the combined use of Geographical Information Systems (GIS) with multiagents for simulation of social or urban environments, which characterizes a geosimulation (Benenson & Torrens, 2004) and (Billari & Prskawetz, 2003). In educational agent-based simulation systems, intelligent agents support the interaction between the simulation model and the user (Gibbons, et al., 2001). Simulation aims to represent one phenomenon via another. In educational terms, simulation is important because it allows learning through the possibility of doing (Piaget, 1976). On the other hand, social or urban environments are dynamic, non-linear, and made of a great number of interacting entities, characterizing a complex system (Wu, 2002). Interactive aspects in these systems (such as precision and realism in simulations) can evoke different emotions from students.

In this article, we propose a multi-disciplinary strategy for associating the concepts of usability, computer education, and affective quality. In this strategy, the interaction between student and teacher is analyzed under the light of learning strategies used in educational geosimulators for defining the main emotional constructs that are involved in this process. This strategy is composed of a set of steps to be followed by developers interested in defining the affective usability design aspects that an interactive educational system must have in order to evoke students' emotional responses in learning experiences. In addition, it is useful to professionals (such as teachers and designers) interested in evaluating the students' satisfaction using an interactive system. We also present, in the final part of the paper, how we evaluated students' emotional responses in learning experiences by affective levels. For this we used an already-deployed system for training police officers.

## THE MULTI-DISCIPLINARY STRATEGY

Figure 1 illustrates the multi-disciplinary strategy we developed to generate a conceptual framework from education, user interaction, and affective quality theories. The framework refers to the association of affective usability design aspects with learning strategies by affective levels. This strategy is composed of three steps. First of all, we identify the learning strategies supported by educational geosimulation systems. Afterward, we analyze the possible emotional responses of affect, activity, and attitude that the stu-

*Figure 1. The multi-disciplinary strategy*

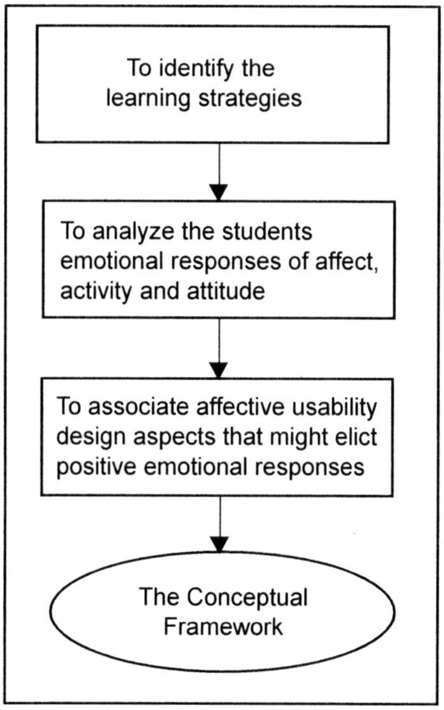


dents can have in the process of interaction with these systems. Then we identify affective usability design aspects in these systems that might elicit positive emotional responses from the students by generating the mentioned framework. Our goal is to provide design concerns elucidated from theoretical literature when appropriate, which can help professionals both to design and to evaluate this kind of educational system. Specifically, we intend to show how attention paid to users' emotional responses can be used as a basis for developing and evaluating these systems in order to elicit positive responses from students.

## Identification of the Learning Strategies

Since we are applying this strategy in a particular domain, i.e., Educational Geosimulation systems, we focused on the study of such systems. Recently (Furtado & Vasconcelos, 2007) brought together the components of what they consider relevant for the development of urban activity training by means of simulation systems, and proposed the Educational Geosimulation Architecture (EGA). EGA follows a traditional architecture of an Intelligent Tutorial System (ITS) in which three main models are distinguished: the student model, the teacher model, and the domain model. However, some particular aspects are present in EGA: the GIS, for the appropriate representation of the simulation environment (domain model), and multiagents,such as domain agents that represent the domain and/or student model, and pedagogical agents, representing the teacher model). Last, but not least important, there are the user interfaces that are the communication channel between the system and the student. The three basic strategies of learning that the architecture entails are: i) learning by instruction, which is obtained from the material (information, examples, and concepts) provided by the teacher; ii) learning by doing, when a student uses the simulation (Piaget, 1976) and; iii) learning by reflection; when the architecture provided students with assistance to help them have a better understanding of the phenomena underlying the simulation.

## Analysis of Students' Emotional Responses in Affective Quality

In this step, we perform two tasks. First the students' possible emotional responses during their learning experiences are analyzed regarding the affective quality concepts. The *concepts of affective quality* are organized in three levels (according to Norman's affect model (2004):

- At the *Affect* level, the interactive experiences of the users are based on their feeling states (such as motivated, enthusiastic, and calm) for using a system that can be beautiful, attractive, and with less constraining interaction. In addition, users have pleasure in having fun (Jordan, 2006);

- At the *Activity* level, the *activities* of users are based on free exploration, context, and participation. When performing simulation activities, the users' emotional responses can be: i) engagement in figuring out simulations (stimulated by perceived ease of use and no feeling of risk and failure); and ii) involvement in understanding the simulation content (stimulated by the perceived ease of content localization and trust);
- At the *Attitude* level, the *attitudes* of the users are supported by assistances. When analyzing simulation results, the users' emotional responses can be perceived: After superficial involvement, they can have a sense of true commitment in understanding the simulation context that can lead to the formation of an attitude.

Second, the possible emotional responses are associated with the identified learning strategies by considering the three affective levels. For instance, the affect (such as feeling states) is a neurophysical state that is consciously accessible as a simple, non-reflective feeling (Zhang & Li, 2004). This is the reason we associated it with the learning-by-instruction strategy. The simulation and explanation learning strategies, in turn, raise emotional responses related to the behavioral (activity) and reflective (attitude) levels of a person, respectively. For instance, a learning system may elicit enjoyment (e.g., pleasure). Then, the users may continue using it for a long time and become emotionally absorbed (e.g., engagement by content that matches their objects of interest, preferences, and restrictions). Finally, the users may decide that they like a specific learning task, which leads to the formation of an attitude (e.g., a summary evaluation of an experience supported by explanations).

Identification and Association of the Affective Usability Design Aspects. Finally, affective usability design aspects that might elicit positive emotional responses from the students are identified and associated with a specific learning strategy by affective levels. Figure 2 depicts three different parts: In the left part, we can visualize the affective levels of the student when interacting with an EGA system, whose components are illustrated in the right part. In the central part, we can see several aspects associated to each affective level and that can allow learning through a certain strategy.

At the affect level, affective usability design aspects related to *look, sound, and feel* dominate. They are being associated to the learning-by-instruction strategy because an instruction-based system can be designed with sophisticated technological resources (as tangible interfaces) that positively influence users' experience on an emotional level (such as having fun) when interacting with it. Design aspects can refer to the characteristics

*Figure 2. Affective usability aspects in educational geosimulation systems*

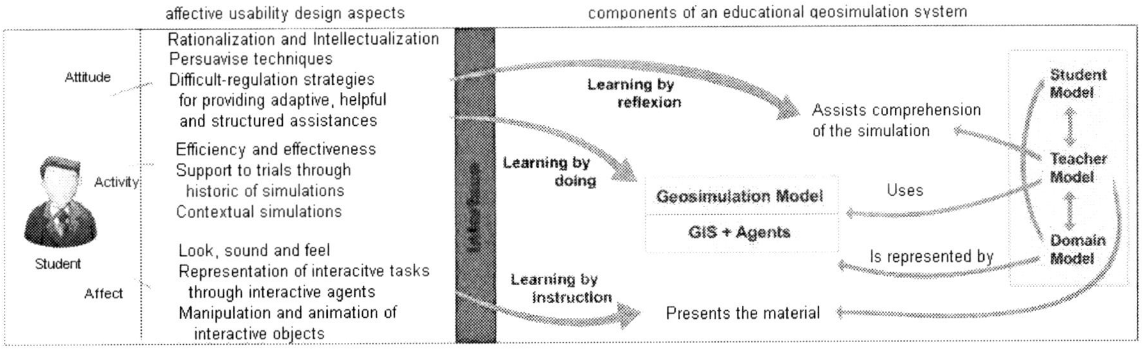

of an educational system's interactive agents. An agent can be modeled anthropomorphically. It can be seen through synthetic characters in the user interfaces (UI) of affective systems. A design concern about this characteristic is that designers should know that anthropomorphic effects can cause interaction problems when considering the fact the user can expect the system to be intelligent and cognitively potent. This expectation may lead to frustration in the user when the system cannot meet it (Höök, 2004). Design aspects can lead to enhancing the users' satisfaction when interacting with the interactive objects that represent agents, users' action, and/or context of study. Here are two examples of these aspects: i) direct manipulation, whereby users need to have complete control over the system and it can be hard to accept characters in the interface that run outside their control (Höök, 2004); and ii) animation, which—through several effects (such as color changes and panning)—can make a UI more enjoyable to use, leading users to have fun.

At the activity level, the affective usability design aspects related to *efficiency and effectiveness* dominate and make it possible to get emotional responses in simulation experiences. In a simulation process, the users' possible emotional responses can be elicited from their engagement in figuring the simulations out. The design aspect refers to the support that an educational system gives to trials through history of simulations, free exploration, and treatment of errors. For instance, if the users can make trials as many times as they want, with no feeling of risk and failure, then they will have more intentions to do so. The number of times the users figured the simulations out and the time spent in this process are important criteria to measure their engagement in interactive simulation experiences.

At the attitude level, the affective usability design aspects related to *rationalization and intellectualization* dominate and can lead students to have emotional responses in Assistance Experiences. Like teachers, educational systems can provide students with different modalities of assistance, such as: a) explanation—this refers to the act or process of explaining something; b) hint— this is a brief or indirect suggestion, a tip; and c) tutorial—this refers to instructions describing how the users can proceed at a certain moment. In (McGuinness, et al., 2006), some explaining techniques show users the provenance of the explanations. Persuasive technologies, studied in the area of Captology, can interactively manipulate what people think and do. Examples of these technologies are the following: persuading through customization, simplifying, and guided persuasion (Fogg, 1006). Tailored information is more effective than generic information in changing users' attitudes and behaviors. "Hint" is a kind of simplifying persuasive technique, because users know they will find brief help. "Tutorial" is a guided persuasion technique that provides opportunities to persuade users along the way.

The affective usability design aspects illustrated in Figure 2 are admittedly incomplete: They are meant to spawn new ways of thinking about the users' satisfaction with these systems by only focusing on the way in which the interactive objects that represent the learning strategies should be perceived by the users. Characteristics about the users and their environment of interaction are not considered.

## EVALUATION OF THE EMOTIONAL RESPONSES IN LEARNING EXPERIENCES

In this section, we describe an educational geosimulator, called the ExpertCop system, that follows EGA architecture. Then we describe our experiment of evaluating the affective usability design aspects of this system when used in learning experiences.

## System Overview

In brief, the ExpertCop system supports learning by means of simulating phenomena that provoke crime in an urban area. The goal is to lead the students to understand the consequences of a police resource allocation plan as well as the cause-and-effect relations. In ExpertCop, the simulations occur in a learning environment along with graphical visualizations that aid the users' learning. The domain agents are the police team, the criminals, and the targets (notable points). Criminals are the most important agents in the simulation process. Their behavior is based on a rule base which orients them to look for targets and to commit crimes (see Furtado & Vasconcelos (2007) for a detailed description). The Pedagogical Agent (PA) represents the teacher model and is aimed at helping the users to understand the phenomena represented by the simulation.

Interaction with the domain agents is done at two moments. First, before the simulation, the user must allocate the police in the areas to be patrolled and available on the geoprocessed map. Crimes are represented on the map as points. The goal of the user is to provide a good allocation, which prevents the occurrence of crimes to the greatest extent. Second, during the simulation, the movement of the police patrol routes is shown. The user can follow the simulation process in the simulation interface. At the end of the simulation process, the user accesses the system's pedagogical tools. Upon each new allocation performed, the system can comparatively evaluate the simulated moments, showing the user whether the modification better effected the crime rate or not. After the simulation, interaction is possible via queries to crimes that occurred. It is up to the PA to answer these queries, providing students with several assistances (such as the reasons that lead agents to perform the crimes).

## Evaluation

ExpertCop was used to support a course on Information Technology for Police and was intended to help police officers reflect on the forms of treatment and analysis of information, and how these influence the understanding of crime. The audience was made up of thirty professionals in the area of public safety: civil police officers, chiefs of police, and military police (which are the majority). A quantitative analysis of the effectiveness of ExpertCop in the learning process is discussed in Furtado and Vasconcelos (2007). In this paper, we will concentrate on the description of an empirical qualitative analysis of the affective usability design aspects of this system that could evoke students' emotional responses in learning experiences at the *affect*, *activity*, and *attitude* levels.

The testing session was composed by the following scenarios of the training process: familiarization, resource allocation, simulation, and evaluation of simulation results. It took place in 30 to 40 minutes. Initially, students made use of the tool in an illustrative simulation to familiarize themselves with the functionalities. In the resource allocation scenario, training was carried out by a set of at least two simulations in city areas. In the first simulation, the students had to create and configure a certain number of teams (according to the size of the area), allocate them on the map, and activate the simulation. At the end of the first simulation, we asked the students to identify, according to their beliefs, factors (concepts) that influenced the occurrence of the crimes. They did so by observing the map of the crimes that occurred and those that were avoided. After collecting the students' concepts, we allowed them to use the system's pedagogical support. After this moment, the collection of factors influencing the crimes was carried out again. In the subsequent simulation, we repeated the same area to serve as a comparison with the initial simulation already completed. At the end of the process we also applied a questionnaire asking about how much fun the system was and how difficult learning its use was. We participated in the testing session as experimenters.

## User Test Results

The results refer to the analysis of the affective usability design aspects of the system that can evoke students' emotional responses. A discussion about these emotional responses evoked by the system is done as follows and is classified into three groups related to the *affect*, *activity*, and *attitude* levels.

### Emotional Responses Due to the Representation of the Interactive Elements

We evaluated here what interactive aspects can bring aesthetics and beauty to the system, and consequently pleasure to the students in using it. The affective usability design aspects of the system considered were: the choice of significant symbols, the definition of how these symbols were used during the communication of the users with the system, and the ergonomic and graphical design aspects used to create the artistic UI design. The users had the following emotional responses: i) during the initial contact with the system, the emotional response of the users was one of complete pleasure, due to the aesthetic aspect of the interface with interactive objects and colors; ii) The entertaining aspect of the game-like structure was potentially a responsible factor for elicitingthe pleasure of having fun from the students. Being fun is particularly delicate in game-based ITS systems. Sometimes the system interface is so much fun that it leads the user to game with the system without any intention of learning the process behind the game. It is important, then, to design the ITS to be resistant to the game factor, i.e., the user obtaining good scores in the game by just guessing. Moreover, the game aspect cannot bypass the main goal which is learning from the use of the system; and iii) Several graphical characteristics elicit good feelings from the users, leading them to use the system comfortably. Assistances are provided in several ways, and are associated with the agents allocated on the map. One of these assistances is the possibility of visualization on a map of notable points such as squares, gas stations, drugstores, schools, etc. Doing so, the user can allocate the police agents by directly drawing on the map the routes that must be followed by these agents while taking into account the hot spots to be monitored. Even the task of police allocation, that our observations and questionnaire responses have captured as being the most difficult thing to do in the system, was considered agreeable to do. Our analysis is that the direct manipulation of the agents representing the police officer in the map is the main factor that influenced this feeling.

### Emotional Responses During the Simulation

We evaluated how the system provided students with support to trials in simulation experiences. The affective usability design aspects of the system considered were: history of simulations, free exploration, treatment of errors, and the simulation content's characteristics. The emotional responses of the users observed were the following: i) Involvement in understanding the simulation content: all the users found the results of the simulation very interesting, in particular, six users who lived in the same city represented in the GIS. They were eager to plunge into allocation and then anxious to see the results. They likewise took far longer than the others for performing the allocation process. We think this was provoked by the contextual identification aspect of the users with their area of work. In learning theories, it is known that users are more involved with the content when it represents objects of their own interest; and ii) Engagement in figuring the simulations out: The process is friendly and a history of simulation results can be accessed allowing the users to follow their performance. During the iterative process of allocation-simulation-visualization of results, users applied some spontaneous collaborative practices to identify similar strategies among themselves. The possibility of accompanying their evolution in the simulations by means of bar graphs helped this collaboration process.

### Emotional Responses Due to Assistances

Here we evaluated how the system motivated students to try out one or more assistances. The affective usability design aspects of the system considered were: persuasiveness, trust, and credibility in assistances to obtain. The users had the following emotional responses: i) They reacted most positively to the hint explanation (showing crime patterns) that made them reflect on their allocation, commenting that they understood why some crimes have not been avoided. A hint button persuaded users to "click on it" and the messages could affect the users' intentions of use (in other words, motivate them to learn by exploring the assistances); and ii) In the first simulation, users trusted in the system's accuracy but not in its results. They trusted more in the system when an explanation of the occurrence of the events was given. They could interpret the model behind the simulations.

## RELATED WORKS

There are some works that make the relation between affect and usability in several domains (see examples in Zhang & Li, 2004). However, there was no mention of any works in the educational domain. We found in the literature of applications for Digital Television, the work of Chorianopoulos and Spinellis (2006) that associates design features to affective constructs and defines an evaluation model of the interaction of the viewers with the interactive applications for Digital Television. This model presents a clear separation among the UI and the TV content. In this paper we followed a different approach by considering the system as a whole. As we cannot measure a system independently, we preferred to associate these concepts by integrating them into the generic learning strategies.

In learning systems, the association of these concepts is still an open issue. In Perez, et al. (2004), an affective model is included in the tutorial system architecture. Their idea is to identify the students' emotional status for providing adaptive assistances. Affective relations of power and identity among students have been explored in collaborative virtual learning systems, but there is no relation between these concepts and the usability in these systems (Mattos, 2005). Several works in games and entertainment (Galvão, et al., 2000; Leemkuil, et al., 2003) use simulation with an educational propose. Game simulators have a different pedagogical strategy because they focus on the results of the simulation, emphasizing only the fun aspect of the interface. In learning systems, the most important aspect is the process itself, and it should also include the formation of attitude. In our previous work (Furtado, et al., 2007), we showed the relationship existing among the usability and affective quality concepts in which both must be employed to evaluate the users' satisfaction with EGA systems. However, we did not describe a strategy that professionals can follow for identifying the affective usability design aspects in educational systems, as we have done here.

## CONCLUSION

The strategy proposed in this paper suggests grouping affective usability design aspects into affective levels and associating them into learning strategies. Then it suggests considering the aspects associated with a certain learning strategy when designing the interaction of an educational system that intends to assure such learning. This strategy is based on the fact that users have emotional responses when manipulating some interactive aspects in learning experiences. This was the reason that we illustrated this strategy with the evaluation of an educational system deployed for the area of law enforcement. A training course with police officers interacting with this system was observed and three different scenarios were taken into account to discuss the results. The perceived behavior of students in ExpertCop revealed the association of their satisfaction in using the system with the HCI concepts described in Figure 2.

If this strategy goes beyond what the traditional usability concepts offer for designing and evaluating systems, it has some limitations as well. We intend to complete the identified aspects that do not address, for instance, characteristics of context in which the interaction should take place. They refer more to the way in which the interactive objects that represent the learning strategies should be perceived by the users. Furthermore, other interactive learning processes and their students' emotional responses must be studied. Finally, we intend to conduct another training course in order to evaluate the subjective user satisfaction by applying a verbal decision analysis. Our goal is to better understand the subjective questions that influence the decision of the most appropriate interaction design solution for a specific learning strategy. The idea is to structure the perceived aspects as criteria to obtain a ranking of interface solutions.

## REFERENCES

Benenson, I., Torrens, P.M. (2004). Geosimulation: object-based modeling of urban phenomena: Computers, Environment and Urban Systems. *Computer, Environment and Urban System, 28*, 1-8.

Billari, C.F., Prskawetz, A. (2003). *Agent-based computational demography: Using simulation to improve our understanding of demographic behaviour.* Germany: Phisica-Verlag.

Chorianopoulos, K., Spinellis, D. (2006). User interface evaluation of interactive TV: A media studies perspective. *Univiversity Access Information Society, 5*, 209-218.

Fogg, B.J. (2003). *Persuasive technology.* Morgan Kaufmann.

Vasconcelos, E., & Furtado V. (2006). Educational geosimulation. In Brasileiro, F., & Mendes, M. (Eds.), *Advances in computer-supported learning.* IGI Global: Hershey.

Furtado ,V., & Vasconcelos E. (2007) Geosimulation in education: A system for teaching police resource allocation. *International Journal of Artificial Intelligence in Education, 17.*

Furtado, E., Furtado V., & Vasconcelos E. (2007). A conceptual framework for the design and evaluation of affective usability in educational geosimulation systems. *Proceedings of Interact.* Rio de Janeiro.

Galvão, J. R., Martins, P. G.,& Gomes, M. R. (2000). Modeling reality with simulation games for a cooperative learning. In *Proceedings of the Winter Simulation Conference*, (pp. 1692-1698). Orlando, Florida: Society for Computer Simulation International.

Gibbons, A. S., Lawless, K. A., Anderson, T. A., & Duffin, J. (2001). The web and model-centered instruction. In B.H. Khan (Ed.), *Web-based training.* Englewood Cliffs, NJ: Educational Technology Publications, 137-146.

Höök, K. (2004). User-centred design and evaluation of affective interfaces. In Zs. Ruttkay, & C. Pelachaud (Eds.), From brows till trust: Evaluating embodied conversational agents. Kluwer. http://www.sics.se/~kia/papers/hook1.pdf

Jordan, P. (2002). *Designing pleasure products.* Taylor & Francis.

Leemkuil, H. H., de Jong, T. , de Hoog, R. , & Christoph, N. (2003). KM quest: A collaborative internet-based simulation game. *Simulation & Gaming 34*, 89-111.

Mattos, F. L. (2005). *Concepção e desenvolvimento de uma abordagem pedagógica para processos colaborativos a distancia utilizando a internet.* PhD. UFC. McGuinness, D., Li Ding, Furtado, V., Glass, A., Zeng, H., & Chang, C. (2006). Explanation interfaces for the semantic web: Issues and models. In *Proceedings of SWUI,* V.1. Berlin: Springer Verlag.

Norman, D. (2004). *Emotional design.* Basic Books.

Perez,Y., Gamboa R., & Ibarra, O. (2004). Modeling affective responses in intelligent tutoring systems. In *Proceedings of IEEE ICALT.*

Piaget, J. (1976). *Le comportement, moteur de l'évolution.* Paris: Ed. Gallimard.

Wu, F. (2002). Complexity and urban simulation: Towards a computational laboratory. *Geography Research Forum 22*, 22-40.

Zhang, P., & Li, N. (2004). The importance of affective quality. *CACM'04.* 2004.

*This work was previously published in International Journal of Information and Communication Technology Education, Vol. 4, Issue 1, edited by L. Tomei, pp. 15-24, copyright 2008 by IGI Publishing (an imprint of IGI Global).*

# Chapter 4
# Creating High Quality Learning Object Metadata Based on Web 2.0 Concepts

**Daniel Dahl**
*European Research Center for Information Systems (ERCIS) Westfälische Wilhelms-Universität Münster, Germany*

**Gottfried Vossen**
*European Research Center for Information Systems (ERCIS) Westfälische Wilhelms-Universität Münster, Germany*

## ABSTRACT

*When introducing the metadata standard LOM, objectives such as the ability to find or to reuse learning objects were followed. These objectives are actually achieved in LOM to a limited degree only, despite the designation as de-facto standard for description of electronic learning content. Based on the complexity of the standard, a high theoretical potential faces rejection in practice. One reason for this is that the process of metadata generation—for example, who creates which metadata attributes—is not defined in detail yet. This paper illustrates an approach which guarantees a high quantity as well as a high quality of learning object metadata records, bringing together known ways of metadata creation and the new paradigm of users describing content as implemented in recent Web 2.0 applications. In the context of a concrete e-learning platform, we exemplarily illustrate who creates which metadata records of LOM in which way at what time. Finally, we show why this approach of creating metadata matters as we measure our metadata quality and compare it with other's findings.*

## INTRODUCTION

Electronic Learning, in particular in the form of Blended Learning, is applied by a rapidly increasing number of universities and companies. Realizing the concept of learning objects (Wiley, 2002) the ability to find and reuse content is generally based on the use of metadata. Due to its wide dissemination IEEE LOM (http://ltsc.ieee.org/wg12/20020612-Final-LOM-Draft.html) can be considered as de-facto standard: With more than 40 attributes, subdivided into 9 main categories, a broad description of learning objects is enabled. Metadata is collected and stored in a central place, making

DOI: 10.4018/978-1-60566-936-6.ch004

content available for potential users. In this way transparency of existing e-learning content and its integration within varying context is enabled (Dahl and Vossen, 2007).

While the great number of attributes enables a detailed description of learning objects, in practice a comprehensive usage of these is rare. Studies show that common attributes like *title* or *format* are filled quite often, while fields like *difficulty* or *structure of learning object* receive only little attention (Friesen, 2004). As long as metadata is only used in a single context respectively in a single system, a reduction of the attribute amount might even be reasonable, as the focus can be set regarding the specific end user (Dahl, Vossen and Westerkamp, 2006); by doing so, complexity is decreased and usability increased. Problems arise if repositories communicate and interact with each other, for example when querying distributed e-learning catalogues: While on the one side metadata records might be considered as crucial and obligatory, the same attributes might never be used on the other side as they are only optional. With a small intersection of filled metadata records the primary objectives like finding and reusing learning objects become impossible to achieve. Furthermore, if metadata is created the way it is mostly today a high risk for superficial records arises when a single person tries to fill as many metadata fields as possible: In result a high quantity might face a low quality. In order to enable cross-system finding and cross-system reusability of learning objects, a high quantity along with a high quality of metadata must be guaranteed, which actually seldom is the case.

Thus, the core dilemma of learning object metadata creation is derived from the discrepancy between the high potential of LOM in theory and the rare implementation and usage of the complexity in practice. We put this down to the aspect that a crucial question is not answered yet:

***Who*** *creates* ***whenwhich metadata records*** *in* ***which way****?*

Though it is obvious that a single person is hardly predetermined to fill in all metadata records (e.g., presented in some kind of list with empty text fields) this approach can indeed be found in practice. However, we often see different sources interacting within the process of metadata generation. In order to be able to find and reuse, it has to be defined in detail which records are generated by whom at which time and in which way. Only in this way a high quantity along with a high quality of metadata can be achieved.

With the objective to define the process of learning object metadata generation for a concrete learning context at a university, the remainder of this paper is structured as followed: In Section 2 we examine in which way metadata for learning objects can actually be created. Furthermore the Web 2.0 tagging approach introducing the user of a system as metadata creator within a community is analyzed. Section 3 brings together the different ways of metadata creation in a single model and draws first conclusions regarding actors within the process of learning object metadata creation (answering *who?*). Furthermore, we disengage the well known structure of LOM with its nine main categories; instead we introduce a view founded on a more classical metadata perspective. This view, breaking up the original LOM hierarchy, reveals groups of metadata records that might be generated together in the same way (answering *which metadata records?*). Section 4 describes in a real world scenario the use of a learner-centered e-learning platform; in a process model it is shown where metadata is created before and during the usage of learning objects (answering *when? / in which way?*). Section 5 then illustrates technical aspects behind this process model, as the 'big picture' of the interacting applications as well as the modular core component of metadata creation are discussed in detail. Finally, Section 6 focuses on measuring the created metadata: Applying the LOM quality metrics we show why our approach described in this paper really matters. Concluding, Section 7 summarizes the paper and outlines future work and research.

## RELATED WORK

Learning object metadata can be created in many ways, each of which has its individual advantages and disadvantages. Two different approaches can primarily be identified: Following a top down approach, metadata records are filled purposefully and explicitly, while following a bottom up approach information that was collected over time is analyzed and processed in order to gain relevant metadata. In the following some top down approaches as well as one bottom up approach will be discussed.

It is obvious to delegate the task of metadata creation to an expert of the respective learning context, for example the author of a learning object or another explicitly selected person. In order to simplify the process of metadata creation so called LOM editors, which represent the standards hierarchical model for example in a tree view (Cebecci and Erdogan, 2006), can be used. Filling text fields and selecting drop down lists step by step the full complexity of LOM can be enabled. Some editors support template creation, storing attributes that are used quite often. As mentioned before one notable disadvantage of this approach is founded on the amount and complexity of LOM: If only for reasons of time and costs it is nearly impossible for a single person to create high-quantitative as well as high-qualitative metadata for a set of learning objects. A collaborative approach of metadata creation is usually neither defined nor explicitly supported.

Another approach is using computer support to a much higher degree: Automatic metadata generation extracts relevant information from learning objects and the context they are stored or used in (Cardinaels, Meire and Duval, 2005). While on the one side costs for staff and effort are reduced on the other side imprecise up to erroneous metadata records have to be taken into account. For example, as the generation of usable metadata from text-based learning objects is relatively reliable this is usually not the case when analyzing multimedia content. Even extracting the title of a Power Point Slide is an enormous challenge (as long as the title is not already set explicitly within the slides´ metadata). Concluding automatic metadata generation can be designated as a promising way that should not be used exclusively without human interaction because of a certain degree of error-proneness; however, very simple metadata attributes might be generated this way to almost 100% correctness (e.g. format or size of learning object).

Hybrid Systems strike a balance between the ways of automatic and human metadata generation (Motolet and Baloian, 2006). Based on an automatic analysis of a learning object three groups of information can be build: Very probable values, probable values and restrictions of possible values. While the first group, including the format or the size of a learning object, usually does not require a human revision the second groups´ values are not that reliable and need to be verified. The third group consists of restrictions not suggesting concrete metadata records but reducing the scale of possible values. The approach using Hybrid Systems simplifies the process of metadata generation for persons like authors or experts integrating aspects of automatic analysis techniques. However, a detailed answer for the crucial question posed in the introduction as well as a collaborative approach is still missing.

A collaborative way of human metadata generation was introduced by Or-Bach in 2005: Following the pedagogical objective of learning content reflexion students are the ones to create learning object metadata. To give an example, programming units are annotated and described which results in an abstraction and recognition of high level concepts as smaller units, concrete examples and exercises are brought into relation. Rethinking and reinterpreting learning objects within the process of metadata generation a learning progress is achieved. This approach is the first one among activating users of a system as metadata creators. However it is not the main objective to generate preferably complete and high-quality metadata sets in the sense of e-learning repositories (ability to find, reusability). For example the idea to combine user-driven metadata generation and the expert knowledge of teachers is not pointed out and it does not become clear how collaboration at this level might look like.

The challenge of making multimedia-based content searchable is also a relevant topic independently of the learning context: For example, how can photos be labelled in order to discover them more easily? Within the scope of the recent Web 2.0 phenomenon (O´Reilly, 2005) social tagging of content emerged as a promising new way for discovery and categorization. The creators of metadata are longer experts or authors. Instead, the creation of metadata is done by system users (Mathes, 2004), who might primarily see individual reasons like an easy discovery of personally interesting photos. Based on personal metadata, interesting information for an entire community can be collected, provided the metadata of many users is brought together. Thus under certain conditions the user of a system acts as a metadata creator without being explicitly aware of his role. While in a collaborative way of metadata creation as mentioned before when students describe learning objects metadata records are filled purposefully, users of a Web 2.0 tagging application primary see personal advantages in their acting. It must be pointed out that in this case the source of metadata is not necessarily an expert of a domain wherefore a high quality of metadata is not guaranteed.

To conclude, different ways of metadata creation characterized by individual advantages and disadvantages have been discussed. In our opinion the bottom up approach as it can be found in Web 2.0 has a high potential for the context of electronic learning and the creation of learning object metadata. As this potential has not been analysed, our objective is a combination with the top down approaches mentioned before in order to accentuate the advantages and to attenuate the disadvantages.

## ANALYSING LOM FROM A CLASSICAL METADATA PERSPECTIVE

In Section 2 approaches from different scenarios of metadata generation have been introduced. In **Figure 1** these are brought together into a comprehensive model. At the first level a classical subdivision into automated and human metadata creation is made (for example cf. Gill, Gilliand-Swetland and Baca, 1998). Additionally Hybrid Systems as introduced before mediate between the two main categories.

Following Cardinaels, Meire and Duval (2005), automatic metadata generation is broken down into 4 aspects: content analysis, context analysis, usage analysis and structure analysis. While in a content analysis information is extracted from the learning object itself (for example language, size), a context analysis involves the environment the object is currently used in. This way relevant information from user or course profiles can be gathered for metadata creation. A usage analysis for example evaluates how often a learning object was viewed by users or how long it took a learner to solve a special exercise. Conclusions regarding special metadata records can be drawn. A structure analysis involves relationship amongst objects: For example course metadata might contain relevant and applicable records for its single learning units.

Human metadata creation firstly can be subdivided into experts, mostly professional metadata creators, and authors (Greenberg et al, 2001). While experts, known from libraries, for example, are predetermined for a categorization of content, authors create abstracts or lists of keywords. Additionally one more human metadata creator can be

*Figure 1. Ways of learning object metadata creation*

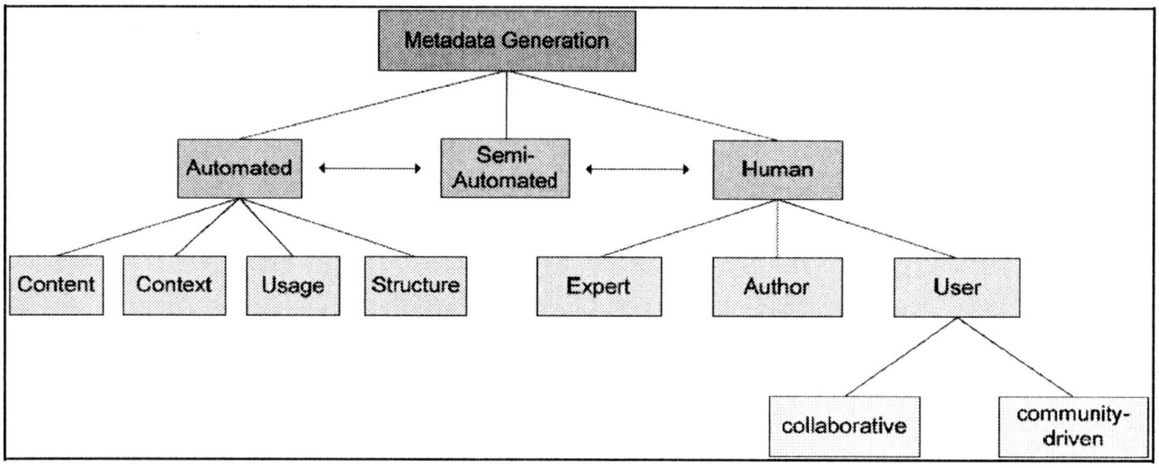


identified: the user of a system. In Section 2 students creating metadata records with a pedagogical objective have been mentioned. A collaborative, explicit approach has been identified. On the other hand user-driven generation as to be found in Web 2.0 tagging applications is characterized by users annotating content of their own accord with primary individual reasons. Bringing together all community metadata interesting information can be gathered. In this way of metadata creation is denoted as community-driven.

Which of these ways are interesting in combination for the challenge to describe learning objects? First of all combining automatic and human approaches seems to be reasonable. Regarding the quality of metadata Lux, Klieber, and Granitzer (2006) differentiate manually generated high level metadata and automatically generated low level metadata. The fuzzy border between these two is known as the semantic gap; to bridge this gap an enormous effort has to be made if human interaction is excluded. This view, confirmed by practical problems as mentioned in Section 2, justifies a combination of automatic and human metadata creation. All four automatic analysis techniques can be used to gain information from learning objects. Referring human metadata creation the role of an expert ought to be discussed within the learning context. While in a library this job might be explicitly filled by a person who is trained to categorize content, at a university to a high probability this is not the case. However at a university due to his expert knowledge a teacher can fill this role. As the teacher is normally also the author respectively the person in charge for a learning object the two roles merge in this context. More than that, he or she is also kind of an observer, conducting the learning process and enriching the learning object with additional information whenever this is required. Users of a learning object can be denoted as learners; this might be students as well as lecturers doing further studies. The collaborative approach introduced before is rather a special case: Especially the aspect of motivation for users to create metadata is missing here as long as there is no external pressure. As will be shown later, this is not the case when following the community-driven approach.

After having considered which ways of metadata generation are relevant in the context of electronic learning, the question arises of what an assignment of concrete metadata attributes might look like. Breaking up the hierarchical model with its nine subcategories a new view on LOM, based on a classical metadata perspective, gives a first impression on groups of metadata that should be

*Figure 2. LOM from a classical metadata perspective*

| | Discovery | Use | Authentication | Administration |
|---|---|---|---|---|
| **objective** | General:Title<br>General:Coverage<br>General:Identifier<br>Lifecycle:Contribute | Rights:Copyright<br>Rights:Description<br>Rights:Costs<br>Edu:Inter.Type<br>Edu:Res.Type<br>Edu:Inter.Level<br>Technical:Duration<br>Technical:Location<br>Technical:Size<br>Technical:Req.<br>Technical:Platf.Req.<br>Technical:In.Remarks<br>Technical:Format<br>General:Language<br>General:Aggr.Level<br>General:Structure | Lifecycle:Status<br>Lifecycle:Contribute<br>Lifecycle:Version<br>Annotation:Date<br>Annotation:Entity | Lifecycle:Status<br>Lifecycle:Contribute<br>Lifecycle:Version<br>Rights:Costs<br>Rights:Copyright<br>Rights:Description<br>Metameta:Contribute<br>Metameta:Language<br>Metameta:Scheme<br>Metameta:Identifier |
| **subjective** | General:Description<br>General:Keyword<br>Edu:Description<br>Class.Description<br>Class.Keyword<br>Class.TaxonPath<br>Class.Purpose<br>Annotation:Description<br>Relation:Resource<br>Relation:Kind | Edu:AgeRange<br>Edu:Context<br>Edu:LearningTime<br>Edu:Sem.Density<br>Edu:Language<br>Edu:Difficulty<br>Edu:EndUserRole<br>Edu:Description | Relation:Kind<br>Relation:Resource<br>Annotation:Description | |

generated in the same way. In 2001 Greenberg has analysed common metadata standards for digital objects, concerning a subdivision of their attributes into different metadata classes. Amongst others Dublin Core (DC), which is much more general than LOM as it allows a description of any digital objects, was examined. As in practice also learning objects are annotated with DC it is obvious to adopt the applied classification schema for LOM, too. In doing so we have categorized the single LOM attributes into four subclasses:

*Discovery Metadata* includes all attributes that support the ability to find of learning objects. For the purpose of restriction – as LOM is a descriptive standard mostly all attributes more or less help to find an learning object - only attributes are denoted as Discovery Metadata if they enable finding an object for themselves, thus without being combined with other attributes. Accordingly the records title or keywords are included, while the format, which might be helpful in combination with the first ones but regarded as a single entry does not enable a purposeful search, is not listed. *Use Metadata* contains all attributes that are meaningful while a learning object is used. This includes technical information like format or system requirements as well as intellectual characteristics as property rights or restrictions regarding the usage. *Authentication Metadata* involves attributes that guarantee the integrity and the overall trustfulness of a learning object. Attributes like the source of a learning object, its version or the relation to other objects are grouped here. *Administration Metadata* includes attributes supporting the management of a learning object as information about ownership or all meta-metadata (e.g. who created the metadata records).

In order to achieve a finer subdivision additionally each of the four metadata classes is split up into Objective and Subjective Metadata (Duval et al, 2002). Speaking of Objective Metadata involves facts as the date of creation or the current version number. Subjective Metadata may vary depending on the person who is annotating. Examples are abstracts, summaries or keyword lists. Accordingly we assign the LOM attributes a total of 8 categories. Analysed are all 45 records of LOM level two, which means all attributes below the nine main categories. Each of these is assigned to a minimum of one and a maximum of four categories; classifying a single record subjective and objective at the same time is excluded.

**Figure 2** illustrates the assignment of the single LOM attributes. The classification is based on different sources (like Greenberg) and explicit proposals (e.g. made by Duval or Cardinaels) on the one hand and a detailed analysis accomplished by the authors on the other. Discussing each record would go far beyond the scope of this paper; it should be mentioned that the figure only shows one - as we think quite reasonable - possibility of an assignment.

Referring back to the ways of metadata generation identified as relevant prior in this section, groups of attributes can be assigned. Objective Discovery Metadata is predestinated for creation by a teacher as the values are unambiguous (for example the title or the coverage of a learning object). By contrast Subjective Discovery Metadata involves individual records: Depending on the annotating person the structure of a hierarchy or the listed keywords may vary. Accordingly these attributes should be generated by learners in order to get an impression of the system users´ language. The group of Use Metadata seems suitable for automatic generation. While Objective Use Metadata typically can be generated by content, context and structure analysis for Subjective Use Metadata usage analysis seems to be applicable. Objective Authentication respectively Administration Metadata partially can be generated automatically with a high degree of correctness (for example metametadata records); however, partially an interaction with a teacher is absolutely necessary. In a semi automated approach costs of a learning object or copyrights can be proposed analysing related objects (for example examining releases of the same author or hierarchically higher respectively lower structured content). The metadata attribute describing the relationship can also be filled this way. Completing Subjective Authentication metadata as well as Subjective Discovery Metadata again seems to be predestinated for a creation by learners. A concluding appraisal in either case should be the job of a teacher as only he or she has expert knowledge in the respective domain

## METADATA GENERATION IN A REAL WORLD SCENARIO: THE PROCESS VIEW

The conclusions obtained above are now brought together in a real world scenario. The e-learning platform Learnr (Dahl, Lechtenbörger, Sieberg and Vossen, 2008), developed and in use at the University of Muenster in Germany, not only gives learners a central point of access for actual learning objects, but also enables online annotation of content as known from Web 2.0. Especially activities experienced in the real world during intensive learning activities (for example prior to exams) are virtually supported. For example, learners can arrange digital post-its, tag content with keywords for the purpose of categorization and retrieving, build relationships amongst learning objects, or create summaries and file cards for recapitulation of important aspects.

**Figure 3** illustrates the combination of top down and bottom up approaches in the context of an e-learning platform. Initially specified metadata is declared by a teacher while he or she uploads a learning object to the platform. Of course, this metadata is not entered into a complex LOM editor, but into single text fields that are in line with the learning platform. Furthermore, setting up the learning object in the context of the platform, content, context and structure analysis deliver a remarkable amount of attributes automatically.

Entering the phase of learning objects usage on one side algorithms can analyse the usage behaviour and generate metadata automatically, on the other side users individually begin to work with the content that is offered: Analogous to Flickr (http://www.flickr.com) tags are set to categorize objects. Structures are defined just like in RawSugar (http://www.rawsugar.com) bringing tags into relation. Well known from Delicious (http://www.del.icio.us) references can be defined, including internal as well as external resources. Digital post-its, as found on the MyStickies-webpage (http://www.mystickies.com), can be used to annotate or to create summaries. In consequence of all these activities following the bottom up approach a comprehensive amount of potential metadata accumulates. In the following the e-learning platform collects and preprocesses all accumulated metadata. Thus first of all data produced by learners is analysed and selected using data mining techniques (first filter). Then relevant information is presented to the teacher in a dialog (along with the already automatically generated metadata). The teacher might finally decide which

*Figure 3. Metadata generation process in the context of an e-Learning platform*

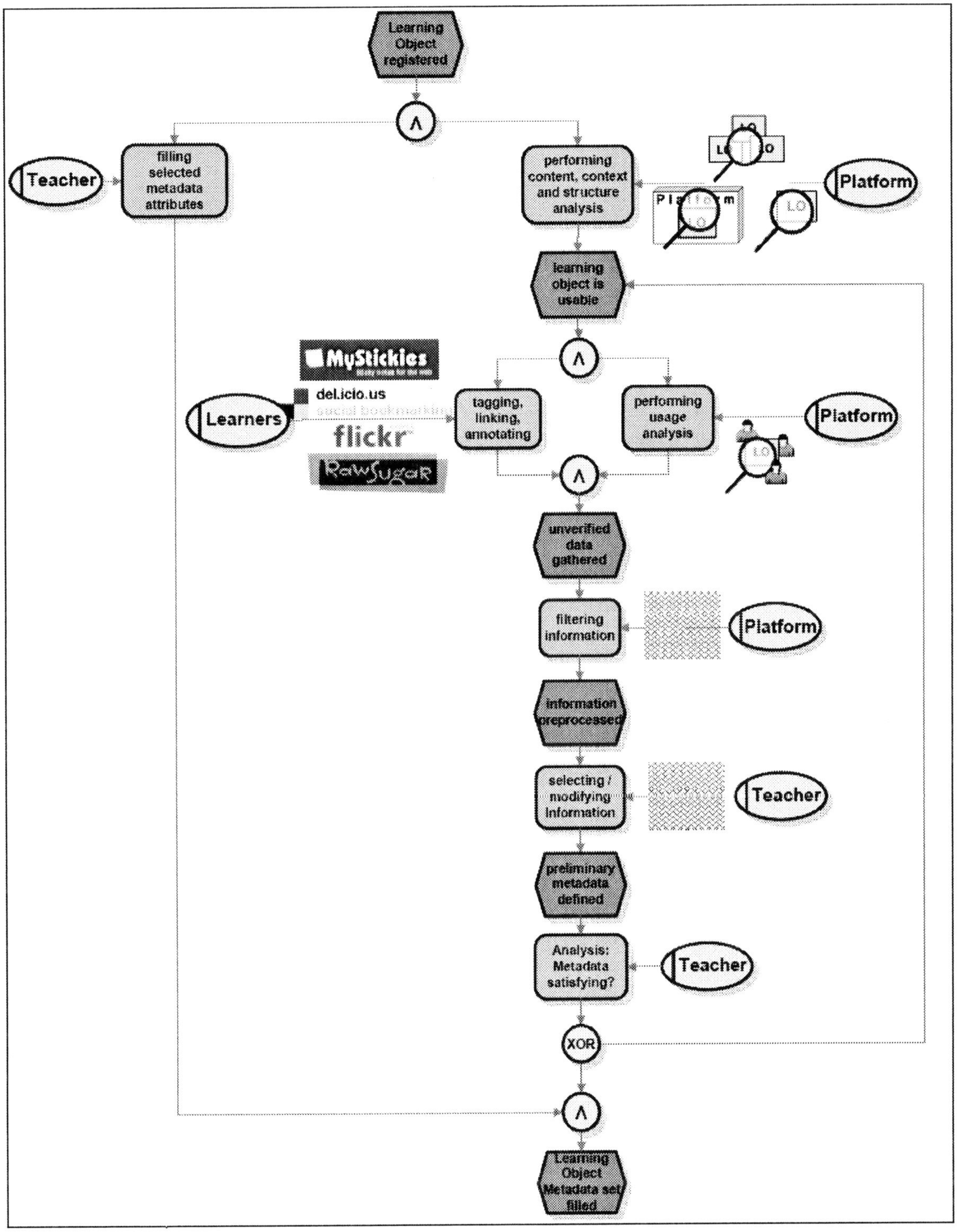

data is relevant enough to become representative learning object metadata (second filter). By and by based on automatic analysis (context, usage) as well as activities by learners more relevant information accumulates; in consequence, the preprocessing of data by the platform and the selection by the teacher is not a single task. Rather it is a periodic, iterative process. An applicable timeframe in the context of a university could be a semester for example; from the beginning to the end quality of metadata will increase as more and higher quality data will be gathered.

Summarizing our findings we can explicitly answer the crucial questions of metadata generation (**Figure 4**). Initially the expert of the domain, the teacher (cf. Objective Discovery Metadata in Figure 2), fills records concerning the *title* and the *coverage* of a learning object. Likewise the records *creator*, *version* and *status* are set. In the scenario described above the *educational description* reflects the idea of blended learning (accompanying lectures), a variation would be a consequence of using self assessment units in distance learning for example. Metadata attributes describing *rights* aspects are also filled by the teacher.

At this stage the platform already supports the teacher (cf. Objective Use Metadata in Figure 2). Performing structural analysis, learning objects that are similarly structured or derived from the same author can be identified. Accordingly, the expert will receive proposals based on this information (*Rights:Costs, Educational:Description*). The *educational description,* for example, should be identical to a high degree concerning a course and its single units. At the same time the learning objects′ *relationship* to already existing learning units can be specified (for example isPartOf-relationship). Content analysis primarily delivers technical information like the *format* or the *size* of a learning object. Also the *language* used, the

*Figure 4. Defining the creation of learning object metadata*

| Who | When | How | LOM-Attributes | |
|---|---|---|---|---|
| Teacher | initial | LOM-Editor | General:Title<br>Lifecycle:Contribute<br>Lifecycle:Status<br>Rights:Description<br>Rights:Copyright | General:Coverage<br>Lifecycle:Version<br>Educational:Description<br>Rights:Costs |
| Platform | initial | Structure-Analysis | Relation:Kind<br>Educational:Description<br>Rights:Copyright | Relation:Resource<br>Rights:Costs<br>Rights:Description |
| | | Content-Analysis | General:Language<br>Technical:Size<br>Technical:Requirement<br>Technical:PlatformRequirements<br>Educational:InteractivityType<br>Educational:ResourceType | Technical:Format<br>Technical:Location<br>Technical:InstallationRemarks<br>Technical:Duration<br>Educational:InteractivityLevel |
| | | Context-Analysis | General:Identifier<br>General:AggregationLevel<br>Lifecycle:Date<br>MetaMetadata:Contribute<br>MetaMetadata:Language | General:Structure<br>Lifecycle:Contribute<br>MetaMetadata:Identifier<br>MetaMetadata:Schema<br>Educational:Context |
| Learners | use | Tags, Notes, Links | General:Keywords<br>General:Description<br>Annotation:Description<br>Classification:TaxonPath<br>Classification:Description | Relation:Resource<br>Educational:Difficulty<br>Classification:Purpose<br>Classification:Keyword |
| Platform | use | Context-Analysis | Annotation:Entity | Annotation:Date |
| | | Usage-Analysis | Educational:SemanticDensity<br>Educational:AgeRange<br>Educational:Language | Educational:EndUserRole<br>Educational:LearningTime |

*resource type* (e.g. slide, exam), the *interactivity type* (active, expositive) and the *interactivity level* will be identified with a high probability. Performing context analysis an unambiguous *identifier* is generated and information based on user profiles is extracted; for example, conclusions regarding *publishing person* and *timestamp*, *educational context* or *meta-metadata* in general are drawn. As already mentioned, learners are annotating learning objects (cf. Subjective Discovery Metadata in Figure 2). Based on tags keyword lists can be deduced, bookmarks to internal (other learning objects) or external (e.g., Wikipedia) resources deliver relationships, and notes on virtual post-its can be stored as LOM-annotations. Using appropriate algorithms and bringing together individual tag hierarchies a classification going far beyond flat folksonomy namespaces can be inferred. The general description of a learning object can also be created processing user generated metadata (analysing summaries). While the learning object is used context analysis delivers information about annotation creators (profile analysis) and corresponding timestamps. A usage analysis involving user profiles enables conclusions regarding end-user role (teacher, learner), age range or learning time: how long did it take until learners marked a learning object as "understood" in order to assign a lower priority concerning the future learning process (cf. Subjective Use Metadata in Figure 2).

## METADATA GENERATION IN A REAL WORLD SCENARIO: THE TECHNICAL VIEW

Now that we have illustrated the process of metadata creation in detail, we can look at selected technical aspects. As we have seen in the previous section, the process of metadata creation is initiated within a selected learning platform (in our example: Learnr). So, as it can bee seen in the left part of **Figure 5**, we have implemented a modular library that enhances the functionality of Learnr. The ACE package (the name refers to the three sources of metadata creation: automatic analysis, the community of learners, and experts) controls the entire process of metadata creation; however, not everything takes place within the ACE module. In Section 4 we have pointed out that some of the LOM attributes can be created by more or less simple automatic analysis. Especially pure content analysis, like extracting the format, the size, or the duration of a learning object, does not need to be implemented over and over again, as the algorithms that are used always are the same. Being aware of this, Cardianaels, Meire, and Duval (2005) published the SamgI Web services (Semi automatic metadata generation Interface). So we simply use the SamgI services by uploading a learning object and receiving the metadata we are interested in.

However, of course most of the metadata is created within the Learnr platform itself. As the ACE package is put into the application context of Learnr, it can use all the programming routines, for example, for connecting to the underlying relational database and query its tables. In this way it is possible to process context and structure analysis as described in the previous section. But more than that, all information that accumulates within the learning community (tags, notes, ratings, summaries, etc.) is also available for further analysis.

Additionally, we deliver one more modular component with our ACE-library: A complete LOM editor. The LOM editor developed at RWTH Aachen[1] is implemented as a very simple Java applet that runs within any Web browser. So, as described in the process model in Figure 3, concluding the process of metadata creation the expert can validate and modify all LOM attributes that have been generated as we load all of them into the editor. To make this clear once more: We only use a LOM editor to validate and modify metadata, not to create it. The expert does not need to add any more attributes in this stage; more than that,

*Figure 5. Technical view on creating and storing metadata*

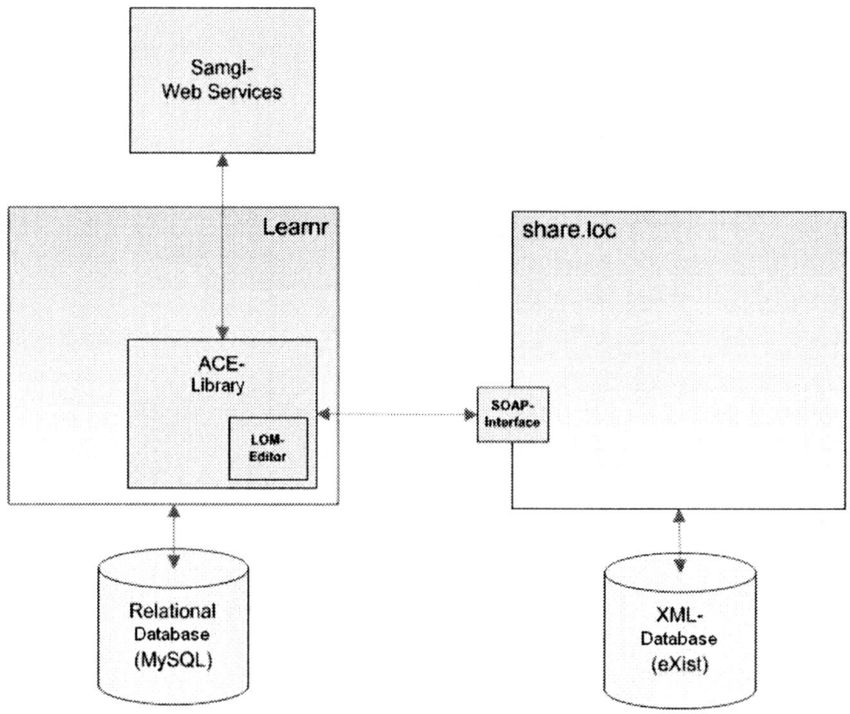


he or she is not even forced to use the LOM editor to finish the metadata creation process at all; if he does not use the editor, the created metadata will be sent to our metadata repository share.loc automatically.

In order to store and administrate the metadata we use our e-learning metadata repository share.loc (Dahl, Vossen and Westerkamp, 2006). The objective of share.loc is to provide a single point of access to learning objects available at the University of Muenster. However, we do not store all learning objects themselves but only their metadata (including explicit references) in order to provide some 'Yellow Pages' where teachers and students can search for content they are interested in. As share.loc provides service-oriented interfaces resp. Web services based on REST (connector to database tier), SOAP (connector to application tier) and WSRP (connector to visualization tier, e.g., to integrate search functionality and GUI into university portals) we use these to access and store metadata from Learnr. According to the LOM XML standard (http://standards.ieee.org/reading/ieee/downloads/LOM/lomv1.0/xsd), a learning object metadata set consists of a single XML file which is stored in an underlying XML database. The XML file consists of several XML fragments representing the nine LOM main categories, which in turn store the metadata attributes that have been created. To make this structure clear we have a closer look at one of the main ACE classes within the Learnr platform, the GeneratorHandler (see **Figure 6**). This class is used to initiate the process of metadata creation.

In order to start the creation of metadata, the GeneratorHandler class needs to be invoked by the method 'createMetadata'. According to the LOM tree, the top class administrates nine LOM secondary level classes, like the LifecycleGeneratorHandler or the TechnicalGeneratorHandler, which in turn will be invoked. Each of the nine sub classes in turn invoke sub classes they

*Figure 6. Metadata generation class model*

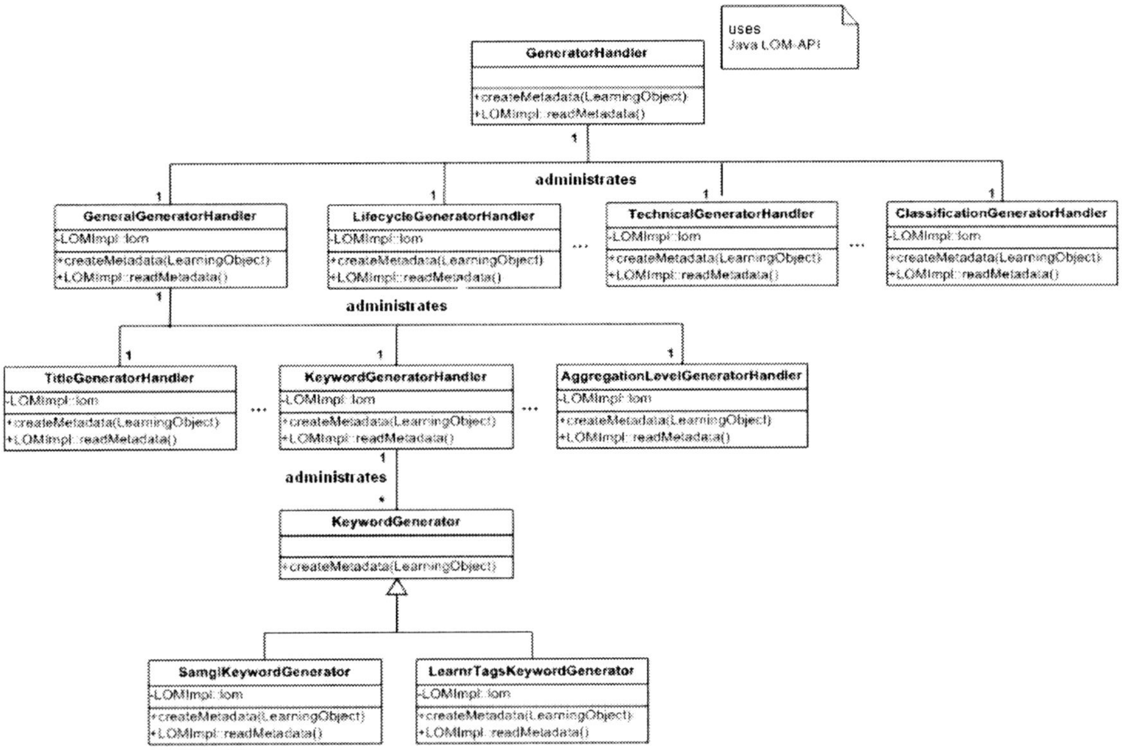


administrate; so the GeneraleGeneratorHandler invokes, for example, a TitleGeneratorHandler and a KeyWordGeneratorHandler. The KeyWordGeneratorHandler administrates one or more concrete keyword generators, which inherit methods and attributes from KeywordGenerator. So, finally in this example keyword metadata is created within the two classes which access the SamgI Web services and the Learnr database in order to query the corresponding table that contains user tags. Then it is the task of the KeyWordGeneratorHandler class to mix the metadata that is received from the sub classes in order to create small objects that hold the metadata. These objects refer to the Java-LOM-API (http://sourceforge.net/projects/lom-j/) as they can easily be transformed into small, LOM compliant XML pieces.

In order to obtain a coherent, LOM compliant XML file that can be stored in share.loc, the main class GeneratorHandler and its method readMetadata need to be invoked. Performing a tree traversal, each node that contains a LOMImpl representation will be visited. This way, step by step a coherent LOM XML file will be created.

More information on the entire technical infrastructure that has been implemented is given in Dahl (2008). However, in this section we have already pointed out that we do have more in mind than just creating metadata in a single learning platform. More than that, we have created components and interfaces that can be reused in varying learning platforms at our university. We can reuse the modular ACE library and adapt it for additional learning platforms that are in use. And of course, we can connect these platforms to the share.loc repository in order to store the metadata that is created.

## MEASURING LEARNING OBJECT METADATA QUALITY

Although we have suggested that the metadata we created was good concerning attribute quantity and quality, we have been looking for a proof of our concept; and we have found it based on the quality metric for learning object metadata as published by Ochoa and Duval (2006). On an abstract layer, the metric consists of the dimensions *consistency*, *conformance*, *completeness*, *connectedness*, *readability*, and *timeliness*. In order to apply this metric we have analysed all the metadata sets that were created based on our process model (Section 4) and the implementation of the ACE library in Learnr (Section 5) automatically. For this we used a Java-implementation based on the algorithms as published by Ochoa and Duval, which was connected directly to the share.loc repository.

To give the reader an impression of how these quality dimensions are calculated, we provide a short introduction here; further information can be found in the publications of Ochoa and Duval (2006) or in Dahl (2008). Concerning *consistency* we have to disclose inconsistencies amongst different metadata attributes in one and the same LOM set. For example, the predefined LOM attributes structure and aggregation level should as well be in line as interaction type and interaction level. One indicator to calculate *conformance* of the given metadata is the entropy term, which discloses whether there is any significant unique description compared to the other learning objects within the repository; the more unique the metadata, the better a learning object can be found. For example, a keyword that is listed in 50% of all metadata sets is clearly not a unique and significant metadata attribute. The *completeness* algorithm simply analyses how many metadata attributes of each LOM set are filled with values other than 'null'; so this dimension is in-line with the quantity-term we used throughout this paper. *Connectedness* is calculated as ingoing and outgoing links to LOM sets are measured; the more dense the network of LOM sets is, the better it can be interpreted by a system or a user. In order to calculate *readability*, we analyse free text fields like the title or the description with the flesh-index-algorithm (long sentences with long words are harder to understand than short ones with short words) and grammar and misspelling checks. Finally, the *timeliness* dimension simply indicates whether the metadata is up-to-date: In our example all of our metadata was created recently. Finally, all dimensions are scaled to a range of 0-10 (10 means best) except the connectedness dimension which represents an absolute value.

**Figure 7** illustrates the average results as we have applied the learning object metadata quality metrics to our share.loc repository; furthermore results as published by Ochoa and Duval (2006) are listed. The table shows that we were indeed able to generate a high completeness rate of almost 85%. The missing attributes are related to missing functionality in our Learnr system, as timestamps are not created with every user interaction and roles like learning object editor are not available. Concerning consistency, no major inconsistencies were disclosed; at least based on the algorithms that Ochoa and Duval used as well. The conformance result is only satisfying; we refer this to the fact that we only analysed LOM sets that were created within Learnr. Of course, in this context some attributes like structure (linear) and resource type (slide or video) are mostly the same in our data sets. While the readability results are very satisfying, the connectedness analysis is even more pleasant: Especially based on our Learnr context analysis (is-based-on-/is-basis-for-relations) and the social

*Figure 7. Results of Metadata Quality Analysis*

| | share.loc | Ariadne | OCW |
|---|---|---|---|
| Completeness | 8,48 | 6,43 | 4,03 |
| Consistency | 10 | 10 | 10 |
| Conformance | 3,24 | 2,08 | 4,76 |
| Readability | 6,44 | 0,99 | 5,42 |
| Connectedness | 7,32 | 0 | 0 |
| Timeliness | 10 | 5,53 | 10 |

tagging concept (references-relations) we could create a high density LOM network. Finally, as mentioned before, the timeliness result is obvious as all of our metadata was created recently.

In conclusion, we were happy with the results of our learning object metadata quality analysis. Quality dimensions such as completeness and connectedness already benefit to a high degree from our Web 2.0-driven approach of metadata creation. We expect to follow the conformance dimension after we have set up the ACE library in further learning platforms, as then the overall accuracy of the metadata attributes stored in share.loc will rise. On the other hand, we will have to keep track of the good results concerning consistency, readability, and of course timeliness.

## CONLCUSIONS AND FUTURE RESEARCH

In this paper a Web 2.0-inspired way of metadata creation has been described. Answering the crucial question of *who creates when which metadata records in which way,* it has been shown that high quantity as well as high quality of metadata can be achieved. Breaking up the classical LOM hierarchy with its nine main categories a subdivision of the single attributes into classical metadata classes has been performed; first conclusions regarding groupings of metadata records could be drawn. A real world scenario has illustrated practical impact in a learner-centered e-learning platform. Finally, a quality analysis has confirmed the approach that has been introduced in an impressive way.

Future work will focus on implementing ACE modules for further learning platforms. Thanks to the modularity and the clear interfaces it should be easily possible to enrich systems that were developed at our university during the past years (e.g., xLx, OpenUSS) as well as Open-Source applications (e.g., Moodle) with metadata creation libraries. We envision a star federation with all platforms loosely connected to share.loc via Web services. Although we are aware that the less 'Web 2.0 features' any of these platforms offers the lower the metadata quality might be, we are very optimistic that in any case we will be able to increase the overall metadata quality significantly.

In this new context with a considerable amount of high quantity and high quality metadata for the first time, we already see the rise of new challenges: How can we maintain the quality of our metadata over time, and can we keep track of metadata quality, i.e., watch over its evolution? A person in charge might be interested in questions such as: Which applications deliver satisfying metadata in general? Which applications deliver a high quantity and high consistency of metadata? How does metadata change in a certain application along selected dimensions like readability or connectedness? At first glance, these questions seem to concern a local problem that needs to be solved within the University of Muenster. However, we imagine that in the future multiple repositories of learning content will need to be able to talk to each other, and that users (students) will want to search for learning content across repository boundaries. To this end, we need to look at the metadata problem at a more global scale, since the questions above might be of high interest, for example, in an SQI-based repository foundation (simply replace the term *application* by *repository* in all these questions). Towards this future scenario of real learning object metadata quality control one of the next steps we will address is the definition of a multi-dimensional data model for learning object metadata quality results that helps to answer the pragmatic questions stated above.

## REFERENCES

Cardinaels, K., Meire, M., & Duval, E. (2005). Automating metadata generation: The simple indexing interface. In *Proceedings International Conference on World Wide Web* (pp. 1-8).

Cebeci, Z., & Erdogan, Y. (2005). Tree view editing learning object metadata. *Interdisciplinary Journal of Knowledge and Learning Objects*, *1*, 99–108.

Dahl, D. (2008). *Integration of learning objects on metadata level – A Web 2.0-coined approach (German)*, Akademische Verlagsgesellschaft AKA, Heidelberg.

Dahl, D., Lechtenbörger, J., Sieberg, J., & Vossen, G. (2008). Learnr – Web 2.0-Driven Learning. In *Proceedings International Conference Web Based Education* (pp. 136-141).

Dahl, D., & Vossen, G. (2007). share.loc: Transparenz und Integration von Lernobjekten an der Universität Münster. *Working Paper E-Learning Competence Center Münster.*

Dahl, D., Vossen, G., & Westerkamp, P. (2006). share.loc – A Mulit-Tiered Interoperable E-Learning Metadata Repository. In *Proceedings IEEE International Conference on Advanced Learning Technologies (ICALT)*(pp. 891-895).

Duval, E., Hodgens, W., Sutton, S., & Weibel, S.-L. (2002). Metadata principles and practicalities. *D-Lib Magazine*, *8*(4). doi:10.1045/april2002-weibel

Friesen, N. (2004). *International LOM Survey: Final Report*. Retrieved January 1, 2009 from http://jtc1sc36.org/doc/36N0871.pdf

Gill, T., Gilliland-Swetland, A., & Baca, M. (1998). *Introduction to metadata: Pathways to digital information*. Getty Research Institute.

Greenberg, J. (2001). A quantitative categorical analysis of metadata elements in image-applicable metadata schemas. *Journal of the American Society for Information Science and Technology*, 917–924. doi:10.1002/asi.1170

Greenberg, J., Patuelli, M. C., Parsia, B., & Davenport Robertson, W. (2001). Author-generated Dublin core metadata for web resources: A baseline study in an organization.In *Proceedings International Conference on Dublin Core and Metadata Applications* (pp. 38-46).

Lux, M., & Klieber, W., & Granitzer. (2006). On the complexity of high level metadata. *Journal of Universal Knowledge Management*, *1*(1), 54–58.

Mathes, A. (2004). *Folksonomies – Cooperative classification and communication through shared metadata*. Retrieved January 1 2009 from Retrhttp://www.adammathes.com/academic/computer-mediated-communication/folksonomies.html

Motolet, O., & Baloian, N. (2006). Hybrid systems for generating learning object metadata. In *Proceedings IEEE International Conference on Advanced Learning Technologies (ICALT)* (pp. 563-567).

O'Reilly, T. (2005). *What is Web 2.0*. Retrieved January 1, 2009 from http://www.oreillynet.com/pub/a/oreilly/tim/news/2005/09/30/what-is-web-20.html

Ochoa, X., & Duval, E. (2006). Quality metrics for learning object metadata. In *Proceedings World Conference on Educational Multimedia, Hypermedia and Telecommunications* (pp. 1004-1011).

Or-Bach, R. (2005). Educational Benefits of Metadata Creation by Students. *inroads – . SIGCSE Bulletin*, *37*(4), 83–97. doi:10.1145/1113847.1113885

Wiley, D. A. (2002). The instructional use of learning objects. *Agency for Instructional Technology.* [1]http://www-i5.informatik.rwth-aachen.de/i5new/staff/chatti/LOMEditor/index.html

# Chapter 5
# Web 2.0:
# A Vehicle for Transforming Education

**Julia Gooding**
*Robert Morris University, USA*

## ABSTRACT

*Web 2.0: A Vehicle for Transforming Education includes practical and accessible overviews of some of the most commonly used and most useful technologies. The article serves as an idea generator, especially for teachers looking for ways to update their courses or to explore new concepts in learning. Technologies once only imagined are now opportunities to be implemented in the classroom. Audio and video conferencing, blogs, podcasts, RSS feeds, social bookmarking, and wikispaces are popular means of communicating in today's society. However, Web technology is developing at such an exponential rate that even the newest of these technologies, like Web 2.0, may one day soon be a footnote in computer history. Once these newer technologies are better understood and appreciated, educators can evolve their teaching strategies to help their students remain competitive in the global society.*

## INTRODUCTION

It seems that every institution, academic or corporate, is eager to gain access to anything involving Web 2.0. Although this term was coined by Timothy O'Reilly in 2004, it has evolved into a colloquialism that refers to the current state of the ever changing World Wide Web. Formerly, as amazing as the Web was, it was mostly a static entity on which programmers posted information in a specific format which others could simply view. Or, as Baumann (2006) asserts, "Before Web 2.0, programmers posted Internet content, and the exchange of information was only one way" (p. 38). However, society was pleased because it had information at its fingertips that previously required much time and labor to access.

As is generally the case, consumers demanded more. Perhaps this was, in part, due to the advances in the computer-animated graphics used in movie making, the highly interactive nature of the newest video games, or the increased dependence on e-mail and instant messaging that have permeated society. Whatever the case may be, computer users insisted that the newest technologies be made available to them. Fortunately, Web technology is developing at such an exponential rate that what we now know as Web 2.0 may one day soon be a footnote in computer history. In the meantime, what do *you* know about Web 2.0?

Although definitions abound which attempt to describe this phenomena, a prevailing theme of collaborative interaction arises. Web 2.0 provides "... ubiquitous access to data, an architecture of participation, and distributed independent developers 'playing well together.' Most importantly, everything is 'always in beta'—that is, constantly open for improvement by user feedback" (Umbach, 2006, p. 192). Hauser (2007) echoes this assertion of Web 2.0 as being "... an environment filled with opportunities to not only create content in new ways but also to share information, communicate differently, collaborate easily with the rest of the world, and self-publish" (p. 27). Coombs (2007) provides yet another interpretation. "It [Web 2.0] is often defined by the technologies that are part of it: social software, Weblogs, linklogs, folksonomies, wikis, podcasts, RSS feeds, and Web services" (p.17). Other applications include tagging, social bookmarking, online learning communities, and online office applications. All of the aforementioned share the characteristics of being user-centered, user-generated, and user-controlled. Or, as Breeding (2007) suggests, it involves

*... building an environment that's more focused on the user, that embraces the dynamic content over static pages, that not only delivers content to users but also seeks content from users, and that fosters engagement, participation and collaboration.* (p. 22)

Many teachers are taking notice of these new technologies because many are inexpensive and easily accessible methods to incorporate technology, to increase critical thinking, and to promote substantive conversation in the classroom. According to Driscoll (2007), teachers are "... discovering many cross-curricular projects such as conducting interviews, creating classroom news broadcasts, recording class discussions and explorations, sharing feedback about books, or discussing papers they have written" (p. 12). Podcasts, wikis, and blogs seem to be at the forefront of their efforts; however, audio and video conferencing, RSS feeds, and social bookmarking activities are also increasing in popularity. This is, in part, due to the ease of use and affordability of these tactics; and, as Driscoll contends, "Teachers can now focus on the important question, 'Why do I want to use this technology?' instead of, 'How do I use this technology?'" (p. 10). The following examples are merely a few of the technologies available as part of Web 2.0, but are the most pervasive at this juncture.

## Audio and Video Conferencing

With academic standards and 21st century skills emphasizing reasoning, communication, and technology, it is vital that educators are finding more expedient ways of communicating with others outside the classroom. Newer forms of telecommunicating, which were previously available to only wealthy corporations, are now available at little or no cost to everyone.

*Using technology to communicate and collaborate across different countries can create a more global learning environment, can allow for cross-cultural studies, and can enhance understanding and appreciation of education in contexts other than your own.* (Driscoll, 2007, p. 12)

The easiest and perhaps the most commonplace of such communications are e-mail and instant messaging (IM). Instant messaging allows for online synchronous discussions, however, most schools block access to IM technology for fear of its misuse. E-mailing, although it is still too site regulated, is a form of asynchronous communication which allows for collaboration around the globe.

Audio and video conferencing are becoming more popular and only require a simple computer with a microphone, and webcam—standard equipment for today's computer systems and available as an inexpensive upgrade for older models. Services like MSN Messenger or AOL provide video links as part of their IM services at no charge.

Equally exciting is the opportunity to skypecast. A skypecast is actually a large-hosted call for groups of up to 100 people from anywhere in the world. By accessing www.skype.com, business leaders, teachers, and students can host or participate in a skypecast. This service is also free, and can be used for online lectures, class discussions or even guest speakers.

## Blogs

*A blog is a personal diary. A daily pulpit. A collaborative space. A political soapbox. A breaking-news outlet. A collection of links. Your own private thoughts. Memos to the world. Your blog is whatever you want it to be.* (http://www.blogger.com/tour_start.g)

Driscoll (2007) adds, "Blogs are simple online journals primarily used to support communication in the form of presentation, and they provide a great tool for class interaction" (p.10). These journals take many forms, but they all contain entries by a variety of persons, with the most recent information posted first. They may be considered similar to reverse threaded discussions, in which each new entry is a response to previous ones. To blog or not to blog? That is the question.

Creating a blog is as simple as writing an e-mail, and the services are free through many providers including www.weblog.com, www.blogger.com, Bravenet.com, edublogs.org, and 360.yahoo.com. Each of these providers offers brief online tutorials to assist the user in creating his or her own free blog. Freyer (2006) suggests that teachers set up blogs for use with their students, or even to have students create their own. Many are still reluctant to do so, for fear of their misuse. However, Freyer argues that blogs are safe.

Teachers can set up classroom blogs so that only students' first names are used, and blog visitors cannot reach out and touch a student. Similarly, teachers can moderate comments posted to classroom blogs to ensure that inappropriate or offensive content is not published (p. 30).

Blogging allows students to create, publish, and share their thoughts. They provide the opportunity for critical thinking and collaboration. This collaboration can be among the students in a particular classroom, in a particular school, or even another class somewhere else in the world. Support for the use of blogs is also presented by Thornburg (2007) as follows:

Often students create documents that only a teacher will see. But when students create online reports of their work, they can hear from others who have an interest in their work. This give-and-take provides tremendous incentive for students to share their perspectives with clarity and vision (p. 21).

## Podcasts

The word podcasts is another morpheme of two words describing audio or video broad*casts* that can be played on an i*Pod*. Because of the abundance of iPods and the advent of iTunes, most people are familiar with podcasting. When implementing iTunes, the user can subscribe to hundreds of free podcasts or download music, videos, or audio books for a small fee. Though this is the most prominent site for podcasting, one can access podcasts at a variety of individual Web sites, as well.

Aside from merely accessing podcasts, individuals or groups can also design, create, and publish their own. This process can also be done for little to no cost by using a computer, a microphone, and free online studio services like http://www.

audacity.com. And once podcasts are produced, they have to be uploaded and hosted somewhere on the Web. Sites that will facilitate this process include iTunes, OurMedia, and podcastpeople, among others.

Why go through all of this trouble? Richardson (2007) contends, "When groups of learners coalesce around shared passions online, they experience something that is difficult to replicate in a physical space" (p. 151). Support for this position is found in Driscoll's (2007) statement, "Podcasting gives teachers and students an audio distribution syndication ability to share their research, perspectives, and stories with an audience beyond their classroom" (p. 12). Similar to blogs, they could describe mathematical processes, interpret laboratory data, describe how they would conduct experiments, or even share their results. Other ideas for student podcasts, as presented by Hauser (2007) include the following: "interview visiting authors, teachers, and other students; record morning announcements; practice foreign languages; record their own stories or poems; record comments during field trips; [and] discuss topics taught in class" (p. 47).

## RSS Feeds

RSS is an acronym for "really simple syndication," which is basically a language for publishing informational feeds and distributing them throughout the Web. These feeds, according to Perkins and Pfaffman (2006), ". . . are the type of newsfeeds most people are familiar with as part of their Web homepage where they appear as breaking news, sports, [or] entertainment" (p. 35). RSS documents can contain anything the author chooses, from a brief summary, or abstract, to a full-text document.

Subscribing to an RSS feed is similar to subscribing to the print equivalent of a podcast. Individuals can access these feeds by clicking on the RSS icons which are located on Web pages, blogs, newsletters, and so forth. But knowing that this service is available is not enough. The feeds require special software in order for the user to view them in a traditional format rather that in the language, itself. This software is:

A 'feed reader' or an 'aggregator.' The user subscribes to a feed by entering the feed's link into the reader or by clicking an RSS icon in a browser that initiates the subscription process. The reader checks the user's subscribed feeds regularly for new content, downloading any updates that it finds (http://en.wikipedia.org/wiki/RSS_(file_format)).

Free feed readers are available from www.feedreader.com, www.illumio.com, or www.newsgator.com, among other sites. "Once you get the idea and have chosen your preferred RSS reader you can subscribe to dozens of different feeds...news, newsletters, articles, blogs and more" (http://www.xml.com/pub/a/2002/12/18/dive-into-xml.html). "Useful information lies buried in data streams. All you have to do is find it. And that's where RSS comes to the rescue" (Tebbutt, 2007, p. 17).

## Social Bookmarking

Clicking on *bookmarks* or *my favorites* is commonplace for individual computer users as a means of storing links to sites on their own computers for future access. However, the newest trend of social bookmarking may soon replace the conventional method of categorizing personal information. Social bookmarking includes tagging, linklogs, and folksonomies. It is "designed to keep found things found, identify new communities, discover new websites, make us more productive, and allow us to create new tools to push the frontiers of the Web's utility" (Gordan-Murnane, 2006, p. 27).

The philosophy behind social bookmarking is that of all of the new Web 2.0 technology: user-created with community access. In short, users identify items or sites of interest by *tagging* them.

Tagging is a term used in a number of contexts for different purposes, mostly referring to adding a tag of some form. Tags are best considered as a keyword that refers to an 'ad hoc' classification and sorting of information. Within Web 2.0 applications, tagging can be applied to the URLs of Web pages, to photographs and images as well as ideas, concepts and various projects (http://recap.ltd.uk/Web2/bookmarking.php).

These tags and categories are stored online, and thus can be retrieved, shared, and used by anyone at any computer with Internet access; hence, the "social" in social bookmarking. This social aspect promotes countless educational opportunities.

Examples [in which] children, young people and educators could derive benefit from this technology include:

- Finding and creating new learning communities of users based around a certain topic
- Sharing access to categorized resources in an efficient way
- Developing new insights about a topic by discovering the views and perspectives of others
- Creating a range of contextual taxonomies that carry a specific meaning for a learning community (http://recap.ltd.uk/Web2/bookmarking.php)

Social bookmarking can be done individually, as described previously, or as a group. When done collectively, it is often referred to as a folksonomy. "The term folksonomy is a portmanteau that specifically refers to the tagging systems created within Internet communities. A combination of the words *folk* and *taxonomy*, the term...literally means 'people's classification management'" (http://www.answers.com/topic/folksonomy?cat=technology). Gordon-Murnane (2006) implies that the "development of folksonomies can be seen as a value-added feature of social book marking [because they] connect different groups of people together, and the more people that use them, the better the services become" (p. 29). Free sites providing this service can be found in Table 1.

*Table 1. Free social bookmarking services*

| Site | Web Address |
|---|---|
| BlinkList | http://www.blinklist.com |
| del.icio.us | http://del.icio.us |
| Furl | http://www.furl.net |
| Magnolia | http://ma.gnolia.com |
| RawSugar | http://rawsugar.com |
| Scuttle | http://scuttle.org |
| Shadows | http://www.shadows.com |
| Simpy | http://www.simpy.com |
| Spurl | http://www.spurl.net |

## Wikis

Anyone who has researched a topic on the Internet is familiar with Wikipedia. What that researcher may not realize is that it is one of the largest wikispaces on the Web. Huffman (2006) describes wikispaces (or wikis) like this one, as ". . . online collaborative communities that lend themselves to continuous editing and refinement of content. They work best at aggregating and distilling shared knowledge and include the ability to track article evolution so that content often reflects a blend of voices" (p. 16). The beauty of a wiki is that it not only promotes collaboration among individuals from around the world, it depends on it. The interactive nature of its design allows individual users to enter or edit entries in real time and to have those contributions published immediately. Wikipedia began with a small number of entries and has grown into one of the most popular and widely used source of information in the world. Because of the ability to constantly edit and update, users are kept abreast of current and accurate information without having to wait for the next edition of a book to be published. In other words, the more wikis are used, the better they become—

in theory. The drawback here, since anyone can contribute, is that wikis should be continually monitored to make sure the information presented is, in fact, factual and accurate; hence, the belief is by some that Wikipedia, as well as other wikis, are not actually reliable reference sources.

Educators may be interested in the idea of wikis because they are easy to create and are free to publish. The only equipment needed is a classroom computer. Sites like www.wikispaces.com, www.wetpaint.com, or pbwiki.com provide free platforms on which to build and publish wikis. Driscoll (2007) suggests that wikis can be used for "group-based writing projects, collaborative note taking, or brainstorming" (p. 11). She further asserts, "The capabilities of wikis in the classroom can be a broadening learning experience, as student groups build rich, deep content over time" (p. 11). Organizations might appreciate the wiki because, as Huffman (2006) argues, "A wiki could easily be developed for distance learning or enhancements to in-class work and/or project collaboration and team experiences" (p. 18). This is bad news for expensive learning communities like E-College, WebCT, and Blackboard.

## SUMMARY

As society changes, so do the skills that are necessary for our survival in it. The Partnership for 21st Century Skills (2004) outlines five key learning and thinking proficiencies that are considered to be the cornerstones of that success. Those skills include: critical thinking and problem-solving skills, communications skills, creativity and innovation skills, collaboration skills, contextual learning skills and information and media literacy skills. "Web 2.0 is transforming the Web into a space that allows anyone to create and share information online—a space for collaboration, conversation, and interaction; a space that is highly dynamic, flexible, and adaptable" (Coombs, 2007, p. 17). Richardson (2007) implies that Web 2.0 technologies help address those 21st Century Skills because they:

Require the ability to find relevant sources of information, to assess the trustworthiness of those sources, to coherently engage with the ideas those sources offer, and to make transparent our own experiences and ideas in ways that leave opportunities for others to engage (p. 150).

Many of the technologies presented can be exploited by teachers to increase substantive conversation in the classroom. But in today's educational system, substantive conversation is not limited to just the classroom. As the *National Science Education Standards (NSES)* (1996) maintain, "Good science programs require access to the world outside the classroom" (p. 220). The challenge lies in the assertion that instructional technology has grown almost exponentially in the past decade alone, and it continues to do so at an amazing rate. Now, however, students can communicate with other classrooms around the world, take virtual field trips, and even talk directly with scientists in real time who are working in the field. Krueger and Sutton (2001) assert:

As classrooms become more science-like, teachers will provide students with activities that differ from those now typically in the curriculum. Rather than reading a text and answering written questions, students will be out in the field using probeware to collect data. They will use computer software to model or graph the data . . . The Internet and other communication technologies will provide opportunities for students to collaborate as most scientists do—not by doing experiments together, but by sharing data and hypotheses, and building on results from other groups. E-mail and videoconferencing link student scientists anywhere in the world so that science becomes a global behavior. (p.75)

This technology provides incredible opportunities for collaboration and conversation that must be incorporated into all classrooms if they intend to become and remain progressive. Support for this position is found in Rockman's

(1998) *Leader's Guide to Education Technology*, in which he claims, "Students who use computers in schools demonstrate improved motivation and enthusiasm for school; critical thinking, problem-solving, and independent learning; skills and content knowledge; and ability to compete in the workforce" (p. 3).

Most teachers have experienced search engines, e-mail, chats, databases, or even online public access catalogs (OPACs) for personal or professional use. For example,

Sixth-graders...[from] Alaska used the e-mail based KIDLINK project to correspond with peers around the globe. Even though these children of fishermen were puzzled by strange-sounding careers like 'orthodontist' and 'seismologist' and had to describe what it meant to slice muktuk with an ulu, they discovered, as one said, 'when you look at people, they look real different, but when you look at their words, you realize we're all alike inside' (Rockmann, 1998, p. 8).

Finally, these and new technological opportunities which are surfacing should all be explored. Inventions and discoveries throughout history did not permeate society as quickly as these technologies, which will continue to do so as they evolve. So what does all this mean for the corporate or academic world? It means change. In order to compete, we must embrace change, if for no other reason than it is inevitable. As Freyer (2006) asserts, "Why focus on content transmission in the classroom when we can help students become content creators as well as consumers?" (p. 32). "We must be readers and writers, editors and publishers, to maximize the benefits of our participation; and we must be willing to collaborate and cocreate with others, working closely together to learn even more in the process" (Richardson, 2007, p. 150). Once we understand the newer technologies, become familiar with them, and appreciate their ever-changing nature, we can truly evolve with them and stay competitive in our global society.

## REFERENCES


Answers.com. (2007). *Folksonomy*. Retrieved from http://www.answers.com/topic/folksonomy?cat=technology

Baumann, M. (2006). Caught in the web 2.0. *Information Today, 22*(8), 38.

Breeding, M. (2007). We need to go beyond web 2.0. *Computers in Libraries, 27*(5), 22-25.

Coombs, K. (2007). Building a library Web site on the pillars of web 2.0. *Computers in Libraries, 27*(1), 16-19.

Driscoll, K. (2007). Collaboration in today's classrooms: New Web tools change the game. *MultiMedia & Internet@Schools, 14*(3), 9-12.

Excess Voice. *RSS Explained.* Retrieved from http://www.excessvoice.com/rssexplanation.htm

Freyer, W. (2006). To blog or not to blog? You decide. *Interactive Educator, 2*(2), 28-33.

Gordon-Murnane, L. (2006). Social bookmarking, folksonomies, and web 2.0 tools. *Searcher: The Magazine for Database Professionals, 14*(6), 26-38.

Hauser, J. (2007). Media specialists can learn web 2.0 tools to make schools more cool. *Computers in Libraries, 27*(2), 6-8, 47-48.

Huffmann, K. (2006). Web 2.0: Beyond the concept-practical ways to implement RSS, podcasts, and wikis. *Education Libraries, 29*(1), 12-19.

Krueger, A. & Sutton, J. (Eds.). (2001). *EDThoughts: What we know about science teaching and learning.* Aurora, CO: Mid-continent Research for Education and Learning.

National Research Council. (1996). *National Science Education Standards.* Washington, DC: National Academy Press.

Partnership for 21st Century Skills. (2004). Assessment of 21st century skills. Retrieved from http://www.21stcenturyskills.org

Perkins, M. & Pfaffman, J. (2006). Using a course management system to improve classroom communication. *The Science Teacher, 73*(7), 33-37.

Richardson, W. (2007). Teaching in a web 2.0 world. *Kappa Delta Pi Record, 43*(4), 150-151.

Rockman, S., et al. (1998). *Leader's guide to educational technology.* Alexandria, VA: EDvancenet.

Russell Educational Consultancy and Productions. (2007). *Linklogs and social bookmarking.* Retrieved from http://recap.ltd.uk/Web2/bookmarking.php

Tebbutt, D. (2007). RSS is the new radar. *Information World Review*, Retrieved from the Library, Information Science & Technology Abstracts database.

Thornburg, D. (2007). Connected learning: Changing education through technology use. *Interactive Educator, 3*(2), 20-21.

Umbach, J. (2006). Web 2.0: The new commons. *Feliciter, 52*(5), 192.*What is a blog?* Retrieved from http://www.blogger.com/tour_start.g

*This work was previously published in International Journal of Information and Communication Technology Education, Vol. 4, Issue 2, edited by L. Tomei, pp. 44-53, copyright 2008 by IGI Publishing (an imprint of IGI Global).*

# Chapter 6
# Factors Encouraging or Discouraging Students from Taking Online Classes

**Chuleeporn Changchit**
*Texas A&M University, Corpus Christi, USA*

**Tim Klaus**
*Texas A&M University, Corpus Christi, USA*

## ABSTRACT

*Advances in communication technologies, such as widespread use of the Internet, have opened new avenues for continuing higher education. These advances have allowed educators to provide for and satisfy individual variations in learning. Generally, online courses are adaptations of traditional courses; some courses are more suitable than others for such online instruction. As the trend to offer online courses continues, understanding the factors that lead to students' preference can be useful. Online courses can be costly to develop and to implement, and inappropriately designating courses for online participation can lead to lower student retention rates. This study focuses on students' perceptions of online courses. The results identify issues that affect students' perceptions, and this study concludes by suggesting ways for universities to design online programs that better suit the desires of students.*

## INTRODUCTION

Advances in technology have enabled instructors to design online courses that better meet the needs of students. Colleges and universities around the world are now able to offer higher education courses to students online. This capability can address certain issues students may develop in pursuit of their education goals, such as resolving issues with commuting to and from campus, and communicating with others. Because of the advantages to both students and the university, enrollment growth is considerable, from 1.6 million students (9.6% of total enrollment) in 2002 to 3.9 million students (21.9% of total enrollment) in 2007 (Allen & Seaman 2008).

The increase in online course offerings and enrollment can be tied to necessity. Higher education institutions have faced changes in their student demographics in recent years as more and more students no longer fit the traditional profile of a young, full-time, in-residence student. As the demograph-

DOI: 10.4018/978-1-60566-936-6.ch006

ics change, so do the education needs. There is a higher demand for more flexible and convenient methods in obtaining a higher education. Also, there is a demand by contemporary society for a more technologically savvy workforce. Even if traditional methods are preferred, there is a need to teach students to incorporate technological proficiency into their everyday education.

With these types of considerations, educational institutions are looking into incorporating online capabilities into their courses. However, there are questions regarding the ability of institutions to afford the cost of successfully implementing and coordinating online courses, and the appropriateness of online learning in meeting institutional goals. Furthermore, the benefits that online courses can provide may be dependent on the individual; what may be beneficial to one student may be a hindrance to another. For example, some students may not find online courses beneficial to their education needs and prefer the person-to-person interaction of a traditional course. Others may find the convenience of online courses as an important factor in their course format decisions. These and other factors can be highly influential to decisions.

Since students have perceptions about online courses that influence their subsequent decisions whether or not to take online courses, it is important to understand the factors that surround perceptions of benefit toward an online course setting. Designers of higher education courses can better create course options and curriculum for their students that address higher education needs. The purpose of this paper is to understand the factors that affect a student's perception of online courses as well as the factors that are perceived as important for online courses. In pursuit of this objective, this paper first discusses prior studies that address issues related to online courses and students' perceptions. This is followed by the study's methodology. Next, an analysis of the factors that were found to be significantly different between subjects who prefer online and those who prefer traditional class setting are discussed. In conclusion, the usefulness of the significant factors which can be utilized by those in higher education is discussed as well as future research development in this area.

## LITERATURE REVIEW

### Online Courses

Online learning has been recognized as a technique that concentrates more on students than traditional classroom learning as students are expected to actively involved in building their learning process (Benefits & Compensation Digest 2008). As such, online courses demand that students take more charge in their learning, and as it is an individual determinant, this increases the individual learning experience level. The main concern of the higher learning institution is to ensure that they offer sound online courses so that students and faculty are best able to participate and gain an appropriate learning experience (Lam 2005).

Online courses are growing in number, both in the number offered in universities and in the number of students participating in the classes (Lee, Tan, & Goh 2004). These courses can be tailored in several ways, with varying degrees of video, web-material and participation of faculty members based on the student population, the course, and the time-frame. No one format for a course is best for every participant, students and educators, or for every course. Prior studies have shown that online courses must meet certain standards like those required for in-class courses. For instance, all online courses should initiate with clear learning objectives that will give shape to the development of the course in which all the listed expectations for the attendants are clearly defined (Dykman & Davis 2008). Other important standards are the constant interactions that must be enabled, and therefore take place, among students and professors. This constant interaction helps to increase the effectiveness of the online course (Dykman & Davis 2008). In addition, other factors must be considered such as the size of the class. If an instructor is handling a large class, the possibility of interaction with all students will be minimized, therefore affecting the necessary connection that an instructor must have in order to succeed in the quality of the process (Dykman & Davis 2008). Some case studies have clearly shown the effectiveness of incorporating online tools to assist in facilitating the classroom instruction and discussion (Tennent, Windeknecht, & Kehoe 2004; Hong 2002; Lee, Tan, & Goh 2004).

Beyond the necessary elements that compose the design of an online course, there are many additional features that an instructor can use in an online course, such as the personal profiles of instructors and students. The use of this profile gives the instructor an opportunity to present to students their professional background, past experience in online teaching, if any, and further describe their expected results from teaching the course (Dykman & Davis 2008).

Typically, students who better handle online course formats are those who fit a more independent, self-motivated profile, and have clear career goals. Motivation and discipline seem to play an important role in all online classes. Prior studies have demonstrated that both highly disciplined and motivated students are best able to complete the required self-learning assignments and online lectures by minimizing the distractions (Anderson 2008).

Lower retention rates for online courses have been attributed to reasons such as lack of personal interaction, inexperienced faculty, students unaware of the expectations, and students with multiple obligations. Students' willingness and ability to adapt to the environment and work within the online environment are usually found to be the major determinants of satisfaction with the learning experience (Stokes 2001). Motivation can also be maintained or increased through the constant communication between an instructor and students in which the instructor demonstrates interest in teaching and helping students with their learning process (Dykman & Davis 2008). It is especially important that institutions take these characteristics into consideration when tailoring their courses with online formats. Understanding the characteristics of students who are attracted to online courses can help determine the format and options inputted in course design so that they can learn with as much efficiency and effectiveness as possible.

As institutions continue to consider the formatting of their offered courses, it is important for universities to assess what types of technology best suit their courses and institutional goals. Not all options are necessary for classes to be effective. Teleconferencing via telephone and video may be effective in one course but not another (Hazari 2004). It is important that institutions assess what is effective for their students and the nature of a course before structuring the course's online aspects. Deciding whether a traditional, in-class or a non-traditional, technology enhanced format for a course would best suit their students should not be determined solely by an administration, but through soliciting student input. Hence, the overall balance between the appropriate technology integration and the academic strategy is more important than just the focus of applying sophisticated technologies (Shank 2004).

Even if financial data may support the use of online courses in terms of saving money, it would not be beneficial to implement the online courses if there was resistance by students (Wang 2004). There is still a need for traditional, face-to-face course design, in addition to online environments. Designing and implementing online courses may actually add the costs of online infrastructure and course maintenance to an institution rather than reducing their costs (Banas & Emory 1998). It is important that institutions conduct a prior assessment of students' attitudes toward online courses in order to make the optimum course design for the institution as well as its participants. This study looks at student attitudes towards online courses overall rather than examining the contents of a specific class, examining the characteristics of student preferences toward online courses. The following section describes the factors that may affect students' preference towards online courses.

## Proposed Factors Affecting Online Course Preferences

Based on the Theory of Reasoned Action (Ajzen & Fishbein 1980), the Technology Acceptance Model (TAM) was developed (Davis 1989; Mathieson 1991; Taylor & Todd 1995). There have been many studies which have examined TAM, with results consistently showing a significant relationship between the two independent variables: perceived usefulness and perceived ease of use, and the dependent variable: attitude towards use. In regards to online courses, students must interact with a web-based technology in order to complete the course. Thus, based on TAM, the following two variables will be examined: 1) Perceived Usefulness of Online Course Technology; 2) Perceived Difficulty of Online Course Technology. Based on TAM, following are the first two hypotheses:

H1: Perceived usefulness affects a student's preference towards online courses.
H2: Perceived difficulty of an online course affects a student's preference towards online courses.

Besides students' perceptions, certain characteristics of students might affect their preferences for classroom settings. Older, experienced students with family and work commitments are joining the student population. This group has been found to be the most prolific in utilizing online or web-based courses (Marks, Sibley, & Arbaugh 2005). Furthermore, age has been found to significantly distinguish between successful and unsuccessful completion of online courses (Muse 2003). Two variables will be examined based on these student demographics: 1) Age; 2) Employment Status.

H3: Age affects a student's preference towards online courses.
H4: Employment status affects a student's preference towards online courses.

In addition, online learning expands the pool of students available for university or college level courses. The student profile for online courses is not limited to the traditional, dorm-resident college students. Other demographics, such as gifted students in rural areas and students who live at a distance that makes in-class participation a difficulty, make use of the opportunity to expand their learning environment and take college classes. The online course format can address the issue of commuting and the difficulty with attending classes for some students. Online courses provide them with an opportunity to advance their education in a non-traditional setting (Parmar 2005). Thus the fifth variable that will be examined is Distance from Home.

H5: The distance a university is from a student's home affects a student's preference towards online courses.

Computer self-efficacy has been studied in various publications and has been defined as "an individual's judgment of efficacy across multiple computer application domains" (Marakas, Yi, & Johnson 1998, p. 129). Furthermore, Marakas et al. (1998) points out that there is a difference between task-specific and general computer self-efficacy. Even for users with general computer self-efficacy, they may lack task-specific computer self-efficacy, which can affect perceptions of online course formatting. One study that examined computer self-efficacy found that it affects the perceived ease of use towards new systems (Agarwal, Sambamurthy, & Stair 2000). It is an understandable human nature that people may not desire a technology in which they feel unskilled at using or feel that their abilities are lacking. In some cases, computer self-efficacy can be improved by either training or by providing user support mechanisms (Bendoly 2000). Familiarity with the technology is important and thus the following two variables will be examined: 1) Online Course Experience; 2) Currently Taking an Online Course.

H6: Students who have taken an online course will prefer online courses.
H7: Students who currently are taking an online course will prefer online courses.

Another area that affects perceptions of online courses is its convenience, which may be an important reason for taking Online Courses (Medlin et al. 2004). Perceived Ease of Access has been found to contribute to student satisfaction (Dreenan & Kennedy 2005). Convenience was operationalized in this paper through two variables: 1) Own a computer; 2) Internet Access at Home.

H8: Students who own a computer will prefer online courses.

H9: Students who have Internet access at home will prefer online courses.

One other independent variable that will be examined is Gender. Previous studies have had mixed results in regards to significant differences between genders in learning environment. For example, Shea et al. (2006) did not find any significant difference among genders in learning environments while Williams and Subich (2006) found that men and women differed in their preference for learning experiences. In regards to online versus traditional learning situations, gender may impact the preference. Thus the final hypothesis proposes that gender will impact course preference.

H10: Gender is significantly related to online course preference.

## Research Model and Variables

Based on the hypotheses described in the previous section, the model used to guide this research is illustrated in Figure 1. Ten measurement variables and one response variable were measured as follows:

*Figure 1. Research model*

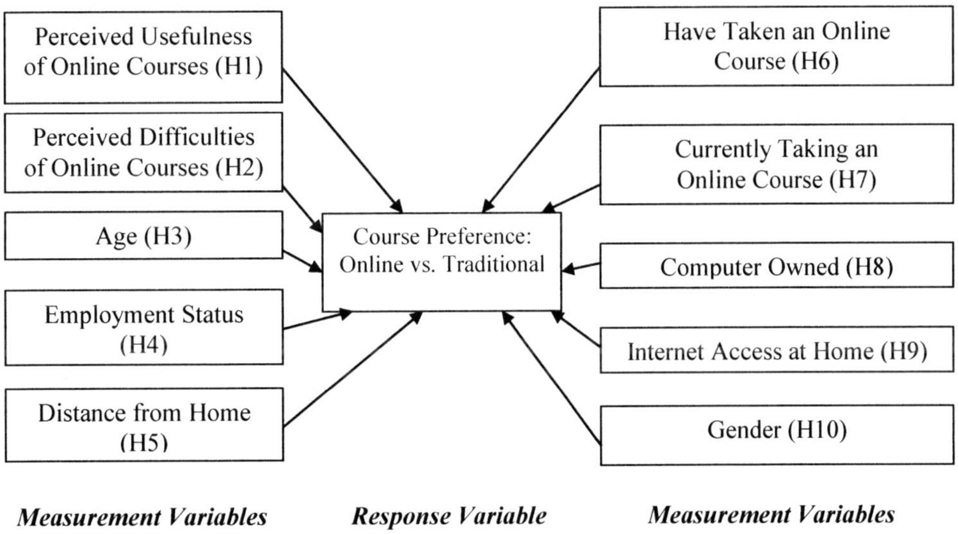

## METHODOLOGY

A direct survey was used to collect the data for this study (see Appendix A). The survey questions were compiled from questions from previous studies pertaining to online courses as well as suggestions from students and researchers (Changchit et al. 2006; Demb et al. 2004; Luarn & Lin 2004; Moore & Benbasat 1991). These questions were designed to gather data on students' perceptions of online courses, as well as their demographics. To validate the clarity of these questions, three professors and three students read through the survey questions. Revisions to the survey were made based on the feedback received.

A total of 32 items were used in the survey. The first 14 questions measured students' general perceptions of online courses, the next four questions gauged the importance of four predetermined factors, the other 13 questions collected demographic data, and survey item 32 measured students' preference for either online or traditional course design.

Surveys were distributed to 225 students enrolled in a mid-sized university. The participants were given the survey and allowed class time to complete the survey. All participants were informed that participation in the study was voluntary and that individual responses would be kept anonymous. The students were asked to rate survey items on a Likert-scale from 1 to 5 with 1 being "strongly disagree" and 5 being "strongly agree" and then to self-classify on survey items concerning specific demographic characteristics. Two hundred and eighteen (218) participants completed and returned the survey instruments. Approximately 21.1% of the respondents preferred traditional classes while 78.9% preferred online courses. Table 1 summarizes additional demographics of the respondents.

*Table 1. Subjects 'demographics*

| | | | | | |
|---|---|---|---|---|---|
| **Age (in years)** | | | | | |
| Under 18 | 18-29 | 30-41 | 42-49 | Over 49 | No Answer |
| 0(0.00%) | 200(91.74%) | 14(6.42%) | 3(1.38%) | 0(0.00%) | 1(0.46%) |
| **Gender** | | | | | |
| Male: 103(47.25%) | | | Female: 115(52.75%) | | |
| **Own a Computer** | | | | | |
| Desktop | Laptop | Both | Neither | | |
| 75(34.4%) | 72(33.03%) | 65(29.82%) | 6(2.75%) | | |
| **Internet Access at Home** | | | | | |
| Dial-up | High speed (i.e., DSL,) | None | No Answer | | |
| 18(8.26%) | 184(84.4%) | 15(6.88%) | 1(0.46%) | | |
| **Distance from Home** | | | | | |
| <10 min. | 10-30 min. | 30-60 min. | 1-2 hours | >2 hours | No Answer |
| 86(39.45%) | 87(39.91%) | 31(14.22%) | 9(4.13%) | 4(1.83%) | 1(0.46%) |
| **Employment Status** | | | | | |
| Full Time: 72(33.03%) | | Part Time: 83(38.07%) | | Unemployed: 63(28.90%) | |
| **Take an Online Course Before** | | | | | |
| Yes: 99(45.41%) | | No: 118(54.13%) | | No Answer: 1(0.46%) | |
| **Currently Taking an Online Course** | | | | | |
| Yes: 37(16.97%) | | No: 179(82.11%) | | No Answer: 2(0.92%) | |

## ANALYSIS AND DISCUSSION

In order to determine the grouping of the survey items, a factor analysis was conducted on the items in the instrument against the two constructs initially presented in the research model which are: (1) perceived usefulness, and (2) perceived difficulty of use. The result from the factor analysis confirms the grouping of these two constructs. The factor matrix is presented in Table 2.

Factor loadings over 0.5 on one factor and less than 0.5 on all other factors produce a clean loading (Hair, Anderson, Tatham, & Black, 1995). As a result of the factor analysis, three survey items (Q3, Q4, and Q10) were discarded due to cross loadings between factors or no apparent loading. The remaining items loaded cleanly onto the two constructs.

*Table 2. Factor analysis*

| Component Matrix(a) | | | | |
|---|---|---|---|---|
| | Component | | | |
| | 1 | 2 | 3 | 4 |
| **Q1** | 0.290 | **0.610** | 0.064 | -0.560 |
| **Q2** | 0.183 | **0.590** | 0.213 | -0.576 |
| **Q3** | 0.211 | 0.153 | 0.663 | -0.026 |
| **Q4** | 0.124 | 0.498 | 0.358 | 0.117 |
| **Q5** | -0.114 | **0.617** | 0.156 | 0.515 |
| **Q6** | -0.187 | **0.677** | 0.185 | 0.411 |
| **Q7** | **0.681** | -0.349 | 0.262 | -0.067 |
| **Q8** | **0.756** | -0.277 | 0.362 | 0.060 |
| **Q9** | **0.796** | -0.178 | 0.211 | 0.096 |
| **Q10** | 0.590 | -0.197 | 0.517 | 0.196 |
| **Q11** | **0.670** | 0.259 | -0.473 | 0.113 |
| **Q12** | **0.755** | 0.209 | -0.429 | 0.122 |
| **Q13** | **0.699** | 0.237 | -0.483 | 0.115 |
| **Q14** | **0.683** | 0.040 | -0.200 | -0.067 |

### Differences between Online Group and Traditional Group

In order to test the hypotheses stated in the prior section, two groups were created (Online and Traditional), based on survey item Q28 – "On the average, I prefer .......... classes", in which subjects marked either "online" or "traditional". T-tests were then conducted on the means of these two groups. The results of the tests are shown in Table 3.

The major findings in testing Hypothesis H1-H10 are summarized in Table 3, which shows t-test results pertaining to the hypothesis as follows:

- **Hypothesis H1:** The t-test confirms that there is a significant difference in students'

*Table 3. Group differences*

| Hypotheses | p-value |
|---|---|
| H1: Perceived usefulness affects a student's preference towards online courses. | 0.017** |
| H2: Perceived difficulty of an online course affects a student's preference towards online courses. | 0.000*** |
| H3: Age affects a student's preference towards online courses. | 0.081* |
| H4: Employment status affects a student's preference towards online courses. | 0.008*** |
| H5: The distance a university is from a student's home affects a student's preference towards online courses. | 0.023** |
| H6: Students who have taken an online course will prefer online courses. | 0.000*** |
| H7: Students who currently are taking an online course will prefer online courses. | 0.024** |
| H8: Students who own a computer will prefer online courses. | NS |
| H9: Students who have Internet access at home will prefer online courses. | NS |
| H10: Gender is significantly related to online course preference | NS |

NS – Not Significant, * Significant at $p < 0.1$, ** Significant at $p < 0.05$, *** Significant at $p < 0.01$

perceptions on the usefulness of online courses. The result indicates that students in the Online Group perceive a higher usefulness in online courses than those in the Traditional Group.

- **Hypothesis H2:** The t-test reveals a significant difference in students' perceptions on the difficulty of online courses. It is evident that students in the Online Group perceive less difficulty in online courses than those in the Traditional Group.
- **Hypothesis H3:** The t-test shows that there is a significant difference in students' ages between the Online Group and Traditional Group. The result indicates that on the average, students in the Online Group are older than those in the Traditional Group.
- **Hypothesis H4:** The t-test reveals a significant difference in students' employment status. The result indicates that a higher percentage of students in the Online Group is employed than in the Traditional Group. It is apparent that students who are employed prefer online course formats as they perceive more flexibility in terms of accessing class materials and studying at their own pace.
- **Hypothesis H5:** The t-test confirms that there is a significant difference in the distance from students' homes to their universities between students in the Online Group and the Traditional Group. The result reveals that students in the Online Group are living farther away from their universities than those in the Traditional Group.
- **Hypothesis H6:** The t-test indicates a significant difference between the level of online course experience. Students who prefer online courses have more experience with taking online courses than those who prefer traditional courses.
- **Hypothesis H7:** The t-test shows that there is a significant difference between students' status in the Online Group and those in the Traditional Group in whether they are currently taking online courses. More students in the Online Group are currently taking online courses than those in the Traditional Group.
- **Hypothesis H8-H10:** The t-test reveals no significant differences between students in the Online Group and those in the Traditional Group on the following issues:

*Figure 2. Students' perceptions on factors important for online courses*

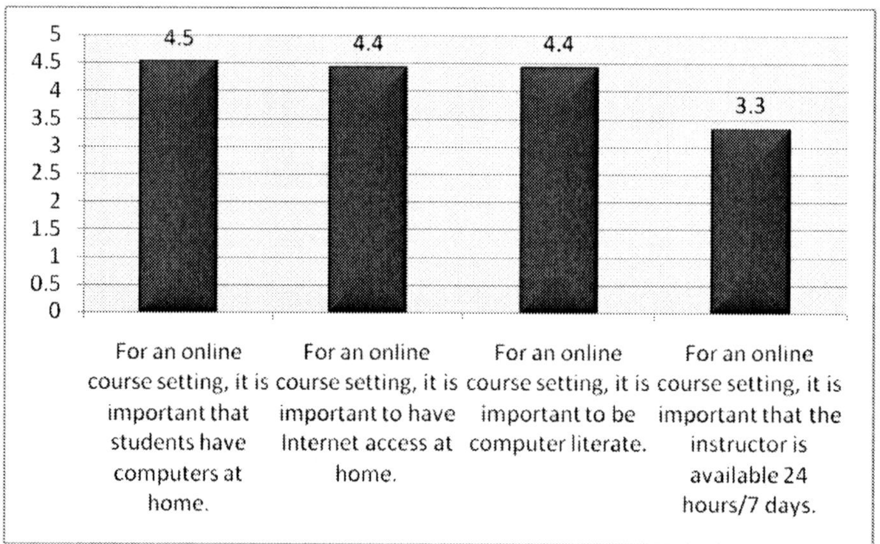

(1) own a computer, (2) have Internet access at home, and (3) gender.

In addition to the hypotheses, a post-hoc analysis was performed on additional questions to examine which factors are important to students when considering online courses. As shown in Figure 2 below, four factors were examined. The results of this post-hoc analysis provides an empirical glimpse into the minds of students as to what they perceive as important to an online course setting. The findings reveal that students, regardless of their preferences of online or traditional classes, place a very high level of importance on the following factors: (1) students have computers at home, (2) students have Internet access at home, and (3) students should be computer literate.

## DISCUSSION AND CONCLUSION

In order to remain competitive and responsive to advances in technology, the number of universities offering online courses has increased rapidly over the past decade. However, this type of learning evolution creates a higher risk for students, faculty, and universities. In order to smooth the transition, the factors critical to successful online courses have been identified and described. As shown in the results section, there are various significant factors that affect student preference.

Based on these factors, there are several areas that a university can focus on. Since the first two factors, perceived usefulness and perceived difficulty, affect students' preferences, course designers should focus on these two areas when designing a course, making the online course comparable to a traditional course. Second, a university should promote online courses in such a way that students understand that the online courses are useful and that online courses do not have a significantly different level of difficulty than traditional courses. The less difficulty students perceive with online courses, the more they tend to support the online option.

The other five factors are important demographics for a university to consider: 1) age; 2) employment status; 3) distance from home; 4) have taken an online course previously; and 5) currently take an online course. Since all of these factors can positively impact a student's preference towards online courses, it is important to know to whom the courses should be promoted. For example, potential non-traditional students can be contacted to increase enrollment through publicizing the flexibility available with online courses. Furthermore, current students could be contacted to let them know what previous students of online courses think about the courses. By reducing uncertainty about expectations, it is likely that a higher number of students will enroll.

These initial findings warrant further investigation. To achieve a better understanding of all of the critical factors in online courses, future research should gather more samples and may also include the perceptions of faculty, administrators, and staff as well as those of students. Furthermore, investigating the characteristics of courses that make them better suited to an online format would benefit higher education institutions striving to meet the needs of students.

## REFERENCES

Ajzen, I., & Fishbein, M. (1980). *Understanding attitudes and predicting social behavior*. Englewood Cliffs, NJ: Prentice-Hall.

Allen, I.E., & Seaman, J. (2008). Staying the course: Online education in the United States. *The Sloan Consortium*, 1-28.

Anderson, W. (2008). Online degrees: Are they right for you? *Certifications Magazine*, *10*(6), 18–23.

Banas, E. J., & Emory, W. F. (1998). History and issues of distance learning. *Public Administration Quarterly, 22*(3), 365–383.

Benefits & Compensation Digest (2008). Tools and techniques, *45*(8), 10.

Bocchi, J., Eastman, J. K., & Swift, C. O. (2004). Retaining the online learner: Profile of students in an online MVA program and implications for teaching them. *Journal of Education for Business, 79*(4), 245–253. doi:10.3200/JOEB.79.4.245-253

Changchit, C., Cutshall, R., & Elwood, S. (2006). Students' perceptions of the laptop program: What factors should be considered before implementing the program? *International Journal of Information and Communication Technology Education, 2*(2), 53–61.

Davis, F. D. (1989). Perceived usefulness, perceived ease of use, and user acceptance of information technology. *MIS Quarterly, 13*(3), 319–340. doi:10.2307/249008

Demb, A., Erickson, D., & Hawkins-Wilding, S. (2004). The laptop alternative: student reactions and strategic implications. *Computers & Education, 43*(4), 383–401.

Drago, W., & Peltier, J. (2004). The effects of class size on effectiveness of online courses. *Management Research News, 27*(10), 27. doi:10.1108/01409170410784310

Drennan, J., & Kennedy, J. (2005). Factors affecting student attitudes toward flexible online learning in management education. *The Journal of Educational Research, 98*(6), 331–338. doi:10.3200/JOER.98.6.331-338

Dykman, C., & Davis, C. (2008). Online education forum - part three a quality online educational experience. *Journal of Information Systems Education, 19*(3), 281–289.

Hazari, S. (2004). Strategy for assessment of online course discussion. *Journal of Information Systems Education, 15*(4), 349–355.

Hong, K. (2002). Relationships between students' and instructional variables with satisfaction learning from a Web-based course. *The Internet and Higher Education, 5*(3), 267–281. doi:10.1016/S1096-7516(02)00105-7

Lam, W. (2005). Teaching e-business online: The Universitas 21 global approach. *Journal of Electronic Commerce in Organizations, 3*(3), 18–41.

Lee, C. S., Tan, D. T. H. T., & Goh, W. S. (2004). The next generation of e-learning: Strategies for media rich online teaching and engaged learning. *International Journal of Distance Education Technologies, 2*(4), 1–16.

Luarn, P., & Lin, H. H. (2005). Toward an understanding of the behavioral intention to use mobile banking. *Computers in Human Behavior, 21*(6), 873–891. doi:10.1016/j.chb.2004.03.003

Marks, R. B., Sibley, S. D., & Arbaugh, J. B. (2005). A structural equation model of predictors for effective online learning. *Journal of Management Education, 29*(4), 531–563. doi:10.1177/1052562904271199

Mathieson, K. (1991). Predicting user intentions: comparing the technology acceptance model with the theory of planned behavior. *Information Systems Research, 2*(3), 173–191. doi:10.1287/isre.2.3.173

Medlin, B. D., Vannoy, S. A., & Dave, D. S. (2004). An Internet-based approach to the teaching of information technology: A study of student attitudes in the United States. *International Journal of Management, 21*(4), 427–434.

Moore, G. C., & Benbasat, I. (1991). Development of instrument to measure the perceptions of adopting an information technology innovation. *Information Systems Research*, *2*(3), 192–222. doi:10.1287/isre.2.3.192

Muse, H. E. (2003). The web-based community college student: An examination of factors that lead to success and risk. *The Internet and Higher Education*, *6*, 241–261. doi:10.1016/S1096-7516(03)00044-7

Parmar, N. (2005). Gifted students connect online with colleges. *Wall Street Journal* (Eastern Edition), B1.

Shank, P., (2004). When to use instructional technology and when not to? *T+D, 58*(9), 30-37.

Shea, P., Chun, S. L., & Pickett, A. (2006). A Study of teaching presence and student sense of learning community in fully online and web-enhanced college courses. *The Internet and Higher Education*, *9*, 175–190. doi:10.1016/j.iheduc.2006.06.005

Stokes, S. P. (2001). Satisfaction of college students with the digital learning environment: Do learners' temperaments make a difference? *The Internet and Higher Education*, *4*(1), 31–44. doi:10.1016/S1096-7516(01)00046-X

Taylor, S., & Todd, P. A. (1995). Understanding information technology usage: A test of competing models. *Information Systems Research*, *6*, 144–176. doi:10.1287/isre.6.2.144

Tennent, B., Windeknecht, K., & Kehoe, J. (2004). Teaching with technology: Value-added innovation or necessity? *Campus-Wide Information Systems*, *21*(4), 144. doi:10.1108/10650740410555016

Wang, W. (2004). How university students view online study: A PCP perspective. *Campus-Wide Information Systems*, *21*(3), 108. doi:10.1108/10650740410544009

Williams, C. M., & Linda, M. S. (2006). The gendered nature of career related learning experiences: a social cognitive career theory perspective. *Journal of Vocational Behavior*, *69*, 262–275. doi:10.1016/j.jvb.2006.02.007

## APPENDIX A

Please circle the answer that best represents your opinion.

For an online class setting (compared to a traditional class setting)
*Table 4.*

| ①=Strongly Disagree | ②=Disagree | ③=Uncertain | ④=Agree | ⑤=Strongly Agree |
|---|---|---|---|---|

| | | |
|---|---|---|
| 1 | …., students should spend less time to study for an online class. | 1 2 3 4 5 |
| 2 | …., student should get a better grade in an online class. | 1 2 3 4 5 |
| 3 | …., there will be fewer group projects in an online class. | 1 2 3 4 5 |
| 4 | …., the tuition for an online class should be less expensive. | 1 2 3 4 5 |
| 5 | …., my study habits should be improved with an online class. | 1 2 3 4 5 |
| 6 | …., my time will be spent more efficiently with an online class. | 1 2 3 4 5 |
| 7 | …., it will be difficult to participate in the class. | 1 2 3 4 5 |
| 8 | …., it will be difficult to communicate with peers. | 1 2 3 4 5 |
| 9 | …., it will be difficult to communicate with the instructor. | 1 2 3 4 5 |
| 10 | …., it will be difficult to participate in a group project. | 1 2 3 4 5 |
| 11 | …., it will be difficult to turn in an assignment. | 1 2 3 4 5 |
| 12 | …., it will be difficult to access class materials. | 1 2 3 4 5 |
| 13 | …., it will be difficult to take examinations. | 1 2 3 4 5 |
| 14 | …., it will be difficult to learn the technologies required for the online class. | 1 2 3 4 5 |
| 15 | …., it is important that students have computers at home. | 1 2 3 4 5 |
| 16 | …., it is important that the instructor is available 24 hours/ 7 days. | 1 2 3 4 5 |
| 17 | …., it is important to have Internet access at home. | 1 2 3 4 5 |
| 18 | …., it is important to be computer literate. | 1 2 3 4 5 |

Please circle your answer for the following questions.

*Table 5.*

| | |
|---|---|
| 19 | I own: 1. Desktop computer 2. Laptop computer 3. Both 4. Neither |
| 20 | My ethnicity: 1. African 2. Anglo 3. Asian 4. Hispanic 5. Native American |
| 21 | My computer knowledge is: 1 (Very poor) 2 3 4 5 6 7 (Excellent) |
| 22 | My classification is: 1. Freshman 2. Sophomore 3. Junior 4. Senior 5. Graduate |
| 23 | My college is: 1. Arts/Humanities 2. Business 3. Education 4. Nursing/Health Science 5. Science/Technology |
| 24 | My age: 1. Under 18 2. 18-29 3. 30-41 4. 42-49 5. Over 49 |
| 25 | It takes me ............ to travel from the place I currently live to the University:<br>1. < 10 Minutes 2. 10 – 30 Minutes 3. >30 Minutes – 1 Hour 4. >1 Hour – 2 Hours 5. >2 Hours |
| 26 | I commute to the University by: 1. Car 2. Bus 3. Walk |
| 27 | What Internet access do you have from home? 1. None 2. Dial-up 3. High-speed (i.e., DSL, Cable) |
| 28 | My gender: 1. Male 2. Female |
| 29 | Did you take a web-based course before? 1. Yes 2. No |
| 30 | Do you currently take a web-based course? 1. Yes 2. No |
| 31 | My current employment status is: 1. Full-time 2. Part-time 3. Unemployed |
| 32 | On the average, I prefer ........... classes. 1. Online 2. Traditional |

-------- Thank You --------

# Chapter 7
# Enhancing Scholarly Conversation Through an Online Learning Community

**Brian Thoms**
*Claremont Graduate University, USA*

**Nathan Garrett**
*Claremont Graduate University, USA*

**Terry Ryan**
*Claremont Graduate University, USA*

## ABSTRACT

*This paper reports on action research (AR) that implements online learning community (OLC) software to foster conversation and community at a specific graduate school. Informed by theories of conversation, online learning, and social networking we incorporate Web 2.0 technologies in the creation of a user-centric OLC. A distinguishing feature of our software is that, rather than being centered on courses like traditional course management software (CMS), our software is oriented towards and controlled by individuals. Results indicate that stakeholders—graduate students and faculty—appreciate and find value in the OLC we implemented.*

## INTRODUCTION

The leaders of our school, a graduate school in the U.S., believe that its continued success depends on the existence of a vibrant intellectual conversation among its stakeholders—students, faculty, staff, and alumni. Unfortunately, recent trends (primarily a decrease in the presence of stakeholders on campus) have led to a reduction in the vitality of this essential conversation. To help foster and revive it, these same leaders have asked our research group to design and implement an information technology (IT)-based solution.

Our guiding philosophy has been to allow people to say what they want to say, to listen to what they want to listen to, to increase their understanding both of themselves and of their fellow community members, and to do it all without having to spend an increased amount of time on campus. To this end, we have implemented software designed to

DOI: 10.4018/978-1-60566-936-6.ch007

promote free expression of identity and ideas, by and between individuals. Our intent has not been to replace face-to-face interactions, but to supplement them with a persistent virtual component.

We believe that we have achieved a measure of success in improving scholarly conversation at our school and also in learning how conversation can be promoted by IT. This paper presents the nature of our school's problem, what we did to design and implement a solution, the impact of the solution on our school and what we plan for the future.

## BACKGROUND

### Bringing Conversation Online

Etched on the perimeter wall of our school there is a phrase, "The center of a college is in great conversation and out of the talk of the college life springs everything else." This observation is widely accepted at our school, especially by our project stakeholders, but it has some problems as a compass for taking action.

Conversation (to say nothing of great conversation) is difficult to define, even when viewed in terms of a specific population, in our case graduate students, faculty, administrators and alumni. To make progress, we adopted a simple initial working definition of conversation as purposeful (that is to say, related to graduate school activities) peer-to-peer talk. More formally, conversation is a speech exchange system that is structured around turn-taking, a sequential organization of who gets to say what and when (Sacks et al., 1974). Because speech acts need not be oral or face-to-face, conversation need not be, either. Within this view of conversation as structured speech, not necessarily involving face-to-face talking, it is possible to distinguish a number of purposes for conversation in graduate education. Jenlink and Carr (1996) identify four types of conversation with varying degrees of applicability in a graduate setting:

- Conversation as *dialectic* with focus on logical argument and distilling truth.
- Conversation as *discussion* where many people advocate for their own individual positions.
- Conversation as *dialogue* with focus on constructing meaning through multiple perspectives.
- Conversation as *design* with goals and a focus on creating something new.

Although these notions of conversation are not void at our school, they have been largely confined to the on-campus setting. As IT researchers in a largely brick and mortar academic institution, we looked for ways where software has been and can be used to bring conversation into the 21st century. In any acceptable solution, stakeholders would have to be more involved in all types of conversations, both face-to-face and online.

Most prior research about online conversation focuses on problem solving, decision-making and discussion depth (Sherry, 2000). Although these matters are important, they do not directly address the goals of our project. In a graduate school setting, conversation must extend beyond simple interactions, allowing people to wrestle with complex problems from multiple perspectives. In considering what to do to help our school, we concluded that a more novel approach would be needed to support conversation.

Inspired by outstanding examples of online conversation in a number of very popular online social networking (OSN) applications—including Facebook™, MySpace™, LinkedIn™, and Classmates™—we focused our efforts on the Web 2.0 technologies that they use. Some Web 2.0 technologies, such as blogs, wikis and peer-to-peer networking, provide the capabilities for users to participate in online conversations, as exemplified by popular OSNs.

Online Social Networks offer an intriguing solution because they employ social ties among friends, and friends-of-friends, to aid in information exchange. OSNs are also intriguing because of their distinction from more traditional course management systems (CMSs). Unlike CMSs, OSNs let individuals own their space, develop their own communities, and control their own participation. In CMSs, control of most aspects of usage rests with course instructors, not students. Furthermore, OSNs make it easy to allow non-student stakeholders to participate, unlike CMSs, which are generally restricted to students enrolled in the particular course.

With these considerations, we decided that some kind of educational OSN application, one that could support stakeholder-defined online learning communities (OLCs), would be our best option for increasing conversation at our school. The largest differentiator between an OSN and OLC is the focus OLCs place on learning. Although learning can occur in an OSN, the primary objective of an OLC is to foster learning through community.

## Online Learning and Online Community

Studies have shown that interaction and dialogue are essential for productive learning (Nicol et al., 2003; Cook, 2002; Sorenson and Takle, 2002). Additionally, Internet and multimedia technologies play increasing roles in reshaping the way knowledge is delivered, providing valid alternatives to traditional classroom learning (Zhang, 2004; Garrison, 2002). Ultimately, online learning offers just one more approach to learning in addition to new ways of collaborating (Haase, 2005).

Wu and Hiltz (2004) used a questionnaire to examine students' perceptions of their learning after using online discussions. They found that students who were more motivated and enjoyed their experiences also reported higher perceptions of their learning achievements, although they recognized that there may be a difference between perceived and actual learning. Webb et al. (2004) have also noted a positive association between participation in what is termed 'e-learning dialogues' and learning outcomes.

In addition to learning, an OLC must also foster community, often measured in terms of social capital. In a virtual setting, social capital is a common social resource that facilitates information exchange, knowledge sharing, and knowledge construction through continuous interaction, built on trust and maintained through shared understanding (Daniel et al., 2003). Huysman and Wulf (2005) propose that the higher the level of social capital, the more members are stimulated to connect and share knowledge. This sharing aspect challenges an individual to draw upon and provide value for themselves and the community. Social resources often include common identity, familiarity, trust, and a degree of shared language and context among individuals (Lesser and Prusak, 1999).

## Theory

The primary objective of research is to promote stakeholder-defined OLCs, which amount to computer-based and learning-oriented communities of practice. A community of practice (CoP) is a group that works together towards common goals, collaborating on common problems, sharing best practices, supporting one another and sharing a common identity (Lave and Wenger, 1991). Critical for the specific knowledge needs of a CoP is that knowledge is self-generating and perpetuating, and the transfer of knowledge is an intrinsic aspect of its functioning (Adams and Freeman, 2000). In creating a sustainable CoP at our school, we draw on theories of learning and community to guide our research. The following diagram shows activities in a CoP (represented by activity theory) as the driving force that enables 1) individual learning (represented by constructivism) and 2) community interaction (represented by social presence).

*Figure 1. Theoretical model for OLCs*

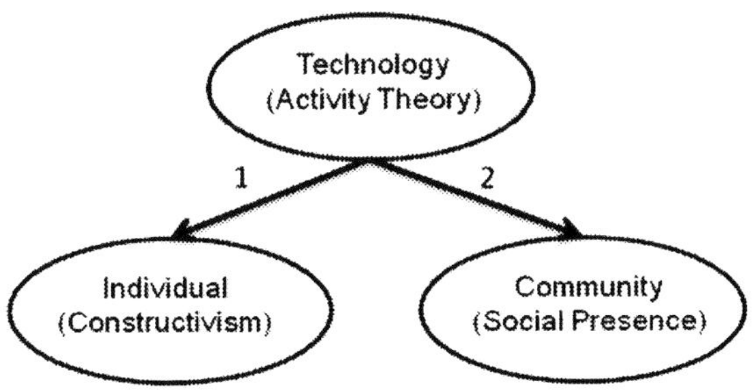


## Activity Theory

Activity Theory can be used as a lens for understanding sociotechnical interactive networks as a function of technology, community development and user interaction between the two. In Activity Theory, activities are goal-directed, where there exists multiple ways to achieve those goals, oftentimes through adaptation (Bødker, 1989). As users begin to adapt to new technologies, they adapt their activities in order to meet the required objectives. Consequently, when individuals use the Web 2.0 technologies comprised in our OLC, they will also adapt their activities to meet course and personal objectives. In a study on higher education, Issroff and Scanlon (2002) found that Activity Theory forces instructors to consider multiple factors that can impact a specific technologies usage. Furthermore, when able to choose educational activities from both online and face-to-face mediums, instructors can potentially select the activity that provides the best fit with any particular learning objective (Heckman and Annabi, 2006).

## Constructivism

Online learning theories are used to understand ways in which individuals learn. At the graduate level learning is largely self-directed. Constructivism views each learner as a unique individual with unique needs and backgrounds, both complex and multidimensional (Gredler, 1997). Constructivism also encourages, utilizes, and rewards these characteristics throughout the learning process (Hagstrom and Wertsch, 1997).

Although constructivism began as a theory of learning, it has progressively expanded its realm, becoming a theory of teaching, a theory of education, a theory of the origin of ideas, and a theory of both personal knowledge and scientific knowledge (Matthews, 2002). The roots of a CoP can be traced to constructivism (Johnson, 2001; Palloff and Pratt, 1999; Savery and Duffy, 1996). Consequently, as a specific type of CoP, an OLC is a user-centric and user-driven and offers each user control over his or her own space. For our purposes, constructivism helps place the individual at the forefront of our OLC design allowing individuals to manipulate the OLC, as they wish, to maximize their learning experiences. As a user-centric and user-driven tool, a stakeholder-defined OLC offers individuals ***complete*** control over their own space and encourages users to explore and take ownership of the OLC.

## Social Presence Theory

Individuals are also influenced to a great extent by their surrounding environment. Therefore, an equally critical component for a successful OLC involves a user's perception of community within the OLC. Social Presence Theory looks at the degree to which an individual's perception of an online community affects his or her participation in that community. Several factors come into play when measuring social presence centered on social context, online communication, and interactivity (Tu, 2002). These components range from group attitudes towards online communication to more personal attitudes on privacy and risk.

Research has shown that social presence is one of the most significant factors in improving instructional effectiveness and building a sense of community. Research by Stacey (2002) discovered that establishing social presence is an important aspect for effective online interaction and learning and found that a high quality of electronic communication helps to engage students and aids in their learning of the course material.

For our specific study, an OLC will be used to compliment face-to-face classroom experiences therefore some degree of social presence, prior to individuals using the tool, may already exist. Our goal will be to extend this classroom experience into an asynchronous online environment.

## RESEARCH DESIGN

### Action Research (AR)

The Action Research (AR) approach is highly appropriate for implementing and measuring our software since a sustainable OLC will involve improvements and modifications in order to realign the software with stakeholder objectives. Modifications are derived from the attitudes of course instructors, students, as well as trends in research and popular online networks. This paper details the planning (i.e. analysis and design), implementation and evaluation of our OLC.

Additionally, we focus our study on a specific population at our school, students taking transdisciplinary courses (t-courses). T-courses are semester-long course required by all second-year doctoral students. Each t-course is comprised of students from different disciplines. T-courses provide an interesting population since these courses, in the past, have relied primarily on traditional instructor-facilitated lecture and face-to-face discussion with little to no virtual collaboration. Where collaborative work was assigned, students would identify their own methods for collaborating, which, again, may or may not have consisted of a virtual component. Ultimately, there was no formal meeting place, virtual or real-life for students to meet and collaborate outside the classroom. While we do not wish to supplant these valuable face-to-face interactions, we hope to provide students and instructors with a blended learning approach, consisting of both face-to-face and virtual components.

Our high-level research question asks how technology can promote conversation. More specific questions focus on the roles learning and community play in conversation at graduate school, and primarily:

1. Can an OLC enable learning among doctoral students?
2. Can an OLC foster community at our school?
3. Will our implementation of an OLC be accepted at our school?
4. Will our OLC implementation be able to sustain an ongoing user population?

### Building the Claremont Conversation Online (CCO)

Members of an OLC should be able to state what they think, comment on what others have said, collaborate on common statements, and share information in many forms. Web 2.0 technologies, such as blogs, wikis, and peer-to-peer file sharing support these activities. And increasingly, individuals are becoming more familiar with these technologies, making their introduction into the classroom more-or-less seamless. A survey of U.S. internet-using teens showed 57% having had created online content including blogs, artwork, video, and content remixing with 19% of U.S. youth ages 12-17 having created a blog, and 38% of U.S. youth reported reading them (Lenhart and Madden, 2005). Furthermore, a 2007 survey states 55% of all American teens having at least one online profile on a social networking site (Lenhart and Madden, 2007). However, the blogosphere is not exclusive to digital natives (individuals who have always had access to the Internet). In fact, Technorati has indexed over 70 million international blogs, and records almost 120,000 new blogs created daily (Sifry, 2007). Thus, as users of advanced learning environments begin to expect such capabilities, a number of solutions that integrate blogs, wikis and file sharing have become available.

During the planning stage of our research we evaluated a variety of proprietary and open source software solutions. We compared technologies based on cost, usability, extensibility, customizability and the range of features each offered. We ultimately decided on the Elgg platform, which labeled itself as *Elgg: the Online Learning Landscape*. In addition to the common Web 2.0 utilities such as blogging, file sharing and peer-to-peer networking. Also an essential feature, and illustrated in Figure 2, is the ability for individuals and communities to restrict access to data across a number of levels allowing members of a community to keep content public (across the Internet) or restrict it to various levels (such as the individual, community, logged in users or custom levels).

Extensibility was another critical factor in our choosing the Elgg platform. As open source, Elgg (shown in Figure 3) provided us with the freedom to develop new components and to customize the software to meet the needs of our school, as well as our research objectives. As one example, the requirement to support individual portfolio building and group writing led to the design construction and implementation of a wiki add-in.

Another important feature of Elgg was that it was easy to integrate with our schools IT infrastructure. Open source LDAP code was modified to use our school's student database, allowing users to log in with their school email id and password, eliminating the burdensome task of having an additional login id and password. Additionally, we were able to customize the look and feel of Elgg to match CGU (also illustrated in Figure 2). Thus, the tool software transformed from Elgg, to the Claremont Conversation Online, or CCO as it is more commonly referred.

The ability for users to customize the look and feel of their personal space was also important. While a number of standard templates exist, Figure 4 showcases how individuals, with basic skill in CSS and HTML coding could customize their respective environments.

A central aspect of Elgg is the ability for users to build sub-communities, a feature we deemed essential for collaboration across multiple projects. For each course implementing our OLC, a specific sub-community was created. In each community, users would have access to the same set of features they would have in their personal space. Consequently, community members could restrict access to content solely to the community (as shown in Figure 2). Figure 5 illustrates an example of a community wiki page, where multiple

*Figure 2. Content restrictions*

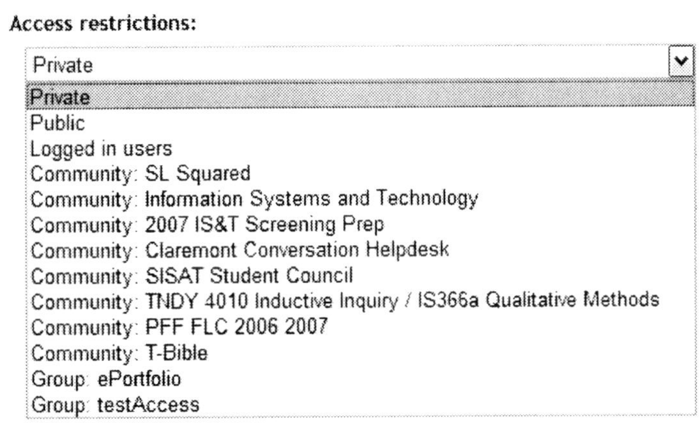

Figure 3. CCO environment

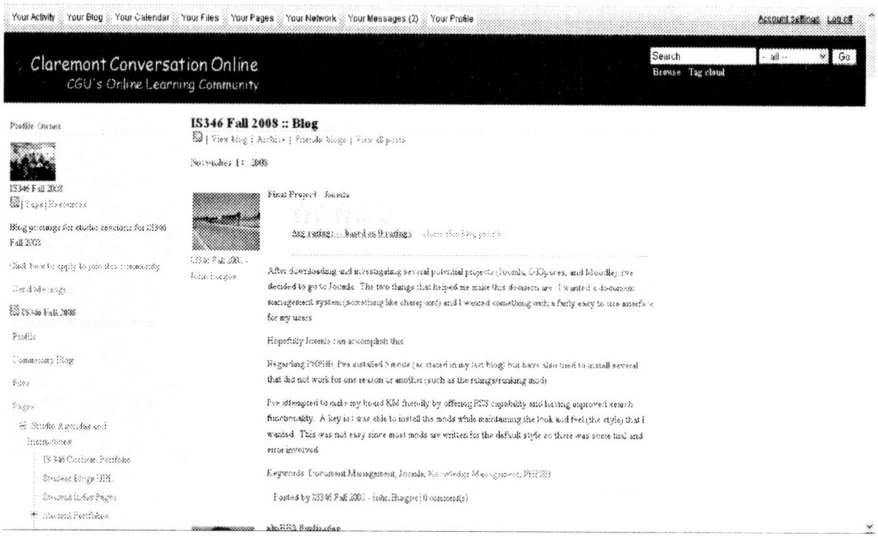


Figure 4. CCO customized look and feel

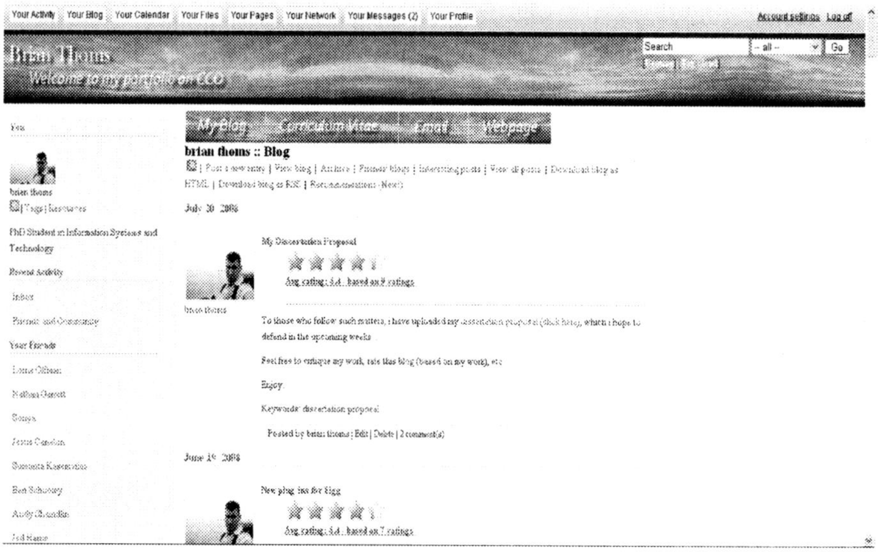


users could collaborate on a single document, keeping it restricted until they wanted to publish it. Figure 5 also shows the use of multimedia for more dynamic content.

## IMPLEMENTATION

During the analysis and design phase of our project, we met regularly with project stakeholders and course instructors for training purposes. During this time we also engaged in regular discussions on how an OLC could augment classroom experiences. Although course instructors were not mandated to use our software, we worked closely to showcase the benefits of using a combination of traditional classroom techniques with a virtual component to further extend the classroom conversation. More specifically, we focused on how

*Figure 5. CCO collaborative writing*

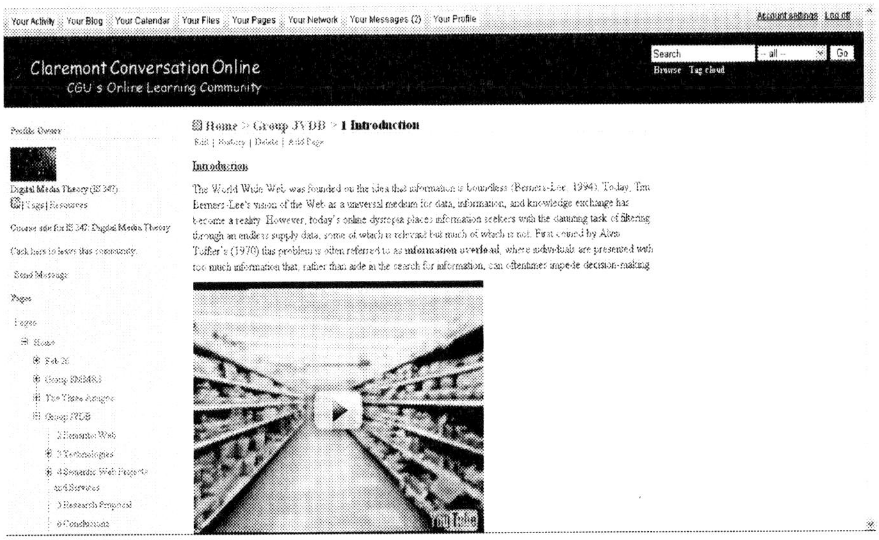


course syllabi could be aligned with our software. For example, if a course required weekly assignments based on selected course readings, or guest lectures, we recommended the community blog as a viable option for individuals to express themselves. For more collaborative projects, we recommended that students use the wiki. In the end, it was left up to the instructor to decide how best to use the technologies and our role was to help facilitate instructors in achieving their specific objectives. The result was 100% adoption by t-course instructors.

Our group sessions also provided feedback on what additional features should be incorporated and helped to identify system bugs as well. After three months of development, our tool was a hybrid of in-house and Elgg-bundled software, hosted on a local web server and customized for our specific school. Table 1 provides a list of Web 2.0 technologies that comprised our tool. Each technology, with the exception of the chat feature, used sparingly, played an integral part in learning and forming community.

*Table 1. CCO features*

| Bundled with Elgg | Add-in |
|---|---|
| Profiles | Wiki |
| Blogging | Enhanced blog commenting |
| File repository | Enhanced wiki commenting |
| Networking capabilities | Chat Room (added post-release) |

## SPECIFYING LEARNING

During the first semester of implementation we focused on providing our school with a stable OLC. During this time we also measured acceptance of the CCO software, how it aided in learning, and how it helped in building community. Although we released the tool to a wider population of courses, we focused primarily on t-courses for analysis, which by their nature place a high emphasis on collaboration and discourse.

### Pretest Analysis

We conducted an initial pretest to capture general information about our user population prior to their using the CCO. The pretest, a closed ended questionnaire, yielded 51 usable responses (62%) from our initial pool of 82 t-course participants.

## Predicting Online Success, Technology Familiarity and Social Tendencies

The first measurement looked to predict the success of our OLC through a general online compatibility scale. The instrument was designed primarily to assess the willingness of an individual to use online methods of learning and communicating. Developed by Smith et al. (2003) and consisting of 22 items, each scored on a 5-point scale, the average rating was 2.16 indicating that, on average, users would have minimal resistance to learning and communicating through online methods.

In addition to online readiness, we measured how familiar individuals were technologies incorporated in our software. Detailed in Table 2, with the exception of RSS and wiki technologies, the majority of users were familiar or somewhat familiar blogging, file sharing and social networks.

As one last pretest measure, we included a general sociability index based on an introversion/extroversion assessment technique developed by an in-house expert. The questionnaire measures an individual's sociability based on responses to 17 questions, each with two choices (one indicating an extroverted action, the other an introverted response). The average rating of 1.54 indicated a balance of introverts and extroverts.

## Usage Data

To get a general indication of the impact the CCO was having, we used indirect data collection based on database updates. Table 3 shows side-by-side comparison of technology usage. In this research we focus on t-course usage, comprising roughly two-thirds of the site participants. These users accounted for the majority of file uploads (60%), wiki edits (64%) and wiki comments (62%). However, the same two-thirds accounted for less total blog additions (42%) and blog comments (43%) with less unique wiki page contributions as well (36%).

## Posttest Analysis

The posttest questionnaire focused on assessing online community, social learning and specific technologies, in addition to overall CCO satisfaction. Closed- and open-ended questions, distributed to the same population of t-course participants (81 individuals), resulted in 50 (61%) usable responses, 35 of which also responded with qualitative data.

*Table 2. Technology familiarity (n=51)*

| Technology | Familiarity (in %) | | |
|---|---|---|---|
| | Very Familiar | Somewhat Familiar | Not Familiar |
| Blogging | 33% | 53% | 14% |
| E-Profiles | 20% | 43% | 37% |
| File Sharing | 31% | 41% | 28% |
| RSS | 12% | 26% | 63% |
| Social Networks | 39.5% | 29.5% | 31% |
| Wiki | 23.5% | 27.5% | 49% |

*Table 3. Site usage: T-course vs. all*

| Category | T-Courses | All | As % |
|---|---|---|---|
| Users | 82 | 130 | 63% |
| Communities | 16 | 29 | 55% |
| Blogs | 486 | 1154 | 42% |
| Blog Comments | 422 | 986 | 43% |
| Wiki Pages (All) | 907 | 1427 | 64% |
| Wiki Pages (Unique) | 127 | 356 | 36% |
| Wiki Comments | 23 | 37 | 62% |
| File Uploads | 115 | 193 | 60% |

## Overall User Experience

A 5-point scale measured the impact of our tool. Overall, the CCO was positively received and the majority of participants agreed that the tool was an important edition to their course. Table 4 provides a snapshot of our data, where only 12% of respondents indicated that they were dissatisfied with the tool. 58% were satisfied or strongly satisfied with the software, while 30% remained neutral. Although 40% of the respondents felt they had less interaction with their instructor, 55% reported more interaction with peers than in other classes. An important aspect of the t-course is in building peer-to-peer relationships and 47% reported a strengthening in peer-relationships compared with other courses. It was largely encouraging to find that only 3% of respondents disagreed with the notion of an OLC at our school, and 67% agreed or strongly agreed.

Lastly, over 50% felt that our software was preferred over traditional CMSs and that an academic OLC is an important addition to the graduate experience. Along these lines, the general consensus was that our software should be hosted under an EDU domain (71%), as opposed to our choice for an ORG extension, implying that an institutionally backed tool would be more appropriate.

*Table 4. Overall user experience (n=50)*

| Survey Item | 1 | 2 | 3 | 4 | 5 |
|---|---|---|---|---|---|
| Overall software satisfaction | 22% | 36% | 30% | 12% | 0% |
| Achieved class objectives | 19% | 41% | 30% | 10% | 0% |
| More interaction (professor) | 14% | 24% | 22% | 16% | 24% |
| More interaction (peers) | 25% | 30% | 27% | 13% | 5% |
| Strengthened relationships | 22% | 25% | 30% | 14% | 8% |
| Should have an OLC | 24% | 43% | 30% | 3% | 0% |
| Prefer over traditional CMS | 24% | 30% | 32% | 12% | 0% |
| Consider use outside the course | 8% | 39% | 31% | 14% | 8% |
| Importanc of a .DU extension | 31% | 40% | 22% | 3% | 3% |

1=Strongly Agree, 2=Agree, 3=Neutral, 4=Disagree, 5=Strongly Disagree

## Community Assessment

We also measured how effective the CCO was in building community and fostering social interaction. The responses we received were encouraging and largely positive. Based on a 6-point scale, 82% of respondents believe the software was an excellent medium for social interaction. Additionally, the majority of respondents were comfortable introducing themselves (88%), conversing (84%) and interacting with others (92%) through the CCO. The CCO also helped individuals form distinct impressions of one another (86%). Additionally, respondents felt that the community helped to improve their learning (78%). Table 5 provides a complete breakdown of these responses.

## Technology Assessment (Quantitative Data)

Quantitative and qualitative data measured how specific technologies enhanced learning, social interaction and community. Using the same six-point scale, users ranked each technology on the criteria of how well each fostered these three areas. Overall, users agreed (average rating of 2 and 3) that the ability to blog and comment on other blogs helped across all areas. Similarly, wiki writing and sub-communities also helped to improve learning, social interaction and community building (average rating of 2). Table 6 provides a detailed breakdown across these survey items.

## Technology Assessment (Qualitative Data)

We also asked a number of qualitative questions allowed individuals to elaborate on various aspects

*Table 5. Community and site effectiveness (n=50)*

| Survey Item | 1 | 2 | 3 | 4 | 5 | 6 | NA |
|---|---|---|---|---|---|---|---|
| Was an excellent medium for social interaction | 8% | 36% | 38% | 4% | 10% | 2% | 2% |
| Felt comfortable conversing through this medium | 14% | 50% | 20% | 12% | 2% | 2% | 2% |
| Felt comfortable introducing myself in this course | 16% | 42% | 30% | 10% | 0% | 0% | 2% |
| The medium helped me feel like part of the community | 12% | 22% | 50% | 6% | 6% | 12% | 0% |
| Was comfortable participating in course discussions | 26% | 38% | 24% | 4% | 8% | 0% | 0% |
| Was comfortable interacting with others in this course | 32% | 34% | 26% | 4% | 4% | 0% | 0% |
| My point of view was acknowledged by others in the course | 20% | 38% | 24% | 12% | 4% | 0% | 2% |
| Was able to form different impressions of course participants | 20% | 40% | 26% | 8% | 2% | 0% | 4% |
| A strong sense of community existed in this course | 24% | 20% | 28% | 10% | 12% | 4% | 2% |
| Course communit improve my leaning | 20% | 28% | 30% | 8% | 6% | 8% | 0% |

1=Strongly Agree, 2=Agree, 3=Agree Somewhat, 4=Disagree Somewhat, 5= Disagree, 6=Strongly Disagree, NA=Not Answered

of the system and will, in turn, help us improve our software going forwards. When asked what features worked best, individuals responded closely stating that the ability to view peer work and work collaboratively were strong points of the system. One response stated, "My favorite part was that I could see other people's work, which allowed me to learn from them." Another response stated, "I think it's beneficial to share knowledge & opinions w/ classmates and the [software] helped support these sharing activities." Responses on site recommendations and improvements ranged broadly. Some recommended more customizable profile pages, similar to popular social networking websites, while others requested interactive online tutorials. Lastly, a theme developed on fixing and improving existing site navigation and functionality for greater ease of use.

## DISCUSSION

Prior to implementing our software there existed no persistent space for learning and community building outside the classroom. In this research we remedied this problem with social software and implemented a stakeholder-defined OLC. For several reasons we believe courses implementing the OLC design achieved significant advantages over t-courses from the past. T-courses are inher-

*Table 6. Technology assessment (n=50)*

| Statement | Average Rating for Course Component | | | |
|---|---|---|---|---|
| | Blogging | Commenting | Wiki Creation | Sub-communities |
| Improved learning quality | 2 | 3 | 2 | 2 |
| I felt comfortable using | 2 | 2 | 2 | 2 |
| Excellent for social interaction | 3 | 2 | 2 | 2 |
| Improved sense of community | 2 | 2 | 2 | 2 |
| Facilitated by the instructor | 2 | 2 | 2 | 2 |
| People responded to | 2 | 3 | 2 | 2 |
| Create distinct impressions | 2 | 2 | 2 | 2 |

1=Strongly Agree, 2=Agree, 3=Somewhat Agree, 4=Somewhat Disagree, 5=Disagree, 6=Strongly disagree, or NA

ently collaborative and require individuals to communicate with one another outside the classroom. Therefore providing students with access to a virtual space from the onset helped to streamline asynchronous collaboration.

Furthermore, the persistent nature of the OLC provides students with access to a breadth of knowledge available during and after the course. Although only 48% indicated interest to continue interacting through the software (Table 4), nothing prevents an individual from returning to formulate their own knowledge communities and/or recruit new members later on. The bottom-up (student-owned) nature of the software provides all members with complete capabilities to establish new communities and also explore the over 114 communities that currently exist across the OLC.

Additionally, using the OLC provided students with the experience of using Web 2.0 technologies in a blended-learning environment. As identified in the background, Web 2.0 technologies are fast-becoming fused into various aspects of higher education and exposing a predominantly doctoral population early on will better prepare those students for teaching and research careers that may incorporate similar tools in the future. This was particularly important, considering that our population was found to be less familiar with popular technologies as such as wikis (Table 2).

We also feel that our research provides generalizations outside our respective institution. In higher education, instructor controlled learning environments are often the norm (i.e. traditional CMS tools). Our decision to use and extend the Elgg open source software offers a unique model that emphasizes students' autonomy within the larger community.

## CONCLUSION AND FUTURE WORK

In this research we expand scholarly conversation at our school through the addition of a virtual component. Our tool has achieved a measure of success in providing an asynchronous online component for students and faculty to collaborate and share knowledge in courses and on external projects. Guided by high-level theories of learning and social networking, direct and indirect data indicates a user-driven OLC to be beneficial to learning at the graduate level. Consequently, the OLC approach may provide a solution to increased scholarship at our school as well as a valid alternative to more traditional CMS solutions. Our initial success has provided a proof-of-concept for a more integrated installation of the software and subsequent financial backing from our school.

As we enter subsequent phases of our research we look to build on and improve numerous aspects of the system. Additionally, although the initial acceptance of our tool is a critical issue in determining whether it is successful, its continued use is at least as important (Bhattacherjee, 2001). Therefore, we will continue to monitor how our tool is adopted and used across our school, looking for ways to extend its use, including to alumni who are also interested in continuing the scholarly conversation.

## REFERENCES


Adams, E. C., & Freeman, C. (2000). Communities of practice: Bridging technology and knowledge assessment. *Journal of Knowledge Management*, *4*(1), 38–44. doi:10.1108/13673270010315939

Bhattacherjee, A. (2001). Understanding information systems continuance: An expectation-confirmation model. *MIS Quarterly*, *25*(3), 351–370. doi:10.2307/3250921

Bødker, S. (1989). A human activity approach to user interfaces. *Human-Computer Interaction*, *4*, 171–195. doi:10.1207/s15327051hci0403_1

Cook, J. (2002). The Role of dialogue in computer-based learning and observing learning: An evolutionary approach to theory. *Journal of Interactive Media in Education.*

Daniel, B., Schwier, R. A., & McCalla, G. (2003). Social capital in virtual learning communities and distributed communities of practice. *Canadian Journal of Learning and Technology, 29*(3).

Garrison, D. R. (2002). Cognitive presence for effective asynchronous online learning: The role of reflective inquiry, self-direction and metacognition. *Fourth Annual Sloan ALN Workshop.*

Haase, A. (2005). Trends in online learning communities. *ACM SIGGROUP Bulletin . Special Issue on Online Learning Communities, 25*, 2–6.

Hagstrom, F., & Wertsch, J. V. (2004). The social construction of identity: The clinical challenge. *Topics in Language Disorders, 24*(3), 162–173.

Heckman, R., & Annabi, H. (2006). cultivating voluntary online learning communities in blended environments. *Journal of Asynchronous Learning Networks*, 10.

Huysman, M., & Wulf, V. (2005). Information technology in building and sustaining the relational base of communities. *The Information Society, 21*, 81–89. doi:10.1080/01972240590925285

Issroff, K., & Scanlon, E. (2001). Case studies revisited- What can activity theory offer? In *Proceedings of First Euro-CSCL Conference, Maastricht.*

Jenlink, P., & Carr, A. A. (1996). Conversation as a medium for change in education. *Educational Technology*, 31–38.

Johnson, C. M. (2001). A survey of current research on online communities of practice. *The Internet and Higher Education, 4*, 45–60. doi:10.1016/S1096-7516(01)00047-1

Lave, J., & Wenger, E. (1991). Situated learning: Legitimate peripheral participation. *Cambridge University Press.*

Lenhart, A., & Madden, M. (2005). *Teen content creators & consumers* [White paper, electronic version]. Retrieved October 8, 2007 from Pew Internet & American Life Project at http://www.pewinternet.org/PPF/r/166/report_display.asp

Lenhart, A., & Madden, M. (2007). *Social networking websites and teens: An overview* [White paper, electronic version]. Retrieved October 8, 2007 from Pew Internet & American Life Project at http://www.pewinternet.org/PPF/r/198/report_display.asp

Lesser, E., & Prusak, L. (2000). Communities of practice, social capital and organizational knowledge. *Knowledge and communities* (pp. 123-131).

Matthews, M. R. (2002). Constructivism and science education: A further appraisal. *Journal of Science Education and Technology*, 11.

Nicol, D., Minty, I., & Sinclair, C. (2003). The social dimensions of online learning. *Innovations in Education and Teaching International, 40*(3), 270–280. doi:10.1080/1470329032000103807

Palloff, R., & Pratt, K. (1999). *Building learning communities in cyberspace: Effective strategies for the online classroom.* San Francisco: Jossey-Bass.

Sacks, H., Schegloff, E. A., & Jefferson, G. (1974). A simplest systematics for the organization of turn-taking for conversation. *Language, 50*(4), 696–735. doi:10.2307/412243

Savery, J. R., & Duffy, T. M. (1996). Problem based learning: An instructional model and its constructivist framework. In B. Wilson (Ed.), Constructivist learning environments: Case studies in instructional design, educational technology publications. Englewood Cliffs, NJ: Prentice Hall.

Sherry, L. (2000). Good online conversation: Building on research to inform practice. *Journal of Interactive Learning Research, 11*(1), 85.

Sifry, D. (2007). *The state of the live Web, April 2007.* Retrieved October 8, 2007 from http://www.sifry.com/alerts/archives/000493.html

Smith, P. J., Murphy, K. L., & Mahoney, S. E. (2003). Towards identifying factors underlying readiness for online learning: An exploratory study. *Distance Education, 24*(1), 57–67. doi:10.1080/01587910303043

Sorenson, E. K., & Takle, E. S. (2002). Collaborative knowledge building in Web-based learning: Assessing the quality of dialogue. *International Journal on E-Learning, 1*(1).

Stacey, E. (2002). Social Presence Online: Networking Learners at a Distance. *Education and Information Technologies, 7*(4), 287–294. doi:10.1023/A:1020901202588

Tu, C. H. (1997). The measurement of social presence in an online learning environment. *International Journal of eLearning, corporate, Government, Healthcare, & . Higher Education, 2*(1), 34–45.

Webb, E., Jones, A., Barker, P., & van Schaik, P. (2004). Using e-learning dialogues in higher education. *Innovations in Education and Teaching International, 41*(1). doi:10.1080/1470329032000172748

Wu, D., & Hiltz, S. (2004). Predicting learning from asynchronous online discussions. *Journal of Asynchronous Learning Networks*, 139–152.

Zhang, D. (2004). Can e-learning replace classroom learning? *Communications of the ACM, 47*, 75. doi:10.1145/986213.986216

# Chapter 8
# Learning Business Law Online vs. Onland:
## Student Satisfaction and Performance

**Louis B. Swartz**
*Robert Morris University, USA*

**Michele T. Cole**
*Robert Morris University, USA*

**Daniel J. Shelley**
*Robert Morris University, USA*

## ABSTRACT

*This article reports on two follow-up studies to "A Comparative Analysis of Online and Traditional Undergraduate Business Law Classes" (Shelley, Swartz and Cole, 2007) designed to further examine two critical areas of e-learning, that is, student satisfaction with, and student learning in, an online environment as compared with an onland, or traditional classroom environment. While the initial study found no significant difference between the two, the second study did find statistically significant differences between the online and the onland course formats with regard to two elements of student satisfaction: (1) student satisfaction with the instructor, and (2) student satisfaction with the course structure. The second study went further to look at the effects, if any, of gender, age and nationality on student satisfaction. There was no significant difference found with age or nationality. There was a significant difference between males and females with regard to two of the research questions. The third study focused on student satisfaction and performance in two onland courses. In both areas, results indicated lower overall means for each of the four central research questions.*

## INTRODUCTION

Braun (2008) notes that more than ever before, increasing numbers of colleges and universities are embracing online degrees and courses. As more institutions of higher education offer more online courses – to remain competitive, to expand access to education, and to facilitate student learning- two questions have been raised: Are students learning as well online as they are onground? Has student satisfaction with onland instruction changed as insti-

DOI: 10.4018/978-1-60566-936-6.ch008

tutions are turning to online platforms? Numerous authors have addressed those issues (Keegan, D., 1996; Russell, T., 1999; Schulman, A.H. and Sims, R.L., 1999; Harasim, L. 2000; Ryan, R.C. 2000; Allen, M., Bourhis, J., Burrell, N. and Mabray, E.; Rivera, J.C. and Rice, M.L., 2002; Bernard, R.M., et al, 2004; Frantz, P.L. and Wilson, A.H., 2004; Suanpang, P., Petocz, P. and Kalceff, W., 2004; Fjermestad, Hiltz, S. and Zhang, Y. 2005; Jang, K.S., Hwang, S.Y., Park, S.J., Kim, Y.M. and Kim, M.J., 2005; Tastle, W.J., White, B.A. and Shackleton, P., 2005; Weaver-Kaulis, A. and Crutsinger, C., 2006; Crawford, S.Z., Greenwell, C. and Andrew, D.; Mentzer, G.A., Cryan, J.R. and Teclehaimanot, B., 2007; Pillay, H., Irving, K. and Tones, M., 2007; Bassili, J.N. and Joordens, S., 2008; Braun, T., 2008).

In follow-up studies to "A Comparative Analysis of Online and Traditional Undergraduate Business Law Classes" (Shelley, Swartz and Cole, 2007), the authors found mixed results when examining two critical areas of e-learning: student satisfaction with, and student learning in, an online environment as compared with an onland, or traditional classroom environment. While the initial study found no significant difference between the two, the second study did find statistically significant differences between the online and the onland course formats with regard to two elements of student satisfaction: (1) student satisfaction with the instructor, and (2) student satisfaction with the course structure. The second study went further to look at the effects, if any, of gender, age and nationality on student satisfaction. There was no significant difference found with age or nationality. There was a significant difference between males and females with regard to two of the research questions (Shelley, Swartz and Cole, 2008).

In a third study of two onland sessions of the same undergraduate business law course completed in 2008, researchers found significantly less satisfaction with the instructor and with the structure of the course when compared with prior studies' onground classes (a mean of 4.2050 to 4.6375 and 4.6154, and a mean of 3.5275 to 3.7500 and 3.8846). Results were more consistent with prior years' studies with regard to student satisfaction with the course overall (mean of 4.02 compared with means of 4.1481 and 4.6154) and with regard to learning as measured by grades (mean of 2.4470 compared with means of 2.4500 and 2.7609). (It should be noted that in the 2008 study the grades were raised by 2 percentage points.)

In each of the three studies the primary focus was three measures of student satisfaction, with the course itself, with the instructor and with the structure of the course, and with one measure of learning, grades.

The course that was the subject of each study is an undergraduate business law class required for all business students at Robert Morris University in Pittsburgh, Pennsylvania. Legal Environment for Business (BLAW 1050) is designed to enable students to develop an understanding of the American legal system and to attain a working knowledge of ethics, contract law and consumer protection to a degree sufficient to be useful in business and consumer transactions. At the conclusion of the course, students have learned their legal rights and responsibilities and have gained the ability to apply legal principles to help solve business and consumer problems.

Since its first online offerings in 1999, Robert Morris University has added 246 new online and partially online course offerings. In academic year 2006-07, there were 145 totally online courses university–wide. Of these, fourteen were offered in the School of Business. In that year, there were an additional 136 courses partially online, forty-three of which were in the School of Business. As the University expands its offerings and more and more instructors and students become involved in online education, ensuring instructional quality and learning effectiveness assumes the central role in course planning. In academic year 2008-09, the University added two fully online degree programs in the School of Education and Information Sciences.

Fundamental differences between teaching online and teaching in the traditional classroom pose major challenges and concerns for course instructors and educational institutions. Chief among these is student learning and perhaps to a lesser degree, student satisfaction as it affects learning in an online environment.

## WHY THESE STUDIES?

As the University, and its School of Business in particular, expand online course offerings, measuring student learning in an analytical way that affords objective comparisons between platforms and offers hypotheses about the more specific factors that influence the broad outcome of "student learning" is critical. The School of Business is in the final stage of AACSB (Association to Advance Collegiate Schools of Business) accreditation. Measurement of student learning is central to the review of current course offerings and to the development of new ones and is a critical component of the accreditation process. Student satisfaction with the learning environment not only contributes to student retention, but it also serves as a measure of faculty performance and teaching effectiveness.

In a follow-up study to one reported in 2006 comparing student satisfaction and learning in the online and traditional classroom environments (Shelley, Swartz and Cole, 2007), the authors replicated and expanded upon the first. The second study was conducted by the same instructor, using the same survey instrument as well as the same instructional materials, course outline and assignments for the *Legal Environment of Business* (BLAW 1050) which he taught in both formats in 2006-2007 (Shelley, Swartz and Cole, 2008). The third study was conducted in fall, 2008 with two onland classes taught by the same professor and using all of the same materials and the same survey instrument as in the earlier studies.

## RESEARCH QUESTIONS

The second study looked at the same four research questions as the original study did and added nine more research questions:

- **Research Question 1:** Does student satisfaction with the course overall differ significantly between the online format and the traditional class format?
    - **Research Question 1a:** Does student satisfaction with the course overall differ significantly between males and females?
    - **Research Question 1b:** Does student satisfaction with the course overall differ significantly between the online format and the traditional class format in the combined study group?
    - **Research Question 1c:** Does student satisfaction with the course overall differ significantly between Study Group I and Study Group II?
- **Research Question 2:** Does student satisfaction with the instructor differ significantly between the online format and the traditional class format?
    - **Research Question 2a:** Does student satisfaction with the instructor differ significantly between males and females?
    - **Research Question 2b:** Does student satisfaction with the instructor differ significantly between the online format and the traditional class format in the combined study group?
    - **Research Question 2c:** Does student satisfaction with the instructor differ significantly between Group I and Group II?

- **Research Question 3:** Does student satisfaction with the course structure differ significantly between the online format and the traditional class format?
    - **Research Question 3a:** Does student satisfaction with the course structure differ significantly between males and females?
    - **Research Question 3b:** Does student satisfaction with the course structure differ significantly between the online format and the traditional class format in the combined study group?
    - **Research Question 3c:** Does student satisfaction with the course structure differ significantly between Group I and Group II?
- **Research Question 4:** Does student learning differ significantly between the online format and the traditional class format?

## METHODOLOGY

### Course Design

The online sections of BLAW 1050 were developed using the eCollege™ format. In 2008, the University changed its online provider from eCollege to Blackboard. All students taking an online course at Robert Morris University are required to complete the Online Learning Training Module prior to being registered for the class.

All online sections of the course were developed and maintained by the instructor involved in this study. The online format employed available instructional tools, including digital drop boxes, document share areas, synchronous and asynchronous dialog, e-mail and online assessment. The textbook readings were enhanced and supplemented with lecture notes and illustrations of key points.

The classroom sections of BLAW 1050 used the same syllabus as the online course and had the same assignments and assessments. The topics used in the threaded discussions in the online format were used in real time in the onland format.

### Sample/Participants

In the first study, comparative data was drawn from four online sections of the course (two in 2004, one in 2005 and one in 2006) and two traditional sections in the spring of 2005. Fifty-eight of the 64 enrolled students completed the online sections of BLAW 1050 (N=58) or 90.6%. Forty-six of the 49 enrolled students in the traditional sections completed the course (N=46) or 93.8%. The total number of students receiving grades for BLAW 1050 during the study period was 104 (N=104) or 94.5%.

In the follow-up study, comparative data was drawn from two online sections of the course (fall, 2006 and spring, 2007) and from one onland section in the spring of 2007. Forty students from the online courses responded to a web survey which duplicated the paper surveys (N=40). Twenty-seven of the students from the onland class participated in the study (N=27).

Thirty–nine of the online students completed the course and received a grade (N=39). Thirty-five of the students in the onland class completed the course and received a grade (N=35).Thirty-nine of the 44 enrolled students completed the online sections of BLAW 1050 (N=39) or 88.6%. Thirty-five of the 39 enrolled students in the traditional section completed the course (N=35) or 89.7%. The total number of students receiving grades for BLAW 1050 during the study period was 74 (N=74) or 89% of those who enrolled.

Of the 64 students enrolled in the online sections of BLAW 1050, six withdrew, for a retention rate of 90.6%. The retention rate for the traditional sections was higher, at 93.8%; of the 49 who enrolled, three withdrew. Of the 44 students enrolled in the online sections of BLAW 1050, five withdrew, for a retention rate of 88.6%. The retention rate for the traditional sections was slightly higher, at 89.7%; of the 39 who enrolled, four withdrew.

In the third study, all of the participants were enrolled in one of two onland sections of BLAW 1050. The retention rate in the two classes was 100%. One student was missing in action, but did not formally withdraw.

## Instrumentation

In the first study, a 24 question satisfaction survey with a five-point Likert Scale was distributed in each class. The survey was administered by the instructor after grading was completed. Participation was voluntary. Thirty-three of the 58 online participants responded, for 56.9% return rate. Thirteen of the 46 students in the traditional courses completed their surveys for a return rate of 28.2%.Total number of students participating in the survey was 46.

The same process was followed in the second study for the students in the onland course. For the students in the online courses, the identical survey was uploaded as a web-based instrument. This was done to facilitate both student participation and accuracy of data conversion for analysis. Forty of the 44 students enrolled in the online course responded using websurveyor. Twenty-seven of the 35 students who completed the course onland participated for response rates of 90.9% and 77.1% respectively.

The same web-based survey administered in Vovici was used in the third study. Eighty-four of the 92 students who completed the course and received grades, participated in the survey for a response rate of 91.3%.

Questions one - thirteen applied to students in both the online course and in the classroom course and were answered by both groups in both studies. A comment section was provided on the survey itself for qualitative input. In both studies, question one asked if the student felt he/she had learned the subject material. Questions two and ten focused on the performance of the course instructor. Questions three and four focused on the quality of the selected textbook. These were not used for the analysis. Questions five to nine and eleven to thirteen dealt with issues involved directly with the course structure.

Participant responses from the online and classroom sections were aggregated and compared for both studies. Responses to question one formed the basis for comparison for Research Question 1. Responses to questions two and ten formed the basis for comparison for Research Question 2. Responses to questions five to nine and eleven through thirteen formed the basis for comparison for Research Question 3. Questions 14 -25 were designed specifically for the online format and thus, were not used for the analysis. Final grades from the online and onland courses formed the basis for comparison for Research Question 4 in each study.

Demographic data was collected for research questions one through three in the second study to allow for additional analysis by gender, age and nationality ("international"/ "not international").

In each of the three studies, the structure of the survey allowed for both quantitative and qualitative data to be analyzed. Each study used SPSS for data analysis. In the first two studies, independent-samples t-tests were run for each research question. ANOVAs were run to determine if there was a significant difference between the two studies. In the third study, one sample t-tests were used to analyze results.

Within the context of each study, "satisfaction" is defined as having met expectations as demonstrated by the student responses. "Learning" is defined as having acquired knowledge of the subject matter as evidenced by the course grades. The studies controlled for what Benbunan-Fich, Hiltz and Harasim (2005) refer to as moderating factors that influence the outcomes when measuring learning. These are technology, course, instructor characteristics and student characteristics.

Results from Second Study for RQ 1-4:

**Research Question 1:** Does student satisfaction with the course overall differ significantly between the online format and the traditional class format?

Aggregated mean score for the online sections 3.9500

Aggregated mean score for the onland section 4.1481

**Research Question 2:** Does student satisfaction with the instructor differ significantly between the online format and the traditional class format?

Aggregated mean score for the online sections 4.1310

Aggregated mean score for the onland section 4.6375

**Research Question 3:** Does student satisfaction with the course structure differ significantly between the online format and the traditional class format?

Aggregated mean score for the online sections 3.4125

Aggregated mean score for the onland section 3.7500

**Research Question 4:** Does student learning differ significantly between the online format and the traditional class format?

Aggregated mean score for the online sections 2.6859

Aggregated mean score for the onland section 2.4500

To examine the effect, if any, of the demographic variables on student satisfaction (RQ1-3), independent-samples t-tests were run to test for statistical significance of age, gender and whether or not the participant was an international student. Gender proved to be the only significant factor with regard to student satisfaction (Tables 5-7). There were 36 males and 31 females who participated. There was no significant difference in student satisfaction with the course overall, the instructor, or the course structure with regard to the student's age (18-21, 22-30, 31-38, 39-50, over 50) or whether or not the student was an international student. It should be noted that there were only six international students in the sample, all of whom were in the online courses.

One-way ANOVAs were run to determine if there was a significant difference between online and onland formats when the study group responses were combined (Tables 8 and 9) for research questions 1-3. One-way ANOVAs also were run to determine if there was a significant difference between the two study groups' results (Tables 10

*Table 1. Student satisfaction with the course overall*

| | t-test for Equality of Means | |
|---|---|---|
| | t N=67 | Sig. (2-tailed) N=67 |
| VAR0002 Equal Variances Assumed | - 1.146 | .256 |

*Table 2. Student satisfaction with the instructor*

| | t-test for Equality of Means | |
|---|---|---|
| | t N=225 | Sig. (2-tailed) N=225 |
| VAR0002 Equal Variances Assumed | -4.673 | .000 |

*Table 3. Student satisfaction with the course structure*

| | t-test for Equality of Means | |
|---|---|---|
| | t N=536 | Sig. (2-tailed) N=536 |
| VAR0002 Equal Variances Not Assumed | -3.424 | .001 |

*Table 4. Student learning*

| | t-test for Equality of Means | |
|---|---|---|
| | t N=74 | Sig. (2-tailed) N=74 |
| VAR0002 Equal Variances Not Assumed | .912 | .365 |

and 11) for research questions 1-3. There were 73 students responding who were in the online courses and 40 who were in the onland courses. Study Group I was composed of 46 students; Study Group II was composed of 67 students.

There was no significant difference between the online and onland responses to research question 1, student satisfaction with the course overall (.336) or for research question 3, student satisfaction with the course structure (.092). There was a significant difference between the online and onland responses to research question 2 on student satisfaction with the instructor (.000).

There was a significant difference between Study Group I and Study Group II found with regard to each of the three of the research questions measuring student satisfaction. In each case, the results from the first study were more positive. Significance levels were .001, .001, and .000 for research questions 1, 2 and 3 respectively. Survey comments for the second group were mixed.

Learning, as measured by grades, was higher for online students overall and higher in Study Group I overall. The mean score for the online students was 2.8365. For onland students, the mean score was 2.60545. The mean score for Study Group I was 2.874. For Study Group II, the mean score was 2.56795.

Independent-sample t-test on Demographic Variable

**Research Question 1a:** Does student satisfaction with the course overall differ significantly between males and females?

Aggregated mean score for male students 3.81

Aggregated mean score for female students 4.19

**Research Question 2a:** Does student satisfaction with the instructor differ significantly between males and females?

Aggregated mean score for male students 4.3194

Aggregated mean score for female students 3.9355

Research Question 3a: Does student satisfaction with the course structure differ significantly between males and females?

Aggregated mean score for male students 3.5278

Aggregated mean score for female students 3.5726

Oneway ANOVA Results Comparing Combined Online and Onland Responses

Research Question 1b: Does student satisfaction with the course overall differ significantly between the online format and the traditional class format in the combined study group?

Mean score for students (N=73) in the online courses 4.16

Mean score for students (N=40) in the onland course 4.30

**Research Question 2b:** Does student satisfaction with the instructor differ significantly between the online format and the traditional class format in the combined study group?

**Research Question 3b:** Does student satisfaction with the course structure differ significantly between the online format and the traditional class format in the combined study group?

RQ2: Mean score for students (N=73) in the online courses 4.1370

*Table 5. Student satisfaction with the course overall*

| | t-test for Equality of Means | |
|---|---|---|
| | t N=67 | Sig. (2-tailed) N=67 |
| q1 Equal Variances Assumed | - 2.058 | .044 |

*Table 6. Student satisfaction with the instructor*

| | t-test for Equality of Means | |
|---|---|---|
| | t N=67 | Sig. (2-tailed) N=67 |
| mean2.10 Equal Variances Assumed | .411 | .033 |

*Table 7. Student satisfaction with the course structure*

| | t-test for Equality of Means | |
|---|---|---|
| | t N=67 | Sig. (2-tailed) N=67 |
| mean5to9.11to13 Equal Variances Assumed | .411 | .033 |

RQ2: Mean score for students (N=40) in the onland course 4.6375

RQ3: Mean score for students (N=73) in the online courses 3.6293

RQ3: Mean score for students (N=40) in the onland course 3.7938

Oneway ANOVA Results Comparing Study Group I to Study Group II Responses

Research Question 1c: Does student satisfaction with the course overall differ significantly between Study Group I and Study Group II?

Mean score for Group I (N=46) 4.48

Mean score for Group II (N=67) 4.03

Research Question 2c: Does student satisfaction with the instructor differ significantly between Group I and Group II?

Research Question 3c: Does student satisfaction with the course structure differ significantly between Group I and Group II?

RQ2: Mean score for Group I (N=46) 4.5652

RQ3: Mean score for Group II (N=67) 4.1418

RQ2: Mean score for Group I (N=46) 3.8899

RQ3: Mean score for Group II (N=67) 3.5485

Results from the third study which focused on the core research questions measuring student satisfaction and student performance (RQ1-RQ 4) demonstrated an overall decrease in satisfaction and in performance. It is possible that as the School of Business has raised its acceptance requirements for students, that the student profile has changed in ways that might have affected

*Figure 1. Student satisfaction with the course*

**ANOVA**

| | Sum of | | | | |
|---|---|---|---|---|---|
| Between Groups | .475 | 1 | .475 | .935 | .336 |
| Within Groups | 56.427 | 111 | .508 | | |
| Total | 56.903 | 112 | | | |

*Figure 2.*

**ANOVA**

| | | Sum of | | | | |
|---|---|---|---|---|---|---|
| mean2.10 | Between Groups | 6.473 | 1 | 6.473 | 15.750 | .000 |
| | Within Groups | 45.624 | 111 | .411 | | |
| | Total | 52.097 | 112 | | | |
| mean5to9.11to13 | Between Groups | .699 | 1 | .699 | 2.880 | .092 |
| | Within Groups | 26.942 | 111 | .243 | | |
| | Total | 27.641 | 112 | | | |

*Figure 3. Student satisfaction with course overall*

**ANOVA**

| | Sum of | | | | |
|---|---|---|---|---|---|
| Between Groups | 5.484 | 1 | 5.484 | 11.839 | .001 |
| Within Groups | 51.419 | 111 | .463 | | |
| Total | 56.903 | 112 | | | |

*Figure 4.*

**ANOVA**

| | | Sum of | | | | |
|---|---|---|---|---|---|---|
| mean2.10 | Between Groups | 4.890 | 1 | 4.890 | 11.498 | .001 |
| | Within Groups | 47.207 | 111 | .425 | | |
| | Total | 52.097 | 112 | | | |
| mean5to9.11to13 | Between Groups | 3.180 | 1 | 3.180 | 14.429 | .000 |
| | Within Groups | 24.461 | 111 | .220 | | |
| | Total | 27.641 | 112 | | | |

their course expectations. All other elements for the course remained constant through the four years, 2004-2008.

RQ1: Mean score (N=84) 4.0200

RQ2: Mean score (N=84) 4.2050

RQ3: Mean score (N=84) 3.5275

RQ4: Mean score (N=84) 2.4470

## DISCUSSION

Several studies comparing online with onland instruction and learning have appeared in the literature. Fjermestad, Hiltz and Zhang (2005) reviewed thirty published empirical studies which compared the effectiveness of course delivery, the authors conclude that the evidence is overwhelming: online delivery is at least as effective as traditional classroom delivery (p.39). The authors looked at access, faculty and student satisfaction, student learning and cost effectiveness. With regard to student learning, their results are consistent with other studies that have found online instruction to be equal to or better than face-to-face instruction. With regard to student satisfaction, the results were mixed, with the "no significant difference" being the overall conclusion (p.48).

Of the twelve studies on student satisfaction, 41.6% were positive for online, 25% were negative. In a third of the studies, student satisfaction as measured yielded no differences between the two modes. With regard to objective measures of learning, 61.7% resulted in a finding of "no difference." 34% positive for online learning mode and four percent negative for online learning. The sample size was 47 (pp 45-46).

Thomas Russell's *The No Significant Difference Phenomenon*, published in 1999, summarized 355 research reports, papers and summaries on the subject of online versus traditional learning. He found no significant difference in grades, satisfaction or effectiveness when "e-learning" was compared to traditional teaching. Other studies have supported Russell's findings. Taking additional factors into consideration, Navarro & Shoemaker (2000) found little or no difference between online and classroom learning when such issues as race, gender, technological and academic backgrounds, and socioeconomic status were taken into account.

In their study of the empirical literature comparing student satisfaction with distance education to traditional classroom instruction in higher education, Allen, Bourhis, Burrell and Mabray (2002) found that the level student satisfaction with instruction is not diminished by online platforms when compared to classroom instruction.

The findings of Shelley, Swartz and Cole (2007) were consistent with those of earlier studies (Schulman and Sims, 1999, Navarro & Shoemaker, 2000, Suanpang, Petocz and Kalceff, 2004, Bernard, Abrami, Lou, Borokhovski, Wade, Wozney, Wallet, Fiset, and Huang, 2004) and supported the proposition that a course provided online would offer a comparable, if not better, learning environment for students than the same course presented in the traditional format. Braun's 2008 study also found overall student satisfaction with online instruction when compared with the traditional classroom instruction with regard to academic course content and instruction.

Bassilli and Joordens (2008) were able to demonstrate increased satisfaction and enhanced learning with an online platform where certain features were used, specifically, media players. Using the media player tool allowed the student to pause and replay course lectures as needed which appeared to enrich the learning experience.

Yet, Rivera and Rice (2002) reported that while several studies (including Russell's 1999 work) have demonstrated that online and traditional courses were comparable with regard to cognitive factors (learning, performance and achievement), the same could not be demonstrated consistently with online learning with regard to student and instructor perceptions and satisfaction. Rivera and Rice did a comparative evaluation of one course offered in three formats: online, traditional classroom and web-enhanced classroom. Using questionnaires to evaluate student satisfaction, grades to evaluate student performance, and discussion and anecdotal references to evaluate instructor satisfaction with teaching online, Rivera and Rice compared the efficacy of the three formats. They found that the exam score averages were close in all three, thus supporting the finding by others that online and traditional classroom courses are comparable with regard to the cognitive factor, in this case student performance.

However, their results showed significant differences in levels of satisfaction among all three formats, including the web-enhanced hybrid (onland with an online component). The 100% online web-based instruction was the least satisfactory to students. As the authors point out in their discussion of instructors experiences with the different formats, there are a number of factors that might be influencing results, such as the students' comfort level with technology, varying level of instructional support and instructors' familiarity with the course material delivery platform.

Results from other studies of student satisfaction and performance are mixed as well. Bernard et al. (2004) concluded that the differences between the two modes of instruction were not significant. Their study was a meta-analysis of the empirical literature comparing distance and classroom instruction in which they analyzed 232 studies measuring student achievement, attitude and retention. They found the effect sizes to be basically zero on all three measures and found wide variability in the results due in part to the disparity in the degree of rigor in the studies analyzed. Some applications of distance education were more successful with regard to student learning than classroom instruction; some were less successful than classroom instruction. Jang, Hwang, Park, Kim and Kim (2005) also had mixed results in their comparative study of undergraduate nursing students' learning.

Mentzer, Cryan and Teclehaimanot's 2007 study had different results. They found that final grades in the online course were lower than in the onland course; and that student satisfaction with the course and with the instructor was more positive in the classroom setting than in the online course. Researchers controlled for bias by using a modified interaction analysis instrument. The same evaluation instrument was used to evaluate student satisfaction. Evaluating participation in online courses and the relationship to grades, Davies and Graff (2005) did not find that greater student participation online resulted in higher grades. They did see however, that where grades were low, online participation was low as well.

Our original study found no statistically significant difference between the online and traditional instructional/learning formats with regard to any of the four research questions on student satisfaction and student learning. These results were consistent with earlier studies (Schulman and Sims, 1999, Navarro & Shoemaker, 2000, Suanpang, Petocz and Kalceff, 2004, Bernard, Abrami, Lou, Borokhovski, Wade, Wozney, Wallet, Fiset, and Huang, 2004) and supported the proposition that a course provided online would offer a comparable, if not superior, learning environment for students than the same course presented in the traditional format.

While the results from the first study clearly fall into the "no significant difference" category and support the majority of the earlier studies, the results from the second and third studies are more mixed. For example, although there was no significant difference in student satisfaction with the course overall or in student learning between the online and onland formats, there was a significant difference found in student satisfaction with the instructor and with the course structure. In both cases, the mean scores for the onland course students were higher than for the online course students. Yet, student learning, as measured by final course grades, was higher for the online course students. Since none of the elements from the first study were changed in the second, i.e., the same instructor, course materials, course structure, and exams, the results are puzzling. Nor were there any changes in the third study of the two onland classes.

In the first study, student satisfaction with the course structure was slightly higher in the online format as opposed to the onland format. Student learning in the online courses were slightly higher than for those in the traditional classes. Those results reinforce Russell's "no significant difference" phenomenon. In the second study, student satisfaction with the course overall, with the instructor, and with the course structure was higher for students in the onland course than it was in the first study.

The first study's survey results also supported findings in the earlier work by Schulman & Sims and by Ryan with regard to research questions 1-13 on student satisfaction with the course, the instructor and the course design of BLAW 1050, Legal Environment of Business.

In the earlier study, student input under "Comments/Suggestions" was comparable, with the difference that students in the online courses also referenced the online features (positively) and that students in the traditional class setting commented on the outside assignments and exams. Seventy-two percent of the online students who participated in the study also added comments compared with 69% from the students in the traditional classroom setting. This feature was used less in the second study (two comments from those in the onland course, fifteen from those in the online course). Comments ranged from enthusiastic about the experience to some complaints about the text and the delivery platform. Comments in the third study were consistent with earlier studies, but as in the second study, this feature was not used very much.

Study limitations in the first study- sample sizes and the difference in participation rates- were ameliorated in the second study. In the first study, 59.6% of the students in the online courses participated while only 28% of the students in the traditional courses participated. Participation was higher in the second study, in part due to the ease of use of web-based survey instrument. Forty of the 44 enrolled students responded to the survey for a 90.9% response rate. At least one of these withdrew prior to the end of class however. Participation was higher in the onland course as well, due in part to the presentation of the survey during class time. Twenty–seven of the 35 students who completed the course completed surveys for a response rate of 77.1%. Response rate in the third study was the highest, at 91.3%.

## CONCLUSION

Fundamental differences between teaching online and teaching in the traditional classroom pose major challenges and concerns for course instructors and educational institutions. Chief among these is student learning and perhaps to a lesser degree, student satisfaction as it affects learning in an online environment.

Fjermestad, Hiltz and Zhang (2005) concluded their study with the observation that more methodologically rigorous studies need to be done before the "Which is better" question can be answered. (p. 49). Arbaugh and Hiltz (2005) would concur. They too discuss the difficulty in reaching definitive conclusions when measuring learning because of variations in measurement tools and methodologies. The majority of the published work to date has found that either there were no significant differences between the two delivery vehicles or that if there were significant differences between the two; learning was greater in the online format. Why was that? Trying to answer that question, the authors looked at the variety of tools used to measure learning, such as grades, collaborative exams, projects and portfolios, course outcomes, as well as attitudinal surveys to measure satisfaction with the learning process. From their review of quantitative methodologies used to measure learning and satisfaction, Arbaugh and Hiltz conclude that for such studies to be useful they need to be more rigorous investigations of learning effectiveness, employing more "valid and pedagogically sound" methodologies (p. 97).

Rivera and Rice's (2002) results illustrate the need to improve the technology and course delivery aspects of online instruction in order to improve student satisfaction. Their study also points out the need for research methodologies that can uncover the answer to the question of effectiveness of online education.

Rudestam and Schoenholtz-Read (2002) suggest that access to, and use of, the internet for knowledge transfer present challenges and opportunities for creating new paradigms for learning and arguably for creating new philosophies and theories of learning. To go further with online instruction, they state, requires a reexamination of core beliefs about teaching and about learning (p.4). This being true, fundamental differences between online and traditional instruction pose major challenges and concerns for course instructors and educational institutions. Measuring effectiveness of teaching platforms and learning models using student outcome measures is one of those challenges/opportunities.

Pillay, Irving and Tones (2007) warn that as increasing numbers of educational institutions move to online instruction, there needs to be a corresponding level of attention paid to the personal and technical qualities required for success in the online environment. Their work examines diagnostic tools available to accurately assess student readiness for online learning.

A similar argument might be made for developing appropriate tools to assess instructor readiness and competencies. While Tastle, White and Shackleton (2005) also found that students tended to learn somewhat more in an online course, academics teaching these courses were not as satisfied with the experience as might be expected from the enhanced student learning results.

Our studies broadly support the conclusions drawn by others. However, with regard to the comparative effectiveness of online learning, a more nuanced study of online learning compared with classroom learning of business law as taught in BLAW 1050 is needed. It would be important to explain the differences between student satisfaction with the instructor and with the course structure with in the online platform and with the onland platform found in the second study that were not present in the first. Additional research, to determine why overall student satisfaction and student learning appears to be declining in the onland courses, may also be warranted.

## REFERENCES

Allen, M., Bourhis, J., Burrell, N., & Mabray, E. (2002). Comparing student satisfaction with distance education to traditional classrooms in higher education: A meta- analysis. American Journal of Distance Education, 16(2), 83–97. doi:10.1207/S15389286AJDE1602_3doi:10.1207/S15389286AJDE1602_3

Arbaugh, J. B., & Hiltz, S. (2005). Improving quantitative research on aln effectiveness. In S. Hiltz & R. Goldman (Eds.), *Learning together online: Research on asynchronous learning networks* (pp. 81-102). London: Lawrence Erlbaum Associates.

Bassilli, J. N., & Joordens, S. (2008). Media player tool use, satisfaction with online lectures and examination performance. Journal of Distance Education, 22(2), 93–108.

Benbunan-Fich, R., Hiltz, S., & Harasim, L. (2005). The online interaction learning model: An integrated theoretical framework for learning networks. In S. Hiltz & R. Goldman (Eds.), *Learning together online: Research on asynchronous learning networks* (pp. 19-37). London: Lawrence Erlbaum Associates.

Bernard, R. M., Abrami, P. C., Lou, Y., Borokhovski, E., Wade, A., Wozney, L., et al. (2004). How does distance education compare to classroom instruction? A meta-analysis of the empirical literature. Review of Educational Research, 74(3), 379–439. doi:10.3102/00346543074003379doi:10.3102/00346543074003379

Braun, T. (2008). Making a choice: The perceptions and attitudes of online graduate students. Journal of Technology and Teacher Education, 16(1), 63–92.

Crawford, S. Z., Greenwell, C., & Andrew, D. (2007). Exploring the relationship between perceptions of quality in basic instruction programs and repeat participation. Physical Educator, 64(2), 64–71.

Davies, J., & Graff, M. (2005). Performance in e-learning: Online participation and student grades. British Journal of Educational Technology, 36(4), 657–663. doi:10.1111/j.1467-8535.2005.00542.xdoi:10.1111/j.1467-8535.2005.00542.x

Fjermestad, H. S. & Zhang, Y. (2005). Effectiveness for students: Comparisons of "in-seat" and aln courses. In S.Hiltz & R.Goldman (Eds.), *Learning together online: Research on asynchronous learning networks* (pp. 39-79). London: Lawrence Erlbaum Associates.

Fowler, D. (March 2005). Are on-site courses as effective as online? *Online Cl@ssroom.*

Frantz, P. L., & Wilson, A. H. (2004). Student performance in the legal environment course: Determinants and comparisons. Journal of Legal Studies Education, 21(2), 225. doi:10.1111/j.1744-1722.2004.tb00318.xdoi:10.1111/j.1744-1722.2004.tb00318.x

Harasim, L. (2000). Shift happens: Online education as a new paradigm in learning. The Internet and Higher Education, 3, 41–61. doi:10.1016/S1096-7516(00)00032-4doi:10.1016/S1096-7516(00)00032-4

Jang, K. S., Hwang, S. Y., Park, S. J., Kim, Y. M., & Kim, M. J. (2005). Effects of a Web-based teaching method on undergraduate nursing students' learning of electrocardiography. The Journal of Nursing Education, 44(1), 35–39. PubMed

Keegan, D. (1996). *Foundations of distance education* (3rd ed). London: Routledge.

Mentzer, G. A., Cryan, J. R., & Teclehaimanot, B. (2007). Two peas in a pod? A comparison of face-to-face and Web-based classrooms. Journal of Technology and Teacher Education, 15(2), 233–246.

Navarro, P., & Shoemaker, J. (2000). Policy issues in the teaching of economics in cyberspace: Research design, course design, and research results. Contemporary Economic Policy, 18(3), 359–366. doi:10.1093/cep/18.3.359doi:10.1093/cep/18.3.359

Pillay, H., Irving, K., & Tones, M. (2007). Validation of the diagnostic tool for assessing tertiary students' readiness for online learning. Higher Education Research & Development, 26(2), 217–234. doi:10.1080/07294360701310821doi:10.1080/07294360701310821

Rivera, J. C., & Rice, M. L. (2002). A comparison of student outcomes and satisfaction between traditional & web based course offerings. Online Journal of Distance Learning Administration, 5(3).

Roblyer, M. D., & Wiencke, W. (2003). Design and use of a rubric to assess and encourage interactive qualities in distance courses. American Journal of Distance Education, 17(2), 77–98. doi:10.1207/S15389286AJDE1702_2doi:10.1207/S15389286AJDE1702_2

Rudestam, K. E., & Schoenholtz-Read, J. (2002). The coming of age of adult online education. In K.E. Rudestam & J. Schoenholtz-Read (Eds.), *Handbook of online learning: Innovations in higher education and corporate training* (pp. 3-28). Thousand Oaks, California: Sage Publications.

Russell, T. (1999). *The no significant difference phenomenon*. Office of Instructional

Schulman, A. H., & Sims, R. L. (1999). Learning in an online format versus an in-class format: An experimental study. T.H.E. Journal, 26(11), 54–56.

Shelley, D. J., Swartz, L. B., & Cole, M. T. (2007). A comparative analysis of online and traditional undergraduate business law classes. International Journal of Information and Communication Technology Education, 3(1), 10–21.

Shelley, D. J., Swartz, L. B., & Cole, M. T. (2008). Learning business law online vs. onland: A mixed method analysis. International Journal of Information and Communication Technology Education, 4(2), 54–66.

Suanpang, P., Petocz, P., & Kalceff, W. (2004). Student attitudes to learning business statistics: comparison of online and traditional methods. Educational Technology & Society, 7(3), 9–20.

Tastle, W. J., White, B. A., & Shackleton, P. (2005). E-learning in higher education: The challenge, effort, and return on investment. International Journal on E-Learning, 4(2), 241–251.

Telecommunications, North Carolina State University Chapel Hill, N.C. Ryan, R.C. (2000). Student assessment comparison of lecture and online construction equipment and methods classes. *THE Journal, 6*(27).

Weaver-Kaulis, A., & Crutsinger, C. (2006). Assessment of student learning outcomes in fcs programs. Journal of Family and Consumer Sciences, 98(1), 74–81.

# Chapter 9
# Fostering Meaningful Interaction in Health Education Online Courses:
## Matching Pedagogy to Course Types

**Richard G. Fuller**
*Robert Morris University, USA*

**Gary Kuhne**
*Penn State University, USA*

## ABSTRACT

*This research study examined the best interactive practices of effective health care education faculty from six major universities that offer online health care programs. Program directors from six major universities identified effective faculty, from which twelve faculty members were interviewed to uncover effective practices and an additional thirty faculty participated in a Delphi study to identify and prioritize effective practices. The findings for this study indicate that different types of facilitation approaches are needed to generate adequate interaction in four distinct types of health care courses, i.e., foundational classes, skills classes, analysis/synthesis classes, and hybrid type courses.*

## INTRODUCTION

Wlodkowski (1999, 1985) suggests that effective instructors must have expertise, empathy, enthusiasm, and clarity, a conclusion that is relevant whether the teaching takes place in traditional face-to-face settings or in online formats. Interaction with the student is a central factor in demonstrating each of these elements. Effective instructors have discovered interactive practices that work well in face-to-face classrooms. The challenge for online instructors is to discover how to replicate effective interaction practices within the online, asynchronous learning environment.

As with traditional face-to-face teaching, there are methods and techniques that work in some venues and not in others. Strategies that foster effective interaction in an engineering classroom may not offer the same efficacy in a healthcare program. Effective teachers are willing to explore why certain interaction techniques work and don't work in order to discover the most effective techniques for their particular educational programs.

Knowles (1999, 1980) and Rogers (1969) argue that adult education teachers serve as facilitators, providing the resources to enhance and facilitate the self-directed learning opportunities of their students. Such an understanding of the role of instructors is particularly pertinent to the online asynchronous arena of higher education systems, where educators must design a variety of courses for a variety of learners. Effective instruction requires that teachers understand the changing needs of their learners based upon the nature of the educational program. In other words, an effective instructor does not approach each learning situation with the same pedagogy and style. Different styles of course require different techniques to facilitate success learning. Knowles (1980) suggests the specific learning needs of the particular participants of a given learning activity must be diagnosed. Understanding the unique needs of different university and college programs at graduate and undergraduate levels in different disciplines will go far toward enhancing interactive teaching practices online. Regardless of the mediated nature of the communication, "It is the teacher's responsibility to precipitate and facilitate learning that has purpose and is focused on essential concepts and worthwhile goals" (Garrison & Archer, 2000, p 48.). Adults and distance-education students relate in an interactive collaborative construction of knowledge, a system that typifies many of the concepts of adult education theory (Anderson, et al., 2002). The dilemma facing online instructors is how best to accomplish the designing, facilitating, and guiding of a predominantly text-based learning arena to best foster the different levels of interaction required for learning success.

## BACKGROUND

In conducting online teaching, interaction needs to be planned to facilitate learning. Vrasidas and McIsaac (1999) found that structure can affect interaction, and concluded that educators need to design courses to foster learner-to-learner interaction and dialogue. Kozma (1991) agrees with the need for less structure and more dialogue and suggests that learners should actively collaborate in order to construct knowledge rather than relying solely on knowledge gained from direct instruction. For such collaboration to occur, learners must feel a sense of connectedness with the group (Gibb, 1995).

Howland and Moore (2002) found that when students initiated interaction with instructors and other students, knowledge was often built spontaneously, such as through students guiding the direction of discussion-board threads. Such student leadership then led to positive results for others. One student said, "Several times, I have seen questions asked by others that had not even occurred to me to ask and the answers benefited me" (p. 188). Swan (2001) found that students with higher levels of interaction with their classmates through online discussion also reported higher levels of learning and satisfaction from courses. Rovai and Barnum (2003) also found evidence that student perception of learning from online courses was positively related to course interaction, lending further support to the need to provide opportunities for online students to learn by active interaction with each other and with instructors. Effective online instructors develop highly interactive material and facilitate participation in online discussions. Rovai and Barnum also suggested that passive interaction, analogous to listening to,but not participating in, discussions, was not a significant predictor of perceived learning in the present study. Consequently, using strategies that promote active interaction leads to a greater perception of learning and higher levels of learner satisfaction.

Beaubien's research (2002) described instructor characteristics that contribute to effective online courses. Students need to feel that the instructor is online regularly. The instructor does not need to be intrusive to the online dialogue but his/her presence needs to be known. Short postings are good for the most part but the teachable moment should be capitalized upon to provide sufficient information and clarification as is necessary. Sometimes the instructor can pose questions that will stimulate or lead the discussion in a direction. Instructor modeling of a high level of presence sets a positive norm for the class and encourages students to do the same. Moore (2001) suggests that instructor interaction should have the goal of establishing a culture of independent learning and peer participation. Positive instructor feedback tends to bring out the best in people and motivate them to invest discretionary effort (Braksick, 2000; Daniels, 2000). Positive instructor feedback can energize the learning system and increase interaction frequency.

Vrasidas (1999) examined the conceptual framework of interaction in online courses and found that the factors influencing interaction were learner control, social presence, structure, feedback, and dialogue. In a follow-up study, Vrasidas and McIsaac (1999) found that each of these factors has specific implications for teaching practice. For example, activities can be structured to increase interaction with the instructor, other students, and the course content. Discussing a paper outline with an instructor; collaborating activities with peers, and participating in required online discussions were found to increase interaction among participants. This study, not unexpectedly, found that higher-than-expected workloads contributed to decreased interaction. Thus, the appropriateness and on-task time of course requirements must be suitable to facilitate online interaction.

## PURPOSE OF THE STUDY, HYPOTHESES, AND QUESTIONS

The purpose of this study was to investigate the interactive practices of effective online health care graduate and undergraduate instructors as gleaned from the experience of successful faculty. This study focused on understanding how the online teaching technologies in the courseware of an Internet portal system were being employed to promote interaction. The goals of this research study were: (a) to better understand the phenomenon of successful online computer-based education in graduate and undergraduate healthcare education through the identification and description of online educational constructs that exemplify effective interactive practice, (b) to better understand how effective distance educators in health-care education utilize the innate capabilities of online courseware to support interactive constructs, and (c) to better understand what techniques and strategies faculty employ to foster and facilitate the sense of interaction. It was hypothesized that effective faculty employ certain methodologies, practices, and mindsets in planning and active teaching phases to promote interaction when utilizing online courseware. The research asked what successful online distance education faculties do to make their teaching more interactive, and do they use different approaches to generate interactivity in different situations.

### The Study or Methodology

The research design for this study involved a triangulated three-tiered process. The first phase was to identify graduate and undergraduate health-care faculty from major university schools of nursing and health professions who provide effective interactive education that fosters learning. Potential faculty were identified by program directors from six university health education programs (The University of Pittsburgh, George Washington University, West Virginia State

University, West Chester University, University of Maryland, Baltimore County (UMBC), and Drexel University), who based their identification on the criteria that the instructor, (a) promotes a high level of student-to-student interaction through threaded e-mail and discussion-board activities as well as other activities that allow learners to construct or formulate an idea in a deeper sense, and raises the interest and motivation of the students, (b) the instructor promotes a high level of student-to-instructor interaction through both quantity and quality of assignments that maximize the impact of interactions, and (c) the instructor promotes a high level of student-to-content interactions through offering a variety of activities and resources that offer students a variety of alternatives for learning.

The second phase, utilizing the program director's listings, involved interviews with twelve (12) selected faculty from the list developed in phase one. These instructors were requested to participate in phenomenological interviews, either face-to-face or by telephone to establish trends and common themes in effective online instruction.

The third phase of the research involved taking the 12 phenomenological interviews and synthesizing the results to create a Delphi questionnaire for use with an experienced group of 30 faculty members drawn from the list developed in phase one. Trends and common threads were identified to assist in categorizing the data. To assure that all common themes and trends were identified, a second evaluator was utilized to assure inter-rater reliability. The questionnaire was operationalized following the Delphi Technique with the 30 additional faculty. This tool allowed a group of defined experts to come to a consensus of opinion when the decisive factors were subjective, and not knowledge-based. Through a series of questionnaire exchanges, the experience group identified additional ideas through individual brainstorming and communicating ideas with the investigator to clarify and validate the findings from the previous questionnaire. Questionnaires were exchanged through e-mail to maximize efficiency and minimize time associated with conventional mail. A series of three exchanges with progressive fleshing out of ideas as well as generation of new ideas beyond those attained through the phenomenological interviews was facilitated. It was determination after the third questionnaires that no new ideas had emerged.

## Limitations

The attempt of this research was to capture interactive fundamentals of practice among health care online educators. It is acknowledged that in the context of the interview, that some may have issued their espoused theory and not their actual theories in use (Argyris, 1999). Argyris suggests that people consistently act inconsistently, unaware of a contradiction existing between their espoused theory and their theory-in-use (what they actually do in the practice setting). Despite this possibility, participants offered what they envisioned as the most effective pedagogies to maximize interactivity in online health care programs.

It is further recognized that one of the limitations of this study is that the definition of effective practice is based upon the single lens, or perspective, of the identified instructors. This research did not take into consideration student perspectives of effective practice. This research also relied upon the identification of effective faculty from the perspective of program directors who based their recommendations upon a provided set of criteria.

## Interview Findings

The 12 faculty members interviewed suggested that different types of courses require different types of facilitation to generate effective interaction. As one instructor for this study stated:

*There are different types of courses, such as the hard sciences classes versus the social science classes. There may be classes such as skills classes, such as a clinical class, or a research class where students need to leave with a skill. There may be social science or discussion/opinion classes that take information and apply it in different situations. This is both online and in the traditional courses. There is a big difference. At times the conceptual pieces in putting things together as opposed to the nitty-gritty facts and research and data collection are trickier to do. The relating this in the everyday life can be more difficult and more challenging.*

Four types of courses were distinguished, including foundational classes, skills classes, analysis/synthesis classes, and hybrid type courses.

- **Foundational or rote memorization courses:** Such courses provide the foundation or knowledge that will be used to build upon in other courses. The prioritized interactions in such courses focus on helping the student make more "student-to-content" connections. As one instructor from this study stated, "In the foundational course, you either get it or you don't." Foundational courses include courses like anatomy and physiology, pathophysiology, medical economics, and the business of health care. Many of the assignments in such courses are e-mailed weekly based upon module content to assure interaction with the content and understanding of key concepts. Some "student-to-student" is fostered, primarily through having assignments shared and discussed on discussion boards, something that does not routinely happen in a face-to-face classroom. As one instructor stated:

Health Care economics is a foundational course (not all economics courses are this way). In this course it is more of the memorization and regurgitation that is important. I can't ask them to compare and contrast John Maynard Keyes to Karl Marx because they are not there yet. I facilitate this differently than an upper level class. I use more discussions based upon the facts. They come into their first class and think that they understand this stuff. They think that they have all the answers. But they give what I call, "man on the street" answers. I have to remember that what I am doing here is building a "foundation." This type of course and a rote memorization course is foundational for other courses.

Such interactions are not generally open-ended and are designed primarily to reinforce the memorizing of basic concepts that will be applied later in their professional studies and practice. For example, an online nursing student may not understand all the reasons for memorizing anatomy and physiology, but will come to understand better the value of such memorization as the anatomy and physiology facts are applied in future courses.

The faculty interviewed believed that while foundational courses do not innately lend themselves to great online discussions, providing interactive e-mailed assignments, more instructor presence for Q & A, and instructor interaction in discussion boards helps to foster meaningful learning interactions with such courses.

- **Skills based courses:** Such courses require students to gain a particular skill(s) set that is applicable to a specific environment. The prioritized interactions in such courses focus on content, however, and expand beyond "student-to-content" connection, requiring instructor facilitation and presence to promote dialogue specific to the skill and the application of the skill. Instructors facilitate critical thinking and understanding of the concepts through the skillful use of questioning, such as "if we did this what would happen" or "how could we do this if…?" or "Great idea. Does anyone have any other directions?" Examples of such skill-based courses would include Nursing Research, Medical Informatics, and Patient Assessment. Skills classes such as Nursing Research or Health Education Research should be project-based, as students need to flesh out ideas and application of concepts. Interaction focuses on how and where to perform the skills as well as facilitating students toward a finite answer and how they get that answer (example: Identifying research questions). As one instructor stated:

*Part of my goal is that I want them to understand the complexities of research and how you work with other people, so I do include a group project into the course. I believe that you learn from your peers. There is a skill set in how do you develop a proposal, how do you interpret statistics, how do you design research. The goal isn't to make them an expert when they are finished but to give them a set of skills that they can apply.*

Another example drawn from an instructor teaching a Physical Assessment course:

*I found teaching Physical Assessment online a difficult course to teach online because there is, or should be, so much hands on, and if you teach it in the classroom there is. So it is very difficult to deliver the physical assessment content Web-based because there has to be a video portion. The student has to be able to see how you percuss (thumping of the different body areas to determine if air or fluid is present), what the assessment of the abdomen looks like, and one of the big issues that we have is not bandwidth on our end at the university but among the receiving students. So I can decide to send them this fabulous thirty-minute clip, the best I've ever seen, and they are not going to be able to download it if they don't have the technical capabilities. I have to make a decision about how much information I can chunk into a block.*

Assignments that facilitate student-to-content interaction in a skills-based class are weekly assignments that are e-mailed back to the instructor to assure that they are interacting with content and understanding it. Interaction with students and instructor focuses on the processes that they are going through in learning the content.

- **Synthesis/Analysis Courses:** Such discussion-based courses are used to teach students to analyze a situation and engage in problem solving. Instructors find that authoritative postings tend to shut down dialogue or that students simply parrot instructor ideas or postings (most students do not want to challenge instructors and risk receiving a lower grade). Dialogue among participants provides regular opportunities for reflection and inquiry and requires the least intervention in the discussion boards of all the course types. Simulated interaction in this manner through subject matter presentation can subsume part of the interaction by causing students to consider different views, approaches, and solutions and generally to interact with a course.

These are courses where core information is presented but there is not necessarily a right or wrong view. An example may be an Issues in Health Care, Health Care Policy, or Nursing Practice course where a module or lesson would focus on "Compare the value of the Canadian versus the USA health systems." This is the type of course or topic that will prompt many opinions and views. The key is to flesh out all angles of the subject and have the students explore and support the differing viewpoints with the facts. For the instructor in this type of class, the key was not to intervene too much. As one instructor stated:

*I've found that if I post my particular opinions about a topic, then I change the discussion in that students stop posting or they just restate my opinions. I think you have to be very careful. What I do is post more personal experience than personal thoughts and beliefs. You have to control that and it can be very difficult. You have to post more with "what I've found is." Any time you give them a clue on which direction you lean the majority of the class is going to lean that way too because they want a good grade too. So you need to be very careful how you do that. I think that if you come down with a very dogmatic statement then you shut them down as people don't want to be wrong. They are still learning and fleshing out their own thought processes and if you post too much you shut down their thought processes or they may not agree but are not willing to take this instructor on.*

There is a need to establish the culture of independent collaborative learning. Instructors do not respond to the majority of postings in this type of course but read them all and respond to key ideas and elements and, through additional questions, guide the learning process. If the discussion is getting one sided or negative, the instructor can, through a posted thought, direct the dialogue to view all sides. "That is true but what would 'so and so' say regarding this and why?" One technique in this type of course is to create an online debate where students don't get to pick but defend an assigned point of view, which forces them to see all sides. In a Nurse Practice course, one technique reported was that students were facilitated to choose a topic, interview someone, and then write a paper and discuss their issue and what they have found by leading a discussion-board thread.

- **Hybrid Courses:** Such courses have a combination of the above three and require a mixing of techniques to facilitate interaction. An example of this may be a Health Care Management or Leadership course where there is specific theory to understand and employ but also where you want students applying and understanding their own personal leadership style in different situations.

## THE DELPHI FINDINGS

The Delphi process suggested that interaction strategies vary according to the type of course. Those participating in the Delphi component agreed with the majority of the interview findings, and in many cases further elaborated on the four types of course offerings.

### Foundational or Rote Memorization Courses

It was agreed that while these types of courses don't generally lend themselves to great online discussions of the material, the interaction generated was centered on assisting students with learning the content material. It was also believed that a greater instructor presence was necessary so that some students don't have the feeling of learning alone or in a vacuum.

### Skills-Based Courses

It was agreed that the interaction in a skills-based course is centered on the content, and that interaction was again used to make connections between content and skill application. This type of course was again found to require more instructor presence for students to gain the skill, and that the instructor's role was to generate thinking and a better understanding of the concepts. A majority of the Delphi participants utilize the weekly assignments format for a skills class. The concept of forcing a student-to-student interaction just for the sake of having one was not described to be effective. One instructor stated: "Certain topics don't lend themselves to meaningful discussion. Having a discussion assignment because there should be student-student interaction is not effective, does not facilitate mastery of the material, and frustrates the students." Another articulated that: "Instructor interaction in this type of course takes a lot of instructor facilitation and takes on a greater role with telephone follow-up when they are having difficulty in understanding concepts."

### Synthesis/Analysis or Discussion-Based Courses

It was agreed that in these types of courses, the free flow of ideas monitored and facilitated by the instructor is the best technique. This requires that the instructor monitor carefully, and he or she must be diligent in fostering good online dialogue about the topics at hand. In this type of course, a majority of instructors reported using a "search and report" technique, where students go out and research a topic and then report back as a catalyst for generating good discussion boards. Additionally, instructors reported using weekly discussion board topic questions based upon the readings and students' own research to generate discussions. The instructor would monitor and then post as was appropriate to guide, stimulate, and assure good dialogue. As one instructor stated: "In my nurse practice course they will choose a topic and go interview someone and then write a paper and discuss their issue and what they have found by leading a discussion board thread."

### Hybrid Courses

It was agreed that there are these types of courses but not all agreed on which type of course was in which category. Some reported that a research course was felt to be a skills class while others felt that it was a hybrid class where multiple techniques are employed. It becomes less of an interest to this research to categorize courses as it is to understand that there are different courses that may need facilitation using different techniques differently to maximize the online learning.

## CONCLUSION AND DISCUSSION

One of the more significant findings in this research is the identification and confirmation that different types of courses require different types of facilitation to generate interaction. The four types of courses in the health-care educational systems (foundational classes, skills classes, analysis/synthesis classes, and hybrid type courses) require the employment of specific facilitation techniques on the part of online instructors. This finding has implications for interactive design and online teaching.

### Foundational or Rote Memorization Courses

The interaction in this type of course is more focused on providing a greater understanding of material to serve as the basis for future learning. The central focus here is on student-to-content interaction. Health care programs have innately established a framework of knowledge and skill that serves as a foundation for future learning. The information learned in a Medical Terminology course is the building block for an Anatomy and Physiology course. The knowledge gained in these courses serves as the universal knowledge for the rest of the health care clinical education, whether in nursing or any other allied health field. These courses are typically very structured in nature and require the instructor to clearly articulate the material for ease of consumption by the learner. The learner of online foundational material must be independent in learning the material but, as with a face-to-face class, the interaction is designed to assist the learner to that end. Carnwell's (1999) concept of developing internal dialogue is supported here. His research indicated that the students in this situation desire more highly structured materials. Design of text materials is also important since the level of structure within the text may create either independence or dependence in students. Jones and Kemper (1994) suggested that independence can be fostered by requiring students to use self-study packages in an unsupervised manner.

Instructors need to design interaction that is more finite in assisting participants to understand the material. This interaction takes its form in assisting students to learn and memorize the material, to see how it all fits together. The design of the interaction is seen through e-mailing assignments to the instructor for feedback and assistance in learning the key objective elements of the foundational course. While this type of course does not lend itself to great online discussions, but is more focused on learning the concepts, it is also paramount for the instructor to practice a greater presence that provides the student with a sense of contact with the instructor (Townsend, 2002; Kanuka, Collett, & Caswell, 2002; Beaubien, 2002).

## Skills-Based Courses

The implications of the research suggest that instructors need to design their interaction in this type of online course to assist students to gain the skills necessary to function in the particular health-care environment. Effective interaction requires designing discussion board activities to present applications and allow students to see other's work and learn from each other toward the final goal of attaining the skill, whether it is learning the physical skills of chest percussion, or lung sounds (where there are significant limitations), or the mental skills of designing and conducting medical research, or the skill of learning database construction and manipulation in a medical informatics class. The ability to demonstrate competence for the physical skills innate to health-care practice is one that has significant limitations in the current online environment within the present available technological structures, and assessment of competency often requires a face-to-face environment with credentialed professionals. The implications for the design of interaction are for the instructor to use a more facilitative role to assure that student-to-student interaction is being assisted here toward applying the information from the class to the field. Kennedy (2002) supported this premise as he suggested that learner-to-learner interaction is a valuable part of the online learning experience and that the distance education format is particularly well suited to engaging students in this type of interaction. These discussions can be deeper and more reflective, covering a broader range of issues that assists students to gain the skills taught in the course. Encouraging greater interaction among learners not only enhances student learning and application of the new skill set, but also places the instructor in a more supportive, facilitative role, which results in more efficient use of instructor time (Udod & Care, 2002). Instructor interaction in this type of course needs to take on greater monitoring of the learning, and the utilization of personal e-mail contact and telephone conversation may assist with students having difficulty in understanding concepts.

The interaction to facilitate these skills in online teaching needs to be planned to include learner-to-learner contact to assist students to learn from each other in how they design and apply the skills discussed above. Vrasidas and McIsaac (1999) supported this finding, as a need for educators to structure for dialogue including learner-to-learner interactions was found important. Kozma (1991) supported this premise seeing the need for more dialogue by visualizing learners actively collaborating with the medium to construct knowledge and skill.

## Synthesis/Analysis Courses

The implications for these courses are for the instructor to understand that there may be multiple views of a situation that require exploration by the participants to fully realize the depth and breadth of the concepts. The obvious nature of these types of courses is for students to explore all aspects of the topic and then draw conclusions based upon the information presented and explored. The online instructor in this type of course must take care not to be too authoritative, as stated above, otherwise he or she will shut down the exploration and the insights that can be gleaned from students interacting and dialoguing about opposing viewpoints. Instructors will find that taking an authoritative stance and letting participants see their opinions on an issue will close dialogue, or they will find that students will simply repeat or "parrot" the instructor's view on the issue. This is in direct conflict of what the instructor may wish to accomplish by allowing students to analyze and synthesize all views of the problem or issue. Dialogue among participants provides regular opportunities for reflection and inquiry (Wesley & Buysse, 2001) and requires the least intervention in the discussion boards of all the course types. Simulated interaction through subject matter presentation in pre-produced courses can subsume part of the interaction by causing students to consider different views, approaches, and solutions and generally to interact with a course (Holmberg, 1999, 1989).

## Hybrid Courses

Courses that reflect a combination of the above three forms require a mixing or blending of techniques to facilitate interaction. The advantage for an online instructor is to be aware of the first three types of courses and the methodologies that are effective in facilitating those types of courses, and then applying that on the micro-application level for the hybrid course.

If we frame the curriculum in the context of the four different types of courses, then we provide a better opportunity in the planning phases to maximize the learnings, as we are focused on the goals and best structure for interaction given the different types of course limitations and opportunities available in the online courseware. Knowlton (2000) believed that learning and teaching are reconceptualized in the online course to allow maximum independence among students by framing the curriculum and student interactions through the providing of resources and opportunities. Framing is used to facilitate students' desire to develop and implement shared goals in making connections with the curriculum. Students must be able to find space for their own inquires and needs within the assurance of a well-planned, content-rich, and flexible learning environment with adequate navigational tools and support systems (Vandergrift, 2002). This type of in-depth planning is more demanding and time consuming than the traditional classroom planning. A classroom teacher can draw upon his/her innate knowledge and platform skills to provide an impromptu lesson structure that cannot be duplicated in the online text-based arena. Vandergrift demonstrated the need for a deeper understanding of the dynamics that online teachers apply to the deliberate acts of design and interaction that facilitate successful courses.

Visor (2000) supported this as faculty serve to design and conduct a course, which is positive for student learning outcomes and serves to maximize the learning opportunities the same way that faculty foster learning in the traditional setting. They prepare and organize content according to well-established and communicated objectives, consider methodologies, which will assist the student to achieve the objectives in an online format, and be cognizant of time that the student will need to spend on the course. Wright and Thompson (2002) support this, as faculty establish a pedagogical strategy and begin to understand how online activities will occur in their discipline for maximum learning, and that faculty create templates in which they can insert their specific academic content. The four types of courses identified in this study have different goals and objectives for learning outcomes. Instructors need to facilitate interaction specific to the needs of the students in the confines of these course types.

The innate properties of the course shell portal allow options for organizing the pedagogy of an interactive course. This also becomes part of the consistency described to establish a "think-forward" type of lesson. The shell portals have innate properties built in that provide a "think forward" consistency of structure that becomes a comfort for students if interaction is designed consistently by the instructor. If the online course facilitates interactions in the nature described in this study, the constructivist and andragogical models of learning predict that successful learning is likely to result. These models require students to create their own meaning to knowledge in a self-directed manner and take more responsibility for their own learning (Knowles, 1999, 1980). This is not to say that the courseware becomes limiting but, as with the traditional classroom, there are confines of what can and cannot be done successfully. The innate properties of the course shell portal allow options for organizing the pedagogy of an interactive course and provide a level of flexibility to allow instructors and designers freedom to explore a number of pedagogies. Instructors need to be open to exploring new pedagogies that require different thinking.

## FUTURE DIRECTIONS

Insight into establishing some of the best methodologies utilized in each of these types of courses would assist online instructors to establish the interaction that is employed and how best to maximize the competencies attained by students. Applying the practices used in the four types of courses would further establish a number of pedagogies that would work in different online health-care educational settings and may open the door for more nontraditional online courses to be offered. Brooks (1999) suggested that curriculum should be examined to determine how technology fits. Using technology, we should identify content that is technology neutral, technology driven, and technology enhanced. Salmon (2000) identified the teacher in the role of e-moderator as the key ingredient for effective teaching and learning online. It is the instructor who is the driver of the technology, given the differing nature of the four types of courses found in this research. Providing online educators with greater insights as to the application of pedagogies in different courses will assist them to create more effective online learning environments and help their students learn the true benefits of online learning communities as established from interactive practice priorities. Until now, there are those who would only see certain types of courses being taught online. With the identification of these four types, and the methodologies that can be employed, additional insights will assist the less traditional online course (clinical-based skills courses) to be opened to this technology.

Applying Holmberg's (2003) conversational theory to the four types of online health-care courses taught will assist in further understanding the dynamics of the interaction required to facilitate learning. This application will aid in offering insights to the pedagogies which promote learning in each of the four course types and the dialogue necessary to promote the interaction. Carnwell's (1999) dialogue is much more than merely transmitting messages to students about requirements of courses. Curriculum planners of distance education programmers and materials designers need to devise a balance of internal and external dialogue which allow students with different learning style preferences and approaches to gain the maximum from their learning experiences. MacDonald (2001) suggests that there may be particular times when extra formative feedback is of particular importance to students. Additional research is necessary to understand the balance of internal and external dialogue and the feedback necessary to maximize the learning opportunities in the four different types of courses.

## REFERENCES

Anderson, T., Rourke, L., Garrison, D. R., & Archer, W. (2002). Assessing teaching presence in a computer conferencing context. *Journal of Asynchronous Learning Networks, 5*,(2). Retrieved April 20, 2003 from, http://sloan-c.org/publications/jaln/v5n2/v5n2_anderson.asp

Argyris, C. (1999). *On organizational learning* (2nd Edition). Malden, MA: Blackwell Publishers Inc.

Beaubien, J. (2002). Harnessing the power of complexity in an online learning environment. In K.E. Rudestam, & R. Schoenholtz(Eds.). *Handbook of oOnline learning: Innovations in higher education and corporate training.* Thousand Oaks: Sage Publications, 221-236.

Braksick, L. W. (2000). *Unlock behavior, unleash profits.* New York: McGraw-Hill.

Brooks, B. (1999). *Professional development for teaching technology across the curriculum.* Minister of Education, Edmonton, Alberta, Canada. Retrieved June, 23, 2005 from, http://ednet.edc.gov.ab.ca/technology

Carnwell, R. (1999). Distance education and the need for dialogue. *Open Learning, 16(2), 58-60.*

Daniels, A. C. (2000). *Bringing out the best in people: How to apply the astonishing* power of_positive reinforcement (2nd ed.). New York: McGraw-Hill.

Garrison, D.R., & Archer, W. (2000). *A transactional perspective on teaching and learning: A framework for adult and higher education.* Oxford, United Kingdom: Pergamon.

Gibb, J. (1995). *Tribes.* Sausalito, CA: Center Source Systems.

Gillis, A., Winston, J., Braid, A., MacDonald, P., & MacQuarrie, M. A. (2000). The learning needs and experiences of women using print-based and CD-ROM technology in nursing distance education. *Journal of Distance Education, 15(1),*1-20.

Holmberg, B. (2003). A theory of distance education based on empathy. In M.G. Moore, & W.G. Anderson (Eds.), *Handbook of distance education.* London: Lawrence Erlbaum Associates, 79-86.

Holmberg, B. (1999). The conversational approach to distance education. *Open Learning, 14(3)*, 58-60.

Holmberg, B. (1989). *Theory and practice of distance education.* London: Routledge.

Howland, J.L., & Moore, J. L. (2002). Student perceptions as distance learners in Internet-based courses. *Distance Education, 23*(2), 183-195.

Jones, A., & Kemper, D. (1994). Approaches to learning and student acceptance of self-study packages. *Educational and Training Institute, 1*(2), 93-97.

Kanuka, H., Collett, D., & Caswell, C. (2002). University instructor perceptions of the use of asynchronous text-based discussion in distance courses. *The American Journal of Distance Education, 16*(3), 151-167.

Kennedy, D. M. (2002). Dimensions of distance: A comparison of classroom education and distance education. *Nurse Education Today, 22*(5), 409-416.

Knowles, M. S. (1980*). The modern practice of adult education: From pedagogy to andragogy.* Englewood Cliffs, NJ: Cambridge Adult Education, Prentice Hall Regents.

Knowles, M. S., Holton, E. F., & Swanson, R. A. (1998). *The adult learner* ( 5th ed.). Woburn, MA: Butterworth-Heinman.

Knowlton, D. S. (2000). A theoretical framework for online classroom: A defense and delineation of a student-centered pedagogy. In R.E. Weiss, D.S. Knowlton, & B.W. Speck(Eds.). *Principles of effective teaching in the online classroom, 84*. San Francisco: Jossey-Bass, 5-14.

Kozma, R. B. (1991). Learning with media. *Review of Educational Research, 61*(2), 179-211.

MacDonald, J. (2001). Exploiting online interactivity to enhance assignment development and feedback in distance education. *Open Learning, 16*(2), 181-189.

Rovai, A. P., & Barnum, K. T. (2003). Online course effectiveness: An analysis of student interactions and perceptions of learning. *Journal of Distance Education, 18*(1), 57-73.

Swan, K. (2001). Virtual interaction: Design factors affecting student satisfaction and perceived learning in asynchronous online courses. *Distance Education, 22*(2),306-331.

Salmon, G. (2000). *E-moderating: The key to teaching and learning online*. London: Kogan Page.

Townsend, E., Campbell, C., Curran-Smith, J., McGinn, F., Persaud, D., Peters, P., Bower, I., & Sheffield, S. (2002) Accessibility and interactivity in distance education programs for health professionals. *Journal of Distance Education, 17*(2), 1-24.

Rogers, C. (1969). *Freedom to learn*. Columbus, OH: Merrill.

Udod, S. A., & Care, W. D. (2002). Lessons learned in developing and delivering Web-based graduate courses: A faculty perspective. *Journal of Continuing Education in Nursing, 33*(1), 19-23.

Vandergrift, K. E. (2002). The anatomy of a distance education course: A case study analysis. *Journal of Asynchronous Learning Networks, 6*(1). Retrieved January 28, 2003 from, http://www.sloan-.org/publications/jaln/v6n1_vandergrift.asp

Visser, J. A. (2000). Faculty work in developing and teaching web-based distance courses: A case study of time and effort. *American Journal of Distance Education, 14*(3), 21-32.

Vrasidas, C., & McIsaac, M. S. (1999). Factors influencing interaction in an online course. *The American Journal of Distance Education 13*(3), 22-36.

Vrasidas, C. (1999). *Meanings of online and face-to-face interactions in a graduate course*. Unpublished Ph.D. Dissertation, Arizona State University.

Wlodkowski, R. J. (1999, 1985). *Enhancing adult motivation to learn*. San Francisco: Jossey-Bass Publishers.

Wright, T., & Thompson, L. (2002). Cost, access and quality in online nursing and allied health professions. *Journal of Asynchronous Networks 6*(2). Retrieved January 3, 2003 from, http://www.sloan-c.org/publications/jaln/v6n2_wright.asp

*This work was previously published in International Journal of Information and Communication Technology Education, Vol. 4, Issue 1, edited by L. Tomei, pp. 44-55, copyright 2008 by IGI Publishing (an imprint of IGI Global).*

# Chapter 10
# Scenegraph-Based Platform for 3D Computer Graphics Training

**Vincent Muggéo**
*University of Geneva, Switzerland*

**Laurent Moccozet**
*University of Geneva, Switzerland*

**Nadia Magnenat-Thalmann**
*University of Geneva, Switzerland*

## ABSTRACT

*We propose a framework for developing online interactive experiments for training students to master the basic concepts of 3D Computer Graphics. As 3D Computer Graphics has applications in a large range of fields (visual arts, media, geography. etc.), we need to devote particular attention to students that are nonexperts in Computer Science and particularly in programming. We also have to take into consideration the resources and efforts required for the development of online training modules. We describe our approach for designing and implementing accurate and efficient training modules and describe how we have implemented one particular use case scenario.*

## INTRODUCTION

In Cunningham (1999), the author stresses the importance of the development of courseware for online and distance education in Computer Graphics. He particularly points out the costs and human resources required. In this article, we propose a framework for developing online interactive modules for training in 3D Computer Graphics (CG). We target two objectives: to offer valuable educational material that can be valid for non-Computer Science students; and to optimize the time and resources required for developing the modules. In the remainder of this section, we will survey the related works in the area of interactive and online educational material for 3D

CG. In the second section we develop our strategy for designing the training module. The resulting implementation is described in the third section, and a specific use case scenario is proposed in the fourth section.

We first briefly survey the available tools that teachers can use for developing interactive training exercises. The 3D CG field is particularly growing with dedicated tools. The first category of tools is 3D design software. In this category, we find some widely used software such as the 3DS Max Studio suite (3DS Max), which features strong modeling capabilities and a flexible plug-in architecture. It is widely used, for example, by video game developers, TV commercial studios, or architectural visualization studios. It is also used for movie effects and movie revisualization. Other commercial and Open Source solutions are also available. However, all these solutions offer the general 3D CG functionalities and much more. The second category of tools is programming solutions. In this case, a graphic library is used to directly write the complete training application. Currently, the most popular libraries are OpenGL and DirectX. Both provide an API dealing with 3D native functionalities. In this context, the teacher has to write a program that handles these functionalities. We need to find a compromise that directly provides a high-level access to state-of-the-art functionalities in 3D CG as with 3D design software, but also offers the flexibility to fully control the access to these functionalities as with 3D programming. There are two classical approaches to train students to let them experiment by themselves 3D CG concepts. They reflect the two types of tools available for developing interactive training material: (1) with programming exercises (Lewis, 2000; Hitchner, 2000; Cunningham, 2000) usually based on a standard programming language such as C++ or Java and a Graphics library API such as Open GL; and (2) with design exercises (Van Gumster, 2003) usually based on a commercial or open source 3D interactive modeler such as 3DS Max or Blender. Although these training methods are effective, they have the following basic drawbacks that prevent them from extending to interactive online training:

- Programming exercises give in-depth access to the basic concepts in 3D, but they also require as a prerequisite that the students have programming skills. This is usually the case for students in Computer Science but not for students from other disciplines. Unfortunately, the acquisition of sufficient programming skills would require too much overload to consider it as a viable general solution. Moreover, according to the range of various programming languages and 3D graphics library available, even Computer Science students may encounter difficulties mastering the programming environment.
- Interactive 3D modelers offer a complete package of features from the most basic ones to the most advanced ones. Unfortunately, as they are professional or semiprofessional production-oriented tools, they are quite complex to use and master, and they also feature a complex user interface. As a result, students need to spend a large amount of their time learning the software itself before being able to experiment with the 3D concepts. The existing efforts to provide interactive or online educational material in 3D Graphics offer a range of modalities from a simple online version of course notes with static graphics illustration (OCGn, HyperGraph), interactive graphics programming demonstration components (CGT, CGEMS, OpenGL tutors) (mostly in OpenGL) to be executed locally after download, to inline interactive Java applets for demonstration (ILO, AlgoViz, Exploratories, ICG). Most of them are oriented to algorithms demonstration rather than practice examples and to "low-level" (mainly graphics programming) aspects.

The AlgoViz (Ullrich & Fellner, 2005) project provides a software environment focused on the visualization of fundamental Computer Graphics algorithms and geometric modeling concepts. The goal is to provide a platform of components, which easily can be combined to create new applications. Based on this environment, the authors have developed an Introduction to Computer Graphics (ICG) online course. This course provides online interactive modules to learn 3D CG concepts such as the rendering pipeline or related topics such as the colors spaces. The available modules are either specific ad-hoc illustrations (Figure 1a) or programming examples (Figure 2). The same comments apply to other solutions such as the ILO (Hanisch, 1996-2003; Hanisch, 2005) system (Figure 1b). MVisio (Peternier, Thalmann, et al., 2006) is a powerful multidevice and multiplatform framework that provides a set of stand-alone executable compact modules allowing both students and teachers to navigate and experience 2D/3D Computer Graphics concepts. However, this rich content and interactive framework is not optimized for distance learning.

## THE 4-AREAS CONCEPT

The general 3D Computer Graphics framework consists of the following items: a set of data

*Figure 1.*

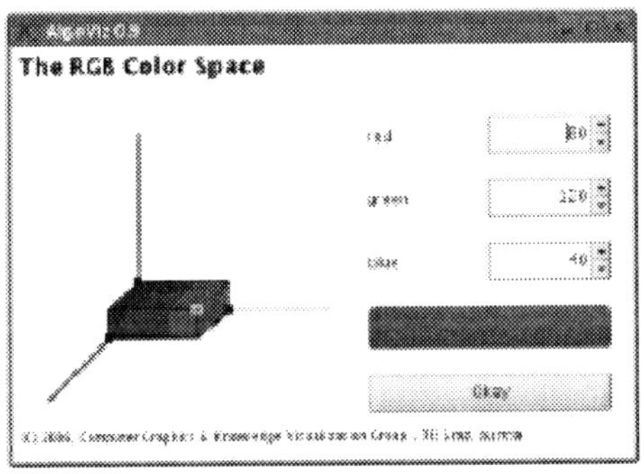


*(a) Ad-hoc demo module for color spaces from the AlgoViz project*

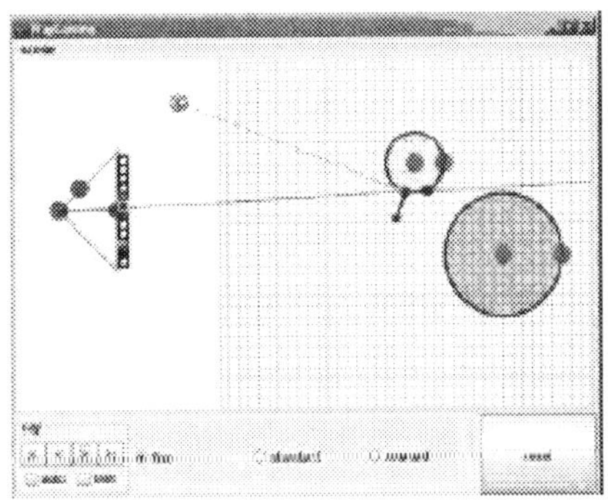

*(b) Ad-hoc demo module for raycasting from the ILO project*

*Figure 2. Online demonstration based on the openGL interpreter from the AlgoViz project*

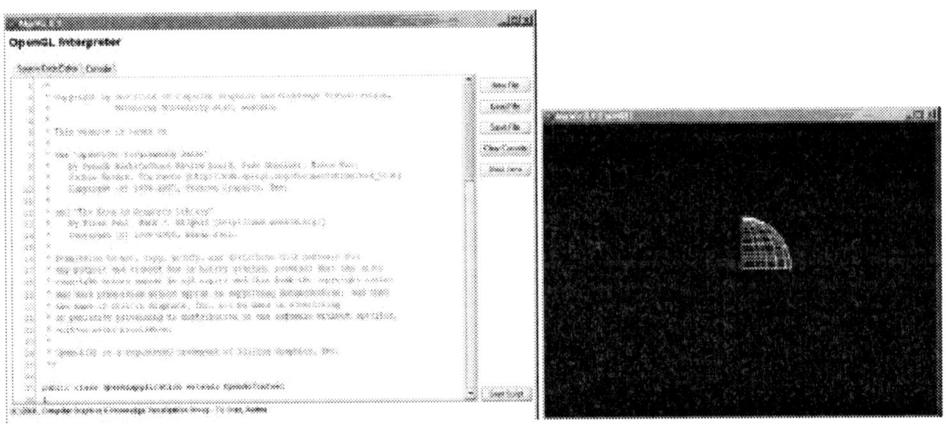

structures describes a virtual scene in three dimensions. The main components are the objects, the materials associated to the objects, the lights, and the cameras. The rendering process then takes the 3D scene description as input and produces a 2D view of the 3D scene that is displayed on the screen as if the 3D scene were photographed with the virtual camera.

Our hypothesis is that students can experiment and further master the various 3D concepts by tuning the parameters of the 3D scenes and interactively simultaneously visualizing the effects onto the data structures, the 3D scene, and the resulting 2D display.

To this end, the resulting design of the user-interface should consider four areas (Figure 3):

1. The control parameters area, where the student can directly change the parameters related to the current illustrated concept (e.g., the location of a virtual camera);
2. The data structures area, where the student can view how the main related data structures are evolving (e.g., the viewing matrix associated to the camera);
3. The 3D scene area, where the 3D scene is visualized together from an extra viewpoint with a combined graphical representation of the involved data structures (e.g., a graphical representation of a camera);
4. The 2D rendering area, where the 3D scene is displayed from the current camera viewpoint. In order to really experiment with the 3D space, the student should be able to navigate in the 3D scene area.

It particularly provides students with a dual view of the virtual scene. When a student modifies the parameter values, he or she can immediately visualize the resulting effects onto the related components: data structure, 3D scene, and 2D rendering. He or she can iterate the same experience with various values in order to assess the effects and correlate them to the influence of each parameter for the current illustrated concept. In order to limit extra unnecessary learning linked to the user interface, the interface must hide the useless features and limit the content manipulation to the ones required to operate the current concept. This approach is close to the one proposed in OpenGL tutors for interactive training in open GL in C and extended for online training in Chen and Cheng (2005); Nate Robins' OpenGL Demos in Ch, and Yang (2002) with GL4Java, an OpenGL Java binding to allow deployment on the Web. However, these interactive environments are more focused to teach 3D programming than for experiencing the 3D concepts.

In addition to this global design, we also consider that students should be faced with real,

*Figure 3. The 4-areas design concept*

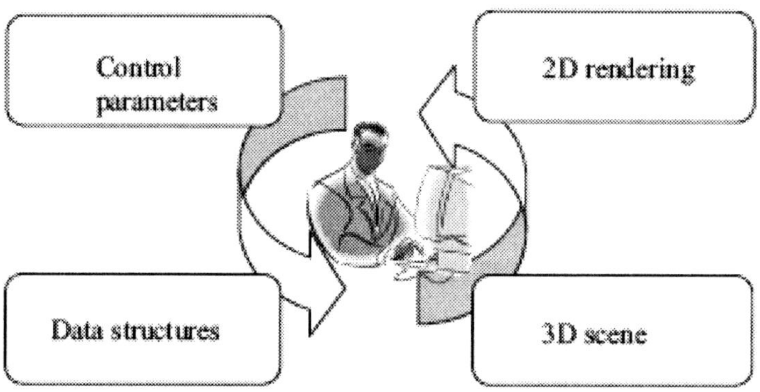

although simple, examples close to the ones they will effectively meet, and not to an ad-hoc illustration of the underlying algorithm that does not correlate the current concept with its effects. Moreover, students should experience the same or similar environment among the different modules. From the development point of view, implementing ad-hoc modules for each concept to illustrate requires a lot of resources even with solutions like AlgoViz. Instead, it would be much more efficient to develop the modules from a global high-level 3D graphics framework.

A scenegraph (Cunningham, 2001) is a high-level data structure that organizes the logical and spatial representation of a graphical scene in a tree structure. Each component of the scene is represented with a node of the tree. Scenegraphs are nowadays commonly available in graphics editing tools, games, 3D applications, graphics libraries, and languages such as Java3D or VRML. They also offer all the state-of-the-art features available in 3D CG. Many open source solutions are available, and many of them are supported by an active community of developers and users that ensure continuous updates that integrate up-to-date 3D features.

## IMPLEMENTATION

The main objective in our application is to propose a complete package integrating all 3D tools that can be manipulated by the Web page. As explained previously, we need to choose a 3D engine that supports scenegraph management. Among the many possible existing scenegraph implementations, we made our choice for OpenSceneGraph because it is cross-platform, it can be transferred easily between various architectures, and it is windowing independent, which is exactly what we need in a Web page. OpenSceneGraph brings all the common 3D CG functionalities required. First, we have to construct our application upon a 3D viewer. This viewer is built with the default OpenSceneGraph viewer, which is optimized for rendering.

According to the C++ technology applied for OpenSceneGraph development, we use an ActiveX control to embed the 3D engine into a Web page. The activeX technology is the only one that provides a complete compatibility with C++ applications and libraries. On the Internet, ActiveX controls can be linked to a Web page, downloaded, and executed by appropriate Web browsers (Internet Explorer, Netscape with plug-in, and Opera with plug-in). Like any executable

*Figure 4. Integration between elements*

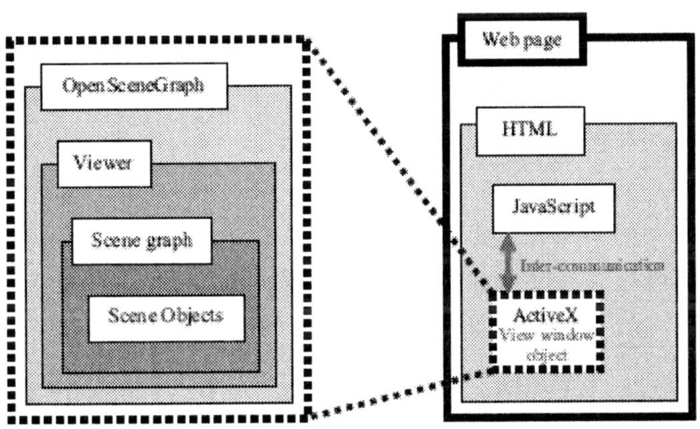

program running on the computer, ActiveX controls can perform any operation on the data, and especially in our case, it can perform OpenSceneGraph operations.

The ActiveX has to interact with OpenSceneGraph functionalities to communicate with the Web page. For this purpose, we developed an API embedding the ActiveX, which is able to interact with the Web page. We use the JavaScript language used to enable scripting access to objects embedded in the Web page.

In Figure 4, the ActiveX is interpreted like an object that could be accessible by the JavaScript language. Finally, our application can be easily deployed through the Internet to the client's computer.

## USE CASE SCENARIO

In this section, we present a complete module (Figure 5) in order to illustrate how we implement training modules based on the framework previously described.

As explained in the previous sections, we build our exercises around precise examples. To understand 3D CG notions, we guide the student to focus on a particular aspect, such as the camera frustum. This example illustrates a precise task related to the virtual camera management. Before showing our concrete example, we briefly summarize what the camera frustum is. A camera frustum represents a delimited volume where we compute the view rendering. Outside of it, we skip the rendering process in order to reduce time computation. This notion of camera frustum comes from the photography field. As shown in the following illustration, a camera frustum is composed of six parameters.

Near clip and far clip parameters define the distance field of view of the camera frustum. Consequently, after the far clip and before the near clip, all information is occulted and, of course, not computed. Top, bottom, left, and right define the height and the width of the near clip. It implies that if we change the distance between eye position and near clip position without changing these four last parameters, we will change the angle of view (Figure 6). The 3D camera management is inspired from the photography field and particularly from lens properties. In this case, we talk about camera perspective with two parameters called vertical FOV (Field of View) and horizontal FOV. These two parameters represent the angle of sight, horizontally and vertically (Figure 7). Both of these representations of camera management

*Figure 5. Camera frustum Web page*

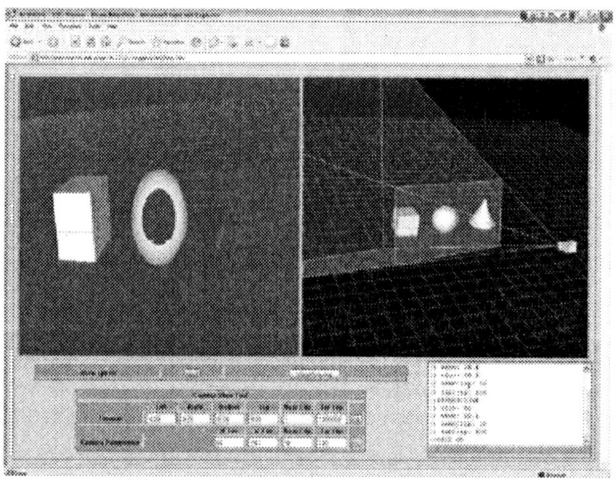

*Figure 6. Camera frustum*

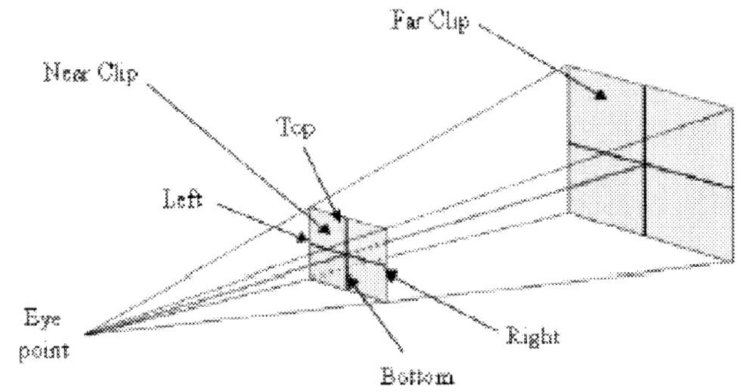


*Figure 7. Camera perspective*

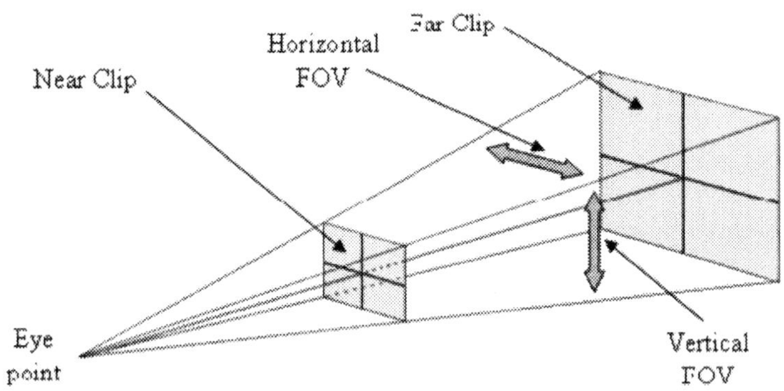


constitute a concrete subject of experimentation for users.

To illustrate the camera frustum manipulation, we create two separate views. The left view is our camera view, which will be manipulated. The right view represents our first camera (left view), viewed from a third-person viewpoint. All the interest lies in this separated representation. Thanks to it, the user can observe in the left view the pyramid, which represents our camera view frustum. This pyramid is composed of a semi-transparent area, which indicates the computation space of the left view. At the top of this pyramid, a camera shape is displayed to indicate the eye point (Figure 8).

In this way, our application displays dynamically the projection view of our camera in the left view and its 3D space representation in the right view. The user gets with the mouse device the handle on both views manipulation. In real time, the user can rotate, translate, and zoom on the scene view. In this situation, our application applies events on both simultaneously. If we assume that the user rotates the left view (our camera frustum projection), in the right view, the pyramid object moves dynamically according of the left view coordinates. On the other hand, if the user rotates the right view, it will have no effect on the left view, but he or she could observe the pyramid object in other points of view. A few objects are added in the scene in order to estimate the view.

*Figure 8. Dual view, original camera view frustum (left view) and its representation (right view)*

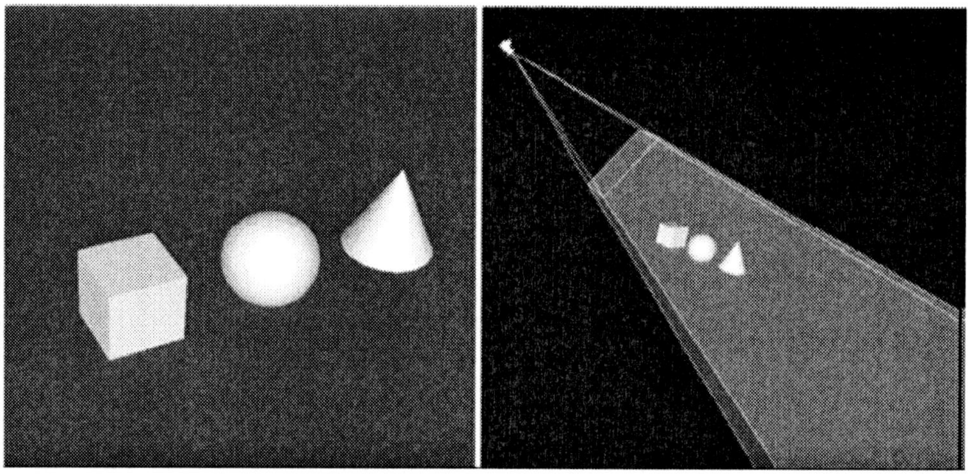

*Figure 9. Scene representation with three shapes and one grid, before and after camera manipulation*

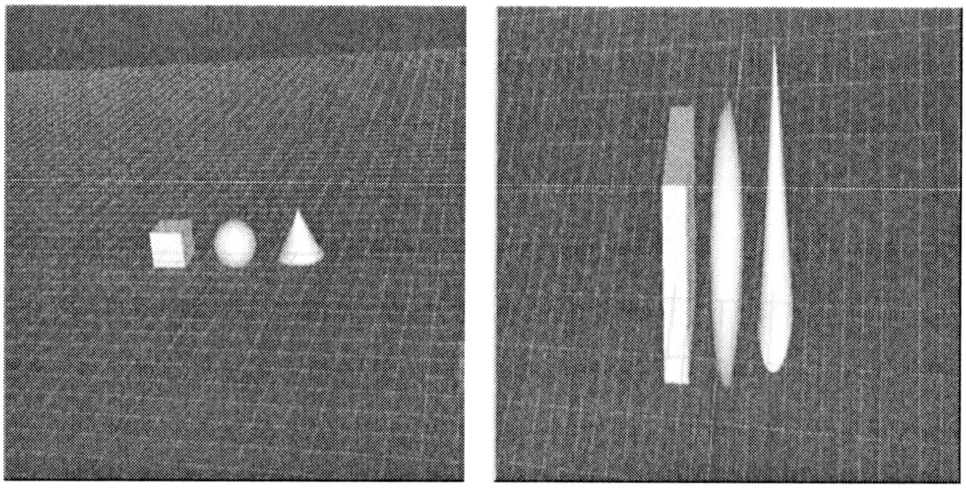

Three simple classical shapes—cube, sphere, and cone—are inserted so they initially appear in the camera view frustum. These are sufficient to observe camera manipulations and effects. Simple shapes often are used to appreciate deformation effects. Moreover, we add a standard grid to underline deformation effects. This grid incorporates x, y, z axis representation (Figure 9). The student can iteratively change the camera view frustum parameters and visualize the effects on the camera object in the 3D space and the resulting effects on the 2D rendered corresponding view, such as in the examples in Figures 10 and 11.

## CONCLUSION

We have proposed a framework for developing am online training module for 3D Computer Graphics based on the 4-areas concept and implemented around a 3D scenegraph viewer. Compared to similar available environments, we propose a complete and global development and design approach

*Figure 10. Dual view with Left=-0.26 Right=0.26 Bottom=-0.26 Top=-0.26 Near clip=1 Far clip=100*

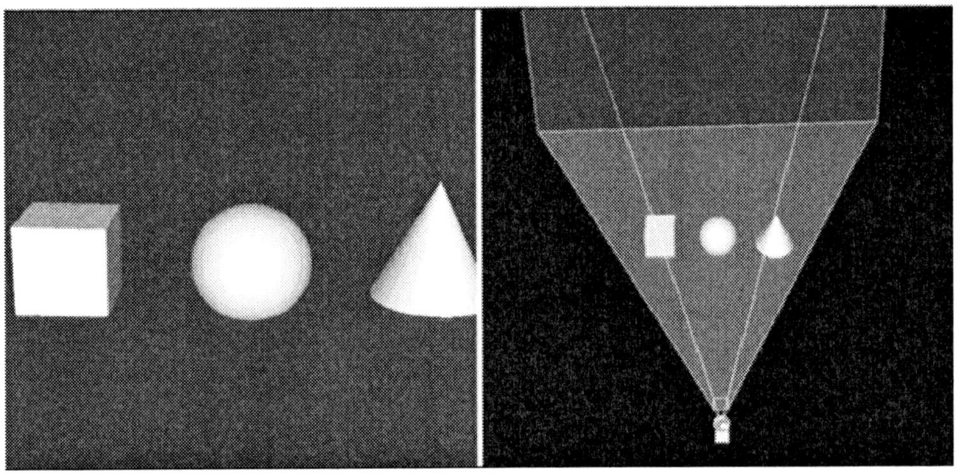

*Figure 11. Dual view with Left=-0.90 Right=0.90 Bottom=-0.10 Top=-0.10 Near clip=1 Far clip=100*

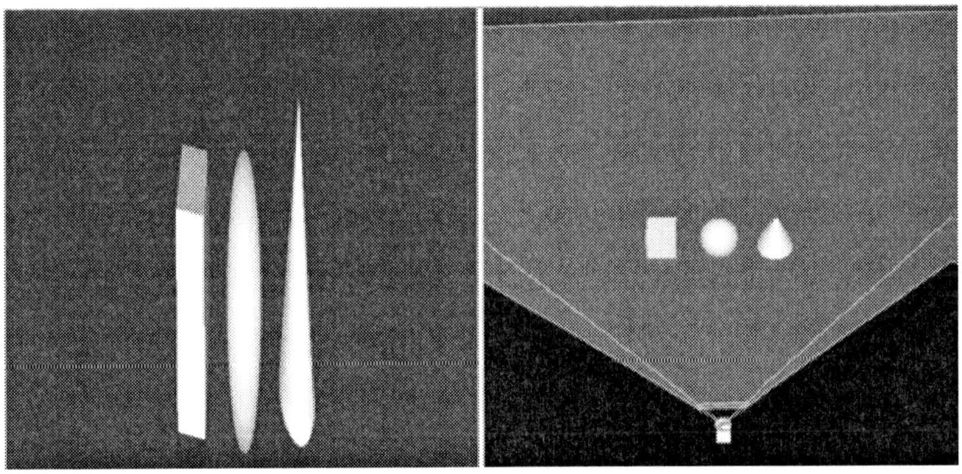

that attempts to address major critical challenges related to distant delivering of rich interactive 3D content for training. From the developer point of view, it provides a good compromise, as the scenegraph scheme offers a high-level and flexible environment for developing training courseware. Moreover, as it provides all the state-of-the-art features available in 3D Computer Graphics, developers can optimize their efforts and devote their attention to the training issues and to the interface with the students. From the student point of view, they can rely on the same interface design and the same 3D environment throughout the various training modules. Moreover, they are faced with a 3D environment similar to the one they will face later when using 3D production or editing software, for example. Each module is dedicated to a single concept, and the module interface is restricting access to the functionalities of the 3D scenegraph to the ones required to operate this concept. Thanks to the 4-areas concept, students can focus their efforts on the concept they have to learn, and by iteratively tuning the control parameters, they can simultaneously visualize the effects onto the 3D scene and the 2D display and correlate these experiences to the theory of the concept.

## ACKNOWLEDGMENT

This work is being developed for the project "Understanding 3D" and is supported by the Swiss Virtual Campus (http://www.swissvirtualcampus.ch/).

## REFERENCES

3DS Max. Retrieved from http://www4.discreet.com/3dsmax/

AlgoViz. Retrieved from http://www.cgv.tugraz.at/people/ullrich/AlgoViz;internal&action=_setlanguage.action

Blender. Retrieved from http://www.blender.org

CGEMS. CG Educational Materials Source. Retrieved from http://cgems.inesc-id.pt/

CGT. Computer graphics tutorials. Retrieved from http://web.sbu.edu/cs/cgtutorials/

Chen, B., & Cheng, H.H. (2005). Interpretive OpenGL for computer graphics. *Computers & Graphics*, *29*(3), 331–339.

Cunningham, S. (1999). Reports of the graphics and visualization education 99 workshop. Retrieved from http://www.eg.org/WorkingGroups/GVE/GVE99

Cunningham, S. (2000). Re-inventing the introductory computer graphics course: Providing tools for a wider audience. *Computers & Graphics, 24*(12), 293–296.

Cunningham, S. (2001). Lessons from scene graphs: Using scene graphs to teach hierarchical modeling. *Computers & Graphics, 4.*

DirectX. Retrieved from http://www.microsoft.com/windows/directx/default.mspx

Exploratories. Retrieved from http://www.cs.brown.edu/exploratories/home.html

GL4Java. Retrieved from http://gl4java.sourceforge.net

Hangzhou, C. (2006). Lecture notes in computer science. Berlin: Springer-Verlag.

Hanisch, F., & Straßer, W. (1996-2003). GRIS/ILO interactive learning objects. WSI/GRIS University of Tuebingen, Germany. Retrieved from http://www.gris.uni-tuebingen.de/projects/ilo

Hanisch, F., & Straßer, W. (2002). *Highly interactive Web-based courseware.* Proceedings of the CGE02 - Eurographics/SIGGRAPH Workshop on Computer Graphics Education, 31–35.

Hitchner, L.E., & Sowizral, H.E. (2000). Adapting computer graphics curricula to changes in graphics. *Computers & Graphics, 24*(2), 283–288.

HyperGraph. Retrieved from http://www.siggraph.org/education/materials/HyperGraph/hypergraph.htm

ICG. Introduction to computer graphics. Retrieved from http://www.cgv.tugraz.at/CGTutorial/CGTutorial.html

ILO. Graphical-interactive systems, interactive learning objects. Retrieved from http://www.gris.uni-tuebingen.de/projects/ilo/index.html

Nate Robins' OpenGL Demos in Ch. Retrieved from http://iel.ucdavis.edu/projects/chopengl/demos.html

OCGn. Online computer graphics notes. Retrieved from http://graphics.idav.ucdavis.edu/education/GraphicsNotes/homepage.html

OpenGL. Retrieved from www.opengl.org

OpenGL tutors. Retrieved from http://www.xmission.com/~nate/tutors.html

OpenSceneGraph. Retrieved from www.openscenegraph.org

Peternier, A.., Thalmann, D., et al. (2006). Mental vision: A computer graphics teaching platform. Edutainment.

Ullrich, T., & Fellner, D. (2005). Computer graphics courseware. Eurographics Education Papers.

Van Gumster, J. (2003). *Blender as an educational tool.* Proceedings of the International Conference on Computer Graphics and Interactive Techniques.

Yang, L., Mustafa Sanver, M. (2002). Web-based interactive 3D visualization for computer graphics education. *ICWL*, 423–432.

*This work was previously published in International Journal of Information and Communication Technology Education, Vol. 4, Issue 3, edited by L. Tomei, pp. 20-30, copyright 2008 by IGI Publishing (an imprint of IGI Global).*

# Chapter 11
# Evaluating WebCT use in Relation to Students' Attitude and Performance

**Lamis Hammoud**
*Brunel University, UK*

**Steve Love**
*Brunel University, UK*

## ABSTRACT

*This chapter presents and discusses the results of a study we carried out to investigate students' attitude and performance to using a managed online learning environment known as WebCT. The chapter starts off with an overview of the literature in this area of research, including a definition of the main technical terms referred to in the research literature. The chapter then goes on to provide a detailed description of the study set-up and presents the main findings obtained from this study. The results are then discussed in relation to previous findings in the research literature.*

## BACKGROUND RESEARCH

The World Wide Web (WWW) provides great opportunities for creating virtual classrooms (Mazza and Dimitrova, 2004) and for building integrated learning and teaching environments (Cheng and Yen, 1998) and it also helps in supporting traditional educational methods (Kalifa and Lam 2002). Teaching and learning are no longer limited by place or time (Kalifa and Lam 2002). The WWW and the internet are considered to be important new methods for delivering online courses (Jiang and Ting, 2000; Lee and Shih, 2001). The interest in web-based learning and technology to support learning is increasing in higher education and this can be seen in the large number of publications in higher education journals in this area (Hoskins and Hooff, 2005; Bower, 2001). Large numbers of educational institutions are offering web-based courses (Owston, 2000) or starting to use course management systems such as WebCT or Blackboard (Mazza and Dimitrova, 2004)

Interaction is central in teaching and learning; the learning process is based on student interaction with instructors, other students, and with the course content (Lei et al., 2003). At the same time, communication and collaboration between the students and instructors can be enhanced by the internet and WWW (Cheng and Yen, 1998).

DOI: 10.4018/978-1-60566-936-6.ch011

## Technical Definitions

A learning platform is defined in a publication from the British Educational Communications and Technology Agency (Becta) as "a generic term to describe a broad range of ICT systems which are used to deliver and support learning. A learning platform usually combines several functions, such as organising, mapping and delivering curriculum activities and the facility for learners and teachers to have a dialogue about the activity, all via ICT. So, the term learning platform can be applied to a virtual learning environment (VLE) or to the components of a managed learning environment (MLE)." (Becta, 2005). The same report defines virtual learning environment and managed learning environment as follows:

*"A virtual learning environment (VLE) is a software tool which brings together resources for curriculum mapping, delivery, assessment, tutor support, communication and tracking."*

*"A managed learning environment (MLE) refers to the whole range of information systems and processes that support learning and the management of learning within an institution. It includes VLEs or other learning platforms, administrative and other support systems." (Becta, 2005) Web-enhanced courses are traditional face-to-face courses which include web-related materials. Web-enhanced courses usually adopt a course management system (e.g. WebCT) (Sivo et al, 2007).*

Web-based learning is a main subcomponent of the broader term e-learning. There are two general types of e-learning which are technology-enhanced learning and technology-delivered learning. Technology-enhanced learning means that students have regular face-to-face meetings with the teacher. Here the traditional face-to-face class is the basic forum for learning, and the technology may make available learning materials online before they are delivered in the class. Technology-delivered learning means students and teachers are not at the same place also it is referred to as distance learning.

## WebCT

WebCT (Web Course Tools) was developed by Murray Golderg, a faculty member at the University of British Columbia (Burgess, 2003; Volery and Lord, 2000). WebCT is an integrated set of educational and management tools and an important provider of e-learning programs. It is specifically used for the design and development of teaching and learning materials. WebCT is mainly used to create sophisticated World Wide Web-based educational environments either by creating entire online courses, or simply by publishing materials that supplement existing courses. Users of WebCT do not need a lot of technical expertise as all content is accessible via a standard Web browser (Volery and Lord, 2000).

Mazza and Dimitrova (2004) stated that course management systems create large log data which contain students' activities in a web-based course. These systems also contain built-in student monitoring features. The instructor can view statistical data about students' use of course pages such as a student's first and last login, the history of pages visited, the number of messages the student has read and posted in discussions, marks achieved in quizzes and assignments, etc. The instructor can use this information to observe students' progress and to discover potential problems.

Students' activities on Web-based course can be measured in the following ways:

- *WebCT* Hits: The number of times each student accessed each page such as homepage and content page.
- *Time:* how much time each students spent exploring a page (such as content page) or using tools (such as quiz or calendar)

- *Communication board* use: This can be measured in two main ways.
    1. Articles Read: The number of articles each student read on the communication board.
    2. Articles Posted: The number of articles each student posted on the communication board.

## Why is Online Education Important?

The rapid development of the internet and WWW technologies enable the building of integrated learning and teaching environments (Cheng and Yen, 1998). Examining contemporary learning theories Cheng and Yen found that the educational focus is shifting from being teacher-centred to being student-centred. They believe that communication and collaboration between students and instructors can be enhanced by the internet and WWW. They also state that interactive and collaborative learning should be the main aim of using the internet and WWW technologies in education. Kalifa and Lam (2002) stated that information and communication technologies help in supporting traditional educational methods and facilitate new methods of teaching and learning. In addition, teaching and learning are no longer limited by place or even time. Universities and organizations use new technology to offer on-line training and courses. A large number of web-based courses are available on the worldwide web (Kalifa and Lam, 2002). The quick development of the internet and WWW provides an important resource for people to easily gain access to various types of information and knowledge. Furthermore the WWW is useful in delivering education because of its use of multi-media, and short response time, etc. (Lee and Shih, 2001). They also said that learners' performance and interest can be improved by using a well-designed World Wide Web (WWW) learning environment. The large number of publications in higher education journals about e-learning and technology to enhance learning indicates the importance of web-based learning (Bower, 2001). The importance of web-based learning has been illustrated in many studies (Hoskins and Hooff, 2005). There are a growing number of institutions offering web-based courses (Owston, 2000) and there are many reasons why online education is important for institutions. These reasons were summarised by Volery and Lord (2000) into four broad categories:

- *Expanding access*: most institutions need to expand access to educational material to meet the needs of individuals in learning and training.
- *Alleviating capacity constraints*: the number of the students is growing more than universities can accommodate. So, online education is seen as a solution to this problem.
- *Capitalizing on emerging market opportunities*: there is a growing acceptance towards lifelong learning among people outside the traditional 18-24 age range. Many institutions are seeking to benefit from the large number of new learners.
- *Serving as a catalyst for institutional transformation*: institutions have a challenge to adapt to a decrease in public funding and increasing competition for students which could be catered for by online education.

## Factors Affecting Students' Attitude and Achievement on Web-Based Courses

Technology has the possibility to enhance and transform teaching, but it can also be used incorrectly or in ways that may interfere with learning so it is important to know how we can achieve effective learning online (Salter, 2003). Different ways can be used to measure the effectiveness of web-based courses. Therefore studies in distance education differ in what they use as evidence of online course effectiveness. For example, Volery and

Lord (2000) collected data from students enrolled in a Global Business course in which WebCT was used to deliver the course materials. They investigated factors that could affect the online course delivery such as: ease of access and navigation, interface, interaction, attitudes towards students, instructor technical competence, and classroom interaction. Then they categorised these under the three success factors as follows:

- *Technology:* WebCT is easy to access and navigate. Students can log in at any time day or night. Also, WebCT is well-designed and structured. This is very important for the students who spend a long time on the site. WebCT has tools which enable students and the instructor to interact and communicate such as the communication board.
- *Instructor characteristics*: there is a significant relationship between teaching effectiveness and instructor characteristics. Instructors' personal approach and their ability to motivate the students in both the classroom and on WebCT are important factors affecting their teaching effectiveness. Furthermore, instructor familiarity with the technology and their ability to use the internet are also considered to be important factors affecting their teaching effectiveness.
- *Student characteristics:* the only student characteristics influencing teaching effectiveness were found to be their previous use of WebCT.

In another study, Jurczyk et al. (2004) found that students' attitude can change during a web-based course. Hisham et al. (2004) stated that there are many factors that can affect student satisfaction with asynchronous e-learning systems (they used WebCT in their study). They said that personalised feedback is an important factor for a successful asynchronous e-learning system. Another factor affecting students' satisfaction is a supportive learning community which can be achieved by the use of tools such as discussion boards. A suitable interface was also found to be another factor which may influence students' satisfaction because a well-designed interface gives students the opportunity to easily access the content. Arbaugh and Duray (2002) found that a large class size has a negative relationship with online learning and course satisfaction. Flexibility of delivery positively affected students' learning and satisfaction. Students who have previous experience in using the internet and on-line courses were found to be more satisfied with the course delivery medium.

One of the factors found to affect students' attitude towards web-enhanced courses is their module leader's attitude towards using the system (Hammoud et al, 2008). The relationship between the students' use of a web-enhanced course (WebCT in their study) and their performance is significant and important. They found that students have a positive attitude towards using WebCT in their courses. They concluded that the module leader's way of using WebCT and their attitude towards WebCT had affected the students' attitudes and performance. However, there is no strong evidence to confirm that the students' marks were affected by their module leader's way of using WebCT.

## Findings from Research on Web-Based Learning

Tian (2001) described the World Wide Web as a vehicle to develop interactive learning and teaching applications. He said that the web is an important tool to facilitate education. Interactive web pages are essential in the learning and teaching process. Therefore, a script language must be used in designing such web pages. Based on his experience and the feedback from multi-choice questions from students, Tian identified six main issues to be considered in designing a computer based course: teachers, students, knowledge, evaluations, communications and the enabling technology.

Rivera et al. (2002) compared students enrolled voluntarily in the same course but different sections. The traditional section is face-to-face lectures and all the assignments and the exams are undertaken in the traditional way. The web-based section depends on the web to deliver the course materials and the assignment and the exams are similar to the traditional way but are all undertaken online. The third section is the hybrid section which is a mixture of the first two sections. They used WebCT to deliver the course materials and at the same time the students attended lectures. For all three sections, the exams questions are taken from the same test bank. By comparing the three groups they found that students' performance was not affected by the way of delivering the materials. However, it has an impact on student satisfaction. They found that the traditional and hybrid students were more satisfied with their courses than the web-based students.

In a study based on postgraduate students at a Malaysian university, Hong et al. (2003) explored students' perception of and success in a web-based learning environment. They chose problem-based learning to implement their study. Problem-based learning is a student-centered instructional approach in which students collaboratively solve problems and reflect on their experiences. In such an approach teachers take on the role as "facilitators" of learning. Hong et al. (2003) compared the differences between the outcomes of a web-based course and a face-to-face version of the same course. They found that most students were satisfied with their web-based learning experience. The students found the web-based course flexible because they could learn anytime and anywhere. A few students felt isolated and needed face-to-face lectures. Developing the students' computer skills was found to be an important aspect supporting students' success and improvement in a problem-based course. Finally the researchers recommended designing clear structures to guide students studying a problem-based module in a web-based environment.

Huifen et al. (2002) studied the development of a web-based course which was an electronic copy of an existing course. Students could choose between a face-to-face learning method and a web-based learning method. The students' results were then compared. In 2000, the web-based group performed better than the face-to-face group. In 2001, although most of students did not have any previous web-based learning experience; they wanted to have more web-based courses. Huifen et al. (2002) stated that the impact of student-student and student-instructor interaction through web-based learning environment is an important issue. The instructor's supervision and communication with students can not be replaced completely by communication and interaction tools through the web-based learning environment. Therefore, the new relationship should be enhanced by the instructors. They should talk with students online more actively and encourage students to participate more in these online discussions. For example, they can respond online to every student's questions which may encourage other students to participate.

In another study, Nageswaran et al. (2000) set out to investigate students' attitude to modules which were supported by WebCT. WebCT was used to enhance and support the traditional classroom. They said that WebCT is a very good supplementary tool for a traditional classroom, especially for courses with large numbers of students. Students in their study considered that supplementing classroom teaching with WebCT is better than replacing it. The researchers found that students have to work collaboratively in order to achieve good understanding of the information on the web which may be promoted by using emails and chat tools between students.

Following on from this, Storey et al. (2002) evaluated the usability of WebCT and blackboard by collecting data from a survey given to set of students during course time. The results showed students' satisfaction with using web-based tools was related to its perceived convenience and flexibility. Students liked being able to access information any time and any place and the way web-based tools supported their learning styles. Storey et al. stated that Web-based learning tools are developing the learning needs and supporting the traditional way of teaching, as well as offering a new way of delivering education.

In another study focussing on student perception of online learning, Hoskins and Hooff (2005) discussed two important questions in examining online learning: "(1) Which students voluntarily utilise web-based learning; and (2) Does this use influence their academic achievement?" p. (177). They observed 110 undergraduate students of different ages and both genders. The students used WebCT to support their course. The results showed that older students accessed WebCT more, spent longer time on it, and used the notice board more frequently than younger students. Males used the chatting dialogue facility more than females. Finally, the results showed that the age and the gender of the learner had a considerable role in determining students' use of web-based learning. They found that there is a relationship between using the discussion board and the students' achievement. Students who posted messages on the discussion board got better grades than those who did not post or posted fewer messages. They considered this finding to be important and they stated that more research is needed to confirm their result and to find the relationship between using specific aspects of an online environment and students' achievement.

Sayers et al. (2004) compared students' performance with and without the support of WebCT. They studied two different groups of students enrolled in the same module in two different years. The authors thought that an on-line assessment could have unfavourable affect on the students' end of semester examination grades; however their results indicated that on-line assessments do not necessarily have a detrimental effect on students' end of semester examination results. In this study students had the traditional end of semester exam and two on-line multiple choice tests delivered by WebCT. The comparison showed that the students who used WebCT achieve slightly better results than the previous year students who did not have WebCT.

## OUR STUDY

This study was conducted at Brunel University. All undergraduate and taught postgraduate courses delivered by the School of Information Systems, Computing and Mathematics at Brunel University are supported by WebCT. Each course consists of a number of modules (a module is an individual unit of study). For each module, students have face-to-face lectures and they can access module resources by using WebCT. WebCT has a communication board that allows the students to communicate with each other and with the module leader to ask any questions they have in relation to the module. Students can also get all the information about the assignments, workshops and making schemes for every module on WebCT. Students are also required to submit their assignments via WebCT.

### Aims and Objectives

The aim of this study was to assess the effectiveness of WebCT in students' academic performance. The study also aimed to assess students' attitude towards using WebCT in their courses.

### Participants

The population of the study consisted of 303 undergraduate students in the Department of Information Systems and Computing. 172 students participated in the first part of the study. 131 students participated in the second part of the study.

Three lecturers in the department also participated in the study. The students participating in the study were between 19 and 23 years-of-age.

## Research Instruments

A mixture of qualitative (e.g. interview) and quantitative (e.g. questionnaire) methods was used in this study. The objective data for this research was collected from the WebCT tracking system data base. There are two main measures of students' use of WebCT; WebCT page hits and bulletin board use. WebCT page hits is the number of times every student accessed each page such as homepage, content page (a module resources page which contains lecture notes). Bulletin board use is the number of messages each student read or posted on the discussion board.

As already stated, this study aimed to assess students' attitude and performance in a web-enhanced course. This information was obtained from a questionnaire specifically designed to measure students' attitude toward WebCT. For the questionnaire, a five point Likert-type scale was used.

The statements were designed to give information about students' general attitude toward WebCT and its effectiveness on their performance. The influence of the communication board was considered to be an important issue that needed to be evaluated in this study because in a similar study Hoskins and Hooff (2005) found that the discussion board can influence students' achievement. Three statements in the questionnaire were designed to assess students' opinion about the effectiveness of the communication board on their interaction with the module leader and with each other and on their achievement in the module.

The questionnaire also contained open-ended questions to gather information about the advantages and disadvantages of using WebCT, WebCT problems, and students' general comments. Some students used the open-ended questions to explain their responses to previous questions. Examples of open-ended questions used in the questionnaire are as follows:

- What advantages do you see in having WebCT in your module? Please provide one or two specific examples.
- What problems have you experienced in using WebCT in this module? Please give specific examples (if any).

This study used the interview as a qualitative method. Interviews were conducted with the module leaders and were designed in a semi-structured format. The first interview was designed to get background information on the lecturer's experience in using WebCT and his expectations of using it in the studied module. The second interview aimed to gather information about three main topics:

- The lecturer's general thoughts, attitude and experience of using WebCT.
- Specific information about the effect of using WebCT on the learning process and its influence on students' performance and on the lecturer's way of teaching.
- The communication and interaction between the students and the lecturer via WebCT communication software.

## Procedure

The study was divided into two parts.

1. Part one: Data was collected from students enrolled on an Information Systems course. In this part of study, students' activities on WebCT were observed in one module. Moreover, students' attitude towards using WebCT in that module was measured using the questionnaire.
2. Part two: Data was collected from students enrolled on an Information Systems course. However, in this part of the study another

group of students was observed. This time the students' activities on WebCT were observed in two different modules. Also, their attitude towards using WebCT to support their modules was measured using the questionnaire.

Both parts of the study followed the following steps:

- Numerical data about students' activities on WebCT was collected weekly.
- The questionnaire was completed by the students at the end of the term during their lab sessions.
- The three module leaders were interviewed twice; at the beginning of the course and at the end of the module.

## Data Analysis

The interviews were taped and transcribed. The students' responses to open-ended questions were analysed and compared to the interviews responses given by the lecturer.

Students' general uses of WebCT were measured by the number of times each student visited WebCT pages or used the discussion board for the observed modules. Students' achievement was measured by their grades. The observed modules were assessed by coursework and written exam. Students' attitudes towards WebCT were measured by using a Likert scale. The data was analysed using SPSS software.

Frequency measures were used to analyse the numerical data which was obtained from the questionnaire. Qualitative data from the lecturer's interview and the students' comments were categorised into main ideas, and analysed by comparing various meanings; then the results were compared with the results obtained from the statistical data which were gathered from the same sample.

The measures of students' academic achievement in the module were correlated (*Pearson's Product Moment Correlation Coefficient*) with the measures of WebCT use. The relationship between the students' achievements and their use of WebCT was also analysed.

## RESULTS

### Results Obtained From the Tracking System During Part One of the Study

Overall, the results from the study indicated that students used WebCT a lot in the observed mod-

*Table 1. Students' visits and use of WebCT pages and tools of the observed module*

| | Number of students | Min use | Max use | Mean |
|---|---|---|---|---|
| Total use of WebCT | 172 | 5 | 1294 | 292.67 |
| Visiting different pages in module resources page | 167 | 0 | 30 | 11.69 |
| Visiting homepage | 172 | 2 | 493 | 70.84 |
| Visiting the organizer | 171 | 0 | 344 | 29.81 |
| Visiting content page | 167 | 0 | 177 | 29.76 |
| Notes | 25 | 0 | 6 | 0.24 |
| Assignment page | 171 | 0 | 150 | 25.52 |
| Calendar | 154 | 0 | 10 | 2.67 |
| Reading mssages | 154 | 0 | 284 | 113.56 |
| Posting mesages | 52 | 0 | 29 | 0.92 |

ules. Most of the students used various pages such as home page, content page, organizer page, and assignment page. The discussion board was also well used; 257 messages were posted on it and 154 students out of 172 visited it regularly, as is shown in Table 1.

Pearson correlations were carried out to investigate the relationship between students' grades in the module and their use of different pages of WebCT for that module.

A positive but weak significant correlation (r=0.236, p =0.002) was found between students` overall grades, and the students' total use of WebCT. Interestingly, no correlation was found between the students' total use of WebCT and their grades for the first task. However a weak positive but significant correlation (r= 0.263, p=.001) was found between the students' total use and their grades for the second task.

The number of different pages each student visited from the module resources was found to have a correlation with students' overall grades. A weak but significant positive correlation (r= 0.181, p= 0.017) was found between the number of different pages each student visited in the module resources and their overall grades. A slightly higher correlation was found between the number of different pages each student visited in module resources and their grades for the second task (r= 0.223, p= 0.003). However, no correlation was found between the number of different pages each student visited in module resources and their grades for the first task.

A weak but positive significant correlation (r= 0.260, p= 0.001) was found between the number of messages each student read via the communication board and their overall grades. There was also a positive significant correlation found between the number of messages each student read via the communication board and first task grades (r= 0.163, p= 0.033), and a slightly higher correlation found for task two grades (r= 0.234, p= 0.002).

There was a weak positive significant correlation (r= 0.174, p= 0.022) between number of messages each student posted on the communication board and their overall grades. However, no significant correlations were found between numbers of messages each student posted on the communication board and their grades for first and second tasks.

In addition, the number of hits for the home page was found to have no significant correlation with students' overall grades. Although, the number of hits for the home page had a weak positive significant correlation (r= 0.215, p= 0.005) with the students' grades on the second task but not with the first task.

## Results Obtained From the Tracking System During Part Two of the Study

The results obtained from the tracking system indicated that students frequently used WebCT on the two modules. Students visited all the main pages such as: home page, content page, organizer, assessment page, and communication board. We referred to the modules as module A and module B. The mean of students' total visits to WebCT of module A is 329 hits, and for module B is 356 hits. The following figure shows the means of the numbers of hits which represent students' total access to each page in each module.

Also figure 2 shows the means number of students' total access to WebCT of each module grouped by weeks.

A positive but weak significant correlation (r=0.39, p<0.001) was found between students' final grades, and "read" for module B. also "read" was found to be significantly correlated with exam grades (r=0.348, p<0.001) and the coursework grades (r=0.294, p=0.001). A positive but weak significant correlation (r=0.237, p=0.006) was found between students' final grades, and "post" for module B. also "post" was found to be significantly correlated with exam grades (r=0.202, p=0.021) and the coursework grades (r=0.197, p=0.024).

*Figure 1. Means of students' total access to WebCT for modules A & B classified by pages*

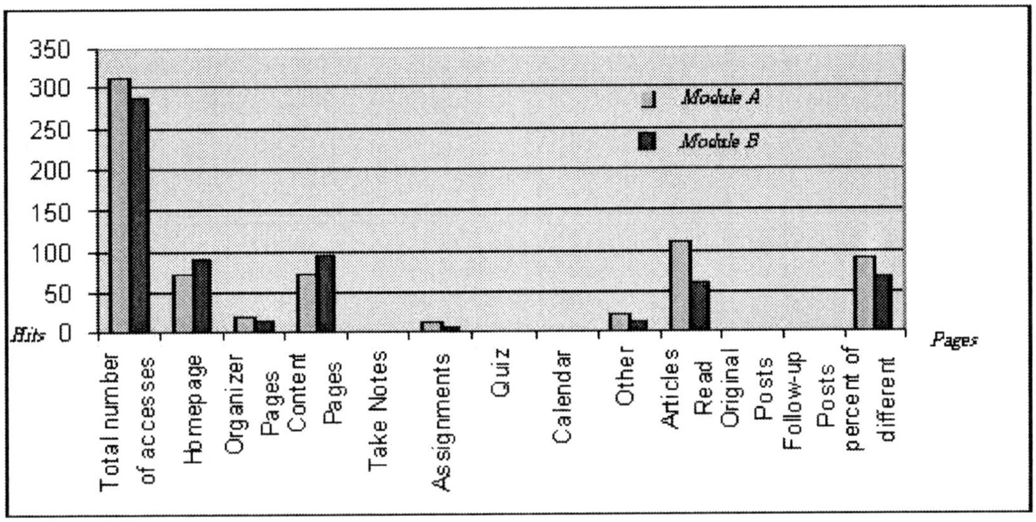


*Figure 2. Students' total access to WebCT for modules A & B divided into weeks*

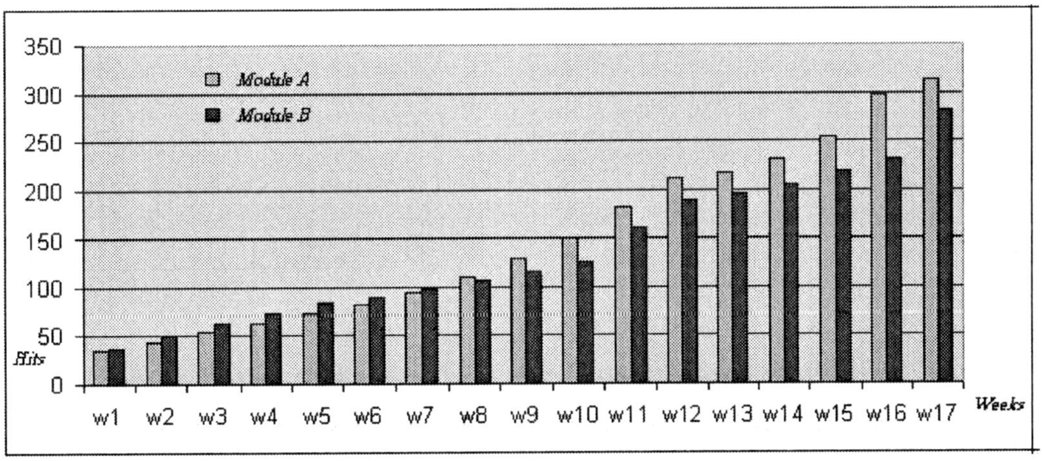


## Questionnaire Results

The majority of the students had experience of using the Web, with 20% percent of students indicating that they had between 3 to 5 years experience. The results obtained from the questionnaire can be summarised as follows:

- The results showed high agreement that WebCT is easy to use.
- There was strong agreement among students that WebCT helped them to cover course content in more detail.
- Students were not sure that WebCT helped them to obtain good grades.
- Students enjoyed using WebCT.
- There was strong agreement amongst students that the communication software in WebCT allowed students to interact more

directly with the module leader and with other students in the module.

- The respondents stated that in comparison to modules that did not use WebCT; they were able to participate more regularly and actively in this module.
- There was agreement among the students that WebCT does not need a lot of improvement.
- The majority of the students did not have technical problems using WebCT.
- Students believed that they were in control of their learning because of the flexibility of using WebCT anytime and from anyplace.
- The students stated that they would recommend using WebCT to support all the courses at Brunel University.

## WebCT Advantages

- WebCT makes the module resource constantly available and up to date.
- WebCT is flexible, students can access to the module resources any time from any place.
- Clearly designed so the student can access different information easily.
- Quick response to all kind of questions via the discussion board.
- Easy to communicate with the module leader and other students in the module.
- Submission of coursework is easier.
- Helps the whole learning approach.

## WebCT Disadvantages

- Students reported their worries about the server slowing down or breaking down if they are submitting coursework at the deadline
- Quality of content depends on what the module leader puts on it.
- Some content is not easily understood
- Face-to-face feedback is quicker
- Discussion board need a lot of improvement
- Content needs to be better organised to give all the modules' timetable and deadlines in one area of WebCT.
- Interface could be improved,
- More tests created by lecturer for students to use are needed.

## DISCUSSION

The results obtained from these studies indicate that students had positive attitudes towards using WebCT as a web-based tool supporting their learning needs. In general, they agreed with statements such as "WebCT helped me to achieve the learning outcome for this module", "WebCT for this module was easy to use". The students' satisfaction and appreciation of web-based course materials can be explained by their familiarity with the technology, and the flexibility of WebCT (i.e. it can be used anytime anyplace). One of the students commented: "I have used WebCT before so I don't need help to use it." The results of students' satisfaction and appreciation of web-based course can be found in previous studies such as Arbaugh (2002).

The questionnaire results showed a high agreement between students that WebCT helped them to cover the module in more detail and also their appreciation of the availability of module resources that were up to date. This corresponds to a module leader stating that his main use of WebCT was to post lecture notes and useful links. This finding is similar to the findings of Motiwalla and Tello (2000) and Wegner et al. (1999); they said flexibility of accessing a Web-based course at anytime and from anywhere is appreciated by the students. One student commented, "*using WebCT keeps me up to date with any new information in the module content.*" Students' satisfaction and appreciation of web-based courses can be found in previous studies such as (Minton and willett 2003; Matuga 2001; Jurczyk et al. 2004; Collins 2000; Hong et al. 2003; Lee and Shih 2001).

All the students in this study are computer science students. They did not mention any technical problems using WebCT. Moreover there was strong agreement from the students that WebCT is easy to use. However, the students reported that WebCT sometimes crashed and they were worried about the server slowing down or breaking down if they are submitting at the coursework deadline. Hong et al. (2003) stated that computer skills are found to be an important aspect for students' improvement in web-based courses. Computer science students accept WebCT more than other students because they are more familiar with the technology and they can solve the technical problem they may face (Burgess, 2003). One of the students stated: "*I like the fact that WebCT has the same layout for all modules. That makes it easy to understand and as a level two student I became used to it after the second or third use.*"

Most of the students believed that they were in control of their learning. The availability of the modules' resources online allowed students to access the learning material anytime from anyplace which is one of the important factors affecting students' learning. One of the students commented: "I am in control of my learning because I can look at lecture slides to prepare for lectures." Another student stated: "WebCT refers to the study guide for learning requirements to pass the module."

The most important point in this study is to assess the relationship between students' use of WebCT and students' achievement. One of the students commented: "*getting good grades depends on what the module leader put on WebCT.*" Another student stated: "*using WebCT does not mean that you will pass the exam.*" At the same time a module leader stated that students who did not attend the lectures may pass the exam but he thinks they will not achieve good grades if they just depend on WebCT alone. In addition, he also stated that WebCT is just a good supporting tool to complement face-to-face lectures and that has no affect on students' achievement. This is similar to Nageswaran et al. (2000) who said that WebCT is a very good supplementary tool for traditional learning. The qualitative data concluded that the materials delivered through WebCT are important and affect the students' learning which is similar to the result of the study by Kalifa and Lam (2002). Similarly, Felix (2001) found that the quality of the delivered information is essential and instructor has to be sure of the level of the material quality going online. Neither the students nor the module leader were sure whether the use of WebCT can affect learning or influenced their achievement. Analyzing the data from the tracking system showed a very interesting result regarding the students' achievement. First of all, a significant relationship was found between the total use of WebCT (hits) and the students' grades, which suggests that students who visit WebCT more get better grades. Visiting different pages within the module resources was found to have a positive relationship with student grades and with their grades for the second task. This result is similar to the findings of Sayers et al. (2004). They found that the students who used WebCT got slightly better grades than those who did not use it and they also found in the same study that WebCT does not have a negative effect on written exam performance.

The only communication tool used in WebCT was the communication board; there was no email tool within WebCT for the observed modules. The data from WebCT tracking system showed that the discussion board was highly used in the modules. Students used the communication board to interact with their module leaders and each other. The questionnaire result showed a good agreement that the communication software enabled the students to interact more directly with the module leader and other students in the module (a similar point was illustrated by the module leaders in the interviews). Many of the students' valued the quick response to their questions. Students appreciated the fact that they could ask any question using WebCT. This can be explained by the fact that that some students feel shy when asking questions face-to-face, due to peer pressure or the large number of the students in the class. The students and the module leaders found the communication board to be a useful tool; however, one of the module leaders stated that the use of discussion board did not have any influence on students' achievement or affect their learning. The module leader commented: "*I think that the availability of the communication board might help the students to ask questions that they will not feel able to ask in face-to-face situations and it may allow people to see the questions that have been asked and the answers that have been given. However, I really don't think it has a huge impact on student learning.*" In contrast to these thoughts, the results from the tracking system showed a positive correlation between the number of messages students read on the communication board and their achievement. Students who read more messages got better grades. These results correspond to the findings of Hoskins and Hooff (2005) who stated that it was extremely promising to find that the use of dialogue can influence the students' achievement in assessed coursework.

More that half of the students agreed that they were able to participate more regularly and actively in this module in comparison to modules that do not use WebCT. This backs up the findings of Kalifa and Lam (2002) who stated that learner interaction with the web-based course is the most important aspect of the learning process.

Some students found it hard to find the information they needed on WebCT and this depended on what the module leader posted in it. One student commented that "*I find it quite easy to use WebCT. However, sometimes it is hard to locate a specific file or to find what you are looking for.*" Also some students stated that WebCT needs some improvement; such as having one file to show all important dates and timetables for all modules they are studying instead of separating the modules, as the current situation holds. One of the students commented: "*I believe nothing is perfect and that there is always room for improvement, so in a sense it can be improved, however I disagree that there needs to be 'a lot' of improvement.*" Also, a module leader found it frustrating sometimes because it looks simple to do some things but when he tried to do it he found it complex. As an example, there is no direct way to post a link in some pages and he must write a script which he copied and pasted every time he wanted to create a link. These comments and the response from students that it was difficult to find wanted information on WebCT emphasises the importance of having a clear structure for all users; which was recommended by Hong et al. (2003). Training for module leaders may also be needed as this affects their way of posting information, which is considered to be important aspect affecting students' achievement.

## CONCLUSION

Overall, the results obtained from this study are encouraging. Finding a positive relationship between students' achievement and attitude and their activities on WebCT can be considered promising. Academically, this study can be used as a first step for similar studies to evaluate the use of WebCT. Practically, the findings that the use of a communication board may influence the students' achievement should be taken seriously. Therefore communication tools should be developed in order to satisfy students' needs and encourage them to use them. Future research can be carried out to evaluate the use of communication tools in relation to students performance; such research should focus on the levels of posted messages not just the number of messages.

## ACKNOWLEDGMENT

We would like to acknowledge the contributions made by our colleagues Sherry Chen and Lynne Baldwin to the journal paper on which this book chapter is based.

## REFERENCES

Arbaugh, J. B., & Duray, R. (2002). Technological and structural characteristics, student learning and satisfaction with Web-based courses. An exploratory study of two on-line MBA programs. *Management Learning, 33*(3), 331–147. doi:10.1177/1350507602333003

Becta (British Educational Communications and Technology Agency). (2005). *An introduction to learning platform.* Retrieved from http://www.becta.org.uk/leas/display.cfm?section=1_2

Bower, B. L. (2001). Distance education: Facing the faculty challenge. *Online Journal of Distance Learning Administration, 4*(2). Retrieved March 22, 2006 from http://www.westga.edu/%7Edistance/ojdla/summer42/summer42.html

Burgess, L. A. (2003). WebCT as an e-learning tool: A study of technology students' perception. *Journal of Technology Education, 15*(1), 6–15.

Cheng, C. Y. Y., & Yen, J. (1998). Virtual Learning Environment (VLE): A Web-based collaborative learning system. In *Proceedings of the Thirty-First Annual Hawaii International Conference on System Sciences, January, 6-9, 1998, Hawaii* (pp. 480-491). IEEE.

Collins, M. (2000). Comparing web correspondence and lecture versions of second-year non-major biology course. *British Journal of Educational Technology, 31*(1), 21–27. doi:10.1111/1467-8535.00132

Felix, U. (2001). A multivariate analysis of students' experience of web-based learning. *Australian Journal of Educational Technology, 17*(1), 21–36.

Hammoud, L., Love, S., & Brinkman, W. P. (2008). The affect of lecturers' attitude on students' use of an online learning environment. In *Proceedings of the 15th European conference on Cognitive ergonomics: the ergonomics of cool interaction ECCE, September 15-19, 2008, Funchal, Portugal.* (ACM International Conference Proceeding Series; Vol. 369)

Hisham, N., Campton, P., & FitzGerald, D. (2004). A tale of two cities: A study on the satisfaction of asynchronous e-learning system in two Australian universities. In *Proceedings of the 21st ASCILITE Conference, December 5-8, 2004, Perth, WesternAustralia* (pp. 395-402).

Hong, K. S., Lai, K. W., & Holton, D. (2003). Students' satisfaction and perceived learning with a web-based course. *Educational Technology and Society, 6*(1), 116–124.

Hoskins, S. L., & Hooff, J. C. V. (2005). Motivation and ability: Which students use online learning and what influence does it have on their achievement? *British Journal of Educational Technology, 36*(2), 177–192. doi:10.1111/j.1467-8535.2005.00451.x

Huifen, L., Xin, H., & Xinyu, Z. (2002). Human communication and interaction in web-based learning: a case study of the digital media Web course. In *Proceedings of the International Conference on Computers in Education (ICCE'02), December 3-6, 2002, Auckland, New Zealand.* (pp. 217-221).

Jiang, M., & Ting, E. (2000). A study of factors influencing students' perceived learning in a web-based course environment. *International Journal of Educational Telecommunications, 6*(4), 317–338.

Jurczyk, J., Benson, S. N. K., & Savery, J. R. (2004). Measuring student perception in web-based courses: a standard-based approach. *Online Journal of Distance Learning Administration, 7*(4). Retrieved March 22, 2006 from http://www.westga.edu/%7Edistance/ojdla/winter74/winter74.htm

Kalifa, M., & Lam, R. (2002). Web-Based learning: Effects on learning process and outcomes. *IEEE Transactions on Education, 45*(4), 350–356. doi:10.1109/TE.2002.804395

Lee, W. I., & Shih, B. Y. (2001). A case study of motivation theory on the web-based learning. In *Proceedings of the ASEE/IEEE Frontiers in Education Conference, October 10-13, 2001, Reno, NV, USA* (pp. 23-27).

Lei, X., Pahl, C., & Donnellan, D. (2003). An evaluation technique for content interaction in Web-based teaching and learning environments. In *Proceedings of the 3rd IEEE International Conference on Advanced Learning Technologies (ICALT'03), July 9-11, 2003, Athens, Greece* (pp. 294- 295).

Matuga, J. M. (2001). Electronic pedagogical practice: The art and science of teaching and learning on-line. *Educational Technology and Society, 4*(3), 77–84.

Mazza, R., & Dimitrova, V. (2004). Visualising student tracking data to support instructor in web-based distance education. In *Proceedings of the 13th international World Wide Web Conference on Alternate Track Papers & Posters. SESSION: Student tracking and personalization, May 17-22, 2004, New York, USA* (pp.154-161).

Minton, T. M., & Willett, L. S. (2003). Student preferences for academic structure and content in a distance education setting. *Online Journal of Distance Learning Administration, 6*(1). Retrieved March 22, 2006 from http://www.westga.edu/%7Edistance/ojdla/spring61/spring61.htm

Motiwalla, L., & Tello, S. (2000). Distance learning on the Internet: An exploratory study. *The Internet and Higher Education, 2*(4), 253–264. doi:10.1016/S1096-7516(00)00026-9

Nageswaran, N., Dreher, H., & Chang, V. (2000). The use of information technology to support a cast of a thousand. In A. Herrmann & M.M. Kulski (Eds.), Flexible futures in tertiary teaching. In *Proceedings of the 9th Annual Teaching Learning Forum, February 2-4, 2000, Perth: Curtin University of Technology.*

Owston, R. D. (2000). Evaluating web-based learning environments: Strategies and insights. *Cyberpsychology & Behavior, 3*(1), 79–87. doi:10.1089/109493100316256

Rivera, J. C., McAlister, M. K., & Rice, M. L. (2002). A comparison of student outcomes & satisfaction between traditional & web-based course offerings. *Online Journal of Distance Learning Administration, 5*(3). Retrieved March 22, 2006 from http://www.westga.edu/%7Edistance/ojdla/fall53/fall53.html

Salter, G. (2003). Comparing online and traditional teaching- a different approach. *Campus-Wide Information Systems, 20*(4), 137–145. doi:10.1108/10650740310491306

Sayers, H. M., Nicell, M. A., & Hagan, S. J. (2004). Supporting and assessing first year programming: The use of WebCT. *Italics eJournal, 3*(1). Retrieved March 22, 2006 from http://www.ics.heacademy.ac.uk/italics/Vol3-1/sayers/Italics.htm

Sivo, S. A., Pan, C. C. S., & Hahs-Vaughn, D. L. (2007). Combined longitudinal effects of attitude and subjective norms on student outcomes in a web-enhanced course: A structural equation modelling approach. *British Journal of Educational Technology, 38*(5), 861–875. doi:10.1111/j.1467-8535.2006.00672.x

Storey, M. A., Phillips, B., Maczewski, M., & Wang, M. (2002). Evaluating the usability of web-based learning tools. *Educational Technology & Society*, *5*(3), 91–100.

Tian, S. (2001). The World Wide Web: a vehicle to develop interactive learning and teaching applications. *Internet research: Electronic networking, applications and policy, 11*(1), 74-83.

Volery, T., & Lord, D. (2000). Critical success factors in online education, the international. *Journal of Educational Management*, *14*(5), 216–223.

Wegner, S. B., Holloway, K. C., & Gordon, E. M. (1999). The effects of Internet-based instruction on student learning. *Journal of Asynchronous Learning Networks*, *3*(2), 98–106.

# Chapter 12
# Examining the Relationship Between Course Management Systems, Presentation Software, and Student Learning:
## An Exploratory Factor Analysis

**Daria C. Crawley**
*Robert Morris University, USA*

**Barbara A. Frey**
*University of Pittsburgh, USA*

## ABSTRACT

*This research examines the relative impact of students' in-class behaviors (i.e., attendance and participation) by assessing student perceptions of the value of instructional technologies, such as eCollege course management systems and instructors' PowerPoint presentations. The results of the study through exploratory factor analyses revealed that 13 items were divided into three factors (electronic presentations, online-course management, and effective classroom behavior) with 53 percent explained variance in instructional technologies' impact on student learning. ANOVA results indicated significant differences in online-course management and perceived impact of electronic presentations on students' classroom behavior among respondents who used the online-course management system. Respondents who used multiple online-course management features viewed it more favorably and did not believe that it had a negative impact on classroom behaviors, such as attendance or class participation compared to those who used fewer features. Implications for construct refinement and future research are discussed.*

## INTRODUCTION

Over the last decade we have witnessed an increase in faculty using a variety of instructional technologies to share and deliver information, including video conferencing, electronic mail, faculty Web sites, and course management systems (CMS). In their 2005 report, *Growing by Degrees*, Educause reported that of the 890 responding colleges and universities in the United States and abroad more than 90 percent used a CMS. Over 65 percent of higher education institutions offering face-to-face courses also offered courses online (Allen & Seaman, 2005). As the interest in distance education and instructional technologies has grown, formats used to reach out to learners unable to attend traditional classes has also grown. Bates (1995) characterizes this growth as a four-generation process, moving from the basic correspondence course to high bandwidth, multi-media courses. Students have gone from interacting with their instructors and classmates via United States Postal Service paper-based mail and the telephone to interacting via Web-based chat, discussion boards, and audio or video connections which allow for both synchronous and asynchronous dialogues. These technologies provide the opportunity to enhance the learning environment in face-to-face courses as well as distance courses.

This study explores the relationship between student learning, classroom behaviors, and the use of eCollege and PowerPoint in an upper-level course required for all business majors. The course, *International Business*, introduces students to the fundamentals of global business operations. During the fifteen-week semester, the instructor and students meet for three 50-minute sessions each week. An interactive approach comprising a combination of classroom lectures, case discussions, and individual and group activities is used to introduce students to the course material.

CMS technology is used to support student learning and deliver course content. According to the report, *Growing by Degrees*, this type of course is considered "Web-facilitated" and delivers up to 29 percent of the course content online, but the course is essentially a face-to-face course (Allen & Seaman, 2005). Given the nature of the global marketplace, it is critical that the latest economic data (i.e., currency exchange, inflation, and stock market rates), cultural and business practices, ethics and social responsibility issues, trade agreements, and business decisions be provided for students in a timely manner. To fulfill this goal, updated material is regularly placed on the eCollege Web site used in this course. The CMS also includes PowerPoint files, readings, homework assignments, links to Web sites, and announcements.

In addition to CMS instructional technology, lectures include PowerPoint presentations created by the textbook publisher, which the instructor regularly updates with current information. Presentations are delivered with the classroom lights on and the instructor moving around the classroom to engage learners. The PowerPoint slides are constantly evolving based on student and peer faculty feedback, best teaching practices, and current business developments.

### Course Management System Literature Review

Many of the same instructional technologies enhance both distance education and traditional classrooms. For faculty, using a CMS frees the faculty member from having to remember to bring additional handouts to class for students absent from the previous session or students from having to seek out a classmate to determine changes in course assignments. Using a CMS expedites many of the mundane administrative tasks of both instructors and students (e.g., posting grades, providing feedback, delivering assignments, and making announcements). Such effective, practical applications were identified as early as 1995 (Campus Computing Survey, 1995). Student learning was enriched through online practice quizzes, links to additional Web sites, and reflective discussions about course concepts. From 1994 to 1995, colleges and universities across the United States reported a 12 percent increase in electronic mail usage and over 50 percent increase in commercial courseware, course management systems (Campus Computing Survey, 1995). These marked increases highlight higher education's expanded use of instructional technologies and, in particular, technologies which enhance accessibility and promote synchronous communication. The adoption level of CMSs has increased dramatically. According to the 2004 Campus Computing Survey results, public university courses using a CMS rose from approximately 11 percent in 1995 to 43 percent in 2004. Private university courses using a CMS increased from 19 percent in 2000 to approximately 47 percent in 2004 (Green, 2004).

The increased use of Web-based instructional technologies has resulted from the need to keep pace with technologically-savvy students with Internet access outside of the traditional campus computer lab. As these structural changes work to decrease the digital divide and instead create "digital unity" (Finn & Inman, 2004) they also create more opportunities for students to engage in learning outside of lectures, classroom walls, and specified class times. The time constraints of asking a question or adding a comment during a class period can now be avoided, being replaced by more thoughtful and relaxed online discussions. Additionally, this framework offers students the chance to engage in learning at the teachable moment when they are most receptive.

As faculty strive to create rich, responsive learning environments, user acceptance of the classroom technology becomes a critical factor, especially when teaching technologically-savvy students. Management education researchers have used Davis' (1989) Technology Acceptance Model (TAM), which addresses how users come to accept and use a technology to examine student acceptance of various instructional technologies. One TAM construct, perceived usefulness, examines the user's perception that the technology enhances performance (Davis, 1989) and the second construct, perceived ease-of-use, examines "the degree to which the user believes that a particular system would be free from effort" (Davis, 1989).

Instructional technology research suggests that students' perception of usefulness of CMS differed by the various features. Students using the Blackboard CMS found that course content features, including course documents, lectures, announcements, and quizzes were more useful than the course support features of the discussion board, external Web sites, faculty information, and e-mail (Landry, Griffeth, & Hartman, 2006). Additional findings indicate that individuals' perception of the usefulness of the Web was positively related to behavioral intentions to use the Web (Liaw, 2002). These results highlight ways faculty positively impact student learning by extending the walls of the traditional classroom to a Web-based environment. Related to this extension is the role of computer-assisted technologies such as PowerPoint, which are used to convey the course content.

## Presentation Software Literature Review

Growing up with television, computers, and video games, many traditional students are used to and even expect technology to be a part of their learning experience. Faculties were continuously challenged with holding the attention of these highly-motivated learners from the high tech generation. With computer presentation programs (such as Microsoft PowerPoint, Asymetrix Compel, Adobe Persuasion, and Gold Disk Astound), faculty can create professional looking presentations to enhance student learning in lectures.

Literature evaluating PowerPoint-assisted lecturing in higher education has increased as instructor use proliferates throughout the college classroom. Ample resources are available to guide instructors and course developers in graphic design for presentation software (Bellamy & McLean, 2003; *Using PowerPoint,* n.d.; *PowerPoint Tips and Techniques,* 2002) but guidelines to enhance the instructional value are lacking. A review of literature on the instructional effectiveness of presentation software in the traditional classroom highlights a few studies which are characterized by the following results. One body of research investigating the efficacy of PowerPoint lecturing in undergraduate classrooms shows that PowerPoint presentations should not be viewed as a replacement for the chalkboard, but rather as an efficient auxiliary medium (Szabo & Hastings, 2000). Chang (2005) emphasized the learning benefits of PowerPoint for international students who struggle with English as their second or third language. Additionally, another stream of research examining the impact of PowerPoint lectures on student grades revealed that PowerPoint lecture groups achieved better grades than the traditional lecture cohort did (Lowry, 1999). Similarly, Harknett and Cobane (1997) reported that students felt that PowerPoint lectures benefited their learning. Some also felt that the visual emphasis in PowerPoint helped them recall the lecture material during exams.

## Research Questions

Examining students' perceptions of instructional technology features that impact their learning may assist faculty in determining how to use these tools to create an engaging learning environment. Faculty appreciation of the relationship between Web-based, computer-assisted instruction, and student learning may highlight the critical factors influencing learning. Hence, this study aimed to examine students' perceptions of the impact of PowerPoint presentations and the eCollege course management system on their learning and classroom behaviors. This study examines student learning by exploring how instructional technologies such as a CMS and PowerPoint influenced student attitudes and behaviors:

- What impact do PowerPoint presentations have on student attendance, attention during lectures, and class participation?
- What is the relationship between retrieving course materials via a CMS and class attendance?
- Do students prefer to access class materials via a shared university sponsored computer drive or a CMS?

## METHODOLOGY

Given that this research was in its exploratory phase, the sample for this study included students from one of the author's classes. The instructor read a statement asking for student participation, which included the confidentiality and anonymity procedures. She left the classroom and a graduate assistant distributed consent release forms and the survey. Students interested in participating were instructed to complete the consent release form and the survey and return it to the graduate assistant. Students who chose not to participate were informed that they could leave the class. Of the 170 students, 134 chose to participate, resulting in a 78.8 percent response rate. Respondents were business majors in a required upper level *International Business* class, and 42.4 percent were female. Analysis indicated no significant gender differences between respondents and non-respondents regarding results.

### Survey

The authors developed a 15-item Likert Scale survey examining student attitudes regarding eCollege, the University-sponsored course management system. Survey questions asked students to assess their (1) likelihood of attending class when the professor posts PowerPoint handouts on eCollege, (2) satisfaction with eCollege, (3) comfort level with using eCollege, and (4) preference for eCollege over the University-sponsored shared computer drive system.

The survey also asked students to assess the following elements regarding the impact of all the PowerPoint presentations they had attended in this *course* on the following: (5) attention to lectures, (6) classroom behavior during lectures, (7) preference for chalkboard or whiteboard lectures, (8) value of handouts for note taking, (9) classroom participation, (10) instructor organization, (11) slide format, (12) student recall of content, (13) student motivation to attend class, (14) identification of key points, and (15) attitude. These 15 items were rated on a five-point scale from strongly disagree (1) to strongly agree (5), with a middle score of neutral (3). A copy of these items is included in Appendix A.

Additional information was collected based on students' self-reported computer technology skills, experience using PowerPoint, and eCollege usage. We created the following independent variables: (1) PowerPoint Experience, (2) Computer Experience, and (3) eCollege Usage.

### PowerPoint Experience

Respondents who attended a PowerPoint training class and/or created a PowerPoint presentation were assigned to group 1 (n=54), and those who had these experiences and/or used animations, sound, or graphics while delivering a PowerPoint presentation were assigned to group 2 (n=79). **Computer Experience***:* Respondents were divided into groups of moderate or expert users, where moderate users (n=42) performed the following tasks: e-mail, word processing, instant messaging, and downloading course materials from eCollege; and expert users (n=90), who performed the above tasks and also used Microsoft (MS) Excel, MS Access, customized PowerPoint presentations with animation, and multiple eCollege features. **eCollege Usage***:* The five-level eCollege usage variable was based on the number of tasks respondents performed with eCollege. Tasks included downloading class materials, checking course grades, turning in assignments, sending e-mail, and participating in a threaded discussion. Respondents performing one task were coded 1, those performing two were coded 2, up to code 5 for five tasks (Mean = 2.6; SD=1.5). The variable correlation matrix and means and standard deviations are summarized in Table 1.

## RESULTS

### Factor Analysis

An exploratory factor analysis was conducted on the 15 Likert scale items listed in Appendix A to determine if the observed correlations could be explained by a smaller number of factors. A principal components extraction was used with a varimax rotation. Using Kasier's (1960) criterion to retain the components with eigenvalues greater than 1.0, a four factor solution accounting for 60 percent of the variance was obtained. The factor loadings exceeded 0.40 and, since items did not cross load, these results demonstrate discriminant validity of three scales. The following variables: (1) the visual images in PowerPoint help recall during exams and (2) preference for advanced PowerPoint with audio, video, or graphics over bullet-point and text only presentations, loaded on the fourth factor which only accounted for 7.2 percent of the variance. Despite high loadings, this factor was eliminated due to a low reliability coefficient ($\alpha$.=.33) between these two variables. A three-factor solution explaining 52.7 percent of the variance was considered the best representation of the data. Table 2 lists the factor analysis results.

*Table 1. Correlation matrix, means and standard deviations (n=134)*

| | Mean | Standard Deviation | Less likely to attend class when slides posted on web | Satisfaction w/eCollege | Comfortable w/eCollege | Prefer eCollege | PPT holds my attention | PPT increases likelihood of inappropriate classroom behavior |
|---|---|---|---|---|---|---|---|---|
| Less likely to attend class when slides posted on web | 2.13 | 0.984 | 1 | | | | | |
| Satisfaction w/eCollege | 3.78 | 0.898 | 0.032 | 1 | | | | |
| Comfortable w/eCollege | 3.99 | 0.888 | -0.016 | .705** | 1 | | | |
| Prefer eCollege | 3.47 | 1.155 | 0.013 | .573** | .538** | 1 | | |
| PPT holds my attention | 3.92 | 0.798 | -0.122 | .271** | .175* | .253** | 1 | |
| PPT increases likelihood of inappropriate classroom behavior | 2.37 | 1.101 | .178* | -0.021 | -0.066 | 0.068 | 0.06 | 1 |
| Prefer traditional lectures | 4.02 | 1.004 | -0.362 ** | 0.014 | -0.017 | 0.049 | .383** | -0.158 |
| PPT helps w/ note taking | 4.3 | 0.835 | -0.157 | 0.041 | 0.095 | 0.12 | .213* | 0.063 |
| Increases classroom participation | 2.26 | 0.904 | .318** | -.215* | -.280** | -.241** | -.173* | .355** |
| Professors organized | 3.92 | 1.045 | -0.166 | .222* | 0.129 | .233** | .495** | 0.053 |
| Prefer PPT w/ text-only over those with audio, video or graphics | 3.28 | 1.120 | -.237** | -0.043 | -0.028 | -0.13 | -0.035 | -0.041 |
| Visual images in PPT help recall | 3.84 | 0.908 | -0.003 | -0.006 | 0.082 | 0.078 | .348** | 0.096 |
| Less motivated to attend when PPT used | 2.11 | 0.898 | .452** | -0.034 | 0.02 | -0.058 | -.367** | 0.147 |
| PPT help emphasize key points | 4.19 | 0.737 | -.188* | 0.143 | 0.105 | 0.17 | .502** | -0.04 |
| Positive attitude toward PPT | 4.17 | 0.786 | -0.169 | .231** | 0.17 | .240** | .630** | 0.046 |

* *Significant at the .05 Level (two-tailed)*
** *Significant at the .01 Level (two-tailed)*

*Table 1. continued*

| Prefer traditional lectures | Note taking | Decrease classroom participation | Professors organized | Prefer PPT w/ text-only over those with audio, video or graphics | Visual images in PPT help recall | Less motivated to attend when PPT used | PPT help emphasize key points | Positive attitude toward PPT | |
|---|---|---|---|---|---|---|---|---|---|
| | | | | | | | | | Less likely to attend class when slides posted on web |
| | | | | | | | | | Satisfaction w/eCollege |
| | | | | | | | | | Comfortable w/eCollege |
| | | | | | | | | | Prefer eCollege |
| | | | | | | | | | PPT holds my attention |
| | | | | | | | | | PPT increases likelihood of inappropriate classroom behavior |
| 1 | | | | | | | | | Prefer traditional lectures |
| 0.167 | 1 | | | | | | | | PPT helps w/ note taking |
| -.274** | -0.097 | 1 | | | | | | | Increases classroom participation |
| 0.428** | .294** | -0.049 | 1 | | | | | | Professors organized |
| 0.129 | 0.086 | -0.116 | -0.025 | 1 | | | | | Prefer PPT w/ text-only over those with audio, video or graphics |
| .203* | .302** | -0.05 | .273** | .206* | 1 | | | | Visual images in PPT help recall |
| -.363** | -.267** | .410** | -.336** | -0.053 | -.181* | 1 | | | Less motivated to attend when PPT used |
| .321** | .305** | -0.12 | .432** | 0.074 | 0.325 | -.338** | 1 | | PPT help emphasize key points |
| .422** | .355** | -.226** | .490** | -0.021 | .325* | -.365** | .514** | 1 | Positive attitude toward PPT |

* *Significant at the .05 Level (two-tailed)*
** *Significant at the .01 Level (two-tailed)*

*Table 2. Exploratory factor analysis-recoded items are presented in bold*

| Survey Questions | POWERPOINT preference | eCollege Preference | Classroom behavior | Delivery Preference |
|---|---|---|---|---|
| PowerPoint presentations hold my attention. | **.779** | .163 | -.023 | -0.01 |
| PowerPoint handouts help me to take better notes during classroom lectures. | **.467** | .072 | .000 | .340 |
| Professors who use PowerPoint presentations are more organized during their presentations. | **.740** | .152 | -.038 | -.070 |
| PowerPoint presentations help to emphasize key points during lectures. | **.687** | .094 | -.0.89 | .133 |
| I have a positive attitude towards PowerPoint presentations. | **.810** | .137 | -.061 | .046 |
| **I prefer traditional lectures using a blackboard or whiteboard to PowerPoint presentations.** | **.562** | -.155 | -.414 | .091 |
| I satisfied with eCollege as a tool to access course materials. | .127 | **.862** | -.008 | -.019 |
| I am comfortable with using eCollege. | .026 | **.871** | -.060 | .098 |
| I prefer using eCollege over the RMU passouts system. | .214 | **.756** | .038 | -.070 |
| PowerPoint presentations increase the likelihood of inappropriate classroom behavior. | .174 | .002 | **.661** | .065 |
| PowerPoint presentations decrease classroom participation. | -.093 | -.301 | **.730** | -.088 |
| I am less motivated to attend class when PowerPoint presentations are used during the lecture. | -.510 | .051 | **.593** | .034 |
| I am less likely to attend class when the professor posts PowerPoint handouts to the Web. | -.211 | .139 | **.658** | -.177 |
| **I prefer bullet-point text-only PowerPoint over more advanced presentations with audio, video and graphics.** | -.097 | -.034 | -.233 | **.839** |
| Visual images presented in PowerPoint presentation lectures help me to recall content during exams. | .469 | .033 | .169 | **.582** |
| **Eigenvalue** | 4.061 | 2.159 | 1.688 | 1.117 |

## Reliability

The reliability dimensions determined from the factor analysis were determined by computing the internal consistency coefficient, Cronbach's alpha coefficient. Nunnally (1967) advised that a magnitude of 0.5 to 0.6 for the Cronbach alpha statistic was sufficient in the early stages of basic research, but that an alpha of 0.8 is more desirable. Cronbach's alpha levels for electronic presentations, online course management systems, and effective classroom behavior items indicated reliability; all three scales showed acceptable alphas (.63-.81). High scores on the electronic presentation scale indicate respondents' positive attitude toward PowerPoint, agreement that PowerPoint holds their attention, emphasizes key points. The scoresportray professors who use PowerPoint as being organized during their presentations, and note preference for PowerPoint over traditional chalkboard or whiteboard lectures. Respondents who scored high on the online course management system scale indicated their satisfaction and comfort using eCollege, and their preference for using eCollege to obtain course materials rather than the University-sponsored shared computer drive. Finally, the effective classroom behavior scale assessed respondents' perception that PowerPoint presentations negatively impact classroom behavior by increasing the likelihood of inappropriate classroom behavior and decreasing participation, increasing students' motivation to attend class, and increasing the likelihood to attend when the professor posts PowerPoint handouts on eCollege. Respondents who agreed that PowerPoint negatively impacted classroom behavior scored high on this item. Table 3 presents the means, standard

*Table 3. Means, standard deviations, alpha coefficient and correlation matrix for all constructs (n=134)*

| | Mean | SD | Alpha Coefficient | POWERPOINT Preference | eCollege Preference | Classroom Behavior |
|---|---|---|---|---|---|---|
| POWERPOINT preference | 4.07 | .65 | .79 | 1 | | |
| eCollege preference | 3.37 | .83 | .81 | .244 ** | 1 | |
| Classroom Behavior | 2.2 | .67 | .63 | -.339*** | -.102 | 1 |

*** Correlation is significant at the 0.01 level (2-tailed).*

*** Correlation is significant at the 0.00 level (2-tailed).*

*Table 4. ANOVA Results*

| | Sum of Squares | df | Mean Square | F | Significance |
|---|---|---|---|---|---|
| PPT Preference | | | | | |
| Between Groups | 1.908 | 5 | .382 | .892 | .489 |
| Within Groups | 54.308 | 127 | .428 | | |
| eCollege Preference | | | | | |
| Between Groups | 12.768 | 5 | 2.2554 | 4.081 | .002** |
| Within Groups | 79.465 | 127 | .626 | | |
| Classroom Behavior | | | | | |
| Between Groups | 4.217 | 5 | .843 | 1.922 | .095* |
| Within Groups | 55.710 | 127 | .439 | | |

** Significant at the .10 Level (two-tailed)*

*** Significant at the .05 Level (two-tailed)*

deviations, alpha coefficients, and correlations at the construct level.

## Comparison of Means

A t-test was done to determine whether there is a meaningful difference between electronic presentation experience, online course management, and effective classroom behavior and respondents' computer experience. We divided respondents into two groups based on their self-reported computer experience. Group 1, moderate users, indicated that they performed the following tasks: e-mail, word processing, instant messaging, and downloading course materials from eCollege; whereas Group 2, expert users, indicated that they performed these tasks and Microsoft (MS) Excel, MS Access, customized PowerPoint presentations with animation, and multiple eCollege features. T-test results indicated no significant differences between respondents who classified themselves as moderate or expert computer users.

To examine differences based on PowerPoint experience, respondents who attended a PowerPoint training class and/or created a PowerPoint presentation were assigned to group 1 and those who had these experiences and/or used animations, sound or graphics while delivering a PowerPoint presentation were assigned to group 2. A second t-test was performed examining the relationship between the dependent variables and respondents' perceptions of classroom behavior. Results indicate a significant difference between respondents on the effective classroom behavior variable. More respondents with limited PowerPoint experience perceived that PowerPoint presentations negatively impacted classroom behavior than those who had more extensive experience *($t=-2.994$ $df=131$ $p< .01$).*

One-way analysis of variance (ANOVA) has been conducted to test whether there is a meaningful difference between the dependent variables and respondents' use of eCollege. ANOVA results, reported in Table 4, indicated a significant difference between respondents based on the extent of their eCollege usage and their online course management preference $F$ (5, 132) = 4.081, $p <$ .01). Respondents who performed more tasks with eCollege were more favorable towards eCollege than those who performed fewer tasks. Additionally, respondents who performed more tasks with eCollege disagreed that the PowerPoint usage led to negative classroom behaviors $F$ (5, 132) = 1.922, $p < .10$).

## DISCUSSION

The primary value of this research is the exploratory factor analyses examining the relationship between CMS, presentation software, and student learning. Factor analysis results suggested a three or four factor solution. The four factor solution produced the following factors: (1) Electronic Presentation component (2) Online Course Management preference, (3) Effective Classroom Behavior, and (4) Delivery Preference. We eliminated the fourth factor after reliability analysis indicated a low Cronbach's alpha coefficient. The final three factor solution offers preliminary support for the electronic presentation, online course management preference, and effective classroom behavior as significant factors when examining student learning. Although previous research examined the relationship between usefulness and ease of use of CMSs (Davis, 1989) and instructional use of PowerPoint on student learning (Harknett & Coban, 1997; Lowry, 1999; Szabo & Hastings, 2000; Chang, 2005) few studies investigate students' classroom behaviors. The inclusion of the classroom behaviors construct provides an opportunity to explore the role of the student behavior component in concert with previously examined concepts.

This research suggests a relationship between students' preference for a CMS and their use of a CMS. Students who performed more eCollege tasks such as downloading course materials, checking grades, and e-mail viewed eCollege more favorably than those who performed fewer eCollege tasks. While previous research (Landry, Griffeth, & Hartman, 2006) highlights the usefulness of CMS content features, it is important to note that this study examines students' perceptions of eCollege assessing their comfort, satisfaction, and accessibility by assessing their preference for using eCollege over the University-sponsored shared computer drive. Faculty use this shared drive to upload course materials (e.g., notes and assignments) which students can only download from computers on campus or by first loading a software program accessible from the University.

Interestingly, we also found a relationship between extensive eCollege usage and classroom behavior. Extensive eCollege users disagreed that PowerPoint usage was related to a decrease in student classroom behaviors such as participation and attendance. This finding is particularly interesting given the anecdotal perception among some faculty that students avoid attending class if they obtain class materials such as handouts, PowerPoint slides, and assignments via a CMS. Our results, though preliminary, suggest that students who performed multiple tasks with eCollege did not perceive a negative relationship between PowerPoint lectures and classroom behaviors, such as attendance and participation. Future research should explore the differences between eCollege users further. Almost half of the students in our study indicated that they preferred using eCollege over the University-sponsored shared computer drive. For example, do students who perform fewer tasks with eCollege experience problems because they are unfamiliar with the CMS features? If this is case, then instructors using a CMS need to ensure that their students are familiar with the CMS features from basic login procedures through participation in threaded discussions. Another potential explanation is that students are unsure how to take notes during a PowerPoint presentation. It's possible that students who bring a printed copy of the slides with them to class and simply follow the instructor from slide to slide without taking notes perceive that class attendance is not a priority or that they have nothing to share since the information is accessible without attending class. Instructors can adapt their PowerPoint presentation to encourage an active two-way dialogue with students with short case studies or slides with discussion questions. International business courses are especially adaptable in this realm given the dynamic nature of the global marketplace.

This study points to several key benefits in using presentation software, such as PowerPoint, and a CMS, such as eCollege, in resident university courses. Students responded positively to PowerPoint as an enhancement and supplement for the delivery of lectures. The handouts printed from PowerPoint files provided outlines for taking notes, identifying key points, and studying content for exams. Furthermore, students perceived the instructor as being organized because the PowerPoint files were developed and posted in the CMS before class. The PowerPoint slides provided an agenda and structure to the class. These results support the findings of Susskind (2005), who noted students attending his *Introduction to Psychology* course when PowerPoint presentations were used had more positive attitudes about the course and greater self-efficacy than when there were no PowerPoint presentations. In similar results, Harding (2005) reported his engineering students found PowerPoint slide-based handouts were a significant benefit for taking quality notes for later reference.

The study did not reveal any deterrents that might discourage the instructor from using PowerPoint. Overall, student respondents did not consider PowerPoint a deterrence to their class participation or attendance. In terms of the design of the PowerPoint slides, a slight majority of learners preferred visual images over text-only content. They indicated that visual images helped them recall course content. This supports one of the principles presented in *Instructional Message Design* (Winn, 1993), "Pictures are usually more memorable than words, and are thus useful when information has to be remembered" (p. 86).

The CMS was also positively perceived by students who indicated a high level of satisfaction and comfort in using eCollege. Currently, this university has a networked Passouts drive, where instructors can post handouts and documents. Unlike the CMS, which can be accessed anytime and anyplace, the Passout system via the University-sponsored shared drive can only be accessed from campus computer labs or after loading software obtained from the University. Apparently, the convenience of accessing files is more important to students than the free printouts offered in the computer labs. Students' positive reactions support an earlier study conducted by Laudato and Nicoll (1999) at the University of Pittsburgh, which noted two-thirds of students (n=1850) commented on the ease of access to assignments, online materials, and communication through their CMS.

## RECOMMENDATIONS FOR PRACTICE

The following recommendations are presented for instructors who are using CMSs in on-campus, face-to-face courses. Professors using instructional technologies such as CMS and PowerPoint may enhance the students' learning experience by demonstrating these tools and explaining their learning benefits. Technology-savvy students do not necessarily have the specific knowledge and skills required for each course and type of technology. Instructors can set the instructional foundation for a course by providing basic training for students on how to use the CMS. Even if students have used the system in previous courses, most courses are organized differently. Students need to understand the structure of the online course materials, such as where to find assignments, announcements, handouts for class, and grades. In addition, students need to know how to print materials from the CMS. This foundation knowledge of the CMS empowers students to be self-directed. Most universities have a help desk, but students cannot ask questions about functions they don't know exist. Online discussions can be especially challenging and unmanageable for students. They can quickly become disjointed and overwhelming. A practice discussion, such as posting student introductions, can orient students to the discussion process. Furthermore, faculty can provide students with clear expectations as to when, how often, and how long postings should be. In a face-to-face course, instructors should integrate and refer to student postings so students know that their contributions are valued.

In addition to being experienced users of the CMS, faculty should be knowledgeable users of the PowerPoint software. They need to consider how to organize slide presentations to engage students and enhance learning. Besides the instructional aspects, basic use of PowerPoint involves some technical skill to add images, charts, and video. In the first or second session of a course, instructors should instruct students on how to retrieve and use PowerPoint slides. In particular, students need to know how to print PowerPoint files as handouts and how to view the slides as a presentation outline. This is also a good time to explain the rationale for providing PowerPoint files and to advise students on how to use them for note taking and studying. Students should understand that these PowerPoint files supplement classroom lectures and activities; they are not a substitute for attending class. Reviewing the slides prior to class serves as an advanced organizer which helps students to prepare for class.

## LIMITATIONS AND FUTURE RESEARCH

New classroom technologies become available at a fast and furious pace. It is sometimes difficult for faculty to know which technologies enhance student learning and make them worth the substantial investment of time it takes to learn and use them. This exploratory study examines the value of the eCollege course management systems and PowerPoint presentations in a face-to-face, upper level business course, *International Business*. Future studies using the same dependent variables can offer support to the link between student learning and electronic presentations, online course management systems, and effective classroom behavior. Additionally, research should be conducted to refine the usage of visual images and animation in electronic presentation items which were eliminated from the analysis due to a low reliability coefficient. As technology continues to advance, inclusion of visual images and animation in electronic presentation will simultaneously occur. Investigation of the impact of these images on student learning is essential to fully understanding the relationship between student learning and electronic presentations. Finally, it would be valuable to determine if student learning differs based on the type of CMS used. Though eCollege shares similar features with Blackboard and WebCT, it is possible that students' comfort level is affected by individual CMS features and/or system design. Overall, instructors who teach with these technologies need to ensure that students know how to efficiently and effectively use them to their full potential.

## REFERENCES


Allen, I., & Seaman, J. (2005). *Growing by degrees: Online education in the United States, 2005.* The Sloan Consortium.

Atkins-Sayre, W., Hopkins, S., Mohundro, S., & Sayre, W. (1998). *Rewards and liabilities of presentation software as an ancillary tool: Prison or paradise?* ERIC Document Reproduction Service; ED 430260.

Bartsch, R., & Cobern, K. (2003). Effectiveness of PowerPoint presentations in lectures. *Computers & Education, 41,* 77-86.

Bates, A. (1995). *Technology, open learning and distance education.* London: Routledge.

Bellamy, K., & McLean, D. (2003). The mechanics of PowerPoint (Electronic Version). *Journal of Audiovisual Media in Medicine, 26*( 2), 74-78.

Campus Computing Survey. (1995). Retrieved May 9, 2006 from, http://www.campuscomputing.net/

Chang, Y. (2005). *The use of PowerPoint as a learning aid for international students.* Research Project Report Presented to the School of Education Indiana University, South Bend.

Cox, J. (2005). Higher ed learns notebook PC lessons. *Network World,* Apr 11, *22*(14), 25-26.

Davis, F. (1989). Perceived usefulness, perceived ease of use and user acceptance of information technology, *MIS Quarterly, 13*(3), 319-340.

DeBourgh, G. A. Simply elegance: Course management systems as a pedagogical infrastructure to enhance science learning. *The Technology Source Archives.* Retreived May 26, 2006 from, http://www.technologysource.org/article/simple_elegance/

Eastman, J., & Swift, C. (2001). New horizons in distance education: The online learner-centered marketing class, *Journal of Marketing Education, 23*(1) 25-34.

Finn, S., & Inman, J. (2004) Digital unity and digital divide: Surveying alumni to study effects of a campus laptop initiative. *Journal of Research on Technology in Education, 36*(3), 297-317.

Fleming, M., & Levie, W. (1993). *Instructional message design: Principles from the behavioral and cognitive sciences* (2nd ed.). New Jersey: Educational Technology Publications.

Frey, B., & Kirby, D. (2002). Student perceptions on the value of PowerPoint. In D.G. Brown(Ed.), *Faculty development programs that work.* Bolton, MA: Anker Publishing Inc.

Gates, K. (1998) Should colleges and universities require students to own their own computers? *CAUSE/EFFECT Journal, 21*(3).

Green, K. (2004). Campus computing 2004:Fifteenth national survey of computing and information technology in US higher education. *The Campus Computing Project,* University System of Georgia.

Hankin, J. (1999) Alice, the college teacher and the rottweiler in wonderland: The prospect and problems of distance education. *Executive Speeches,* 1492: 18-21.

Harding, G. (2005). The students' feedback: A survey of three approached to teaching with modern PowerPoint presentations. *American Society for Engineering Education.* Retrieved April 1-2, 2005 from,.

Harknett, R.J., & Cobane, C.T. (1997). Introducing instructional technology to international relations. *Political Science & Politics, 30,* 496-500.

Journal of Teaching in International Business. (n.d). Retrieved September 25, 2006 from, http://www.haworthpress.com/store/product.asp?sid=91D0P1HS6QEV

Kaiser, H. F. (1960). The application of electronic computers to factor analysis. *Educational and Psychological Measurement, 20*, 141-151.

Kladko, B. (2005). *Lectures have to compete with Web.* Retrieved April 16, 2005 from, http://www.bergen.com/page.php?qstr=eXJpcnk3ZjczN2Y3dnFlZUVFeXkzJmZnYmVsN2Y3dnFlZUVFeXk2NjgwNjU1JnlyaXJ5N2Y3MTdmN3ZxZWVFRXl5NA

Landry, B., Griffeth, R., & Hartman, S. (2006) Measuring student perceptions of blackboard using the technology acceptance model. *Decision Sciences Journal of Innovative Education, 4*(1), 87-99.

Laudato, N., & Nicoll, J. (1999). assessing the impact o n students of online materials in university courses. Retrieved October 3, 2006 from, http://www.educause.edu/ir/library/html/edu9952/edu9952.html

Liaw, S. (2002) Understanding user perceptions of World-wide Web environments, *Journal of Computer Assisted Learning, 18*, 137-148.

Lowry, R. (1999). Electronic presentation of lectures—Effect upon student performance. *University Chemistry Education, 3*(1), 18-21.

Martins, L, & Kellerman, F. (2004). A model of business school students' acceptance of a Web-based course management system. *Academy of Management Learning and Education, 3*(1) 7-26.

Morgan, G. (2003). *Faculty use of course management systems.* 2. Retrieved March 16, 2006 from, http://www.educause.edu/ir/library/pdf/ers0302/rs/ers0302w.pdf

Nunnally, J. (1967). *Psychometric theory.* New York: McGraw-Hill.

Olsen, F. (2001) Getting ready for a new generation of course management systems.*Chronicle of Higher Education*, Vol. 48 (17), A25-A27. Retrieved March 16, 2001 from, http://reddog.rmu.edu:2065/login.aspx?direct=true&db=eric&an=EJ639384

Susskind, J. (2005). PowerPoint's power in the classroom: Enhancing students' self-efficacy and attitudes. *Computers & Education, 50,* 2, 203-215. Retrieved September 25, 2006 from, ACM Portal database.

Szabo, A., & Hastings, N. (2000). Using IT in the undergraduate classroom: Should we replace the blackboard with PowerPoint? *Computers & Education, 35,* 175-187.

Winn, W. (1993) *Instructional message design.* In. Fleming & Levie (Eds.), Englewood Cliffs, NJ: Educational Technology Publications, pp. 55-119.

Using PowerPoint. (n.d.).*PowerPoint tips and techniques.* Retrieved 2002. http://www.pitt.edu/~ciddeweb/fds/tech_presentation_software.htm

## APPENDIX A. SURVEY QUESTIONS

Scale: Strongly disagree (1) to strongly agree (5).

1. PowerPoint presentations hold my attention.
2. PowerPoint presentations increase the likelihood of inappropriate classroom behavior.
*3. *I prefer traditional lectures using a chalkboard or whiteboard to PowerPoint presentations.*
4. PowerPoint handouts help me to take better notes during classroom lectures.
5. PowerPoint presentations increase classroom participation.
6. Professors who use PowerPoint presentations are more organized during their presentations.
*7. *I prefer bullet-point, text-only PowerPoint presentations over presentations with audio, video, or graphics.*
8. Visual images presented in PowerPoint presentation lectures help me to recall content during exams.
9. I am less motivated to attend class when PowerPoint presentations are used during the lecture.
10. PowerPoint presentations help to emphasize key points during lectures.
11. I have a positive attitude toward PowerPoint presentations.
12. I am less likely to attend class when the professor posts PowerPoint handouts to the Web.
13. I am satisfied with eCollege as a tool to access course materials.
14. I am comfortable with using eCollege.
15. I prefer using eCollege over the RMU Passout system.

*Italicized items were reverse scored

*This work was previously published in International Journal of Information and Communication Technology Education, Vol. 4, Issue 1, edited by L. Tomei, pp. 1-14, copyright 2008 by IGI Publishing (an imprint of IGI Global).*

# Chapter 13
# Using a Web-Based System to Support Teaching Processes

**V. Klyuev**
*University of Aizu, Japan*

**G.P. Nikishkov**
*University of Aizu, Japan*

## ABSTRACT

*A platform-independent Java Web application named TSI (Teacher-Student Interaction) that supports communication between an instructor, teaching assistants and students in a traditional on-campus course is presented in this chapter. Using the TSI, the instructor and teaching assistants can handle most of the routine work: upload student personal information, send students personal emails, etc. The system can easily be installed and administered individually by an instructor inexperienced in computers. It is as simple as a pen for students. Students can check their personal data (scores and comments), download educational materials, etc. As part of the TSI, a VBA application is used to analyze the course log files. This tool is helpful in understanding individual and group students' behaviors. The TSI was successfully tested during six years at the University of Aizu (Japan) in an environment where English is one of the working languages and both students and professors are non-native speakers of English.*

## INTRODUCTION

During the last decade, a lot of work has been done in designing new methods and tools to support course management and communication between professors and students in everyday university life. Different solutions have been proposed to simplify student access to class material, to help students submit the results of exercises, to help professors distribute lecture notes, to get student feedback, and to monitor student progress (Llamas-Nistal et al., 2004) In many cases, these solutions are not ideal.

In classroom situations where both teachers and students use their second language, communication becomes a crucial factor. Such a situation is found in some Japanese universities where the official language of instruction is English but both professors and students are non-native speakers of English.

DOI: 10.4018/978-1-60566-936-6.ch013

One solution to improve communication between professors and students is to turn to the Internet and to use modern information technologies. These have impacted the university teaching processes in significant ways.

There are two main directions in designing communication tools:

- universal systems with a large number of features (Angel, 2009; Blackboard, 2009; Top Class E-learning Suite, 2009; WebCT, 2009),
- specialized systems dedicated to specific purposes (Nicenet, 2009; Tuckman, 2003).

The large portion of these systems is designed to support on-line teaching (Pahl, 2003). As it was noticed in (Bonk, 2002), there is a myriad options, making it difficult to determine the right system to apply to university courses. The number of core functions provided by each system is practically the same. For professors, these functions include: a) making lecture notes available to students, b) monitoring student progress, c) analyzing student behaviors, d) obtaining student feedback, e) exchanging e-mails with students, f) providing students with their grades, and g) managing student accounts. For students the set of functions consists of the following: a) accessing class material, b) exchanging e-mails with professors, c) accessing personal grade information, d) submitting the results of exercises, and e) changing their password. The main problem with the tools of the first category i.e., universal systems, is the number of features: The end user does not need most of them (Moodle, 2009; BlackBoard, 2009; Woods et al., 2004; Kaiden, 2002). As a result, the system interface is not intuitive and support from experts and permission for installation from the university network administrator is required (Storey et al., 2002). For example, to learn how to post the basic documents such as a syllabus and a schedule using BlackBoard, the professor should attend a 2-hour workshop (FSU, 2009). WebCT, another famous tool in this category, is equally complicated: Its manual contains 563 pages (Rehberg et al., 2004), which no instructor is inclined to read. Special auxiliary software is introduced to simplify work with BlackBoard and WebCT (Course Genie, 2009). Another drawback of these systems is user frustration with information technology (Lazar, Jones, and Hackley, 2006). Storey et al. (2002) and Masiello et al. (2005) pointed out that when tools are hard to navigate, they not only have a neutral but negative effect on learning.

On the other hand, the tools in the second category, i.e., specialized systems, are difficult to adapt to teaching needs (Yen and Wu-Jeng Li, 2003; Tuckman, 2003).

To summarize, we note that information technologies for teaching processes are growing very fast. The market offers many different tools. To use the most of universal systems, users need external assistance. Instructors and students have to waste time to learn how to apply the features of the tool.

In the traditional on-campus course, instructors prefer to retain their conventional style of teaching even when using computer technologies. They need assistance in the routine. They want to distribute handouts and assignments before the lecture and to post them on the Internet to make materials available only for students enrolled in their class. Communication with students, tracking their progress, and checking grades by students are tasks to put on the Web. In this face-to-face mode, subjects such as math and computer science remain an individual rather than a group activity. This model is distinguished from either distance or traditional instruction. How can it be employed in campus computer labs?

We introduce an approach using the TSI (Teacher-Student Interaction) system that provides assistance in the routine and that may be of interest to teachers and professors. Our aim was to design the tool which can be installed, set up and administered individually even by a professor inexperienced in computers and which is as simple as a pen for the students. The main features of the TSI system are:

- This system is in the public domain.
- It is written in Java: The TSI can run on any computer.
- It is easy to install: An inexperienced computer user can set up this system.
- It is stable: In the six years since we have used the system, it has never crashed.
- An instructor is independent: S/he does not need any external assistance or permission of the network administrator to run the TSI.
- It is safe: The system does not make any security holes in the local network or in the computer where it runs.
- A professor may easily adjust the system to meet the teaching needs.
- The interface is intuitive, which makes it appropriate for both teaching and learning, especially in the case of teaching students who are non-native speakers of English.

In this chapter, we share our experience of employing the TSI and report our findings in teaching courses on programming and software development in an environment where English is one of the working languages, and both students and professors are non-native speakers of English, and student comprehension of spoken English is low.

## FRAMEWORK OF THE TSI

During a semester, interaction between the instructor, teaching assistants (TAs) and students is extremely complex. It includes many verbal and paper-based actions. The set of the core actions was introduced in the previous section. How to achieve the key goal: Both instructors and students have to concentrate on the course and not to waste time in becoming familiar with the communication technology. There are significant issues to be considered.

The classical approach to implement interaction is to involve Web pages, e-mail and CGI scripts. It can be time-consuming to develop and inefficient to use because publishing new materials on the professor's personal Web page implies modifying the page itself. Putting student scores on the Internet makes them available to everybody. Utilizing e-mail lists makes it difficult to send individual messages containing specific information for each receiver. University regulations may forbid CGI scripts on professors' personal Web pages.

Another solution is to use PHP or ASP. This approach is also inappropriate: A standard way to implement a system requires MySQL. An installation of software might be complicated for an inexperienced user. Special measures should be taken to provide the necessary level of security to protect the professor's computer when an application is running.

We used Java to create a Web-based application to implement aforementioned communication functions.

The architecture of the system is presented in Figure 1. The HTTP protocol is used for all communications between users and the TSI, allowing participants to employ standard Web browsers. The TSI supports multiple user access. Since Microsoft Excel tables are widely used by professors to keep student scores, this table is a key data component in the TSI. It is sharable between the professor and TAs for updating. Via this table, the students can check their personal scores. In order to use Excel to work with the table on a personal computer, this table can easily be uploaded to the TSI and downloaded to the professor's machine. It is also possible to edit the score table on the TSI.

After investigating the needs of the instructor, TAs and students, the following scenarios were defined.

*Figure 1. Architecture of the TSI system*

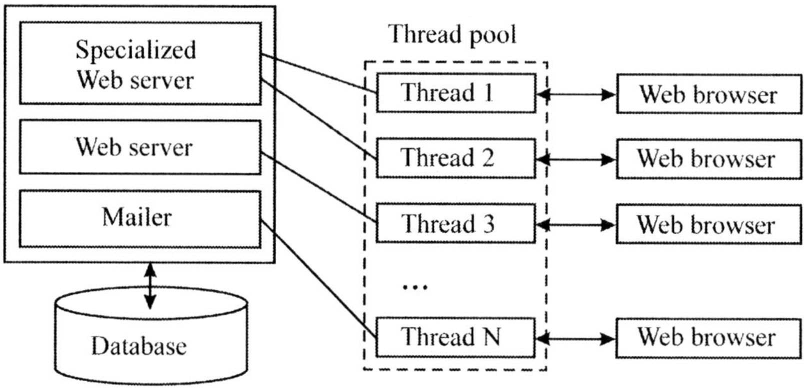


## Initialization

The system does not require installation in the traditional way. A TSI directory with its subdirectories is placed on any networked computer with the Java Virtual Machine, version 1.3 or higher. The running TSI manager displays a window with options to set up the system and to start/stop it. Setup options include a course name and the instructor name. It is possible to specify the common part of allowed IP addresses. The TSI replies only to computers with allowed IP addresses. The given domain name is used to create e-mail addresses for students utilizing their ID. The setup menu is shown in Figure 2.

## Access to the TSI

Any Web browser is used for password-protected communication with the TSI. The instructor has

*Figure 2. Setup Menu*

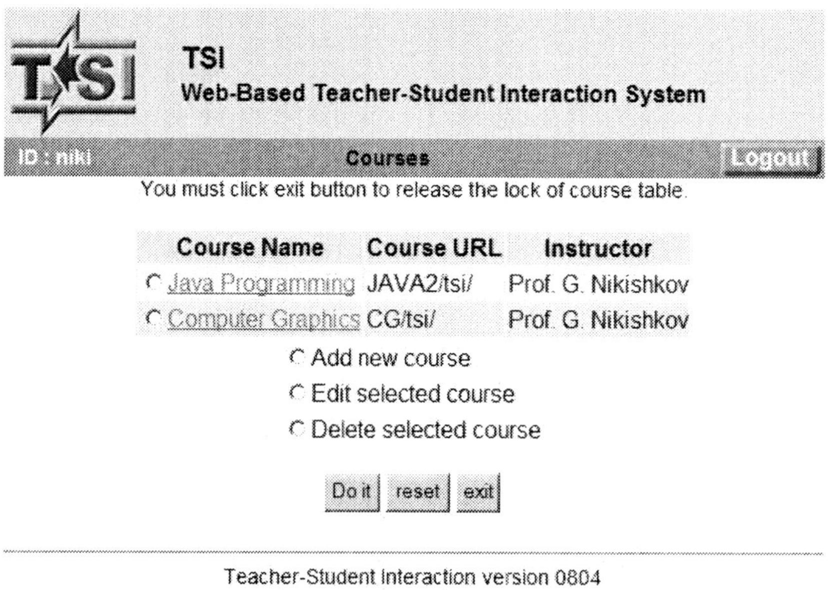

the highest authority in the system access. He or she can look through all stored files. Also, he or she is able to recover lost student and teaching assistant (TA) passwords. A certain level of security is provided to prevent unauthorized access to the system.

## Instructor and TA Scenarios

Scenarios for the instructor and TAs include the following features: table score manipulation, editing course materials, downloading student submissions, sending personal e-mails, tracking student activity.

## Table Score Manipulation

For the instructor and TAs, the main advantage of the TSI system is in the possibility to jointly maintain, modify and distribute a table with personal student information (score table). At the beginning of the semester, the table only contains student names and ID numbers that students use as login names.

During the semester, this table is constantly modified by adding student scores, instructor comments and other personal information. The score table can be modified either in the spreadsheet program or within the TSI using a Web browser. Two strategies can be applied: a) keeping the master copy on the instructor's computer or b) keeping the master copy on the TSI. The second strategy is more efficient if both the instructor and the TAs are going to modify the score table. The current state of the score table is available to all authorized persons (every student can access only its personal data). To prevent any loss of information, there are two possibilities supported by the TSI: a) a backing up of the score table in a text format, and b) downloading the score table to another computer.

## Personal E-Mails

The mail function plays a crucial role in communication with students who are not fluent in the language of instruction, especially in the situations when a student's comprehension of an instructor's spoken language is low.

Information stored in the score table is used to send students personal e-mails. It is possible to specify columns (information content) and students (list of recipients). The TSI provides a default body for the message, which can be edited. Since fields of the score table have no size restrictions, any textual data can be put there. This function is simple to communicate with students and powerful to send them common information (for all students) such as personal comments in the same e-mail. Such a combination is a key feature of the TSI e-mail service.

## Other Functions

Other opportunities for the instructor and TAs are shown in Figure 3. Note that course materials are usually presented as PPT, PDF or PS files. There are several functions available only for the instructor. To analyze student activity, a special VBA application exists.

## Student Scenarios

For a student, a typical scenario to use the TSI is as follows. First, s/he receives e-mail from the instructor with a simple explanation of the TSI, the URL of the login page, and a personal password. Using the Web browser, the student logs into the TSI and changes his or her original password since it was generated automatically. The other student actions can be understood from Figure 3.

The main TSI menus for the instructor and students are shown in Figures 4 and 5, respectively.

*Figure 3. Instructor, teaching assistant and student interfaces to the TSI system*

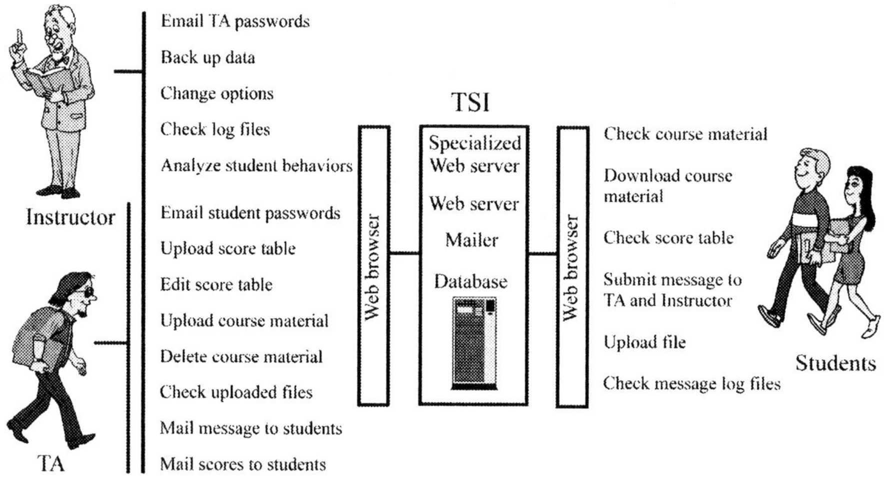


*Figure 4. Instructor Menu*

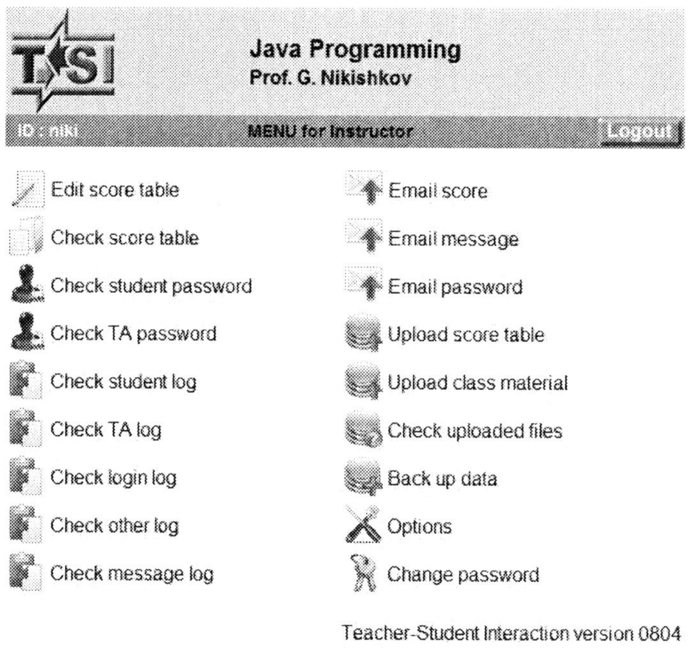


## Our Experience

In a traditional course at the University of Aizu, an instructor and one or two TAs interact with 35 – 96 students. It may include lectures, exercises, a term project and an examination. Exercises are graded on a weekly basis. The semester usually consists of 14 weeks. The TSI was tested by several professors during the last six years to teach different computer science subjects related to programming and software development, such as Programming III, Java Programming, Numerical

*Figure 5. Student Menu*

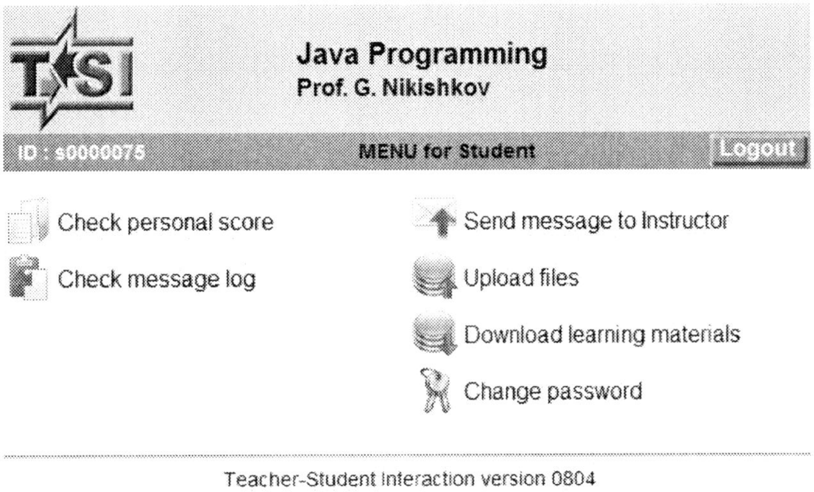


Analysis, Advanced Algorithms, Algorithms and Data Structures, Web Engineering, etc. It was successfully utilized in several English language courses such as Technical Writing, etc. The students were second- and third-year students and classes were held once a week. Each class took two to four academic hours. The total number of student groups was more than 30. The total number of students who used the TSI was more than 1000.

Professors applied the TSI in different ways. Students of several groups were provided with class materials via a different course Web site. In some courses, students uploaded their results to the special server which was separated from the TSI whereas in another courses, students submitted their reports in a traditional hard copy way. Students of one class reported their results orally. The Java Programming and Language Processing Systems courses utilized the full set of the TSI features. These courses also required a mandatory activity: uploading the results of exercises to the TSI.

Figure 6 shows an example of how professors used the main table of the system. Results of exercises of our courses are mainly software source code (C, C++ and Java programs) and term project reports. Most activities are individual exercises except for work on a term project, where students worked in groups of 2 – 4. For English language courses, results of exercises are text files in different formats.

## EFFICIENCY OF THE TSI: AN INSTRUCTOR VIEW

### Methodology

Log files are useful source of information to monitor student activity and progress. Statistics from the logs can help in apprehending student behaviors. This information provides impartial feedback from students.

The TSI keeps the following logs:

- The login log: It keeps track of all successful login operations.
- The student activity log: It records actions such as checking scores, checking and sending emails, downloading materials, uploading results, and changing passwords.

*Figure 6. Variant of a score table*

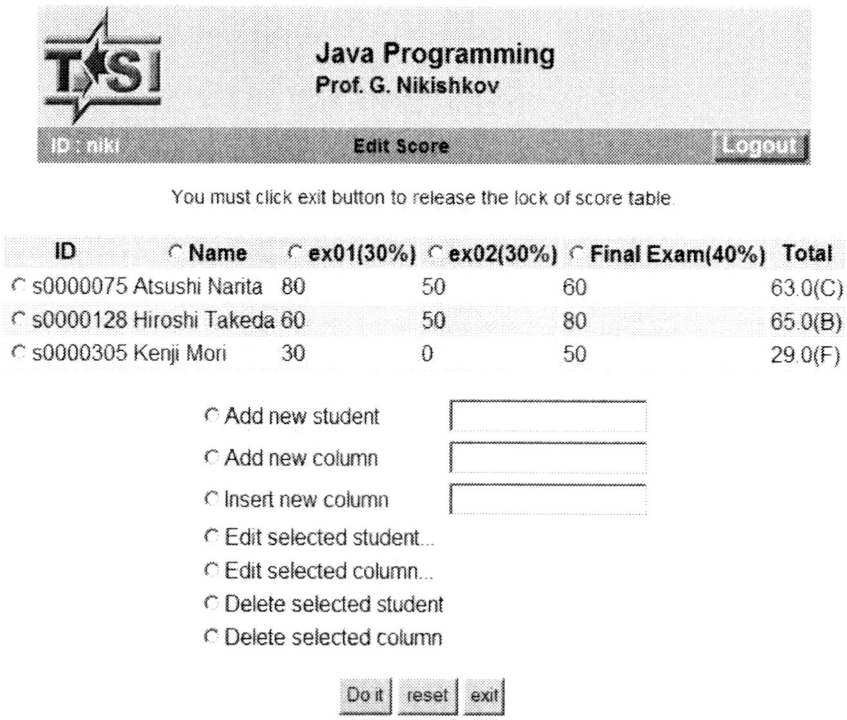


- The professor and TA activity logs: It records all operations done by them.
- The log of unknown users: It shows how often users fail to access the system.

Our VBA application analyzes the logs and different student behaviors (VBA was selected for a statistical analysis because the main table of the TSI is an Excel table):

- Group student behavior: It classifies logs according to the grades received (A, B, C, and D categories) and plots histograms of student activities. These histograms can be populated with data on a daily, weekly, or semester basis.
- Group tendency: To identify possible correlations between student grades and their activities, the application plots the corresponding trend lines. The necessary set of options can be specified.
- Individual student activity and individual tendency: Every aforementioned histogram and trend line can be plotted for each student.

## Results of Statistical Analysis

Log files of all groups were carefully examined using the aforementioned VBA application.

### Examination of Student Behaviors

Students who received an A or B grade used the TSI system significantly more often compared to students who received C or D grades. This result holds across all groups. Figure 7 provides such an example. The common pattern is low student activity in the days following the class, after which activity increases and reaches the maximum on the day of the next class. The number of messages received by the professor or TA is very low.

*Figure 7. Diagram of student activity (for A and D grades) during a week*

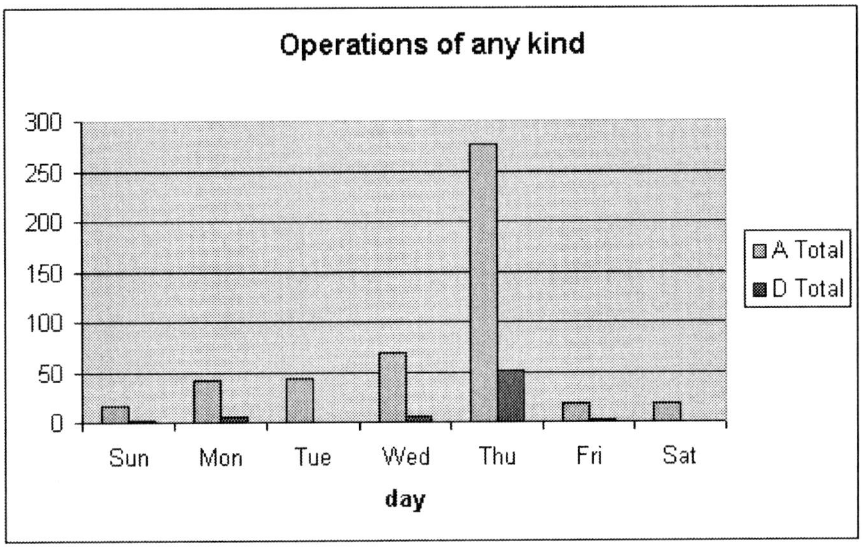


This low activity in sending messages is closely linked to the educational culture in Japanese high schools where students usually do not ask questions (Kluev, 2004). On the other hand, there is no restriction in using tools to send messages: Some students employed the usual mail tool. Students who received an A grade were concerned about changing the password to access the TSI. In contrast, students who received C or D grades did not care about it.

## A Tendency Examination

In programming exercises, students work individually at their own pace and have their own learning styles. We think, this is a reason why we found no correlation between student grades and any type of activity except for mandatory ones, when we analyzed a group student profile.

As we mentioned earlier, some courses such as Java Programming required submission of student results. We analyzed log files of two classes (ONE and TWO) of the second year students enrolled in this course. We found a correlation between this activity and marks obtained by students in the A and B grade groups: $R^2 > 0.74$, there $R^2$ is the coefficient of determination ($\alpha < 0.01$). This is not the case for the C and D grade groups. See Table 1 for details.

The statistical analysis of log files using individual profiles gave promising results: Individual student behavior (in the case of non-mandatory activity such as checking scores) from the A and B grade groups are stable for most students. It is possible to predict their activity during a semester and their marks after four or five classes. If a scale of grades for exercises is the same for each exercise, then such a prediction can be made quite accurately. Figures 8 – 9 illustrate this conclusion for one student who received a B grade. The trend lines for these figures are based on data from the entire semester (Figure 8) and on data from the first five classes (Figure 9). The "score checking" axis presents the accumulated number

*Table 1. Coefficient of determination*

| Class / Grade | A | B |
|---|---|---|
| ONE | 0.81 | 0.74 |
| TWO | 0.87 | 0.82 |

*Figure 8. Individual trend line: Points earned as a function of score checking using data for the whole semester*

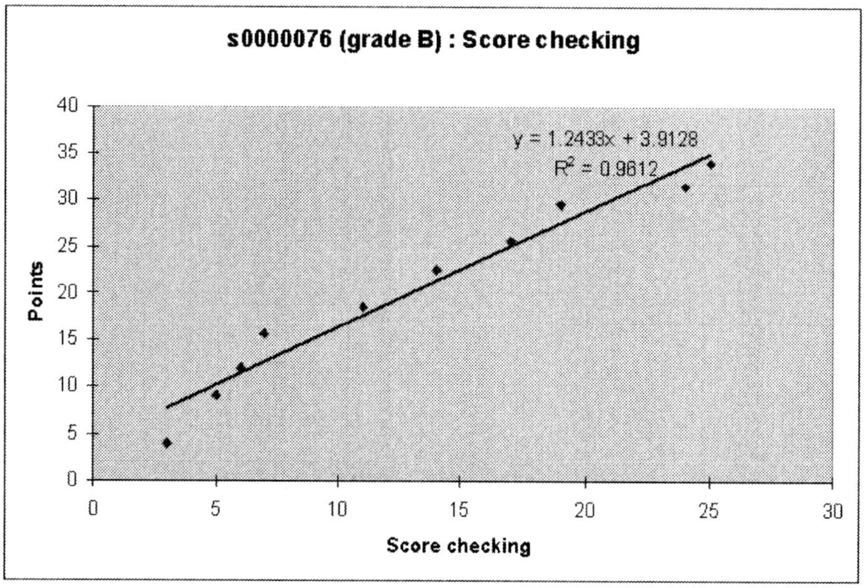


of operations made by the student to check the current score and the "points" axis presents the current score. The following example gives an illustration: At the moment, the student checked his current score 14 times he earned 22.5 points. It happened at the fifth week.

*Figure 9. Individual trend line: Points earned as a function of score checking using data for the first five classes*

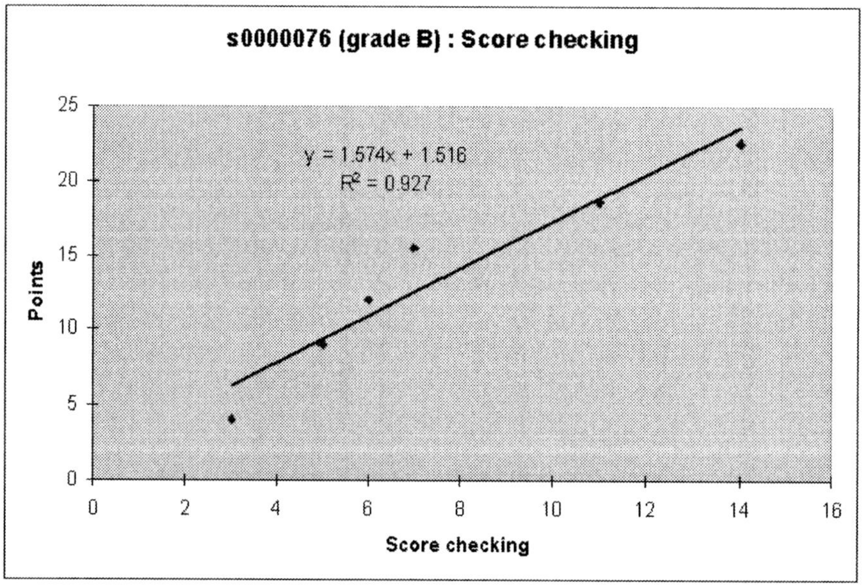

### Practical Issues

Analyzing these figures, the professor can see which students are not doing well in the course. Such a conclusion can be made quite early in the course where low activity predicts that the student will get a C or D grade. The instructor can act to help students. On the other hand, if the majority of students are successfully doing exercises (they have a current score of A), it is worth adjusting assignments to make them more difficult.

## EVALUATION OF THE TSI: A STUDENT VIEW

### Survey Characteristics

A questionnaire is a useful instrument to get users' opinion on tools they utilize (Carrol et al., 2005; Dagdilelis, Evangelidis, Satratzemi, Efopoulos, and Zagouras, 2003; Masiello, Ramberg, and Lonka, 2005). A survey was conducted to understand how easy the system is for undergraduate students, and how helpful it is for them.

The questionnaire consists of 13 questions in which an answer is selected from a prepared list. We distributed the questionnaire among the second year students enrolled in the exercise class of the Java Programming course at the end of the semester in paper and pencil format. Twenty-two students from class ONE and 27 students from class TWO participated in our study. The total number of students enrolled in these classes is 47 and 44 respectively. The number of returned questionnaires was 22 (100%) for class ONE and 20 (74%) for class TWO. The survey form was anonymous and questions were given in English. The answers to some questions can be gained from log files of the TSI. It helped us evaluate the reliability of obtained data. Table 2 presents statistics we got from the questionnaire and from the log files.

Some students provided several answers to Questions 3 and 11. This explains why the total number in the corresponding fields exceeds 100 percent. The "*Other*" items of Questions 5, 9, 10, 11, and 13 include the text in upper case. These are comments from the students.

We analyzed the obtained data from the different angles.

### Easiness Criterion

The majority of our students (88%) were positive about the TSI (Question 1). The key TSI operations were evaluated as follows: 93% of respondents selected a submission via the TSI as their preferred method (Question 2); 76% of participants reported that the feature to download class materials is convenient (Question 5). On the other hand, the number of students who never downloaded class materials is very high (64%). We have to clarify: A separate Web page was setup by the lecturer to provide students with lecture materials and assignment descriptions. The lecturer and the instructor were the different teachers. Teaching materials placed on the TSI were related to exercises. Teaching materials were prepared in English but they included references to Japanese sources on the Internet.

### Convenience Criterion

The majority of participants (67%) were satisfied with getting instructions via the Web (Question 3). Seventeen percent of respondents would like to receive instructional support via e-mail. Practically all participants said that it is convenient to check the current score via the TSI (Question 7). Students regularly received e-mails with their score sent from the TSI once a week, so, they did not need to do that operation (Question 13).

*Table 2. Statistics obtained from the questionnaire and from log files*

| Nr | Question | Possible answers | Class ONE | | Class TWO | |
|---|---|---|---|---|---|---|
| | | | Q'aire | Log | Q'aire | Log |
| 1 | Is it easy for you to use TSI? | Yes | 100% | | 80% | |
| | | No | | | 5% | |
| 2 | How do you prefer to submit results of your work? | a) Electronically via TSI | 100% | | 85% | |
| | | b) Electronically by e-mail | | | 15% | |
| | | c) in hard copy (paper submission) d) Other: | | | | |
| 3 | How do you prefer to get instructions? | a) via Web | 69% | | 65% | |
| | | b) by e-mail | 14% | | 20% | |
| | | c) during face-to-face talk with an instructor (TA / Professor) | 27% | | 15% | |
| 4 | Is English reading related to your course difficult for you? | Yes | 86% | | 75% | |
| | | No | 14% | | 25% | |
| 5 | Is it convenient to get class materials from the Web (from TSI)? | a) Yes | 82% | | 70% | |
| | | b) No | 9% | | 30% | |
| | | c) Other: (I don't know exactly, but I think it is good) | 5% | | | |
| 6 | Do you think TSI helps you to study this course? | a) Yes | 91% | | 60% | |
| | | b) No | 9% | | 40% | |
| | | c) Other: | | | | |
| 7 | Is it convenient for you to check your current results using TSI? | a) Yes | 95% | | 100% | |
| | | b) No c) Other: | | | | |
| 8 | Do you usually check the mail archive on TSI? | a) Yes | 41% | 38% | 30% | 34% |
| | | b) No | 59% | | 70% | |
| | | c) Other: | | | | |
| 9 | Have you ever changed your TSI password? | a) Yes | 27% | 17% | 45% | 34% |
| | | b) No | 68% | | 55% | |
| | | c) Other: (I didn't know about changing the password) | 5% | | | |
| 10 | Have you ever downloaded class materials with comments to handouts and assignments from TSI? | a) Yes | 32% | 36% | 30% | 36% |
| | | b) No | 59% | | 70% | |
| | | c) Other: (I didn't know, I know today) | 5% | | | |
| 11 | What do you usually use (read, study) doing exercises? | a) book, title________ Japanese book recommended by the lecturer | 32% | | 15% | |
| | | Another Japanese book | 17% | | 10% | |
| | | b) Lecture handouts from the Java course Web site | 50% | | 60% | |
| | | c) Comments to assignments from TSI | 9% | 17% | 5% | 10% |
| | | d) Materials from the Web you find on your own | 27% | | 20% | |

*continued on follow page*

*Table 2. continued*

| Nr | Question | Possible answers | Class ONE | | Class TWO | |
|---|---|---|---|---|---|---|
| | | | Q'aire | Log | Q'aire | Log |
| 12 | Have you ever sent e-mails to TA or Professor? | a) No | 64% | | 85% | |
| | | b) Yes, via TSI | 9% | 2% | 5% | 0% |
| | | c) Yes via my regular mail account | 27% | | 10% | |
| | | d) Other: | | | | |
| 13 | How often do you check you personal score? | a) once a week | 32% | | 60% | |
| | | b) once in two weeks | 19% | | 30% | |
| | | c) never check | 31% | 21% | 5% | 10% |
| | | c) Other: SOME TIMES | 13% | | | |
| | | ONES A MONTH | 5% | | | |
| | | THEN I LOGIN | | | 5% | |

## Helpfulness Criterion

Seventy six percent of students involved in this survey are sure that the TSI helps them in studying the course (Question 6). 24 percent oppose this opinion. Question 11 is about materials used by students in direct support of their study: Only seven percent of participants selected comments from the TSI and 36 percent preferred Japanese books. On the other hand, at least 31 percent of respondents downloaded materials from the TSI (Question 10). The data from log files confirm this outcome: 36 percent. Student answers to Question 4 and discussions in the "*easiness*" criterion subsection explain this phenomenon. Some important instructions and comments were sent via e-mail to the students to be sure they got this information because the TSI was accessible only from the university computers; students were able to receive e-mails on their mobile phones. This helps understand the situation with a relatively low activity to download class materials. The mail function as an indirect support feature was used on regular basis. Providing students with multiple choices is one of the key factors to effectively help them.

## Reliability of Our Data

According to the results of the Spearman's rank correlation test (class ONE: $n = 6$, $\alpha = 0.02$, $r_s = 0.9559$; class TWO: $n = 6$, $\alpha = 0.02$, $r_s = 0.9545$), there is a significant correlation between the data from the survey and the log files: Questions 8, 9, 10, 11, 12, and 13. One important outcome was that the use of English for the questionnaire was not a problem for the students; and data we obtained are reliable.

The main result from this survey is that the TSI is easy, convenient and helpful for our students. The same answer, we got from informal conversations with professors and TAs who used the TSI.

## Lessons Learned

The only reliable information we could get about students' group behavior from the log files was that students who work hard usually access the system much more frequently than students who are not successful in the course. The same result was obtained in the study by Grabe and Christopherson (2005).

The results of the statistical analysis are promising from the individual profile point-of-view. For the majority of students who got A and B grades, there is a correlation between a non-mandatory TSI activity (number of operations made by students checking their score, for example) and grades. So in many cases, it is possible to predict a student's grade and upcoming activity quite accurately after four to five classes.

We can also conclude that students with current score C or D (calculated automatically) after 4 classes need extra attention.

## CONCLUSION

The approach described in this chapter is an easy-to-implement alternative to technologies based on general purpose systems that are usually difficult to set up and often use a non-intuitive interface.

The TSI is no substitute for instructors; it supports their communication with students and handles most of the routine work. This system is extremely useful when an instructor works with non-native English speaking students. It will also help users who are not familiar with computers. Professors and students do not need to waste time learning the TSI; instead, they can concentrate on teaching and studying issues. For professors, it is easy to adapt it to their teaching needs. It can be applied as a supported tool for a wide range of teaching subjects. The system is written in Java and may be installed on any machine. The stable nature of this system was proved in the six years since it has been in use at the university.

The VBA application as part of the TSI can help instructors understand individual and group student behaviors. With this information, instructors can make necessary changes to a course during a running semester and help students learn better.

The system was highly evaluated by students as a simple, useful and convenient support tool.

The TSI can be downloaded from the URL: http://www.u-aizu.ac.jp/~niki/tsi/.

## ACKNOWLEDGMENT

The authors thank Dr. Renu Gupta for the critical review of an earlier draft of this chapter. Our specials thanks to our colleagues for providing us with log files of their classes taught with the TSI. Our gratitude is also to our students for their participation in this study.

## REFERENCES

Angel Learning. (2009). Retrieved January 2009 from, http://www.angellearning.com/

Blackboard. (2009). Retrieved January 2009 from, http://www.blackboard.com/

Bonk, C. J. *(2002). Collaborative tools for e-learning.* Chief Learning Officer Magazine, November. *RetrievedJanuary2009 from,* http://www.clomedia.com/

Carrol, J. M., Rosson, M., Convertino, G., & Ganoe, G. (2006). Awareness and teamwork in computer-supported collaboration. *Interacting with Computers, 18*, 21–46. doi:10.1016/j.intcom.2005.05.005

Course Genie. (2009). Retrieved January 2009 from, http://www.wimba.com/products/wimba_classroom/

Dagdilelis, V., Evangelidis, G., Satratzemi, M., Efopoulos, V., & Zagouras, C. (2003). DELYS: A novel microworld-based educational software for teaching computer science subjects. *Computers & Education, 40*, 307–325. doi:10.1016/S0360-1315(02)00133-1

FSU. (2009). Retrieved January 2009 from, http://learningforlife.fsu.edu/ctl/catalog/index.cfm /

Grabe, M., & Christopherson, K. (2005). Evaluating the advantages and disadvantages of providing lecture notes: The role of internet technology as a delivery system and research tool. *The Internet and Higher Education*, *8*(4), 291–298.

Kaiden, R. (2002). A review of WebCT. *The Internet and Higher Education*, *5*(4), 399–404. doi:10.1016/S1096-7516(02)00125-2

Kluev, V. (2004). Experiences teaching Japanese students. *The WSEAS Transactions on Advances in Engineering Education*, *1*(1), 11–14.

Lazar, J., Jones, A., & Hackley, M. (2006). Severity and impact of computer user frustration: A comparison of student and workplace users. *Interacting with Computers*, *18*, 187–207. doi:10.1016/j.intcom.2005.06.001

Llamas-Nistal, M. *Fernández-Iglesias, Manuel J., Anido-Rifon, L.E. (Eds.) (2004).* Computers & education: Towards a lifelong learning society. *Kluwer Academic Publishers.*

Masiello, I., Ramberg, R., & Lonka, K. (2005). Attitudes to the application of a Web-based learning system in a microbiology course. *Computers & Education*, *45*, 171–185. doi:10.1016/j.compedu.2004.07.001

Moodle. (2009). Retrieved January 2009 from, http://moodle.org/

Nicenet. (2009). Retrieved January 2009 from, http://www.nicenet.org/

Pahl, C. (2003). Managing evolution and change in web-based teaching and learning environments. *Computers & Education*, *40*, 99–114. doi:10.1016/S0360-1315(02)00100-8

*Rehberg, S. D., Ferguson, D. M., & McQuillan, J. M. (2004).* The ultimate WebCT handbook, a practical and pedagogical guide to 4.x WebCT.

*Storey, M.-A., Phillips, B., MaczewskiM., & WangM. (2002). Evaluating the usability of Web-based learning tools.* Educational Technology & Society, 5*(3).*

Top Class E-learning Suite. (2009). Retrieved January 2009 from, http://www.wbtsystems.com/solutions/technology

Tuckman, B. W. (2002). Evaluating ADAPT: A hybrid instructional model combining Web-based and classroom components. *Computers & Education*, *39*, 261–269. doi:10.1016/S0360-1315(02)00045-3

Web, C. T. (2009). Retrieved January 2009 from, http://www.blackboard.com/us/index.bbb /

Woods, R., Baker, J. D., & Hopper, D. (2004). Hybrid structures: Faculty use and perception of web-based courseware as a supplement to face-to-face instruction. *The Internet and Higher Education*, *7*(4), 281–297. doi:10.1016/j.iheduc.2004.09.002

Yen, C., & Wu-Jeng, L. (2003). Web-based learning and instruction support system for pneumatics. *Computers & Education*, *41*, 107–120. doi:10.1016/S0360-1315(03)00029-0

# Chapter 14
# The Impact of Examination Software on Student Attitudes and Examination Performance

**Lori Baker-Eveleth**
*University of Idaho, USA*

**Daniel M. Eveleth**
*University of Idaho, USA*

**Michele O'Neill**
*University of Idaho, USA*

**Robert W. Stone**[1]
*University of Idaho, USA*

## ABSTRACT

*The impact of examination software on student attitudes was investigated. The Technology Acceptance Model (TAM) provides the theoretical foundations for studying the completion of examination on student laptop computers. The model applies TAM to link both faculty and technical support for the examination software to student attitudes towards the software, while it is mediated by the perceived usefulness and ease of use of the software. The model is empirically tested using questionnaire responses from 107 students enrolled in sections of a business core course using the examination software. The statistical technique used is structural equations modeling. Empirical results show that perceived usefulness and ease of use of the software have direct, meaningful impacts on developing positive attitudes toward the software. Furthermore, faculty support and ease of system use impact student attitudes in a meaningful fashion indirectly through perceived usefulness. These empirical results are discussed and implications for instructors are offered.*

## INTRODUCTION

Declining technology prices and tight university operating budgets create an opportunity and a need to incorporate computer technology into the learning environment. Furthermore, the mobility of laptops provides a number of teaching and learning-oriented advantages; one advantage that has not been fully explored is using laptops to evaluate and assess

DOI: 10.4018/978-1-60566-936-6.ch014

student performance. The challenge is how to provide a secure examination environment such that students are constrained from accessing notes on their laptop hard drives or the internet, or communicating with other students via email and instant messaging. Another key issue, from both a faculty and student perspective, is the affect on a student's grade. If a student takes an exam using a paper and pencil exam there may be test anxiety, poor handwriting or hand cramps from writing, but most of these do not affect the outcome of the exam. Using technology though to automate the process of the exam, i.e., typed text, allows the student to eliminate the poor handwriting and hand cramping but adds an additional dilemma – what happens if the technology fails during the exam? A laptop could crash, the power could go out forcing battery backup, or the application could fail all these events could affect the outcome of the exam. In addition, as students prepare to enter the work force where there is an expectation of technology literacy, they must be prepared to learn the technology and then adapt to the changing environment.

As part of the student's transition from the university environment to the work force, an important skill they need is confidence in their ability to use technology. Confidence in one's ability can make the difference in a student being hired quickly for their first management position or struggling to find an entry level position. In general, self-efficacy is the individual's perception of possessing the requisite abilities to successfully perform a specific task (Bandura, 1977; 1982; 1986). A companion to self-efficacy is outcome expectancy which reflects the individual's perception regarding the result or gain from successful completion of these tasks. Expectancies have significant influence in a variety of settings involving behavioral and affective outcomes (Henry & Stone, 2001; Stone & Bailey, 2007; Lin, Ko & Wu, 2008).

Understanding self-efficacy and outcome expectancy can help to improve a student's transition from the university environment to the work force (Jenkins & Garvey, 2001). The research presented below focuses on the self-efficacy and outcome expectancy of technology use and its affect on exam performance. The purpose of the research was to evaluate the viability of using examination security software to conduct computer-based in-class examinations. A critical element in this evaluation is the affect on student performance as well as student confidence in their ability to use the technology.

The remainder of this paper is organized into sections presenting the details of this examination of self-efficacy theory and its impact on student's performance. First, a discussion of the theoretical model based on self-efficacy theory is presented. Next, the hypotheses to be tested are derived from the theoretical framework and the methodology used in the research. Finally, based on the empirical results, a discussion of the results is presented followed by conclusions and directions for future research.

## THE THEORETICAL MODEL

Predicting and perceiving student's technology use can be based on self-efficacy theory (Bandura, 1986; Bates & Khasawneh, 2007). Self-efficacy theory has been used to explain user reactions to information technologies (Bandura, 1986; Baronas & Louis, 1988; Hasan, 2003; Havelka, 2003; Martinko, Henry, & Zmud, 1996; Meier, 1985; Potosky, 2002). The theory (Bandura, 1986; 1982) links an individual's cognitive state to a variety of affective and behavioral outcomes and perceptions of future outcomes (i.e., loss of control, low self-confidence, low achievement motivation) (Staples, Hulland, & Higgins, 1998).

Recent research on computer self-efficacy investigated demographic predictors (e.g., academic major, gender, computer-related experience) influence on business student's self-efficacy (Havelka, 2003). Significant differences in self-efficacy ratings resulted for information systems and economics majors compared to management majors as well as those who have greater than five years experience working with computers. Gender differences did not result in any different self-efficacy ratings.

Expectations (e.g., motivation, performance, and feelings of frustration associated with repeated failure) in large part determine affect and behavioral reactions in numerous situations. Bandura (1986) separated expectations into two distinct types, self-efficacy and outcome expectancy. If the system is perceived to be useful, a user is more likely to adopt and use the technology in the future (Henry & Stone, 2001; Martinko, Henry, & Zmud, 1996). The extensive use of technology and information systems in the business education requires many systems to be non-volitional. A system that is mandatory may inflate the system use but the perception of usefulness will still be present (Iivari, 2005). Rai, Lang, and Welker (2002) defined "quasi-volitional IT use" as un- mandated use of the system but not completely volitional because of social pressure and subjective norms in the environment. This means that a student may not be required to use the system but the influences in the college environment encourage it.

An individual's belief that he or she possesses the skills and abilities to successfully accomplish a specific task represents self-efficacy. In addition, an individual's persistence to learn a task impacts his or her perceptions of future outcomes influencing their self-efficacy. Outcome expectancy is an individual's belief that by accomplishing a task, a desired outcome is attained. Outcome expectancy is the consequence of the act and not the act itself. Self-efficacy and outcome expectancy have separate impacts on behavior and affect. However, self-efficacy typically has a larger effect than outcome expectancy (Bandura, 1986) and generally self-efficacy has a direct impact on outcome expectancy (Stone & Henry, 2003).

Four groups of variables or experiences identified by Bandura (1977) impact an individual's expectancy evaluations of a specific task. The strongest is the individual's personal mastery or accomplishments regarding the task. Prior success at performing a task increases self-efficacy and outcome expectancy of that task. On the other hand, failing repeatedly at performing a task lowers these expectations (Gist & Mitchell, 1992). In the context of programming skills, experiences focusing on installation and repairs of computer hardware and software as well as experiences solving computer problems could be viewed as appropriate personal mastery experiences (Coffin & MacIntyre, 1999; Hasan, 2003).

Vicarious experience or modeling the behavior of others who successfully completed the task is the second group of variables. The observer can improve his or her own performance through observing others successfully completing the task (Bandura, 1977; Gist & Mitchell, 1992). In this study, vicarious experience can be viewed as watching teammates or others as they work on similar programming related projects.

Social persuasion is the third group of antecedents to self-efficacy and outcome expectancy. Social persuasion occurs when individuals are led or have it suggested to them that they can successfully complete the task in question and experience the resulting outcome. Common forms of social persuasion are verbal encouragement, coaching, and providing performance feedback (Bandura, 1977). In this study, one form of social persuasion is the encouragement from peers, other students, and faculty to successfully complete a project requiring programming.

The last grouping of antecedent variables is physiological arousal and emotional states. From here after, this will be referred to as physiological arousal. Physiological arousal of the individual impacts his or her expectancy judgments regarding specific tasks (Bandura, 1977). Improving perceptions of self-efficacy and the value of completing the task occur when there is intellectual interest in a task. Negative judgments of one's efficacy and the task outcome can be produced from anxiety regarding a specific task (Bandura, 1986). For the study at hand, the intellectual interest and stimulation of students regarding a project requiring programming efforts can impact the individual's self-efficacy and outcome expectancy about one's programming skills.

Based on the literature presented above, a model was developed. This model relates the antecedents of self-efficacy and outcome expectancy to students' attitudes and behaviors of technology use, mediated by self-efficacy and outcome expectancy. The four antecedents, watching others, experience, faculty support, ease of use, are proposed as directly impacting self-efficacy and outcome expectancy.

## THE EMPIRICAL STUDY

The sample was collected using a paper and pencil survey of undergraduate students enrolled in an integrated business course at the junior year. In selected sections of this course, students were required to complete in-class examinations using their laptop computers. To insure that students completing their laptop examination did not use unauthorized materials during the exam, students were required to use security software. During the exam students had access to a word processing document and a spreadsheet, both developed by the faculty team, once these files were opened the software prevented students from exiting these documents to perform any other actions (e.g., accessing hard drives, the internet, or email).

### Sample Data

A total of 107 students were enrolled in the two sections of the business core, which was taught by a faculty team consisting of five members. There were 63 students enrolled in one section and 44 in the other section. The exam was administered during a common time to 98 students using the secure software. Of the nine students who did not take the exam, five students were excused due to illness and four had conflicts with the common exam time. The 98 students who used the secure software for the exam were given the opportunity to complete a survey five days following the exam experience and prior to receiving their exam scores. Sixty-two students responded, with sixty students fully completing the survey for a 61% response rate. A verbal reminder to return the survey was provided during class prior to the exam scores being released. No additional reminders were provided due to concerns over latency effect and attribution of exam performance.

### Characteristics of the Respondents

Data characteristics of the sample are shown in table 1. The average GPA of 3.06 appears high, but it is the case that all students enrolled in the course must complete several qualifying courses with a minimum GPA. These requirements prohibit some students, who tend to have lower GPAs, from enrolling in the course. The average age of the respondents was 22.2 years old. Both genders were similarly represented to the distribution of the population, at 40% females and 60% males. The percentage of the sample students in each major ranged from a high of 28.33% in Marketing to 3.33% in both Production Operations Management and the Professional Golf Management program.

### Nonresponse Bias

As is the case of any research depending on data collected using a survey, nonresponse is a concern. The issue examined by non-response bias is how representative the sample is of the target population. In other words, the question is do the nonrespondents bias the ability of the sample to accurately represent the target population? To examine the possible presence of nonresponse bias, the sample characteristics were compared to the corresponding values at the College level (i.e.,

*Table 1. Sample and population characteristics*

| Characteristic | Sample | College Population |
|---|---|---|
| Average GPA | 3.06 | 2.98 |
| Average Age (years) | 22.2 | 21.6 |
| Gender: | | |
| Female | 40.00% | 39.00% |
| Male | 60.00% | 61.00% |
| Major: | | |
| Marketing | 28.33% | 22.20% |
| Information Systems | 11.67% | 12.60% |
| Finance | 11.67% | 11.60% |
| Management & Human Resources | 16.67% | 14.00% |
| Production Operations Management | 3.33% | 5.60% |
| Professional Golf Management | 3.33% | 0.70% |
| Accounting | 20.00% | 16.20% |
| Economics-Finance | 5.00% | 6.10% |
| Other/Undeclared | N/A | 9.00% |
| Number of Observations | 60 | 837 |

the target population). The students enrolled in the College represent the population of students that could have been selected into the sample for this study. The average GPA of all students declared as majors within the College was 2.98. Statistical tests indicate that the difference between these two GPAs is not significant at a 20% level in two-tail tests. The average age among all students in the College was 21.6 years. Statistical tests show the average ages are not statistically different at a 5% level of significance. At the College level, 39% of the population was female and 61% male. Tests comparing proportions show no significant differences at a 10% significance level. Also shown in table 1 are the percentages by major for the College. There are no significant differences using a chi-square goodness of fit test and a 10% level of significance. Based on the comparison of these demographic variables between the sample and the College, it is concluded that nonresponse bias does not present a problem.

## THE MEASURES AND THEIR PSYCHOMETRIC PROPERTIES

We modified prior work (Davis, 1989; Agarwal and Prasad, 1999; Stone and Henry, 2003) to develop the research specific questionnaire items to measure the constructs in the TAM. For all measures, students were asked the extent to which they agreed or disagreed with each of the items and were presented with the ordered answer choices of Strongly Disagree, Disagree, Neutral, Agree, or Strongly Agree. To evaluate the appropriateness of the survey instrument generally, we pre-tested an initial questionnaire with 11 students that had completed the course a year earlier and were familiar with the exam software. Based on their feedback concerning the intent of the questions, we eliminated four items. All of these items are discussed below.

### Measures Examined

The external variable measure, Faculty Support, was measured by two items. These items were prompted with, "For each of the following factors, indicate the extent to which you agree or disagree that it helped your use of the software." The items were, "Faculty explaining the advantages of the software" and "Faculty explaining laptop-based exams and related software is inevitable." These items were created by the authors. The other external variable measure, Technical Support, also used an overall prompt, "The IXL Staff..." and referred to the name of the helpdesk for the laptop program that provides technical support. Items for this measure were, "Is helpful when someone has a problem," "Seems knowledgeable about the software," "Is able to identify the source of most of the software problems," and "Is supportive

*Table 2. The questionnaire items, measures, and reliabilities*

| Measures and Indicants | Cronbach's Alpha |
|---|---|
| *Faculty Support* | 0.79 |
| For each of the following factors, indicate the extent to which you agree or disagree that it helped your use of the software: | |
| Faculty explaining the advantages of the software. | |
| Faculty explaining laptop-based exams and related software is inevitable. | |
| *Technical Support* | |
| The IXL Staff… | 0.80 |
| Is helpful when someone has a problem with the software. | |
| Seems knowledgeable about the software. | |
| Is able to identify the source of most of the software problems. | |
| Is supportive when someone has a question or problem with the software. | |
| *Perceived Usefulness* | 0.64 |
| Using the software… | |
| Gives me greater control over my exam performance. | |
| Improves my exam performance. | |
| Saves me time in the exam. | |
| Enables me to accomplish the exam tasks more quickly. | |
| Enhances my effectiveness on the exam. | |
| Improves the quality of the work I do. | |
| Increases my productivity. | |
| *Ease of System Use* | 0.72 |
| I often become confused when using the software. | |
| I make errors frequently when using the software. | |
| *Attitudes Towards Software* | 0.60 |
| Completing exams using the software… | |
| Is a good idea. | |
| Is unpleasant (reverse coded). | |
| Is beneficial. | |
| Fits the way I prefer to work on exams. | |
| Does not fit the way I prefer to work on exams (reverse coded). | |
| Meets my needs. | |

when someone has a question or problem with the software." All of these items were modified by the authors from previously published scales (Stone and Henry, 2003).

For the measure, Perceived Usefulness, the phrase, "Using the software…" prompted its list of items: "Gives me greater control over my exam performance," "Improves my exam performance," "Saves me time in the exam," "Enables me to accomplish the exam tasks more quickly," "Enhances my effectiveness on the exam," "Improves the quality of the work I do," and "Increases my productivity." These items were created from a published scale by Davis (1989). The measure,

Ease of System Use, consisted of two items based on the work of Agarwal and Prasad (1999) phrased as, "I often become confused when using the software" and "I make errors frequently when using the software." Given the phrasing of these items, their values were reverse coded prior to being included in the empirical analysis.

The measure, Attitudes Towards Software, was developed by the authors using items that all began with the phrase, "Completing exams using the software." The specific items were "Is a good idea," "Is unpleasant (reversed coded)," "Is beneficial," "Fits the way I prefer to work on exams," Does not fit the way I prefer to work on exams (reverse coded)" and "Meets my needs." All these items are shown in Table 2.

The measures used in the subsequent analysis were formed by summing the individual questionnaire items. These sums were averages of the responses to each item. This was due to the complexity of the model to be estimated compared to the sample size. The reliability of the measures were evaluated by calculating their Cronbach's Alpha. These reliability coefficients ranged from 0.60 (Attitudes Toward the Software) to 0.80 for Technical Support. The remaining reliability coefficients were: 0.79 for Faculty Support; 0.64 for Perceived Usefulness; and 0.72 Ease of System Use. These reliabilities are marginal values at best and are also displayed in Table 2. The correlations among the measures used in the study were also computed. These values ranged from a high of 0.85 (Perceived Usefulness and Attitudes Towards the Software) to a low of 0.12 (Faculty Support and Technical Support). All the correlations are shown in Table 3.

## THE ESTIMATION OF THE MODEL

As discussed earlier, due to the relatively complex model compared to the small sample size, the measures in the model were formed using the summation of the questionnaire items as the measures. Furthermore, due to the sample size considerations, the model was estimated using three ordinary least squares regressions. Each regression estimated a collection of paths in the model. The regression results are presented in Table 4.

The first regression (Regression One) had Ease of System Use as the dependent variable. The re-

*Table 3. The correlations among the measures*

| Measure Pair | Correlation |
|---|---|
| Faculty Support / Technical Support | 0.12 |
| Faculty Support / Perceived Usefulness | 0.34 |
| Faculty Support / Ease of System Use | 0.14 |
| Faculty Support / Attitudes Toward Software | 0.41 |
| Technical Support / Perceived Usefulness | 0.27 |
| Technical Support / Ease of System Use | 0.17 |
| Technical Support / Attitudes Towards Software | 0.35 |
| Perceived Usefulness / Ease of System Use | 0.59 |
| Perceived Usefulness / Attitudes Towards Software | 0.85 |
| Ease of System Use / Attitudes Towards Software | 0.63 |

*Table 4. The regression results*

| Parameter Estimates | | | | | |
|---|---|---|---|---|---|
| Source | Degrees of Freedom | Sum of Squares | Mean Square | F-Value | Pr>F |
| Model | 2 | 8.73 | 4.36 | 1.27 | 0.29 |
| Error | 57 | 195.87 | 3.44 | | |
| Corrected Total | 59 | 204.60 | | | |

*Figure 1. The significant relationships found for the estimated technology acceptance model*

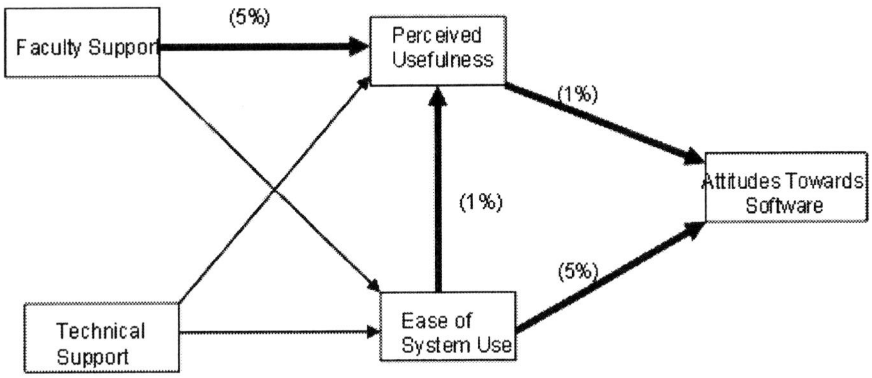


sults indicated no significant impact of the regression equation as a whole in explaining variations in the Ease of System Use measure. Furthermore, only the intercept term had a significant impact on Ease of System Use among the individual explanatory variables. The second estimated regression (Regression Two) used Perceived Usefulness as the dependent variable. The regression equation as a whole explained a significant portion of the variation in Perceived Usefulness. Ease of System Use and Faculty Support had meaningful, significant individual impacts on Perceived Usefulness. The final estimated regression (Regression Three) used Attitudes Toward Software as the dependent variable. For this regression, the estimated equation as a whole explained a significant variation portion of the variations in Attitudes Toward Software. Furthermore, the explanatory variables of Perceived Usefulness and Ease of System Use had individual significant impacts on the dependent variable. All of these results are summarized and presented in Figure 1.

## DISCUSSION

The regression results provided several meaningful results. Faculty Support, such as discussing the advantages of using the software, significantly impacted Perceived Usefulness of the examination security software. Ease of System Use, relating to the interface and minimizing confusion had similar, significant impacts on Perceived Usefulness. These results indicate that faculty actions and software characteristics influence Perceived Usefulness and ultimately students' Attitudes Toward the Software. Furthermore, software characteristics impact Ease of System Use which has direct impacts on students' Attitude Toward Software. Thus, in order to impact current student attitudes and future adoption of examination security software, faculty actions and characteristics of the software such as interface and navigation influence its ease of use and play an important role.

There were also expected results that were not empirically observed. The first of these was that the antecedents of Ease of Use (i.e., Faculty Support and Technical Support) had no meaningful impacts on Ease of System Use. While Ease of System Use impacts students' Attitudes Toward Software, none of the proposed antecedents had any observed impacts on Ease of System Use. Additional research is needed to expand the antecedents to Ease of System Use.

The second predicted result that was not observed was no meaningful impacts from the antecedent of Technical Support. A potential explanation for this lack of meaningful results lies in the value of the technical support for students.

Only students who experienced difficulties made use of technical support. Very few students experienced difficulties and made use of the available technical support. Therefore, this support had little or no value to the majority of students. As a result, these variables had no meaningful impacts in the estimated model.

## IMPLICATIONS FOR INSTRUCTORS

The purpose of the study was to examine students' attitudes toward software providing security in computer-based examinations. The hope was to understand what influences these attitudes have in order to facilitate such software adoption. There are implications for instructors within this context. First ease of use for the software matters. The selection of such examination security software needs to consider the skill set of the students who will use the software as well as the characteristics of the software. The easier a system is to use on the part of students (i.e., a new testing procedure/software), the more accepting they will be of the technology. Also, the type of examinations, essay-based versus multiple-choice, may influence the ease of use. In previous testing of the software using a spreadsheet, certain features are not available. A multiple-choice exam may not need additional features of the software and be considered easier to use. Second, what faculty say and do to support students' use of such software matter. By faculty "selling" the usefulness of the examination software, demonstrating how to use the software, they encourage students' perceptions of system usefulness and their attitudes regarding the software.

Although the Technical Support yielded no significant impact, only two of the 107 students requiring technical staff (extra Ethernet cable and re-booting a locked operating system), the presence and services of the staff are important. If no technical support was available during the examination and a larger number of students experienced problems, the perception of the software being easy to use might be impacted.

As more colleges and universities require laptops, faculty will want to consider using the laptop for evaluation and assessment. Ideally this will further spark software developers to consider all different needs related to evaluation in a secure exam, laptop environment. Moving beyond the needs of multiple-choice exams, incorporating a variety of applications (for example, Microsoft Access) and expanding the tool set within the software would provide a greater understanding of student attitudes towards the software.

## REFERENCES


Adams, D. A., Nelson, R. R., & Todd, P. A. (1992). Perceived usefulness, ease of use, and usage of information technology: a replication. *MIS Quarterly, 16*(2), 227–247. doi:10.2307/249577

Agarwal, R., & Karahanna, E. (2000). Time flies when you're having fun: Cognitive absorption and beliefs about information technology usage. *MIS Quarterly, 24*(4), 665–694. doi:10.2307/3250951

Agarwal, R., & Prasad, J. (1999). Are individual differences germane to the acceptance of new information technologies? *Decision Sciences Journal, 30*(2), 361–391. doi:10.1111/j.1540-5915.1999.tb01614.x

Ajzen, I. (1988). *Attitudes, personality, and behavior*. The Dorsey Press, Chicago.

Ajzen, I. (1991). The theory of planned behavior . *Organizational Behavior and Human Decision Processes, 50*(2), 179–211. doi:10.1016/0749-5978(91)90020-T

Ajzen, I., & Fishbein, M. (1980). *Understanding attitudes and predicting social behavior*. Englewood Cliffs, NJ: Prentice-Hall.

Bates, R., & Khasawneh, S. (2007). Self-efficacy and college students' perceptions and use of online learning systems. *Computers in Human Behavior*, *23*, 175–191. doi:10.1016/j.chb.2004.04.004

Chau, P. Y. K., & Hu, P. J.-H. (2001). Information technology acceptance by individual professionals: A model comparison approach. *Decision Sciences*, *32*(4), 699–719. doi:10.1111/j.1540-5915.2001.tb00978.x

Davis, F. D. (1989). Perceived usefulness, perceived ease of use, and user acceptance of information technology. *MIS Quarterly*, *13*(3), 319–340. doi:10.2307/249008

Davis, F. D., Bagozzi, R. P., & Warshaw, P. R. (1989). User acceptance of computer technology: A comparison of two theoretical models. *Management Science*, *35*(8), 982–1003. doi:10.1287/mnsc.35.8.982

Fornell, C., & Larcker, D. F. (1981). Evaluating structural equation models with unobservable variables and measurement error. *JMR, Journal of Marketing Research*, *18*, 39–50. doi:10.2307/3151312

Gefen, D., & Straub, D. (2000). The relative importance of perceived ease of use in IS adoption: A study of e-commerce adoption. *Journal of the Association for Information Systems*, *1*(8), 1–28.

Ghani, J. A., Supnick, R., & Rooney, P. (1991). The experience of flow in computer-mediated and in face-to-face groups. In *Proceedings of the Twelfth International Conference on Information Systems*. New York, NY.

Hair, J., Jr., Anderson, R. E., Tatham, R. L., & Black, W. C. (1992). *Multivariate Data Analysis with Readings*. New York: MacMillan Publishing Company.

Hasan, B. (2003). The influence of specific computer experiences on computer self-efficacy beliefs. *Computers in Human Behavior*, *19*(4), 443–450. doi:10.1016/S0747-5632(02)00079-1

Havelka, D. (2003). Predicting software self-efficacy among business students: A preliminary assessment. *Journal of Information Systems Education*, *14*(2), 145.

Henry, J. W., & Stone, R. W. (2001). The role of computer self-efficacy, outcome expectancy, and attribution theory in impacting computer system use. *Journal of International Information Management*, *10*(1), 1–16.

Igbaria, M., & Greenhaus, J. H. (1992). Determinants of MIS employee's turnover intentions: A structural equation model. *Communications of the ACM*, *35*, 35–49. doi:10.1145/129630.129631

Igbaria, M., Zinatelli, N., Cragg, P., & Cavaye, A. L. M. (1997). Personal computing acceptance factors in small firms: A structural equation model. *MIS Quarterly*, *21*(3). doi:10.2307/249498

Iivari, J. (2005). An empirical test of the DeLone-McLean model of information system success. *The Data Base for Advances in Information Systems*, *36*(2), 8–27.

Legris, P., Ingham, J., & Collerette, P. (2003). Why do people use information technology? A critical review of the technology acceptance model. *Information & Management*, *40*(3), 191–204. doi:10.1016/S0378-7206(01)00143-4

Lin, M. P., Ko, H. C., & Wu, J. Y. W. (2008). The role of postivie/negative outcome expectancy and refusal self-efficacy of internet use on internet addiction amond college students in Taiwan . *CyperPsychology and Behavior*, *11*(4), 451–457. doi:10.1089/cpb.2007.0121

Martinko, M. J., Henry, J. W., & Zmud, R. W. (1996). An attributional explanation of individual resistance to the introduction of information technologies in the workplace. *Behaviour & Information Technology, 15*(5), 313–330. doi:10.1080/014492996120094

Nunnally, J. (1978). *Psychometric Methods* (2nd ed.). New York: McGraw-Hill.

Rai, A., Lang, S. S., & Welker, R. B. (2002). Assessing the validity of IS Success models: An empirical test and theoretical analysis. *Information Systems Research, 13*, 50–69. doi:10.1287/isre.13.1.50.96

Rainer, R. K. Jr, & Harrison, A. W. (1993). Toward development of the end user computing construct in a university setting. *Decision Sciences Journal, 24*, 1187–1202. doi:10.1111/j.1540-5915.1993.tb00510.x

Rivard, S., & Huff, S. (1988). Factors of success for end-user computing. *Communications of the ACM, 31*, 552–561. doi:10.1145/42411.42418

Shih, H.-P. (2003). Extended technology acceptance model of Internet utilization behavior. *Information & Management, 41*, 719–729. doi:10.1016/j.im.2003.08.009

Shneiderman, B. (1998). *Designing the user interface - Strategies for effective human-computer interaction*. Addison-Wesley.

Stone, R. W., & Bailey, J. J. (2007). Team conflict self-efficacy and outcome expectancy of business students. *Journal of Education for Business, 82*(5), 258–266. doi:10.3200/JOEB.82.5.258-266

Stone, R. W., & Henry, J. W. (2003). The roles of computer self-efficacy and outcome expectancy in influencing the computer end-users' organizational commitment. *Journal of End User Computing, 15*(1), 38–53.

Tractinsky, N. (1997). Aesthetics and apparent usability: Empirically assessing cultural and methodological issues. In *Proceedings of the CHI 97*, Atlanta, GA.

Venkatesh, V., & Davis, F. D. (1996). A model of the antecedents of perceived ease of use: Development and test. *Decision Sciences, 27*(3), 451–481. doi:10.1111/j.1540-5915.1996.tb01822.x

Wiggins, G. (1998). *Educative assessment: Designing assessments to inform and improve student performance*. San Francisco: Jossey-Bass Publishers.

Zhang, P., Benbasat, I., Carey, J., Davis, F., Galletta, D., & Strong, D. (2002). Human-computer interaction research in the MIS discipline. *Communications of the AIS, 9*(20), 334–355.

## ENDNOTE

[1] Authors listed in alphabetical order.

# Chapter 15
# Using PowerPoint to Encourage Active Learning:
## A Tool to Enhance Student Learning in the First Accounting Course

**Elise A. Boyas**
*Robert Morris University, USA*

## ABSTRACT

*Educators have recognized that curricula must become more focused on developing student skills and less focused on memorization of rules. Increasing demands on student time make it necessary for students to access learning support tools on demand. Successful curricula give students opportunities for active learning and opportunities to be self-regulated learners. Teaching students how to be active, self-regulated learners prepares them for careers which will require them to be lifelong learners. The creative use of technology throughout the curriculum can help educators bring active learning experiences to their students. This article describes the development of an interactive PowerPoint module for use in an introductory accounting course in a business school. This use of PowerPoint provides students with immediate, appropriate feedback with explanatory details. This tool is designed to be used outside of the classroom at the student's own pace and can be used in disciplines other than accounting.*

## INTRODUCTION

Educators have recognized that curricula need to become more focused on developing skills which help students to be active, lifelong learners. The creative use of technological tools can assist in achieving this goal by bringing active learning experiences to students. This article describes the need for such tools in the delivery of undergraduate accounting curricula and describes an interactive teaching tool for use in the introductory accounting class.

Over approximately the past fifteen years, the American Institute of Certified Public Accountants (AICPA) (1999, 1998) and the Accounting Education Change Commission (AECC) (1992, 1990) along with the American Accounting Association (AAA) and a host of accounting researchers and educators (for example, Albrecht & Sack, 2000) have focused on the changes in the accounting profession and the need for practicing accountants to be lifelong learners. These reports describe a gap between what accountants do in practice and what accounting educators teach. The overriding conclusion of these reports was that the focus of accounting education must shift to meet practice needs and must teach students the skills necessary to be successful accountants, rather than centering on the delivery of content.

Saunders and Christopher (2003) synthesize the efforts of the AECC, AICPA and AAA and describe the characteristics of successful accounting programs as follows:

1. The student should be an active participant in the learning process.
2. The student should be taught to identify and solve unstructured problems that require use of multiple information sources.
3. Learning by doing should be emphasized.
4. Working in groups should be encouraged.
5. The creative use of technology is essential.

The use of technological tools and platforms can be very helpful in developing these characteristics. But in a survey of 106 accounting chairpersons (at mainly AACSB accredited institutions), Saunders and Christopher find that few accounting programs require students to attend an accounting lab in an introductory course (15.1%), only about half require students to complete a computer-based assignment (54.3%), and few require the completion of a computer-based simulation (20%).

Smith (2001) points out that the best way to prepare accounting students for a career requiring lifelong learning is to show them how to be self-motivated, self-regulated learners while they are pursuing their degrees. A self-regulated learner takes ownership of his or her learning by understanding which learning strategies work best for him or her. This means that a necessary condition for successful self-regulated learning is that the learner should have some free choice over the use of learning tools and that the learner must be actively involved in the learning process.

Chickering and Gamson (2001) describe the relationship between the principles of good teaching practice (as developed by the American Association of Higher Education) and new technologies. They state that good teaching practice should use technology to promote active learning and provide prompt feedback to students about their learning. In addition, technology can be used to increase the value of student time spent on task by making studying and learning more efficient.

Proserpio and Gioia (2007) argue that educators are now teaching a generation of learners who are no longer simply verbal or visual learners. Students now prefer to use interactive media and learn best when presented with learning opportunities based in the technologies they are accustomed to using. Proserpio and Gioia (2007) write, "Students now expect rich, interactive, and even 'playful' learning environments" (p. 73). They suggest that instructors can create a sense of personal involvement and interaction with their students through the use of technological tools. Effective teaching of this generation is dependent on an instructor's creative use of available technology.

Curricula can benefit from the creative use of technology to enhance student learning outcomes. Technological tools can make the student a more active participant in the learning process and can help to enhance student learning by giving students convenient access to review material and immediate, appropriately detailed feedback.

## APPLICATION TO INTRODUCTORY ACCOUNTING

Although there is a perception that multiple choice questions are most suitable for assessing basic student knowledge, properly written multiple choice questions can assess both basic and higher order skills with reliability and validity (Center for Teaching and Learning, 1990). Accounting text materials provide practice quizzes in a multiple choice format to assist student learning and text-related test banks supply a large variety of multiple choice questions. These are often the source of instructor prepared multiple choice assessments linked to course learning objectives.

An important component of effective student learning is appropriately detailed and timely feedback. When multiple choice questions are included in text material as chapter quizzes, students are provided with the correct answer, but are not provided with additional information explaining why wrong answers are incorrect or why the correct answer is correct. Therefore, although multiple choice questions may be an appropriate format for assessing student learning outcomes, their use as a student learning tool is limited since they do not provide appropriately detailed feedback to students. This deficiency can be addressed through instructor-provided resources.

Instructors with direct knowledge of web design can program web-based interactive tools. But many instructors in a content-rich discipline such as accounting do not have the skill set to effectively program an interactive web-based learning tool nor do they desire to invest the time to learn how to do so. They are, however, typically fairly proficient with PowerPoint.

Instructors can use the advanced features of PowerPoint to create interactive multiple choice review tools which provide immediate and appropriately detailed feedback to students. Practice quiz questions from the text material can be used or the instructor can create multiple choice questions. These PowerPoint tools can be hosted on a course management server like Blackboard, or simply e-mailed to students. They can be used in class with a group or as individual assignments to be completed outside of class time.

The Appendix to this article documents the steps necessary to create a PowerPoint tool using multiple choice questions based on material in an introductory accounting course. The student reads a multiple choice question and selects an answer choice by clicking on the choice's letter. A hyperlink takes the student to a slide which will either (1) confirm the choice as correct and give the student the reason the choice is correct or (2) explain why the choice is incorrect, forcing the student back to the question to try again. When the correct answer is chosen, a hyperlink takes the student to the next question. Sound effects are used to reinforce the student's answer and make the learning tool engaging.

This tool can be used in a wide variety of courses where skills can be assessed using multiple choice questions. Student learning in undergraduate courses which are not case-study based can benefit from this tool. While this tool cannot be used to directly measure the achievement of student learning outcomes in assessment reporting, it can be used as a remediation tool for students who are having difficulty mastering course objectives and require extra attention.

The tool described in this article is interactive, provides immediate, appropriately detailed feedback to students and can be worked on by students at their own pace. It gives the student an opportunity to be an active, self-regulated learner and is a creative use of technology to enhance student learning in an introductory accounting course. Initial anecdotal feedback from students in an introductory accounting course who have used this tool is that they find it engaging, easy to use and helpful in their learning.

## CONCLUSION

The typical student, whether a traditional post-high school undergraduate or a non-traditional adult learner, is balancing a great many demands on his or her time. Family, work and school commitments often prevent students from being on campus during faculty office hours or during scheduled tutoring sessions. The new generation of learners is quite comfortable with the use of a variety of forms of technology and may be more inclined to complete coursework in an interactive format than with a book and pencil and paper.

Few accounting programs currently support an organized accounting lab and students may wish to complete their work at times of the day which do not coincide with the availability of the instructor or tutors. Successful accounting curricula give students opportunities for active learning and opportunities to be self-regulated learners. Interactive learning modules like the module described in this article are always available and can be accessed when convenient to the student. These modules can immediately direct the student through hyperlinks to task specific support material, enhancing the value to the student of time spent doing course-related work.

The PowerPoint module presented in this article actively engages students in their learning, gives them opportunities to learn by doing rather than by simply listening to a lecture or reading a book and provides prompt and appropriate feedback to students regarding their work. This module embodies many of the characteristics of good teaching practice, is cost effective to develop and implement and is a creative use of technology to enhance student learning. Although this article focused on the use of this tool in the introductory accounting course, it can be used in other academic areas.

## REFERENCES

Accounting Education Change Commission (AECC). (1990). Objectives of education for accountants: Position statement number one. *Issues in Accounting Education, 5*(2), 307-312.

Accounting Education Change Commission (AECC). (1992). Objectives of education for accountants: Position statement number two. *Issues in Accounting Education, 7*(2), 249-251.

Albrecht, W. & Sack, R. (2000). Accounting education: Charting the course through a perilous future. *American Accounting Association, Accounting Education Series, 16*. Retrieved from http://aaahq.org/pibs/AESv16/toc.htm.

American Institute of Certified Public Accountants (AICPA). (1998). CPA vision project: Focus on the horizon. *Journal of Accountancy, 3*(6), 186.

American Institute of Certified Public Accountants (AICPA). (1999). AICPA core competency framework for entry into the accounting profession. Retrieved from http://www.aicpa.org/edu/corecomp.htm.

Center for Teaching and Learning. (1990). Improving Multiple Choice Questions. Retrieved from http://ctl.unc.edu/fyc8.html.

Chickering, A. & Gamson, Z. (2001). Implementing the seven principles of good practice in undergraduate education: Technology as lever. *Accounting Education News, spring*, 9-10.

Proserpio, L. & Gioia, D. (2007). Teaching the virtual generation. *Academy of Management Learning & Education, 6*(1), 69-80.

Saunders, G. & Christopher, J. (2003). Teaching outside the box: A look at the use of some nontraditional teaching models in accounting principles courses. *Journal of American Academy of Business, 3*(1-2), 162.

Smith, P. (2001). Understanding self-regulated learning and its implications for accounting educators and researchers. *Issues in Accounting Education, 16*(4), 663-701.

## APPENDIX

The example presented below illustrates how to design and structure an interactive PowerPoint slide presentation and includes two questions related to inventory costing methods. The content of the PowerPoint slide is in a box and replicates what the student would see when working through the PowerPoint. The instructions on how to create the slides are written in italics below the box.

### Slide 1

> **Intro to Accounting**
> **Interactive Practice Exercises – Inventory**
>
> Instructions:
> Click on the Blue Arrow or on a word highlighted in Blue to move from slide to slide.
>
> Action Button (Blue Arrow)

*You are going to restrict the student from moving through the slides by controlling the hyperlinks within the PowerPoint. To install the hyperlink action button:*

1. *Select "Auto Shapes" on the bottom left toolbar.*
2. *Select "Action Buttons".*
3. *Select an action button of your choice and insert it into the slide. An arrow pointing to the right is a good choice, indicating that clicking here will move the student to the next slide.*
4. *A window entitled "Action Settings" will appear; select Next Slide on the "Mouse Click" tab under "Hyperlink to" so that the action button is activated when the student clicks on the button.*

### Slide 2

> **Welcome to Interactive Practice Exercises - Inventory**
>
> If you are annoyed by or object to the sound effects in this interactive module, just turn off your speakers.
>
> If you have questions about items included in this module, e-mail your Professor or ask a question during class.
>
> Action Button (Blue Arrow)

*The use of sound effects in this slide show can be very effective but may be objectionable to some students. Placing this disclaimer up front makes students aware that they can turn the sound off without losing content.*

*Follow the instructions in slide 1 to insert the hyperlinked action button to get to the next slide.*

*To insert a hyperlink for students to send an e-mail to you while they are in the PowerPoint do the following:*

1. *Highlight the words "e-mail your Professor".*
2. *Right click.*
3. *Select "Hyperlink".*
4. *Select "e-mail address" at the bottom of the left bar.*
5. *In the window, enter your e-mail address and the subject (for example, Student Question - Inventory Practice Exercises) and click OK.*

## Slide 3

**Question 1**

ABC Company uses FIFO to value its ending inventory of Part R. ABC had the following transactions during January 2007. All sales are at a unit sales price of $20.

| Transaction | Quantity | Unit Cost |
|---|---|---|
| Beginning inventory 1/1/07 | 400 | 10 |
| Purchase 1/10/07 | 200 | 11 |
| Purchase 1/15/07 | 300 | 12 |
| Purchase 1/18/07 | 400 | 12 |
| Sale 1/22/07 | 300 | |
| Sale 1/28/07 | 400 | |

Calculate the value of ending inventory on January 31, 2007.

A $6,200
B $7,200
C $14,600
D $7,400

*Each selection (A-D) will be hyperlinked to a separate slide in the PowerPoint presentation which will immediately tell the student whether the selection is correct or incorrect. In addition, the student will be presented with some information to assist in calculating the correct answer and will be forced to go back to the question to try again if an incorrect choice is selected.*

*The first step is to create four additional slides to follow this question, slides 4-7, and leave them blank for now.*

*To insert the hyperlinks:*

1. *Highlight the "A" on the first choice line.*
2. *Right click.*
3. *Select "Hyperlink".*
4. *From the left menu, select "Place in This Document"*
5. *From the slide choices, select "Slide 4". It will be shown as a blank slide.*
6. *Click "OK" to exit the window.*

7. *Repeat this procedure for the other choices, B, C and D, linking them to slides 5, 6 and 7, respectively.*

*Assume that the student incorrectly selects option A and proceeds to the next slide.*

## Slide 4

| |
|---|
| Sorry, Choice A is not correct.<br><br>HINT: Remember that FIFO (first in, first out) means that the costs associated with the units sold are the earliest or oldest costs in inventory. So, the items sold will be costed out using the unit costs for the beginning inventory and then the next purchase, and so on until all the items sold are costed. This amount equals Cost of Goods Sold.<br><br>This means that under FIFO, the units left in ending inventory at the end of the month are costed out using the newest or more recent unit costs.<br><br>Click here to try again. |

*A behavior reinforcing (and humorous) touch here is to use a sound effect when the student reaches this screen. The fog horn sound clearly communicates that the student has made the wrong choice. To select a sound:*

1. *On the main toolbar, select "Insert".*
2. *Select "Movies and Sounds."*
3. *Select "Sounds from Clip Organizer" (you can also import sound from a sound file but the sound clip organizer has a very wide selection and is easy to use.)*
4. *Search for fog horn or any other sound.*
5. *Click on it and select "Automatically" so it plays when the student reaches the slide.*
6. *A small megaphone icon will appear on the slide to indicate that sound is present.*
7. *To hide the megaphone from the view of the student, click on it.*
8. *Select "Edit Sound Object".*
9. *Under display options, click "Hide Sound Icon during Slide Show."*

*The explanation of why the answer the student chose was incorrect can be as detailed as necessary. If appropriate, you can insert a hyperlink to teaching material on the Web or supported by the textbook to reinforce your explanation.*

*To return the student to the original question to try again:*

1. *Highlight the word "here" in the last sentence.*
2. *Right click.*
3. *Select "Hyperlink".*
4. *From the left menu, select "Place in This Document"*
5. *From the slide choices, select "Slide 3", the slide with the original question.*
6. *Click "OK" to exit the window.*

*You will create similar slides for choices C and D, both incorrect choices. We will proceed assuming the student selects the correct answer, choice B.*

## Slide 5

Choice B is correct!

Here is the calculation. Ending inventory under FIFO means that the items remaining at the end of the month are costed at the most recent unit costs. Notice that the unit selling price of $20/unit is irrelevant in the calculation of ending inventory value, which is based on the cost paid by ABC to purchase the inventory.

| | |
|---|---|
| Units available for sale | 1,300 |
| Units sold | 700 |
| Units in inventory at 1/31/07 | 600 |

400 @ $12 = $4,800
200 @ $12 = 2,400
600 $7,200

Action Button (Blue Arrow)

*Follow the procedure described in slide 4 to import sound onto this slide. Since the student has chosen the correct answer, an appropriate sound effect would be cheering which is available on the sound clip organizer.*

*Follow the instructions in slide 1 to insert the hyperlinked action button to get to the next slide, which would be the next question.*

## Slide 6

**Question 2**

ABC Company uses FIFO to value its ending inventory of Part R. ABC had the following transactions during January 2007. All sales are at a unit sales price of $20.

| Transaction | Quantity | Unit Cost |
|---|---|---|
| Beginning inventory 1/1/07 | 400 | 10 |
| Purchase 1/10/07 | 200 | 11 |
| Purchase 1/15/07 | 300 | 12 |
| Purchase 1/18/07 | 400 | 12 |
| Sale 1/22/07 | 300 | |
| Sale 1/28/07 | 400 | |

ABC pays taxes each year. It is expected that the price ABC will pay for Part R will increase and that inventory levels of Part R will increase. ABC's controller Jane Smith believes it is time to switch to LIFO.

A Ms. Smith is correct. A switch to LIFO will result in higher Cost of Goods Sold which causes Income before Income Taxes to be lower and taxes paid to be lower when compared to FIFO.

B Ms. Smith is incorrect. A switch to LIFO will result in lower Cost of Goods Sold which causes Income before Income Taxes to be higher and taxes paid to be higher compared to FIFO.

C Ms. Smith is correct. A switch to LIFO will result in lower Cost of Goods Sold which causes Income before Income Taxes to be higher and taxes paid to be higher compared to FIFO.

D Ms. Smith is incorrect. A switch to LIFO will result in higher Cost of Goods Sold which causes Income before Income Taxes to be lower and taxes paid to be lower compared to FIFO.

*Although this question is multiple choice, it involves higher order thinking and assesses whether the students understand the theory behind the relationship between choice of accounting method, determination of income before income taxes and taxes paid.*

*Each choice will be hyperlinked to a slide explaining why the answer is correct or incorrect. Follow the procedure outlined in slide 3. In addition, the student can be sent to a Web site for detail.*

*Assume that the student selects choice A, the correct choice.*

## Slide 7

Choice A is correct!

Gross Profit using LIFO and FIFO is calculated below. Other expenses would be subtracted from gross profit to determine Income before Income Taxes.

| | FIFO | LIFO |
|---|---|---|
| Sales Revenue | $14,000 | $14,000 |
| Cost of Goods Sold | 7,400 | 8,400 |
| Gross Profit | $ 6,600 | $ 5,600 |
| | | |
| Taxes Due @ 40% | $ 2,640 | $ 2,240 |

Notice that LIFO saves ABC $400 in taxes because Cost of Goods Sold is higher using LIFO. The calculations for Cost of Goods Sold are below:

COGS – LIFO:
700 @ $12 = $8,400

COGS – FIFO:
400 @ $10 = $4,000
200 @ $11 = $2,200
100 @ $12 = $1,200
$7,400

For more information on the FIFO/LIFO decision, go to:
http://findarticles.com/p/articles/mi_m0DTI/is_12_30/ai_96892300

*A hyperlink sends the student to a Web article for more information on the subject. You can also hyperlink the student to the Web-based support material available with the text.*

At the end of the PowerPoint, you can insert a hyperlink to you, the instructor, asking for feedback on the module. This gives you an opportunity to receive timely comments and suggestions and also enables you to keep track of students who have completed the module. If this is important to you or you are basing a portion of the course grade on completion of these modules, you need to make students aware that they will have to complete the entire module in order to send the e-mail. Unfortunately, students can circumvent this control. A great incentive for them to complete the module, however, is to tell them that course assessments will contain similar multiple choice questions. To insert a hyperlink for students to send an e-mail to you, follow the instructions on Slide 2 and use a descriptive e-mail subject title, such as "Feedback – Interactive Inventory Module".

In order to ensure that the students move through the PowerPoint slides in the order you wish them to, you must set the presentation to "kiosk" mode. From the tool bar select "Slide Show" and then select "Set up Show". Under Show Type select "Browsed at a kiosk" and then click "OK". And to ensure that no one can alter the slide show in any way, it is good to save it as a PowerPoint show. From the tool bar, select "File", then "Save as". At the bottom of the window scroll through the choices in "Save as type" and select "PowerPoint Show (*.pps)." This file can be e-mailed directly to students for their use or set up for their access in a course management system such as Blackboard.

*This work was previously published in International Journal of Information and Communication Technology Education, Vol. 4, Issue 2, edited by L. Tomei, pp. 14-25, copyright 2008 by IGI Publishing (an imprint of IGI Global).*

# Chapter 16
# Building Bridges Online:
## Issues of Pedagogy and Learning Outcomes in Intercultural Education Through Citizenship

**Roger Austin**
*University of Ulster, Northern Ireland*

**John Anderson**
*Queen's University, Northern Ireland*

## ABSTRACT

*This article indicates that there are four key drivers for school-based use of collaborative software;*

a. *The speed at which 'social software' has been taken up by young people outside school which has led some educationists to review the potential of such software in more formal school settings.*
b. *Helping pupils to develop 'knowledge construction skills' which are relevant to a knowledge economy.*
c. *Enabling more pupils to access a wider curriculum.*
d. *The promotion of inter-cultural education through citizenship.*

*In the case of the fourth driver, the article examines in detail the research basis for extending the concept of the 'contact hypothesis' through communication technology. It uses evidence to show that well managed on-line collaboration between school-based students can be a powerful vehicle for intercultural education through citizenship.*

## INTRODUCTION

While there has been extensive research on the role of information and communication technology (ICT) in collaborative learning, there has been relatively little work that explores where ICT has been used explicitly to promote inter-cultural education or citizenship. In a review of recent research on the role of ICT in a range of learning settings, Hartley (2007) refers to what he calls "widening horizons." He comments on one positive account of e-mail used to link students from two geographically-separated primary schools (Ho, 2000). But, he notes that in Fabos and Young (1999), reviews of over a dozen different studies find more limitations. Indeed, they concluded that much of the research was contradictory, inconclusive, and possibly misleading because of its overoptimistic tone.

This article sets out to challenge this somewhat outdated and pessimistic view of the role of ICT in citizenship by referring to a range of international research which has been published since 2001. The authors set out to establish a consensus on the learning gains which arise from well-managed school links and argue that enough is now known about the conditions for success in "e-partnerships" between schools for "e-twinning" to be raised to a level of expectation for all schools. Given the support that is now available from the European Community to promote "e-twinning," the opportune moment for a critical review of learning outcomes that ICT can deliver as well as the conditions likely to enable collaborative learning to emerge may be at hand.

## THE DRIVE TOWARDS COLLABORATIVE WORKING ONLINE

In seeking to explain the significant increase in the use of collaborative software in schools, three "drivers" are proposed. First, the explosion of what Shirky (2003) calls "social software" enabling group communication through social networking has become so prevalent among teenagers and adults (Grant, 2006) that there is, among innovators in the education sector, recognition of the potential for such software in schooling. At the same time, however, negative press coverage surrounding the misuse of social environments leading to cyber-bullying and child abuse creates an environment less supportive of spreading these innovative educational applications.

The second reason for an increased interest in social software arises from a combination of declining student numbers in schools coinciding with pressure on schools to provide a wider and more varied curriculum. School administrators are facing difficult choices about how they can sustain schools, particularly in rural communities and where there are falling enrollments. In Northern Ireland, the Costello report (2004) on post-primary reorganisation claims that e-learning has the potential to make "a major contribution to local partnerships of schools, which could make it possible to provide courses for small groups that would not otherwise be viable". The report calls for further investment in facilities and teacher training "so that they (the teachers) are comfortable with the issues related to teaching in this way" and a development path with clear targets for e-learning emerges to secure significant gains as soon as possible.

It is recognised that technology can have a significant role to play in broadening choice through the online delivery of distance learning to courses traditionally delivered face-to-face to a single class gathered in one room. To reduce the need for learners to travel between school sites during the school day, online technology can support collaboration through the communication tools of text-conferencing, audio and video conferencing, and applications-sharing. And, schools will need open access computer-resource study areas for learners who come and go throughout an extended day.

By carefully targeted pilot projects subject to independent evaluation, much progress has been made over the last four years in understanding what it takes to deliver high quality teaching and high standards of attainment online to school-aged learners, with some of that provision becoming widely accepted.

Over three years, by the summer of 2005 nearly 5,500 pupils with more than 225 teachers in 200 schools had experienced online teaching and learning as part of their curriculum in Northern Ireland. The range across age groups from 11 to 17-year-olds (with the majority focusing on the more mature learners able to regulate their own learning) included advanced vocational education courses and advanced level courses in geography, mathematics, computing, physics, chemistry and biology, and senior courses in ICT. Also involved in these studies was Citizenship teaching for 14-year-olds, Japanese Studies, European Studies, and a range of courses conducted by linking schools from the North and South of Ireland.

The third driver in the use of collaborative learning according to Grant (2006) is the supposition that a knowledge-based economy requires "knowing how to learn and how to participate in creating new knowledge." This argument is reinforced by Austin and Anderson (2008) in their analysis of the relationship between the needs of the economy and the use of ICT in schools. They quote from the European Commission[5] to the effect that young people require the following high-level cognitive skills in a knowledge economy:

- Working safely in teams (whose members may be in different locations);
- Self-reliance and self-management;
- Collaborative problem-solving;
- Creativity and innovation;
- High-level reasoning, analysing, and conceptualising;
- Communicating and understanding within multi-cultural environments; and,
- Autonomous learning

Each of these skills assumes, implicitly, a familiarity with ICT in general and specifically with the use of software that enables communication across sites and probably between countries.

The fourth reason for the growth of collaborative software use in schools is located within a cluster of ideas concerned with citizenship. Ligorio and van Keen (2006) argue for example that:

*the creation of virtual environments was envisioned by the European Commission as a means to foster a European culture. Facilitating exchanges and collaboration among European educational agencies was foreseen to be one of the strategies to consolidate Europe as a geo-political unit.*

Writing in the context of ICT links between schools in Northern Ireland and the Republic of Ireland, Austin, et al. (2003) used evidence from the Dissolving Boundaries programme to state that one of the achievements of this programme was the planting of seeds of "tolerance and respect for other cultural traditions." More recently, Austin (2006) has summarised extensive research on the ways that ICT has been used in the Middle East and elsewhere to promote improved intercultural understanding. Shonfeld, et al. (2006) have provided a compelling case study of how technology has been used in Israel to link a Kibbutzim College of education with an Arab teacher-training college and a religious Jewish women's institute to facilitate "multi-cultural dialogue." The potential of ICT to promote citizenship and intercultural learning is also made clear in the United Kingdom's Department of Education and Skills' review of Citizenship and Diversity (2007). Sir Keith Ajegbo makes the point in the foreword to his report that:

*If children and young people are to develop a notion of citizenship as inclusive, it is crucial that issues of identity and diversity are addressed explicitly—but getting the pedagogical approach right will be critical: the* ***process*** *of dialogue and* *communication must be central to pedagogical strategies for Citizenship.*

And, he concludes in his findings that there is a "new duty on schools to promote community cohesion" recommending that:

*Schools should build active links between and across communities, with education for diversity as a focus. This might range from electronic links (local, national and global). These links should be encouraged particularly between predominantly monocultural and multi-cultural schools. Such links need to be developed in such a way as to ensure they are sustainable. Such work between schools must have significant curriculum objectives.*

Finally, in a study of the challenges in using ICT to improve cross-community relations between Catholic and Protestant schools within Northern Ireland, Austin (2007) argues that the current citizenship curriculum will have a limited impact on young people unless there is far greater cross-community contact, either face-to-face incorporated into courses that pupils are already studying or online. This problem is not restricted to Northern Ireland. Current curriculum reforms globally have some strands in common. Whether called Citizenship or Civics, one of the aims is to disrupt the parochial/insular perspectives of young people and broaden their perspectives so as to help them to develop the competence and confidence to widen their horizons and to think globally.

However, planning for a Citizenship or Civics curriculum often fails to include contact experiences. It has been argued that without contact, learning outcomes will necessarily be impoverished. Curriculum planners and those agencies responsible for the school curriculum need to encourage schools to overcome the obstacles and promote the practice of online approaches to enrich Citizenship studies.

## THEORETICAL MODELS

This second part of the article examines the importance of different learning models in collaborative learning. According to Ligorio and Veermans (2005), research data confirm that international Web-based interaction can work effectively only "when meaningful pedagogical models are implemented." It is argued here that a new paradigm is emerging that can bring together insights from both social learning theories in education and insights from social psychology.

Ligorio and Van Keen (2006) base their work on three theoretical models: theories that see learning as knowledge-building (Scardamalia & Bereiter, 1994), theories based around the idea of a community of learners (Brown & Campione, 1990), and computer-supported collaborative learning (Lehtinen, Hakkarainen, Lipponen, Rahikainen, & Muukkonen, 1999). These models share an approach to learning built on constructivist principles.

More recently, Cochran, Conklin, and Modin (2007) have argued that links between schools can also lead to "higher-level thinking." Using an updated version of Bloom's taxonomy, they argue that a link between students in the U.S. and Finland provided opportunities for metacognition when they had to decide "the best tools to use and how to attack complex problems using them." The authors claim that using the revised Bloom's taxonomy ensured that the students went beyond asking simple questions of each other and "were thinking more deeply." This research is important to sharpen the ways that learning outcomes associated with cognitive interaction might be assessed. The new taxonomy, using the verbs "remember, understand, apply, analyse, evaluate and create," provides a framework that can help teachers think clearly about the way collaborative work should be extended beyond the simple exchange of data. So, learning activities can be planned that explore similarity and difference and, at the highest level, lead to the creation of new knowledge.

While these "models" have gained considerable acceptance, we suggest that in one important respect they do not provide a sufficiently critical framework for the use of ICT in citizenship education. The "missing element" is provided through the contact hypothesis, a construct drawn from the work of social psychologists. Austin (2006) argues that in its simplest form, the contact hypothesis lays down the broad conditions under which contact between two or more groups is likely to be successful. Following Allport (1954) and Pettigrew and Tropp (2004), he says that contact should be cooperative rather than competitive, between partners of equal status, between groups rather than individuals, long term rather than short term, given institutional support, and capable of leading to the development of a "super-ordinate identity" (i.e., one that transcends local, regional, or national identity).

The diagram below illustrates how the aim of promoting citizenship through ICT should take into account the contact hypothesis at the planning stage before considering the "collaborative/knowledge-building" learning activities that lead to the creation of a community of practice. The contact hypothesis is important for three reasons. First, work that involves young people of "equal status," defined here as those of similar age and ability, has been shown to be a significant factor for successful links in the Dissolving Boundaries programme. Successful experience is more likely to encourage schools to work together after their first encounter.

The importance of long-term, rather than short-term contact, should not be underestimated; links of at least one school year are often necessary to allow time for technical difficulties to be resolved and for both teachers and pupils to advance beyond a superficial exchange of pleasantries.

The contact hypothesis has a third significance. We have argued elsewhere (Austin & Anderson, 2008) that institutional support in the form of leadership by head teachers is vital to effectively embed innovation in ICT. This is particularly true when discussing the role of ICT in citizenship, which is disruptive not only in terms of practical issues like timetabling classes to be able to video-conference, but because links to other schools can open up fundamental questions about the function of schooling. Decisions about resource allocation, the role of assessment, teacher service and conditions, styles of learning, and the shape of the curriculum are all loaded with assumptions. While "otherness" and diversity are at the heart of citizenship education, teachers need institutional support to enable them to manage the differences

*Figure 1. A planning and learning model for the use of ICT in citizenship education*

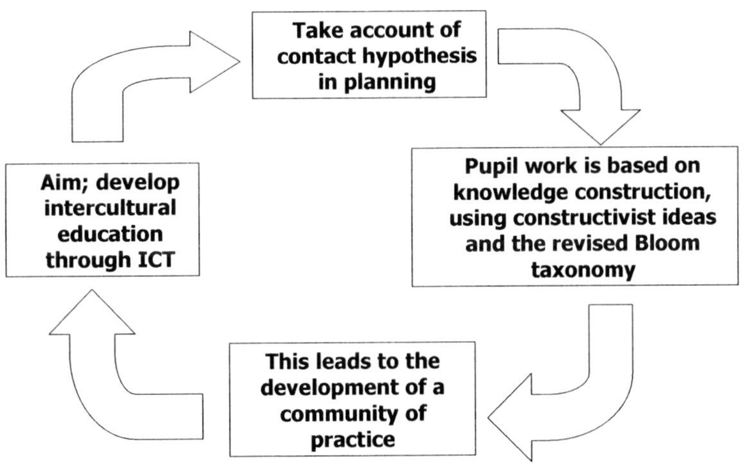

that can emerge when collaborative work is being undertaken with another school. This view is supported by Ligorio and Van Keen (2006) who concluded that the key factors in one of their projects was "not so much teachers'attitudes or expertise but the structural conditions at the level of the school organisation." And, according to Austin, et al. (2003), analysing the work between schools in Northern Ireland and the Republic of Ireland in the Dissolving Boundaries programme, teacher enthusiasm was the single most important factor influencing project work.

Sustainability of this type of work is assisted when teachers have an understanding of the theoretical models outlined, and implies a particular form of professional development that goes well beyond up-skilling in ICT. Davis, et al (2005) suggest that long-term changes to teacher attitudes require an appreciation of inter-cultural competence and a disposition to see schooling in terms of both social inclusion and globalisation.

In concluding this section of the article, we assert that understanding these theoretical models is important not only to ensure sustainability but because they affect every practical outworking of the learning that takes place between schools. In the final part of the article, we analyse evidence from different projects and propose a model of collaborative learning.

## LEARNING OUTCOMES AND COLLABORATION

Analysis of research about the implications of different theoretical models includes the work of Ligorio, et al. (2005) who describe work in which students in Italy and Greece worked together to write fairy tales; each group took turns to start a story and then handed over responsibility to the other group to complete it. The authors use the term "intersubjectivity" to refer to the contributors' ability to "decentralise themselves and to include the interlocutors' perspective within their own view." For them, this concept includes the concept of "constructing new understanding by combining different perspectives." For us, this definition is coterminous with knowledge construction; as Grant (2006) puts it in her study of a school in England that was using collaborative software, there is a principle being used that "knowledge of the group is greater than knowledge of the individual"; in other words, the result of the collaboration between the two groups leads to the creation of new insights.

Two other points about this research are important. The first is the finding that a high degree of intellectual reasoning is needed to accomplish the task in a truly collaborative way. But evidence from Abbot, et al. (2004) derived from the Dissolving Boundaries programme shows that children with special educational needs can also benefit from inter-school collaborative learning when learning is pitched at an appropriate level and when the technology used fits the purpose. In this instance, video-conferencing proved to be a highly successful medium for communication. The point about video-conferencing is highly significant since this area is most likely to see a rapid expansion of low cost opportunities using software like Skype or Marratech, which is expected to benefit all children, both those with special needs and those in mainstream education.

The second point is that of the set task involved the construction of a narrative. The authors assert that this was a "pre-requisite to build an intersubjective space." The issue of defining the exact focus for collaborative work is, in our judgment, critical. Grant, for example, provides an example of pupils in the same school using the collaborative environment Moodle to undertake a project on innovations in technology since 1950. She notes that while the students were keen to publish their own work, once they decided who was responsible for a particular page, the individual or pair's ownership of that page was strongly asserted. She finds evidence of students' reluctance to edit others' work and links this to what she sees as the predominance of "individualised written assessment that pervades school." Where collaboration did take place, it was in the domain of design and technical competence. She argues that students accepted the concept of collaboration in this area precisely because it was on the periphery of the practices of the school community. Her honest judgment was that there was little or no evidence of "knowledge-building" in the short time that this project took place.

The definition of suitable collaborative tasks has also been at the centre of the Dissolving Boundaries programme. Unlike the other evidence quoted thus far, the Dissolving Boundaries approach has stressed the importance of social interaction between pupils as a pre-requisite for curricular work. Since it has been working in primary, special, and secondary schools, it has evolved a distinctive framework for collaboration.

Drawing on the contact hypothesis, links are formed between two schools and, in the classes that participate, sub-groups with some four to six pupils are created. Each of these teams is then paired with a similar team in the other school. This structure is designed to enable contact to be formed within a suitably small but diverse setting where pupils are likely to be working with young people who display a variety of characteristics and personalities. Given that the central goal of this programme is the development of intercultural education through the use of ICT, it is felt to be important to go beyond electronic pen pals to group interaction so that cultural understanding is formed by working with others on a common purpose.

Thus, the first task is for each group to write a summary self-description and post this into shared space that has been created for them and their partner group. While some schools still prefer to exchange individual profiles in the online forum, the inter-group exchanges are seen as being central to the goals of the programme. Current work, which takes account of the restrictions now in place about the sending of personal or group photographs, is designed to encourage the groups to select digital images that have some personal significance and post these into the shared but private space of the group. In some cases, it is the task of the children from the other school to try to identify which members of their linked team selected which images. This work highlights both the need to preserve the distinctive contribution of the individual and the way in which that contribution adds something to the collective enterprise of the group.

A further example of this approach can be seen in a Dissolving Boundaries project that has been built around the study of history. Teachers in two selective secondary schools, one in Northern Ireland and the other in the Republic of Ireland, agreed to investigate the reasons behind the sixteenth century Reformation which led to the English King Henry VIII's break from the Church of Rome. This was an important issue for pupils in both schools since Henry's decision led to conflict between Catholics and Protestants; a conflict that in some respects remains unresolved and very "real" on the island of Ireland. Each of the groups undertook some research and posted their findings in different colours into their shared workspace in Moodle. In this way, they could clearly see the respective input from both schools. This work was then followed by a video-conferencing session designed to explore the differences between the collected information.

This example is a good reminder of the respective contributions that different communication technologies can bring to collaborative learning: where Moodle is asynchronous, allowing all pupils time to research, compose, and present, video-conferencing is "real-time" and well-suited to those occasions when the two groups need to clarify understanding. Schools in Dissolving Boundaries have also used video-conferencing to enable pupils to present work to their partner, to carry out a question-and-answer session with an invited guest (a fireman, an author), and to perform music for each other (Christmas carols).

Since the use of video-conferencing is relatively new in this kind of work, one final point should be made. Teachers and pupils in the Dissolving Boundaries programme agreed that video-conferencing made the links between schools "more real" and this was especially true for pupils in primary schools and special schools. Abbot, et al. (2004) reported a special needs teacher discussing her pupils' attitudes to their weekly video-conferencing session as follows:

*Motivation was fantastic—confidence, yes and self-esteem, yes. They come in on Thursdays all spruced up with their uniforms on and smelling of aftershave and deodorant!*

What this evidence shows clearly is that when we are examining learning outcomes, we need to take account of not just the cognitive domain but the affective and connative ones too. In concluding this section on learning outcomes and collaboration, it should be noted that even when inter-school links through ICT are designed to promote inter-cultural education, the processes involved often contribute to knowledge creation and to the kinds of leaner attributes likely to be valuable in a knowledge-based economy.

## IMPACT ON PEDAGOGY

A final point arising from the Dissolving Boundaries research is the strong evidence emerging with respect to changes in pedagogy. Austin, et al. (2003) reported that the need to integrate ICT into the classroom and use it for communication with another school led to a shift in teaching style from information provider to facilitator. The need for integration must be linked to the ideas emerging from the application of the revised Bloom taxonomy. It is not sufficient to assume that a teacher's change of role to facilitation will provide a strong enough framework to ensure that class work moves beyond "shallow constructivism." While the overall goal of links between schools may be inter-cultural learning, learning must continue to be built on worthwhile social and academic activities that have measurable outcomes.

## CONCLUSION

Pedagogy must be clearly based on theoretical models of learning which we suggest bring together the revised model of Bloom's taxonomy with both the constructivist approach and the insights from the contact hypothesis. As a result, teachers will be able to plan and evaluate work more clearly and systematically.

Second, the use of these new theoretical models is likely to challenge conventional teaching patterns in many schools and in this sense will be disruptive in terms of use of ICT resources, school aims, and outcomes.

Third, learning outcomes should be a judicious blending of both the cognitive, connative, and affective domains, reflecting a wide range of learning that involves selecting which communication technology to use, how to work in a team, how to respect differences, how to extend understanding of a given topic, and how to create new knowledge. Successful assessment regimes require placing a high value on moving beyond recall and comprehension to embrace the kind of learning goals needed for a knowledge-based economy in a multi-cultural world.

Finally, delivery requires a curriculum entitlement entailing political commitment. At a recent international educational ICT conference at the University of Ulster on the theme of *"The role of ICT in Building Bridges and Social Inclusion"* (September 2006), delegates resolved that every school should be entitled to two e-partners, one local and one international, to develop online dimensions for the local and global goals of a Citizenship agenda. A conference resolution was presented to the Minister responsible for Education at the Department of Education in Northern Ireland as advice on policy development. While it may be easy to concur that the purpose of the resolution is highly desirable, it does strengthen the need for the research outlined here into the pedagogic effectiveness of online communications to develop a better understanding and tolerance of difference—local, national, and international—among young people.

## REFERENCES


Abbott.,L., Austin, R., Mulkeen, A., & Metcalf, N. (2004). The global classroom: Advancing cultural awareness through collaborative work using ICT. *European Journal of Special Needs Education, 19*(2).

Anderson, J., & McCormick, R. (2006). Pedagogic quality–supporting the next UK generation of e-learning, Section 26. In U-D. Ehlers & J.M. Pawlowski(Eds.), *Handbook on quality and the standardisation in e-learning.* Springer, Berlin.

Anderson, J., & McCormick, R. (2005). *Pedagogic quality in policy and innovation in education: Quality criteria*, Brussels: EUN Schoolnet

Austin, R., & Anderson, J. (2008). *E-schooling: global messages from a small island.* Oxford: Routledge.

Austin, R., Abbot, L., Mulkeen, A., & Metcalfe, N. (2003). Dissolving boundaries: Cross-national co-operation through technology in education. *The Curriculum Journal, 14*(1), 55-84.

Austin,R. (2007). Reconecting young people in Northern Ireland. In Loader, B. (Ed.) *Young Citizens in the Digital Age*, Routledge.

Austin, R. (2006). The role of ICT in bridge-building and social inclusion: Theory, policy and practice issues. *European Journal of Teacher Education, 29*(2), May, 145-161.

Cochran, D., Conklin, J., & Modin, S. (YEAR). A new bloom: Transforming learning. *Learning and Leading with Technology, 34*(5), 22-25.

Costello Report. (2004). *Future post-primary arrangements in Northern Ireland.*, ¶5.14, 3 http://www.deni.gov.uk/pprb/costello_report.htm (See also the DE pages about the Costello report in http://www.deni.gov.uk/pprb/index.htm)

Davis, N., Cho, M.O., & Hagenson, L. (2006). Intercultural competence and the role of technology in teacher education. *Contemporary Issues in Technology and Teacher Education, 4*(4).Retrieved May 7, 2006 from, http://www.citejournal.org/vol4/iss4/maintoc.cfm

Fabos, B., & Young, M.D. (1999). Telecommunication in the classroom: Rhetoric versus reality. *Review of Educational Research, 68*(3), 322-349.

Grant, L. (2006) Using Wikis in schools: A Case Study. *Futurelab.* Bristol. Retrieved from www.futurelab.org.uk

Hartley, J. (2007). Teaching, learning and the new technology: A review for teachers. *British Journal of Educational Technology, 38*(1), 42-62.

Ho, C.M.L. (2000). Developing intercultural awareness and writing skills through email exchange. *The Internet TSL Journal, VI*(12). Retrieved September 29, 2005 from, http://iteslj.org/Articles/Ho-Email.html

Ligorio, B. & Veermans, M. (2005). Perspectives and patterns in developing and implementing international Web-based collaborative learning environments", *Computers and Education, 45*(3), 271-275.

Ligorio, M.B., & Van Keen, K.( 2006). Constructing a successful cross-national Virtual Learning Environment in Primary and Secondary Education, *Association for the Advancement of Computing in Education Journal, 14*(2), 103-128.

Ligorio, B., Talamo, A. & Pontecorvo, C. (2005). Building intersubjectivity at a distance during the collaborative writing of fairytales, *Computers and Education, 45*(3), 357-374.

Reading.V, (2002). *Knowledge is Power*, European Commissioner for Education and Culture. Extracts from the Public Service Review, European Union, Autumn 2002.

Shirky,C (2003). *A Group is its own Worst Enemy*. Paper given at Etech, April 2003. Available from http://www.shirky.com/writings/group_enemy.html

UK Department for Education andScience (2007). *Diversity and citizenship curriculum review*. Retrieved http://publications.teachernet.gov.uk/default.aspx?PageFunction=productdetails&PageMode=publications&ProductId=DFES-00045-2007

*This work was previously published in International Journal of Information and Communication Technology Education, Vol. 4, Issue 1, edited by L. Tomei, pp. 86-94, copyright 2008 by IGI Publishing (an imprint of IGI Global).*

# Chapter 17
# The Holistic Model for Blended Learning:
## A New Model for K-12 District-Level Cyber Schools

**Alex Stone**
*VLN Partners, Inc., USA*

## ABSTRACT

*Pennsylvania is at the forefront of the public cyber charter schooling movement in America. As more and more students elect to transfer from traditional public schools into cyber charter schools–and their districts of origin are forced to forfeit their tuition allocations–a need for a public school alternative to cyber charter schools has emerged. Using current practices in Pennsylvania's public schools as a backdrop, this article presents a new model for district-level cyber schooling, called the holistic model for blended learning, that public schools in Pennsylvania (and elsewhere) can use to compete with cyber charter schools and meet the growing demand for K-12 online learning.*

## INTRODUCTION

Pennsylvania is at the leading edge of a national movement toward K-12 distance education and this phenomenon is perhaps most apparent in the growing popularity of cyber charter schools across the state. In the 2006-2007 school year, an estimated 15,000 students were enrolled in Pennsylvanis's cyber charter schools. This number represents an increase from approximately 10,000 students in the 2005-2006 school year and 5,000 students in the 2004-2005 school year (Smith, 2005; Chute, 2005; Silver, 2007). This enrollment trend, coupled with the fiscal policies that govern financing cyber charter schools, poses significant threats and challenges to Pennsylvania's public school system. Public school districts need to adopt a model for cyber schooling that they can use to effectively meet their students' demand for distance education so they can retain the tuition allocations that they are *obligated* to forfeit to cyber charter schools (Raffaele, 2004). Put very simply, public school districts need a way to compete with cyber charter schools.

For instructional technology innovators who are devoted to supporting and improving public schools in America, the freedom granted to Pennsylvania's cyber charter schools to meet the demand for K-12 distance education, coupled with the financial burden that has been placed upon school districts to finance their endeavors, introduce enormous challenges and opportunities. There is a real and pressing need to craft a district-level response to cyber charter schools in Pennsylvania and the solution that emerges will undoubtedly have implications on a national scale (NCES, 2003).

While few would argue with the commonly voiced claim that the No Child Left Behind (NCLB) initiative of 2000 offers *choice,* another extremely valuable by-product of this piece of legislation is that it inspires a type of healthy *competition* between public schools and external education providers that will ultimately benefit American students. Currently, public school districts are losing this competition when it comes to distance education–and this stark reality becomes apparent if we examine their response to the cyber charter school movement in Pennsylvania (Knade, 2001). Unlike a regular brick-and-mortar charter school where districts at least have the opportunity to offer a viable classroom-based alternative to the curricular options presented, in Pennsylvania's *cyber* charter schools, districts don't even "field a team." In the vast majority of cases, they are unable to meet their students' demand for distance education because they are not aware that a viable model for K-12 district-level cyber schooling exists. While this is a significant problem now, as enrollment in cyber charter schools continues to increase and the schools begin to present larger and larger bills to public school districts, the situation will become critical for public schools in Pennsylvania in the near future.

Several ill-conceived attempts to craft a public school response to the cyber charter school movement in Pennsylvania have already been attempted in the past few years and each has met with limited success. This article examines one particular attempt in the section that explains deficient models. This attempt is well worth investigating for two main reasons:

- It provides an excellent example that can be scrutinized to identify strengths and weaknesses of different models for developing a district-level alternative to cyber charter schools in a real world setting.
- Its lack of success underscores the need for a categorical shift in thinking that must occur within the field of instructional technology if we are to have any practical and positive effect upon the learning and teaching that takes place in the online environment and in the classrooms of the future.

## WHAT MAKES CYBER CHARTER SCHOOLS WORK?

The term *cyber charter school* is used throughout this article to represent the publicly funded, state approved educational institutions in Pennsylvania that provide home-schooled students with a viable full-time curriculum, support services, and means of access—along with a legitimate diploma upon completion of coursework. While there are other full-time private cyber schools operating in Pennsylvania that offer similar services for a fee, this study focuses upon *public* cyber charter schools for two reasons:

- They each use models for K-12 online learning that have enjoyed at least some degree

*Table 1. Detailed description of rubric for evaluating various models for K-12 cyber schooling*

| Component/Feature | Description |
|---|---|
| **Grouping #1–Vital Ingredients** | |
| Cyber services | YES or NO<br>All the hardware, software, connectivity, technical support, shipping, and maintenance required to access Web-based lessons. |
| Dedicated online teachers | YES or NO<br>Teachers in a remote location who interact with students online via asynchronous and/or synchronous communication technologies. |
| Lessons designed to support on-site facilitators | YES or NO<br>Typically this on-site facilitation comes from a parent or guardian in a home-based cyber school setting but it can also come from a teacher or a behavior support specialist in other blended learning environments. |
| Learning management system used | LMS name<br>While all models use some kind of LMS, it will be helpful to identify which particular LMS is used. |
| **Grouping #2–Utility for K-12 Online Learners** | |
| Engaging instructional content | Rubric Rating Scale: 1-10<br>Score based upon the quality of the media presented, its ease of use and the degree of interactivity.<br><br>Unlike higher education learners who can rely upon threaded discussions and synchronous communication technologies to carry the instructional message in the online learning environment, K-12 online learners need more engaging instructional materials. |
| User-friendly, universal interface | Rubric Rating Scale: 1-10<br>Based upon the ease of use, lack of nested menus, consistent navigation scheme in different lessons, and design features that make the user interface "seamless" and easy to use.<br><br>Any successful model for K-12 cyber schooling must utilize a user interface that is user friendly and somewhat intuitive. |
| **Grouping #3 – Utility for K-12 Public Schools** | |
| Digital resources blend with, and support, existing classroom practices | YES or NO<br>Perhaps the best way to envision whether or not a piece of digital instruction can be blended with classroom instruction is to envision a teacher in a classroom using a projector to display the material to an entire classroom of students. |
| Lessons can be sequenced to mirror classroom instruction | YES or NO<br>This will prove to be a very important feature as public schools attempt to leverage their existing infrastructure and educational practices to create a model that competes with cyber charter schools. |
| LMS supports interoperability | YES or NO<br>While every LMS includes the ability to deliver native digital artifacts or instructional materials from different vendors (learning objects) as "external" supplements to the primary instructional message, only a select few enable native publishers to deliver digital materials as an integral part of the primary instructional sequence. In other words, only a select few LMSs provide the ability to blend off-the-shelf digital materials (LOs) with native artifacts in online lessons. |
| Degree of performance data tracking available in LMS | Rubric Rating Scale: 1-10<br>Score based upon the degree of performance monitoring and individualized data reporting provided. A score of 1 indicates that student login information is available. A score of 5 indicates that there is some way for teachers and facilitators to manually input test scores into a reporting module that is embedded in the LMS. A score of 7 indicates that there is some combination of manual entry and automatic generation of performance data for different question types. A score of 10 indicates that the LMS tracks student browsing behavior and generates individual test score reports for students for multiple choice, fill-in-the-blank, short answer and essay questions. |

of success in meeting the demand for home-based distance education in Pennsylvania
- They are using taxpayer money to introduce products and services that compete with Pennsylvania's public school districts for students and funding

The key advantage that public school districts enjoy at this point in the evolution of K-12 distance education in Pennsylvania is the fact that they can analyze existing practices in the 11 cyber charter schools currently operating in the state and craft a second-generation model that encapsulates the strengths and improves upon the weaknesses of each approach. Ultimately, this analysis–coupled with some visionary thinking–can produce a district-level model for cyber schooling that enables public schools to meet the growing demand for distance education in Pennsylvania's student population and provide classroom teachers the opportunity to embrace instructional technology in their current practices.

## A RUBRIC FOR EVALUATING VARIOUS MODELS FOR K-12 CYBER SCHOOLING

In order to provide a clear picture of each model presented later in this article, it will help to identify the vital ingredients that must be in place in any model used to establish and maintain a successful K-12 cyber school. The ingredients identified can then serve as criteria in a rubric that can be used to evaluate the efficacy of any given approach to K-12 cyber schooling. This rubric will not only help to focus the discussion about each of the models described in this article, it can also have great utility for public school administrators, parents, and students as they consider their options in the K-12 cyber school environment today. Finally, it can also prove to be a valuable tool for technology innovators who are interested in crafting a public school response to cyber charter schools.

## DEFICIENT MODELS

A key attribute that cyber charter schools in Pennsylvania share is that they each provide *all* of necessary ingredients that students and parents need to participate in a complete learning experience at home. While it is conceivable to create an incomplete bundle of products and services for K-12 cyber schooling and then sell it to public schools or intermediate units in the hope of *someday* using it to compete with cyber charter schools, any product offering that does so runs the risk of being inadequate, controversial, or incomplete. Ultimately, any such partial solution can cause more harm than good for public schools because in order to work, it places an additional, and often unexpected, burden upon an already strained infrastructure and technology-support staff.

Perhaps the best example of a deficient model that is currently being introduced to public schools in Pennsylvania is a bundle of software products offered by an organization called BlendedSchools (http://www.blendedschools.com/). For a relatively small fee, BlendedSchools offers public schools access to the Blackboard™ learning management system (LMS) along with access to several off-the-shelf courseware products (from Apex Learning, Class.com, Keystone National High School and Advanced Academics) that each has its own user interface and navigation scheme. They also provide access to videoconferenced learning opportunities and other synchronous technologies like Wimba.

**Critique of Deficient Models**. Aside from all the difficulties that arise from the imposition upon classroom teachers to become native publishers, the many problems that arise from mixing off-the-shelf courseware from different vendors in the same learning space (Friesen, 2003), the less-than encouraging results of a major government study on the effectiveness of off-the-shelf courseware (USDED, 2007), and the inappropriateness of the "master teacher" approach to presenting instruc-

*Figure 1. Completed rubric for BlendedSchools' product offering*

| Component/Feature | Response/Score |
|---|---|
| VITAL INGREDIENTS | |
| Cyber services | NO |
| Dedicated online teachers available | NO |
| Lessons designed to support on-site facilitators | PARTIALLY |
| LMS used | Blackboard |
| UTILITY FOR K-12 ONLINE LEARNERS | |
| Engaging instructional content | 6 |
| User-friendly, universal interface | 3 |
| UTILITY FOR K-12 PUBLIC SCHOOL DISTICTS | |
| Digital resources blend with, and support, existing classroom practices | NO |
| Lessons can be sequenced to mirror classroom instruction | NO |
| LMS supports interoperability | NO |
| Degree of performance data tracking available *in LMS* | 3 |

tion via video (Zugner, 1987), BlendedSchools' product offering is a deficient model for K-12 cyber schooling because it is a software bundle only. It simply does not include the critical elements of cyber services or dedicated online teacher support that *must* be in place in any workable model for K-12 cyber schooling. Public school administrators are attracted to the BlendedSchools' product offering primarily because it includes Blackboard licensing in its bundle, and there is some degree of false reassurance that comes with the associated brand recognition. Few administrators realize that Blackboard was originally designed for a higher education audience and this fact becomes apparent in the K-12 cyber schooling learning environment when learners are confronted with online lessons that are driven primarily by asynchronous threaded discussions. Furthermore, few administrators realize just how much of a strain it would be upon their infrastructure to develop a district-level alternative to cyber charter schools using the BlendedSchools product offering. If such a district level alternative were developed, it would not only require an enormous amount of effort to establish and maintain, it would also produce an end-product that is incompatible with existing classroom practices and significantly inferior to other forms of cyber schooling that are available to students in Pennsylvania's cyber charter schools today.

## Practical Utility for Public Schools

Ultimately, deficient models like the BlendedSchools' product offering may provide some educational benefits by offering students exposure to BlackBoard—an experience that could be valuable if they attend college after they graduate. But when it comes to meeting the need for a public school model for cyber schooling that can be used to compete with cyber charter schools, deficient models simply confuse administrators and give them a false sense of hope. Furthermore, such techno-centric and ill-conceived models for K-12 cyber schooling threaten to invalidate the field of instructional technology as a whole because they introduce products that have little-to-no respect for the systems and/or environments that they are designed to support. While these models can work in theory, when it comes to actual implementation,

*Figure 2. Completed rubric for the patchwork model for distance education*

| Component/Feature | Response/Score |
|---|---|
| VITAL INGREDIENTS | |
| Cyber services | YES |
| Dedicated online teachers available | YES |
| Lessons designed to support on-site facilitators | PARTIALLY |
| LMS used | Blackboard |
| UTILITY FOR K-12 ONLINE LEARNERS | |
| Engaging instructional content | 6 |
| User-friendly, universal interface | 3 |
| UTILITY FOR K-12 PUBLIC SCHOOL DISTICTS | |
| Digital resources blend with, and support, existing classroom practices | NO |
| Lessons can be sequenced to mirror classroom instruction | NO |
| LMS supports interoperability | NO |
| Degree of performance data tracking available *in LMS* | 3 |

they invariably fall short when compared to other models for K-12 cyber schooling that are enjoying a good amount of success today. From a pragmatic point of view, the basic fact that this particular product offering from BlendedSchools, and others like it, simply ignores the need for cyber support services and other critical elements that must be in place to offer a viable model for K-12 online learning epitomizes a type of short-sightedness that has plagued the field of instructional technology for the past several decades (Tomei, 2002; Hodas, 1993). Using the 1960s as the starting point for the field of instructional technology, it is easy to trace the repetitive cycle of innovation, implementation, and obsolescence that has made computer-assisted learning technologies like instructional television, videodiscs, and computer-based training programs of little-to-no use in the K-12 learning environment of today (Reiser, 2001). A key attribute that each of these technologies share is the fact that they each failed to use instructional technology to support current practice in the classroom. Deficient models are yet another instance of this type of myopia that must be overcome if the field is to have any credibility in our collective efforts to leverage technology to increase the efficiency of public schools in America.

## PATCHWORK MODELS

One prominent model currently used in some of Pennsylvania's cyber charter schools is sometimes referred to as a *patchwork model*. In this approach, various communication applications, support services, and off-the-shelf courseware products from several different vendors are pieced together to create a virtual learning environment. The key attribute that differentiates the patchwork model from other models is that students experience different user interfaces when they navigate from one class to another. While all courses are launched from the same centralized LMS, the learning interface, navigation scheme, and "look and feel" changes from course to course.

*Figure 3. Completed rubric for the holistic model for course delivery*

| Component/Feature | Response/Score |
|---|---|
| VITAL INGREDIENTS | |
| Cyber services | YES |
| Dedicated online teachers available | YES |
| Lessons designed to support on-site facilitators | PARTIALLY |
| LMS used | Blackboard |
| UTILITY FOR K-12 ONLINE LEARNERS | |
| Engaging instructional content | 8 |
| User-friendly, universal interface | 8 |
| UTILITY FOR K-12 PUBLIC SCHOOL DISTICTS | |
| Digital resources blend with, and support, existing classroom practices | NO |
| Lessons can be sequenced to mirror classroom instruction | NO |
| LMS supports interoperability | NO |
| Degree of performance data tracking available *in LMS* | 7 |

Cyber charter schools that adopt this model typically use Blackboard, and in some cases, they use additional synchronous communication technologies to carry the instructional message. While several existing cyber charter schools continue to utilize the patchwork model, they represent a decreasing percentage of the total student population in Pennsylvania currently enrolled in cyber charter schools. When students and parents compare the total learning experience presented in cyber charter schools that use this model with the total learning experience offered in cyber charter schools that follow holistic models, there is a general trend (reflected in enrollment numbers) toward the holistic models (Chute, 2005).

## Critique of Patchwork Models

One key problem with patchwork models is that they are essentially an attempt to deliver courseware that has its roots in an antiquated tradition of CBT that is quickly moving toward obsolescence in the K-12 environment. CBT courses are essentially interactive training programs that work best for self-motivated, adult learners who access lessons in an isolated learning environment, and problems arise when you attempt to deliver courses from different vendors in the same learning space. Not only are these courseware products based upon a fundamental set of design assumptions that make them inappropriate for the target audience and incompatible with each other (because they each define a unique user interface), they neglect to capitalize upon the Internet's ability to deliver and retrieve information almost instantaneously, and ultimately, they are incompatible with classroom instruction.

The results of an in-depth study recently conducted by the U.S. Department of education (USDED, 2007) seriously challenge the educational effectiveness of off-the-shelf courseware products in the K-12 learning environment. These results certainly cast a shadow of doubt upon models for K-12 cyber schooling that rely upon these products to carry their instructional message, but they also underscore the need for an evolutionary shift away from the thinking that has dominated

*Figure 4. The collaborative model for distance education*

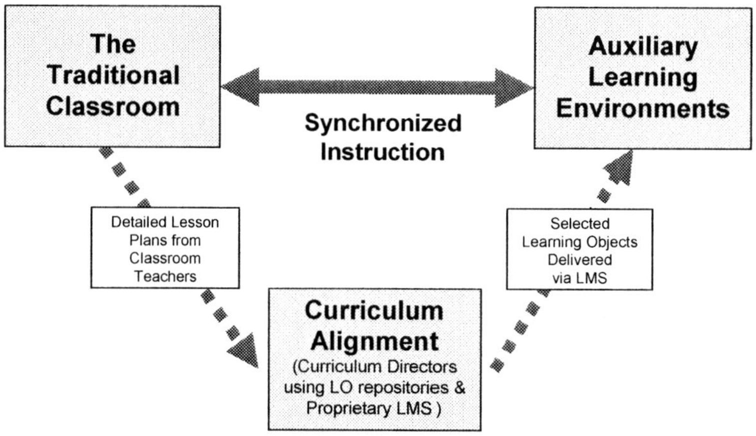


the field of instructional technology for the past several decades.

Patchwork models are a natural, albeit awkward and somewhat adolescent, phase that the field of instructional technology must pass through as it develops a model for K- 12 public cyber schooling that relies less upon once-removed computer programs and more upon products and services that are designed to embrace a more contextual approach to online learning. Future instructional technologists will likely look upon these models as products of a transitional phase between the mechanistic mindset that has dominated the field of instructional technology since the introduction of the personal computer in the 1980s and a more pragmatic perspective that will ultimately emerge as we begin to embrace the power of the Internet.

### Practical Utility for Public Schools

Patchwork models are an interesting phenomenon because they represent a type of evolutionary transition between instructional materials that are delivered in an isolated, PC-based delivery environment and instructional materials that are delivered via the Internet. While patchwork models can conceivably "force-fit" PC-based courseware into a Web-based delivery environment, the problems that arise from such practices prevent them from effectively competing with the holistic models that will be presented in the following sections of this article. Put simply, it may be possible for public schools to mimic cyber charter schools that utilize the patchwork model, but the end result will be a district level cyber school environment that cannot blend with classroom instruction and, ultimately, will not pose any real challenge to cyber charter schools.

## THE HOLISTIC MODEL FOR COURSE DELIVERY

Several cyber charter schools operating in Pennsylvania today use what can be called the *holistic model for course delivery.* The key attributes that differentiate this model from other models are:

- All of the course materials in any given grade are created by the same publishing company.
- The courses materials are designed, from their inception, for K-12 Web-based learning (as opposed to converted or enhanced CBT).
- A universal user interface is employed.

*Figure 5. Completed rubric for the holistic model for blended learning*

| Component/Feature | Response/Score |
|---|---|
| VITAL INGREDIENTS | |
| Cyber services | YES |
| Dedicated online teachers available | YES |
| Lessons designed to support on-site facilitators | YES |
| LMS used | dynamic |
| UTILITY FOR K-12 ONLINE LEARNERS | |
| Engaging instructional content | 8 |
| User-friendly, universal interface | 8 |
| UTILITY FOR K-12 PUBLIC SCHOOL DISTICTS | |
| Digital resources blend with, and support, existing classroom practices | YES |
| Lessons can be sequenced to mirror classroom instruction | YES |
| LMS supports interoperability | YES |
| Degree of performance data tracking available *in LMS* | 7 |

The value of a simple to use, universal user interface cannot be understated. It is perhaps the single most important advantage that this model affords and the cyber charter schools that have adopted this approach are enjoying a great deal of success in Pennsylvania. The success of these cyber charter schools underscores just how important it is to consider the end-user experience when designing a district level model for cyber schooling. Regardless of how instructionally effective any particular piece of stored digital instruction may be, if it is too difficult to access, the educational benefits it offers will be outweighed by the frustration involved with navigation.

## Critique of the Holistic Model for Course Delivery

These course bundles are perhaps the crowning achievement for the mechanistic approach to instructional design that thrives in the isolated, PC-based delivery environment. Even though these courses are designed specifically for Web-based learners, just like the off-the-shelf courseware mentioned in the previous section, they are still greatly influenced by, and beholden to, an antiquated instructional design mindset that can be traced back to the 1980s. Some important features of these course bundles that demonstrate this tendency are the facts that their lesson materials are not easily manipulated, it is difficult to deliver them alongside digital materials from other publishers, they are expensive, and it is not possible to synchronize the sequence of the instruction presented in their lessons with the instruction presented in a district's classrooms. Furthermore, these course bundles are consigned to operate in isolated learning environments that cannot take advantage of open architecture content authoring technologies. The shortcomings of these courseware bundles in the Web-based learning environment arise not because they are in some way instructionally incorrect or faulty, but because they are based upon a fundamental set of design principles that prohibit them from taking advantage the interactivity and interoperability that is now available in the Web-based blended learning environment.

## Practical Utility for Public Schools

Curriculum bundles have been used to set up learning environments that, by design, compete with public schools. Furthermore, they cannot be customized or modified to somehow support existing classroom instruction because they are beholden to an antiquated "all-or-nothing" approach to presenting digital learning materials via the computer that stems from a tradition of designing instruction for delivery on a PC. This exclusionary stance makes blending the instruction presented in these courses with instruction presented in courses produced by other publishing firms, or with instruction presented in the classroom, nearly impossible. Ultimately, it disqualifies this model as a viable public school response to cyber charter schools because, in order to adopt it, schools must establish a learning environment that competes with the instruction presented in their classrooms.

## THE HOLISTIC MODEL FOR BLENDED LEARNING

At least two cyber charter schools and one school district in Pennsylvania today use what can be called the *holistic model for blended learning*. The key attributes that make this model unique are:

- Online lessons utilize learning objects (LOs) to carry a large portion of the instructional message
- A universal user interface, a single login, and some form of LMS-based performance monitoring and assessment capability is provided via a proprietary LMS
- Online lessons can be used to support classroom instruction and they can be influenced by classroom teachers
- Employment of curriculum directors who "piece together" customized online lessons that typically mirror the flow of the instruction presented in the traditional brick and mortar classroom

In the holistic model for blended learning, these attributes work together to offer the critical ingredient of *context* that has long been missing from courseware used in the K-12 online learning environment. Until recently, online courses have been justifiably perceived as a threat to classroom instruction (Hodas, 1993) because they either present an instructional message that directly competes with classroom instruction in an isolated computer-based learning environment or they impose a need for software training upon teachers who are often reluctant to abandon the current paradigms that drive their activity in the classroom. The holistic model for blended learning mitigates this conflict by using current classroom instruction as the starting point, and then employs collaborative development techniques (Stone, 2007) and rapid prototyping processes (Tripp & Bichelmeyer, 1990) to create online lessons that are delivered as whole and complete products that have utility in both the classroom and in auxiliary environments (see Figure 4).

Before describing the nuances of this model, it will help to clear up some language. Rather than delving into the seemingly endless semantic debates that tend to occupy a great deal of time for instructional technology theorists, I will simply state that I advocate the definition for the term "blended learning" presented by Bonk and Graham (2007) that, "blended learning is a combination of online and face-to-face instruction" (p.3). This seemingly innocuous definition poses a very serious challenge to the field of instructional technology because it actually "makes room" for a live person who is expected to participate in the delivery of instruction. In other words, it can be viewed as a move toward using stored media as forms of performance support for teachers and facilitators in the target delivery environment (Geary, 1991; Rosenberg, 2001; Carman, 2002; Voos, 2003).

The holistic model for blended learning offers an alternative to the self-contained and exclusionary off-the-shelf courseware and whole curriculum bundles used in other models for K-12 cyber schooling by leveraging a new form of instructional delivery called learning object-based instruction (LOBI) (Stone, 2007) that has emerged from the learning object movement within the field of instructional technology (Merrill, 2000; Wiley, 2000; Downes, 2001; Freisen, 2003). LOBI is a new form of learning and teaching that harnesses the power of the Internet, allows for open architecture content authoring, and thrives in the holistic model for blended learning. Furthermore, when utilized within the overall theoretical framework established by the holistic model for blended learning, LOBI opens the door to many exciting possibilities for public school districts in Pennsylvania because it provides a means by which they can mount a serious challenge to cyber charter schools and external education providers.

Perhaps the greatest advantage that the holistic model for blended learning affords is the fact that it enables public schools to utilize collaborative development processes to create customized online lessons. These collaborative development processes differ from the systematic design and development processes used to create courseware bundles because they view digital media artifacts (learning objects) as tools that curriculum directors can use to create customized Web-based lessons that have utility across several auxiliary learning environments (Tripp, 1990; Wilson, 1995; Hickman, 2002; Wiley, 2003). In this model, classroom teachers assume a primary role that is somewhat akin to that of a subject matter expert in a traditional courseware development process. Rather than shouldering the burden of producing digital materials, or wading through the seemingly endless sea of digital resources that are available today to create their own online lessons, classroom teachers work with curriculum directors to guide the development process, ensure the quality of the finished product, and verify that the sequence of lessons presented online mirrors the sequence of the instruction presented in their classrooms. While they are certainly encouraged to participate in the open architecture authoring process by creating digital artifacts that can be folded into the final online lessons, it is not necessarily a prerequisite. The result of this collaboration is a customized online curriculum that has practical utility in both the classrooms and in auxiliary learning environments (like cyber schools).

The potential for the future growth of this model becomes apparent when you consider the recent activities of most major textbook publishers. Over the past few years, nearly every popular textbook publishing company has begun to offer libraries of learning objects as a value-added resource that can be used in tandem with their textbook materials. These resources are designed to be free standing and interoperable, meaning that they can be delivered alongside LOs created by other publishers and native materials created by classroom teachers. This trend can help the overall push toward collaborative approaches to blended learning in public schools because it enables classroom teachers to continue to use a familiar resource (textbooks) to guide the flow of instruction in the classroom while curriculum directors piece together learning objects from the same publishing firms (along with learning objects from other publishers that address the same topics) to create customized Web-based lessons that follow the same sequence. Ultimately, the major textbook publishing companies will continue to support this model because it enables them to perpetuate their dominant stance in the educational publishing market.

## Which LMS Works Best with the Holistic Model for Blended Learning?

One of the key challenges that must be addressed for LOBI to work is that of *interoperability* (Friesen, 2001). This is not an issue in the holistic model for course delivery because learners simply need to login once (or actually twice–once for the LMS and once for the courseware bundle accessed) and all of the instructional materials are presented in an enclosed environment with one interface. This works just fine if learners are satisfied with accessing learning materials created by one publishing company only–across all subject areas–at the exclusion of materials from all other publishers. In the holistic model for blended learning, this exclusionary stance is not practical because it forces learners into a closed learning environment. In order to work to its full potential, the holistic model for blended learning requires an open learning environment and an LMS that is customizable on the server-side to support this capability. Currently, this means that some level of customized coding is required so that the LMS can seamlessly "engulf" learning objects from different publishers and eliminate the access difficulties that invariably arise when materials from different publishers are served in a static (as opposed to a dynamic) learning management system.

This need for a dynamic LMS makes the most popular LMS products on the market today like Blackboard, WebCT, and even Moodle, a bad fit for the holistic model for blended learning because it is either impossible (or at least very difficult in the case of Moodle) to perform the customized coding *on the server-side* that must occur to make access of LOs seamless for courseware directors, facilitators, and learners. Without this degree of customization, users encounter numerous password prompts and navigation hurdles that make access an enormous challenge.

At this point in the evolution of LOBI, the best LMS products available are proprietary and dynamic–meaning that they can be quickly modified by programmers who have access to the source code on the server-side to accommodate new types of learning objects as they emerge. Perhaps at some point in the future this requirement will be mitigated by some type of added feature in existing LMS products that streamlines access, or by the ever-elusive LO shared content standard (Verbert & Duvall, 2004), but until those capabilities emerge, the best LMSs for the holistic model for blended learning are proprietary.

## Critique of the Holistic Model for Blended Learning

This model offers an enormous amount of freedom to harness the power of learning objects, a universal user interface, open architecture content authoring, and a dynamic LMS to provide a customized learning experience that takes advantage of many of the opportunities offered in a Web-based delivery environment. Currently, two cyber charter schools in Pennsylvania have adopted this approach and they are enjoying some success. They are, however, underutilizing this model for two main reasons—and these same two reasons can be seen as an opportunity for public schools to use this model to create a district level alternative that will lure their students away from cyber charter schools and back into their student populations:

- They neglect to fully capitalize upon the ability that this model affords them to blend online lessons with classroom instruction because they are not affiliated with any institutions that provide classroom instruction
- They fail to utilize the LOs that are produced by textbook publishers because textbook publishers are, by default, allied with public schools and they are reluctant to support organizations like cyber charter schools that compete with their primary customers

## Practical Utility for Public Schools

Without placing a significant strain upon their infrastructure, the holistic model for blended learning enables school districts to utilize various forms of stored media and instructional technology to support their existing educational practices and leverage the priceless commodity of *context* to extend their influence into the online learning environment. Because it accommodates collaborative development processes and LOBI to carry the instructional message, this model can be used to establish a cyber school environment that utilizes online lessons that mirror both the *scope* and the *sequence* of the instruction presented in designated classrooms in a district, whether or not teachers in those classrooms possess basic digital publishing skills. Finally, because it supports open architecture content authoring, it can be viewed as "the path of least resistance" (Shank, 2007) between where classrooms are now and where they need to be in the future, because it establishes an exemplary online curriculum that teachers can add to, refine, and improve upon each year.

There are several aspects of the holistic model for blended learning that set it apart from other models when it comes to delivering meaningful online lessons:

- Lessons are designed, from the outset, to utilize a dedicated facilitator in the learning that takes place in the ideal target delivery environment
- Curriculum directors work with classroom teachers to create online lessons that match the scope and the sequence of the instruction presented in the traditional brick-and-mortar classroom
- Learning objects used to support online instruction can also be used to complement classroom instruction
- The LMS can be modified at the server side to maintain one universal interface and navigation scheme (i.e., the LMS can be easily modified to address login and navigation difficulties that invariably arise when accessing content from different publishing companies)
- The LMS can accommodate open architecture, and interoperable, content authoring so "native" digital materials can easily be folded into online lessons and delivered alongside content from other publishers.

Each of these characteristics, in and of themselves, offers significant advantages when compared to other models for cyber schooling that are used today. But when these features are combined into one holistic model, they offer a powerful alternative for K-12 distance education that can have a profound impact upon public schools across America

## A Concluding Note about the Holistic Model for Blended Learning and Pragmatism

The situation in Pennsylvania's public schools underscores the assertion that there is a pressing need for a paradigm shift within the field of instructional technology as we collectively address the need for a practical model for K-12 distance education. The earliest attempts to come up with a viable model have relied upon a nominalistic and mechanistic worldview that has dominated the field of instructional technology for the past several decades (Reiser, 2001) and these efforts have met with limited success. A key point that I want to make in this article is that the field of instructional technology can now consider a more pragmatic and contextual worldview when addressing the need for a new model for K-12 distance education. Rather than introducing technologies and learning materials that represent what *can*, or *should* be presented in any given learning environment, it is now possible to use instructional technologies and stored digital media in the form of learning objects to represent what actually *is* happening in the classroom. This shift in perspective places control squarely back into the hands of classroom teachers and enables all interested parties to provide the vital ingredient of *context* that has long been missing from online lessons presented to K-12 learners. This needed transition is perhaps best described by Stephen Pepper in his book *World hypotheses: A study in evidence* (1942) where he explains the root metaphor method and applies it to compare the nuances of the mechanistic world view and the contextual, or pragmatic world view.

It is precisely Pepper's emphasis upon the "real historic event–alive in its present" (p. 232) that makes the holistic model for blended learning so valuable for America's public schools, because this root metaphor encapsulates the need for adaptability in delivery technologies and processes in order to place the online lessons they produce in context. By building upon this root metaphor, the holistic model for blended learning makes it possible for public school districts to establish their own cyber schools and blended learning programs that provide the one key element that has long been missing from K-12 online learning. That element is context – and the holistic model for blended learning not only addresses this need, it does so by placing a minimal strain on district staff and infrastructure.

## REFERENCES


Bonk, C. J., & Graham, C. R. (Eds.). (2007, in press). *Handbook of blended learning: Global perspectives, local designs*. San Francisco, CA: Pfeiffer Publishing.

Carman, J. (2002). *Blended learning design: Five key ingredients*. Retrieved January 26, 2007 from, http://www.agilantlearning.com/instructionaldesign.html

Chute, E. (2005, May 08). Cyber schools spring up in state. *The Pittsburgh Post-Gazette*.

Downes, S. (2001). Learning objects: Resources for distance education worldwide. *International Review of Research in Open and Distance Learning, 2*(1), 29.

Friesen, N. (2001). What are educational objects? *Interactive Learning Environments, 9*(3).

Friesen, N. (2003). Three objections to learning objects and e-learning standards. In R. E. McGreal (Ed.), *Online education using learning objects*, pp. 59-70. London: Routledge.

Geary, G. (1991). *Electronic performance support system* (1st ed.). Tolland, MA: Gery Associates.

Godwin-Jones, R. (2004). Learning objects: Scorn or SCORM? *Language Learning & Technology, 8*(2), 7-12.

Hodas, S. (1993). Technology refusal and the organizational culture of schools. *Education Policy Analysis Archives, 1*(10).

Hickman, L. A. (2001). *Philosophical tools for technological culture: Putting pragmatism to work*. Bloomington: Indiana University Press.

Knade, S.L. (2001). Cyber school litigation expands. *Education Law Focus, 10*(5), 1-4.

Merrill, M. D. (2000). Instructional transaction theory (ITT): Instructional design based on knowledge objects. In *Instructional design theories and models: A new paradigm of instructional theory*. Mahwah, NJ: Lawrence Erlbaum Associates.

NCES. (National Center for Educational Statistics). (2003). *Distance education at degree-granting postsecondary institutions: 2000-2001*. Retrieved November 9, 2006 from, http://nces.ed.gov/pubs2003/2003017.pdf

Pepper, S. C. (1942). *World hypotheses: A study in evidence*: University of California Press.

Raffaele, M. (2004, June 21). Some districts resist paying for cyberschools. The Pittsburgh Post-Gazette.

Reiser, R. A. (2001). A history of instructional design and technology, Part I: A history of instructional media. *Educational Technology, Research and Development, 49*(1), 53-64.

Rosenberg, M.J. (2001). *e-Learning: Strategies for delivering knowledge in the digital age.* New York: McGraw Hill.

Shank, G. (2007). Personal communication. 03/19/2007.

Silver, J.D. (2007, March 18). Cyber-school empire under attack: Beaver County educator fighting grand juries, suits and legislators. *The Pittsburgh Post-Gazette.*

Smith, C. (2005, December 18). State-funded cyber school draws concern. *The Pittsburgh Tribune-Review.*

Stone, A. (in press). Learning object based instruction (LOBI). *Encyclopedia of Information Technology Curriculum Integration.* Hershey, PA: Idea Group Publishing.

Tomei, L. A. (2002). *The technology facade: Overcoming barriers to effective instructional technology.* Boston: Allyn and Bacon.

Tripp, S., & Bichelmeyer, B. (1990). Rapid prototyping: An alternative instructional design strategy. *Educational Technology Research & Development, 38*(1), 31-44.

US Department of Education (USDED). (2007). *Effectiveness of reading and mathematics software products: Findings from the first student cohort.* Online version. Retrieved May 29, 2007 from, http://ies.ed.gov/ncee/pdf/20074005.pdf

Welsch, E. (2002). SCORM: Clarity or calamity? *Online Learning Magazine, 07*(1), 4.

Wiley, D. A. (2000). Connecting learning objects to instructional design theory: A definition, a metaphor, and a taxonomy. In D. A. Wiley (Ed.), *The instructional use of learning objects.* Online version. Retrieved January 26, 2007 from, http://reusability.org/read/chapters/wiley.doc

Wiley, D. (2003). *Getting axiomatic about learning objects.* Retrieved January 26, 2007 from, http://www.reusability.org/axiomatic.pdf

Wilson, B. G. (1995). *Situated instructional design: Blurring the distinctions between theory and practice, design and implementation, curriculum and instruction.* Paper presented at the Association for Educational Communications and Technology, Washington DC.

Verbert, K., & Duval, E. (2004). *Towards a global architecture for learning objects: A comparative analysis of learning object content models.* Retrieved May 24, 2007 from, http://www.cs.kuleuven.ac.be/~hmdb/publications/files/pdfversion/41315.pdf

Voos, R. (2003). Blended learning—What is it and where might it take us? *Sloan-C View, 2*(1), 3-5.

Zugner, L. C. (1987). A reappraisal of instructional television. *ERIC Digest, 12*(00), 1-6.

*This work was previously published in International Journal of Information and Communication Technology Education, Vol. 4, Issue 1, edited by L. Tomei, pp. 56-71, copyright 2008 by IGI Publishing (an imprint of IGI Global).*

# Chapter 18

# Identifying the Risks Associated with Primary School Children Using the Internet

**Derek O' Reilly**
*Dundalk Institute of Technology, Ireland*

## ABSTRACT

*The Internet is becoming widely available and increasingly important in the modern world. Because of this, it is very important that children start to familiarize themselves with the Internet at a young age. As technology is becoming increasingly part of our daily lives, computers and the Internet have been adopted into schools. The sea of information and learning activities available on the Internet has the potential to greatly help in the development of young minds. However, the question remains as to how safe an environment the Internet is for young children. Children might not see any dangers beyond the physical environment where they live. Therefore, the Internet can be an unsafe place for them to venture into. Children can be targeted by a wide number of Internet risks. For these reasons, the problem of how to keep children who are using the Internet safe must be solved. This paper identifies the potential risks associated with primary school children using the Internet. This paper evaluates the level of understanding of Internet safety that children have. This paper identifies what children use the Internet for and what information that they are willing to reveal while online. The findings of this paper are based on analysis of a survey carried out on 645 Irish primary school children in February 2007.*

## BACKGROUND

There is much concern among parents, educators and policy makers regarding the dangers that Internet usage poses to children. The aim of this paper is to identify the actual risks that are associated with primary school children using the Internet. It is important that research be undertaken in this area to allow the various stake-holders the opportunity to separate the actual and perceived risks that exist from children using the Internet. In the absence of scientific research, the information void is being filled by the popular press (Carrington, 2008, pp., p7), who tend to produce inaccurate and overly

DOI: 10.4018/978-1-60566-936-6.ch018

negative accounts of the risks that Internet usage poses to young children. Carrington (2008) states that *"there is a clear positioning of children as innocent and gullible and the construction of a gulf between adult and child in terms of awareness of risk and victim–predator status"* (p. 156). Carrington (2008) reports that this knowledge gap is causing many adults to take a negative approach toward technology and *"the huge issue that teachers, parents, politicians and school administrators make about digital technology is for their own benefit"* (p. 163). Hope, too, believes that the Internet risk debate is being manipulated by certain lobbies. Hope (2006) reports that many older people struggle with the new technologies that children tend to be very comfortable with. Cizek (1999) agrees, stating that *"many teachers feel that their pupils know more about the technology than they do and, as a result, feel threatened by the technology"* (p. 406). Cizek (1999) states that teachers who feel threatened by technology *"are reluctant to adopt and use the technology"* (p. 406). Valentine (2001) states that adult fears about Internet content *"are fears about knowledge rather than violence"* (p. 72). Carrington (2008) states that *"as the technological gap between some adults and children continues to open, there will be increasing amounts of discomfort over young people's use of technologies and the public spaces they make accessible"* (p. 157).

The need for research into the risks that primary school children face while using the Internet has been identified by other researchers. Hope (2006) states that there has only been *"some limited discussion of the issue of children and Internet risks"* and that *"risk arising from school Internet use is a largely neglected issue that needs urgent attention"* (p. 312).

This paper limits itself to the study of the Internet usage behavior of children in the senior four years of primary education. Too many other published studies, such as (NCTE, 2006), attempt to treat children of all ages as being the same. The large range of age groups surveyed in these other studies (the NCTE study deals with children in the age range of nine to 16 inclusive) makes it difficult to draw generalizations from the data. Lawson (2000) argues that discussions about school Internet risks must be age appropriate. Hope (2006) agrees with Lawson, noting that there is a tendency in current research to use the term 'children' in a generic sense, *"treating the group as homogeneous whilst ignoring the issue of different age groups"* (p. 308).

## Context

As technology is becoming increasingly available and progressively more important in our daily lives, the Internet is being increasingly adopted into schools. Condie (2007) states that *"the evidence gathered has shown a steady increase in the number of computers and other technologies"* being used in schools (p. 3). Reflecting upon the importance of computer and Internet usage skills, Carrington (2008) states that the increased usage of computers and the Internet requires *"a concerted effort to ensure that all children in our school systems are provided with opportunities to engage in these new forms of textual and social practice"* (p. 165).

It is very important that children start to familiarize themselves with the Internet from a young age. Carrington (2008) states that *"because young people are born into a world saturated in digital technologies they require literacy practices and skills oriented to multimodality"* and *"to focus on print as an a priori requirement is increasingly a disservice, particularly to those pupils most at risk of poor outcomes from schooling"* (p. 165).

The sea of information and learning activities available on the Internet has the potential to greatly help in the development of young minds. Venezky (2004) states that *"by giving all schools access to an expanding world of resources, both teaching and learning can be advanced to a new level of effectiveness and social importance"* (p. 3). Haugland (1999), citing earlier research, suggests that introducing computers into the classroom can improve pupils' motor skills, enhanced mathematical thinking, increase creativity, increase critical thinking and problem solving, improve language skills, enhance children's self-confidence and communication skills. Haugland also states that children who use computers share leadership roles more frequently and develop positive attitudes toward learning. Whitehead (2002) agrees with this view, reporting that computer usage can lead to increased pupil writing, enhanced cooperative learning, enhanced integration of curriculum, greater application of learning style strategies, increased applications of cross-age tutoring, increased teacher communication, enhanced community relations and enhanced global learners.

However, the question remains as to how safe an environment the Internet is for young children. Wishart (2004) states that there is *"worldwide concern for the safety of young Internet users"* (p. 193). Children can be targeted by a wide number of Internet risks, such as exposure to illegal or harmful material, receiving unsolicited messages and being lured into a physical encounter (NCTE, 2002).

## HOW CHILDREN USE THE INTERNET

Children use the Internet in a variety of ways. These include email, webpage publishing, blogging, instant messaging, chat rooms and social networking.

Email is one of the most well-known and popular communication tools used by children on the Internet. In an educational environment, email can be very beneficial, as it allows children to communicate and share ideas and resources with their teachers and pupils from any location throughout the world. Children can use the Internet to engage in collaborative projects and improve their writing and communication skills. Although email provides many benefits to children, it also provides a gateway for potential risks, such as inappropriate material, cyber bullying and exposure to pedophiles.

Many teachers see publishing class webpages to be a useful exercise. As a result, many children learn the basics of creating webpages as early as primary school. A website can help promote a school within the wider community by publishing information on school activities, policies and events. Habib (1998) states that *"publishing on the Internet benefits both pupils and the audience who access the published information"* (p. 7). Publishing on the Internet gives children a global audience to view their material. Children who publish on the Internet can develop collaborative, creative and critical thinking skills. There are also disadvantages associated with publishing webpages on the Internet. Without proper supervision, children can inadvertently publish personal information, such as their name, address, phone number or a photograph of themselves. In doing so, they risk becoming the target of cyber bullies or pedophiles.

Children are increasingly using blogging as a means of communication. Blogs are the first widely adopted, easy-to-use, publishing tool that allows people to publish on the Internet. Blogs are ideally suited to usage by children, because non-technical people can create blogs. Among other things, children use blogs to create personal journals of their lives. In a classroom environment, blogs can encourage children to write down thoughts and ideas in relation to classroom activities and projects. Richardson (2006) states that *"blogs engage readers with ideas, questions and links. They ask readers to think and respond"* (p. 18). However, blogs that have been published by children present the same risks as webpages do to children.

Instant messaging allows a person to send and receive messages instantly, giving that person the ability to have a real time conversation on the Internet. Real time conversations expose children to a number of risks. Ybarra (2008) reports that children are more likely to be harassed or sexually solicited via instant messaging than by any other electronic means.

Chat rooms use instant messaging to connect people with similar interests to each other. Chat rooms provide an anonymous environment where identities can be kept hidden. Unfortunately, children find the anonymous environment of chat rooms, where people are not talking face to face, to be very appealing. The anonymity of participants may lead children to *"engage in unsafe behavior, or to become susceptible to online grooming"* (NetAlert, 2004, p.10).

Social networking websites combine the features of blogs, instant messaging and chat rooms. Social networking websites allow children to express themselves by sharing their likes and dislikes, opinions and expertise with friends and strangers. Social networking websites allow children to create their own webpages, publish photographs, publish blogs, submit comments, create online polls and upload videos. Social networking websites make it extremely easy for children to share personal information. Zinman (2007) states that *"communication with strangers is often an inherent part"* of social networking websites and that social networking websites *"exist in part to enable unsolicited, yet friendly and welcome communication"* (p. 1). According to Zinman (2007), the *"openness to messages from strangers leaves users of these sites vulnerable to a growing quantity of unwelcome contact"* (p. 1). Of all the current means of Internet communication, social networking appears to present the greatest risk to children. This is because the design of social networking websites actively encourages the sharing of personal information. Gross (2005) states that *"personally identified or identifiable data (as well as contact information) are often provided, together with intimate portraits of a person's social or inner life"* (p. 72). Gross (2005) believes that *"such apparent openness to reveal personal information to vast networks of loosely defined acquaintances and complete strangers calls for attention"* (p. 72).

The Internet has become a crucial part of the daily lives of many children. Children are increasingly using the Internet as part of their social interaction and do not differentiate between online and offline social activities. Valentine (2004) states that, for children, *"the real and the virtual are mutually constituted"* (p. 302).

Carrington believes that a contributing factor to adults' fear of children using the Internet results from the fact that many adults are not proficient with the technology. Carrington (2008) states that *"as the technological gap between some adults and children continues to open, there will be increasing amounts of discomfort over young people's use of technologies and the public spaces they make accessible"* (p. 157). Carrington contends that, rather than basing the debate on scientific fact, many adults engage in reactionary discourse based on a fear of the unknown. Carrington says that this fear is compounded by the popular press, which often depicts children as being at risk while using the Internet.

Lumby (1997) suggests that much parental concern focuses on protecting children's "innocence" by attempting to stop them from gaining access to information about sexual practices and alternative models of sexuality. Valentine (2001) agrees with Lumby, stating that parents fears *"are fears about knowledge rather than violence"* (p. 72).

## ONLINE SAFETY

Personal safety is a major issue when it comes to children using the Internet. According to a report published by the Irish National Centre of Technology in Education, there are three primary categories of risk associated with children using the Internet (NCTE, 2002):

1. Exposure to illegal or harmful material;
2. Receiving unsolicited messages;
3. Being lured into a physical encounter.

In addition to these three categories of risk, consideration should also be given to the ways in which childhood development factors can cause children to be placed at risk when they are online.

Each of these four categories of risk is discussed in detail below.

## Exposure to Illegal or Harmful Material

Children using the Internet face the risk of exposure to *"hate, violence, misinformation, consumer exploitation and sexual predators"* (Berson, 2003, p. 10). Berson (2003) states that *"Without much effort, a child may inadvertently or deliberately be exposed to on-line content that is obscene, pornographic, violent, racist, or otherwise offensive"* (p. 10). Others agree with Benson's argument that the relatively unregulated nature of cyberspace means that sexually explicit discussions, pornography, racial and ethnic hatred, Neo-Nazi groups, and pedophiles can all be found on the Internet (Squire 1996; Whine 1997; Britton 1998; Hope 2006; NCH 2006). Exposure to such material can occur unintentionally when searching for educational content about people or places. Wolak (2007) found that 42% of adolescents surveyed had been exposed to online pornography in the past year. Of those, 66% reported unwanted exposure. Exposure to illegal or harmful website content can have a serious negative effect on a child's behavior or attitude (Ybarra, 2004).

Pupils who deliberately search for sexually explicit material on the Internet will have no difficulty in finding it. In a survey conducted on behalf of the Irish Department of Education and Science, it was reported that 35% of the 848 children surveyed had visited pornographic websites (NCTE, 2006).

## Receiving Unsolicited Messages

Cyber bullying can happen through the use of email, text messages, chat rooms, mobile phones, mobile phone cameras and websites (Campbell, 2005). Although cyber bullying may not be of the physical kind, it is an extreme form of mental bullying, which can lead to the physical harm of an individual. Willard reports that cyber bullying can cause devastating effects to young children (Willard, 2006). Willard reports that the effects of cyber bullying include low self-esteem, poor academic performance, depression, and, in some cases, violence, even suicide. Hinduja (2008) agrees with Willard, stating that *"this negative experience not only undermines a youth's freedom to use and explore valuable on-line resources, but also can result in severe functional and physical ramifications"* (p. 129).

Because of the widespread use of the Internet, cyber bullying has increased dramatically in recent years. The NCH reports that one in every five school children is the victim of cyber bullying (NCH, 2005). Li (2006) reports similar findings, with one in four school pupils saying they are the victim of cyber bullying. A study by Sharriff (2007) found even more worrying results, with 60% of pupils reporting they had been ignored by peers online, 50% saying they had been disrespected, 30% saying they had been called names, and 21% saying they had been threatened. Sharriff (2007) states that *"cyber bullying is especially insidious because of its anonymous nature"* and *"it is dangerous because it most often takes place outside school hours on home computers, making it difficult, if not impossible, to supervise"* (p. 77).

## Being Lured into a Physical Encounter

Internet grooming *"involves a clever process of manipulation, typically initiated through a nonsexual approach, which is designed to entice a victim into a sexual encounter"* with an adult (Brown, 2001, p. 11). Davidson (2005) reports that sex offenders will often socialize and groom children over long periods of time. Davisdon (2005) states that grooming *"involves a process of socialization during which an offender seeks to interact with the victim (and sometimes the victims family), to share their hobbies and interests and to become a part of their life, in order to prepare them for abuse"* (p. 4). Davidson also reports that the "*grooming of a child is often a slow process and certainly a deliberate process, it is much easier to abuse a child when trust is established"* (pp. 4-5). The NCH (2006) agree with Davisdon, stating that *"generally they [sexual predators] will seek to develop a relationship with the child or young person with a view to meeting up with them in real life"* (p. 3).

The anonymity of the Internet makes it is easy for an adult to impersonate a child and lure children into a physical encounter. The most serious risk of Internet usage *"most serious risk involves the possibility of a child being lured into a physical encounter with someone they've met online"* (NCTE, 2002, p. 10). According to Berson (2003)*"The perceived anonymity of the Internet has benefited pedophiles and provided an environment conducive to the exchange of pornography, identification of children to molest, sexual interactions with youth, and support and validation from other adults who share their sexual preferences"* (p. 11). Pedophiles find security by operating within the confines of their own homes, from where they can groom children with the intention of meeting them in real life (Davidson, 2005).

In a survey conducted on behalf of the Irish Department of Education and Science of children's use on the Internet, 7% of the surveyed pupils aged between nine and 16 years old had met someone in real life that they first met on the Internet (NCTE, 2006). Of the children who had met an online contact in real life, 24% said that the person they met had introduced themselves as children on the Internet, but had turned out to be adults in real life.

Not all researchers agree that children are at imminent risk of sexual predators when using the Internet. Carrington (2008), quoting Bob Sullivan, a technology correspondent at MSNBC, states that *"experts could not cite a single case of a child predator hunting for and finding a child through a blog"* (p. 155).

## Childhood Developmental Factors that Contribute to Online Risk

Ybarra believes that children will continue to use the Internet, irrespective of the dire warnings that they receive from adults. Ybarra (2007) states that *"it may not be feasible to change the entire online culture, and the promotion of prevention messages that contradict or fail to recognize widely accepted online behavior may lack credibility with youth"* (p. 142). Ybarra (2007) suggests that *"instead of imparting the message "don't talk to strangers online," a harm reduction approach may be more effective"*(p. 142). Ybarra believes that it can be okay to talk to strangers while online. Ybarra suggests that children only come into danger when they engage in *"risky"* behavior, such as talking about sex or behave aggressively toward others while online.

Oswell (1998) believes that not all children are victims. Oswell categorizes children into three categories: child-as-victim, the child-in-danger and the dangerous child. Hope (2006) states that Oswell's categories *"allows for the idea that young people are not only possible victims but also potential offenders"* (p. 313).

Wolak believes that many young victims of online consensual sexual crimes may not see themselves as being victims. They do not believe that either they or their adult partner are doing any harm. Wolak (2004) states that most victims of sex crimes are *"young teens who are willing to enter into voluntary sexual relationships with adults whom they meet online"* (p. 18). For many troubled children, the Internet provides them with the affection and attention that they lack in the real world. Wolak (2004) states that *"half of the victims"* of online child sexual abuse *"were described as being in love with or feeling close bonds with the offenders"* (p. 11).

Wolak (2004) found that most sex offenders *"did not deceive victims about the fact that they were adults who were interested in sexual relationships"* (p. 11). Willard agrees with Wolak in the belief that many children are willing to have sex with adults. Willard (2007) suggests that predators do not need to be snatching kids by piecing together clues from personal information when they can go for the "low-hanging fruit" - the teens specifically engaging in at-risk behavior, such as posting sexually provocative images in their profiles. Wolak (2004) suggests that educating children about the potential long-term detrimental effects of underage sex might achieve better results than simply advising children not to talk to strangers online.

Peer pressure and group dynamics play a role in children's online behavior. Ybarra (2007) states that *"the role of friends in many online behaviors should be acknowledged. More than 40% of online risky behavior occurred while youth were using the Internet with friends or peers"* (p. 144).

Wolak (2003) reports that having *"home Internet access and high levels of Internet use were related, independently of each other, to forming close online relationships"* (p. 115). Krauss (2008) reports that *"Both males and females with Internet access were also found to report significantly lower ages of first sexual intercourse than participants without Internet access"* (p. 166). This matches previous findings that show how children who are exposed to sexually explicit material on television are more accepting of premarital sex, more likely to misjudge the prevalence of sexual activity, less likely to value the concept of marriage and monogamy, and more likely to consider sexual activity without emotional commitment (Collins R., 2004; Zillman, 1988, 1988b).

The children who are at greatest risk of being bullied or sexually abused online tend to be the same children who are at risk in the real world. Wolak (2003) states that children are more vulnerable to online victimization if they are *"highly troubled, reported high amounts of conflict with their parents, low communication with parents and engaged in high levels of delinquency"* (p. 110). These children are more likely than the general population to seek online friendships or romances.

Ybarra (2005) believes that concerns about young children exposing themselves to pornography on the Internet may be overstated. Ybarra's research shows that 87% of children who seek out pornography are at least 14 years old, which is an age when one would be expected to be sexually curious. Ybarra's results suggest that younger children do not seek out pornography.

## SURVEY

The principal research aim of this paper is to quantify the various Internet usage behavioral patterns that might result in children being exposed to personal safety risks while using the Internet. The instrument used to gather data is a quantitative survey. 645 children from six primary schools in County Louth, Ireland, took part in the survey. The surveyed children ranged in age from nine to 13 years old. Data for this survey was collected in February 2007.

Three factors contribute to the reliability of this survey. Firstly, respondents' confidentiality was ensured. Therefore, there was no reason for the children to give dishonest answers to the survey questions. Secondly, the survey was conducted in ten different schools and the data from all ten schools correlated. Thirdly, both urban and rural schools were represented in the data. This means that the data is more representative of the population. These three factors suggest that the data is reliable.

There are two major weaknesses inherent in the data used in this survey. Firstly, the data collected in this survey was not randomly selected. Instead, access to schools was got via personal contacts. Secondly, the sampled schools were clustered in one county. Both of these weaknesses mean that it is unsafe to make generalizations for the entire population based on the findings of this survey.

The results from this survey show that Internet usage increases as pupils progress through the class stages. Internet usage ranges from a low of 42% in 3rd class and 39% in 4th class up to a high of 62% in both 5th and 6th classes. Other studies, such as that conducted by NSBF (NSBF, 2000) and NCES (NCES, 2003) confirm the increase in Internet usage as children become older.

The results from this survey show that boys and girls are as likely as each other to use the Internet in any year. Weiser (2000) finds similar results. In addition, Weiser states boys tend to use the Internet for entertainment while girls use it for communication.

Three questions were asked of the surveyed children as a means of discovering the Internet usage behavioral patterns that might result in them being exposed to personal safety risks while using the Internet.

Firstly, what do primary school children use the Internet for? Answering this question should allow us to better understand the types of activities that primary school children engage in when they are on the Internet and therefore help us to identify the level of risk that children face when using the Internet.

Secondly, what information do primary school children believe is safe to put online? Answering this question should help to identify the level of knowledge that children have regarding placing personal information online and consequently, it should help to identify the level of risk that children face while using the Internet.

Thirdly, do young people know whom it is safe to talk to online? Answering this question should help to identify any vulnerability that exists due to children being trusting and naïve.

The results from each of these three research questions are discussed in detail below.

## Children's Online Activities

In order to find out what primary school children use the Internet for, pupils were asked which of the following six activities they use the Internet for. Pupils were allowed to select zero or more of the six activities. Pupil responses are shown in Figure 1.

As they get older, children tend to move away from passive Internet activities toward more interactive Internet activities. As they get older, fewer children partake in the passive activity of downloading games. There is also not a significant difference in the number of pupils in each class who participate in the passive activity of downloading music.

On the other hand, interactive activities, such as searching the Internet and emailing friends, all show a gradual increase as children get older. Chat rooms, which are the most interactive activity listed, show the most dramatic increase in rate of usage as the pupils get older. Chat rooms usage rates change from 24% of pupils in 3rd class up to 65% of pupils in 6th class. The trend of pupils to move from passive to more interactive online activities as they get older is also found by (NCES, 2003).

Based on the other graphs, one would expect that the use of blogs would increase as pupils become older. The 4th, 5th and 6th year classes all follow the predicted curve of increasing year on year usage. However, the greatest number of online bloggers is the 3rd class pupils.

The graphs in Figure 1 show that there is significant usage of the Internet by all age groups. The Internet plays a part in the life of a large number of children. Society needs to ensure that children are properly educated in the various areas of Internet usage.

*Figure 1.*

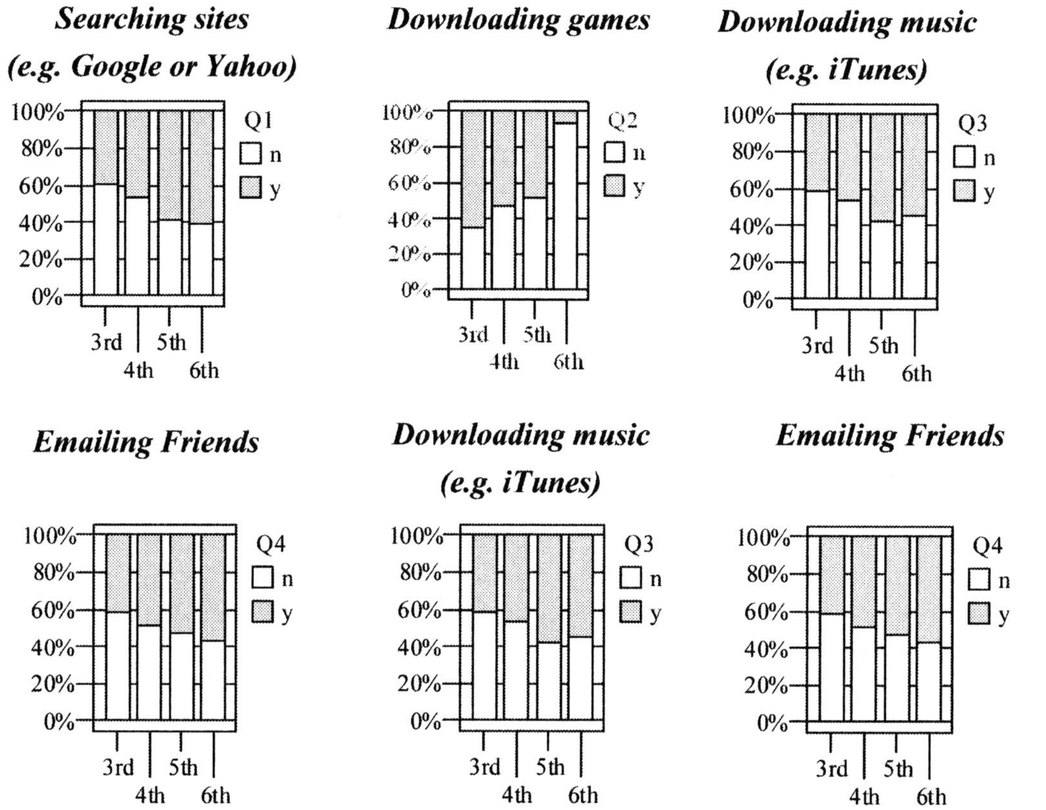


## Placing Personal Information Online

In order to find out what information primary school children believe is safe to put online, pupils were asked to identify which of the following nine items they believed it was okay to put online. Pupils were allowed to select zero or more of the nine items. Pupil responses are shown in Figure 2.

Figure 2 shows that, they get older, children tend to be more aware of the type of information that they should not put online. Carrington (2008), too, finds that older children appear to be aware of what personal information is appropriate to place online. As they get older, there is a gradual decline in the number of children who think that it is correct to put their name, age, address, phone number or email address online. The biggest declines occur in the willingness of older children to place either their address or their phone number online. One must question if this reflects the lessons that children of a very young age receive from their parents and teachers about talking to strangers. Perhaps the figures reflect the fact that very young children are not also told that it is bad to tell strangers their age or email address. This might well be the case, as both parents and teachers are not of the Internet generation. Parents and teachers might not are realize how literally children take their advice.

*Figure 2.*

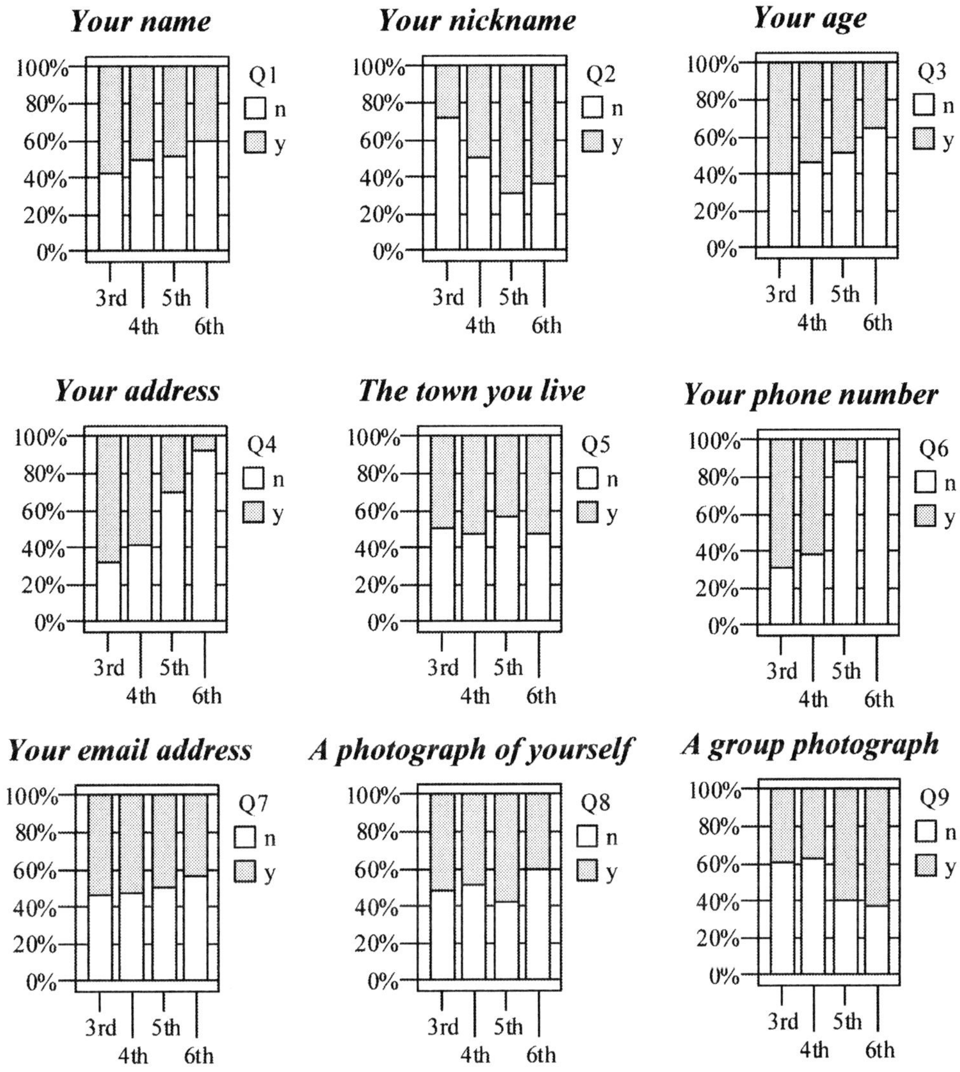


## Online Conversations

In order to find out if young people know whom it is safe to talk to online, pupils were asked to identify which of the following three categories of people it was okay to talk to online. Pupils were allowed to select zero or more of the three items. Pupil responses are shown in Figure 3.

The graphs in Figure 3 show that children of all ages are not at all clear whom it is okay to talk to online. Pupils tend to be too cautious while talking online to people that they know. From a safety viewpoint, this is not a bad thing. However, it might result in pupils not exploiting the full potential of using the Internet.

The graphs show that 4th, 5th and 6th class pupils are too open to communicating with strangers online. 60% of 4th class, 39% of 5th class and 62% of 6th class pupils believe that it is safe to talk online to strangers who the pupils know to be older than themselves. Interestingly, all of the 3rd class pupils said it was not safe to talk to strang-

*Figure 3.*

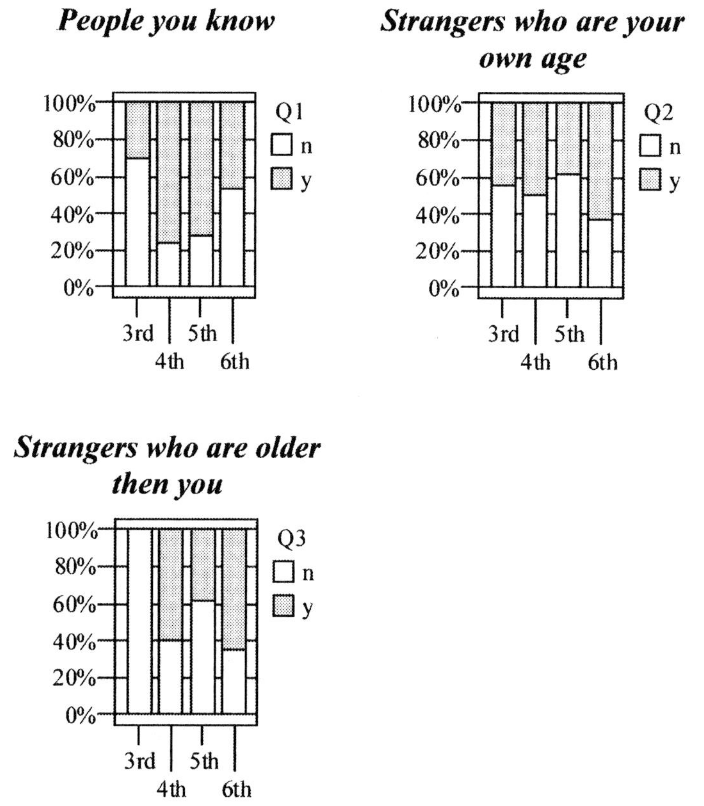


ers online. The majority of 6th class pupils also believe that it is okay to talk online to strangers who are their own age.

Of all pupils, the 6th class pupils feel most safe when talking online to strangers. Over 60% of 6th class pupils feel safe when talking to strangers who are either their own age or are older. Perhaps it is as a result of them having more exposure to online chatting with people their own age that 6th class children feel more comfortable talking online to strangers of all ages. The false sense of security brought on by a familiarity of using the Internet might be the greatest danger that older children face when using the Internet. It only takes one horrible experience to destroy a child's life. Children simply do not understand the very real dangers that are present in the Internet.

## CONCLUSION

Both the literature review and the survey show cause for concern. They suggest that primary school children may be placing themselves at risk as a result of their Internet usage behavior. Children do not appear to be well equipped for dealing with the potential danger that their Internet related activities place them in. It is a fact of life that primary school children use the Internet. Children are not knowledgeable enough to be able to safely use the Internet. Just as they are taught about the dangers that exist in the real world, children need to be taught about the dangers on the Internet.

This research suggests that children's Internet usage is dependent on age. Future research into children's Internet safety should take greater consideration of the differences in Internet usage that exist between various age groups.

It is the role of schools to educate children properly about Internet safety. Wishart (2004) reports that it is generally accepted worldwide that schools have a fundamental role in ensuring the online safety of their pupils. Currently children are not properly taught which information is safe and which information is dangerous to put online. Internet safety should be given as high a profile as is given to education regarding dangers in the physical world. There is no alternative to education. Society has a duty to educate children about the risks that exist on the Internet.

## ACKNOWLEDGMENT

The author would like to acknowledge the contribution of Caroline O' Neill, who was co-author of the conference paper (O' Reilly, 2007), upon which this paper is based.

## REFERENCES

Berson, I. (2003). Grooming cybervictims: The Psychosocial effects of online exploitation for youth. *Journal of School Violence, 2*(1), 5–18. doi:10.1300/J202v02n01_02

Brown, D. (2001). Developing Strategies for collecting and presenting grooming evidence in a high tech world. *National Center For Prosecution of Child abuse . Update, 14*(11).

Campbell, M. (2005). Cyber bullying: An old problem in a new guise. *Australian Journal of Guidance & Counselling, 15*(1), 68–76. doi:10.1375/ajgc.15.1.68

Carrington, V. (2008). 'I'm Dylan and I'm not going to say my last name': some thoughts on childhood, text and new technologies. *British Educational Research Journal, 34*(2), 151–188. doi:10.1080/01411920701492027

Collins, R., E. M., Berry S., Kanouse D., Kunkel D, Hunter S., and Miu A. (2004). Watching sex on television predicts adolescent initiation of sexual behavior. *Pediatrics, 114*, 280–289. doi:10.1542/peds.2003-1065-L

Davidson, J., & Martellozzo, E. (2005, Sept. 8-10). *Policing the Internet and protecting children from sex offending online: When strangers become 'virtual friends'.* Paper presented at the Cybersafety Conference, Oxford, England.

NCES. (2003). *Computer and Internet use by children and adolescents in 2001 statistical analysis report*: National Centre for Education Statistics.

NCH. (2005). *Putting U in the picture: Mobile bullying survey 2005*. NCH, the Children's Charity.

NCTE. (2002). *Be wise on the Web: Information and advice for schools*: Irish National Centre of Technology in Education.

NCTE. (2006). *Survey of children's use of the Internet: Investigating online risk behaviour*: Irish Department of Education & Science.

NetAlert. (2004). *A teacher's guide to Internet safety: How to teach Internet safety in our schools*: Australia's Internet safety advisory body.

NSBF. (2000). *Safe and smart: Research and Guidelines for children's use of the Internet*: National School Boards Foundation. O' Reilly, D., & O' Neill, C. (2007). *An Analysis of Irish Primary School Children's Internet Usage and the Associated Safety Implications*. Paper presented at the e-Learning.

Willard, N. (2006). Flame Retardant: Cyberbullies Torment Their Victims 24/7. Here's How To Stop The Abuse. *School Library Journal*, *52*(4), 55–56.

Ybarra, M. (2004). Sex Differences In Youth-Reported Depressive Symptomatology And Unwanted Internet Sexual Solicitation. *Journal of Medical Internet Research*, *6*(1). doi:10.2196/jmir.6.1.e5

Zillman, D., & Bryant, J. (1988). Effects of prolonged consumption of pornography on family values. *Journal of Family Issues*, *9*, 518–544. doi:10.1177/019251388009004006

Zillman, D., & Bryant, J. (1988b). Pornography's impact on sexual satisfaction. *Journal of Applied Social Psychology*, *18*, 438–453. doi:10.1111/j.1559-1816.1988.tb00027.x

# Chapter 19
# Ethical Issues in Information Technology:
## Does Education Make a Difference?

**Barrie E. Litzky**
*The Pennsylvania State University, USA*

**Effy Oz**
*The Pennsylvania State University, USA*

## ABSTRACT

*This research examines the impact of education on the ethical decision-making outcomes of adult learners in the area of information technology (IT). This study sheds light on the research question "Does IT ethics education make a difference?", and more specifically, "Do ethics courses influence decisions regarding IT ethical issues in adult learners?" In a field study of 78 pre- and post-test surveys, we found that graduate students who took a course in IT ethics made different decisions than those made at the start of the term, for 2 of 6 ethical issues. The ethical issues described in this article are particularly relevant in today's knowledge economy. Implications for IT ethics education and future research in the area are discussed.*

## INTRODUCTION

The corporate scandals of the past several years have sparked intense interest and debate in academic, corporate, and government circles over the topic of business ethics education. These groups agree that more emphasis needs to be placed on business ethics education (Etzioni, 2002; Sims, 2002; Mangan, 2003; Swanson, 2003, 2004; Swanson & Frederick, 2003). This article examines a specific functional area of business ethics education—ethical issues within the information technology field. The ethical issues highlighted in this article have tremendous relevance in the present global knowledge economy, and the findings shed new light on our current understanding of the outcomes associated with business ethics education.

Only a small number of studies have dealt with education of information technology (IT) ethics. Students have been used in measuring attitudes toward IT-related ethical issues (Logsdon, Thompson, & Reid, 1994; Loch &Conger, 1996; Calluzzo & Cante, 2004; Haines & Leonard, 2007), but the purpose of these studies was to measure attitudes at a certain point; they did not focus on the impact of ethics education and therefore did not measure changes in the attitudes. Those that have focused on a single ethical issue usually address only software piracy (Oz, 1990; Logsdon, et al., 1994; Sims, Cheng, & Teegen, 1996; Ramakrishna, Kini, & Vijayraman, 2001; Peace, Galleta, & Thong, 2003), but none have taken a holistic approach to test the effect of IT ethics education on student attitudes.

We conducted a pre-test-post-test study to examine the hypothesis that ethical education in the form of a graduate course makes a difference in students' ethical decision-making. Repeatedly, students in earlier sections of this course said they now discovered "another side to the issue." Thus, we expected that if some students did not recognize the ethical nature of some of the issues at the start of the class, they would by its end; and that if they did not consider all of the relevant stakeholders involved and the entire potential impact of an act, they would by the end of the course. In an effort to understand *whether* and *how* learning takes place, we employed two taxonomies of educational objectives to frame the types of learning outcomes we could expect from a course in IT ethics. A brief discussion of Bloom's (1956) taxonomy of educational objectives and Tomei's (2005) taxonomy for the technology domain follows in the next section.

## Taxonomies of Learning in IT Ethics Classes

Bloom's taxonomy of educational objectives is a widely-used and accepted framework for classifying educational outcomes as a result of instruction. Bloom described three educational domains: cognitive, affective, and psychomotor-skills and the expected outcomes or behaviors associated with each (Reeves, 1990; Krathwohl, 2002). The cognitive and affective domains are relevant to this study. The cognitive domain refers to the development of intellectual competencies whereas the affective domain deals with attitudes, feelings, values, and behaviors. Outcomes associated with teaching ethics in information technology can fall within both domains, although the affective domain outcomes are harder to measure (Reeves, 1990).

In the cognitive domain, learning outcomes fall into one of six hierarchical levels: knowledge, comprehension, application, analysis, synthesis, and evaluation, where each higher category requires the skills and abilities associated with lower categories. Bloom believed educators should focus on developing student problem-solving abilities and higher-order thinking skills. To do this, educators must provide students with methods for application and critical thinking (Reeves, 1990).

To determine cognitive mastery of a subject, many ethics educators use the case method, which lends itself nicely to the development and improvement of critical thinking skills. However, ethics educators often find themselves in a conundrum in terms of measuring learning outcomes. Most likely it is because testing cognitive capabilities only provides insight into one component of a student's learning. The other component focuses on how students internalize what they have learned. Critics of business ethics education suggest that it is difficult if not impossible to determine whether students have become more ethical or whether their moral development has increased because of ethics education. While this may be true, including pedagogies that employ affective domain principles can help to measure outcomes, such as attitude changes that must go along with improved decision-making.

The affective domain refers to the manner in which individuals deal with emotions and includes five outcome levels: receiving, responding, valuing, organization, and characterization (Krathwohl, Bloom, & Bertram, 1973). Bloom suggests the affective domain is a continuum of internalization of values, beliefs, and practices (Reeves, 1990). Tomei's (2005) taxonomy of educational objectives for the technology domain echos Bloom's goals of higher-order learning, including values internalization and organization of values into systems. Specifically, Tomei (2005) suggests that the highest level of learning in the technology domain includes "the ability to judge the universal impact, shared values, and social implications of technology…" (p. 106).

While it is difficult (and perhaps inappropriate) to grade the feelings, attitudes, and moral development of students, business ethics educators must be able to assess whether these student attributes change as a result of coursework, to the degree that better decisions can be made. We argue that the change in students' ethical decision-making behavior over time is an indicator of higher levels of affective learning (Krathwohl et al., 1973) and an indication that students have developed the appropriate tools for understanding the ethical issues in information technology (Tomei, 2005). We employed ethical issues specific to information technology in an effort to gauge affective learning in an IT ethics class. These IT ethical issues are described in the next section.

## IT Ethical Issues

The six issues addressed in this study have been discussed widely in the media and are also addressed in the codes of ethics of IT professionals. These issues include violation of intellectual property rights, violation of customer privacy, violation of employee privacy, spamming, violation of free speech and exaggeration of IT capabilities.

Violation of intellectual property rights is probably the most widely debated ethical issue in recent years because of the widespread use of file-sharing applications such as BearShare, LimeWire, Kazaa, and many others to illegally copy music and movie files. The fight of the Recording Industry Association of America (RIAA) against the practice brought the issue to the courts. Violation of privacy has been an issue of concern for at least a decade. Two types of violations are addressed in this area: violation of customer privacy and violation of employee privacy. The first has to do with collection, use, and dissemination of private information of consumers, and the latter involves use and sharing of information that employers collect about their employees. Spamming has been met with so much disgust by the public that it has been legally restricted in many European countries. There is an ongoing debate in the United States about whether it should be restricted in this country as well.

The ability of millions of people and organizations to publish on the Web almost anything they wish often brings up the question about whether it is ethical to publish anything, including inflammatory comments and incitement. Free speech or violation of it has become an important ethical issue in the Web era. IT professionals are infamous for exaggerating the benefits of hardware, software, and networking technologies. The codes of ethics of IT professional organizations entreat members not to exaggerate the capabilities of information systems. Exaggeration of information systems capabilities is an ethical issue that is of interest for study in students whose education and careers include the development and promotion of information systems.

# METHODOLOGY

## Validation of Research Instrument

Our review of the literature yielded six IT-related issues: violation of intellectual property rights, violation of customer privacy, violation of employee privacy, spamming, violation of free speech, and exaggeration of IS capabilities. In addition to being reported and discussed in the media, all of these issues are addressed in the codes of ethics of organizations of IT professionals. For each issue we composed a brief case. The scenarios are presented in Appendix A.

We sent the scenarios to seven extremely prolific researchers in IT ethics. All have taught IT courses and published articles and books on the topic. Under each scenario, we included two questions. To measure to what degree the phenomenon described in the scenario is of ethical importance, the experts were asked: Regardless of the scenario itself, how strongly would you like to eliminate the phenomenon mentioned in bold-faced letters? (0 = not at all; 10 = very much). To measure to what extent the scenario described the issue, the experts were asked: To what extent does the scenario portray a case of the above (bold-faced) issue? (0 = not at all; 10 = very much). We also asked the experts to provide comments. Based on those comments we made slight changes in the wording of some scenarios.

Prior to sending the scenarios to the experts for validation, we decided that an issue was of ethical importance if the average response to the first question was at least seven, and that a scenario portrayed a case of that issue if the average response to the second question was at least seven. The averages of the responses to all first questions were well above eight. The averages to the second questions were well above nine. Thus, we decided to use all the scenarios. The scenarios in the student questionnaire were not titled to avoid any bias that might be perceived from the titles.

## Participants and Procedure

The participants in this study were graduate students enrolled in five sections of *Ethical Issues in Information Technology* at a large northeastern university. Each section had an average of 25 students enrolled. All students are working adults. All had taken an introductory course in information systems, and most had taken other IT-related courses by the time they participated in the study. Many of the students are IT professionals.

As can be seen in Appendix A, each scenario ends with a question about the likelihood of the respondent to commit the act implied. For each scenario, the respondents were asked to indicate how likely they were to commit the act on a 10-point Likert-type scale (0 = I would not do it, 10 = I certainly would). Respondents were invited to provide written comments or explanations for their answers for each scenario.

Participants were asked to respond twice: once at the start of the course, during the first few minutes of the first class; and again, during the last few minutes of the course in the last class meeting. In the second time, the order of the scenarios was reversed. To ensure anonymity, yet provide a means through which surveys could be paired, respondents were asked to place the last four digits of their student identification numbers in the top corner of the first page of the survey. By comparing the responses from time one at the beginning of the term and time two at the end of the term, we can infer whether the ethics education they received has made any difference on their suggested behavior.

Apparently, the long time between the first and last meetings combined with the reversed order of the scenarios did not raise any suspicions in the students that the scenarios were the same because no student raised any questions about it. Out of a possible 125 students, 78 usable pairs of surveys were obtained indicating a 62% response rate

## Hypothesis

The single null hypothesis of this study was *Ethics in IT education does not change students' ethical decision-making approach.* The hypothesis was tested for each of the six scenarios. It is important to note that this ethics course does not categorically define for the students which specific act is right and which is wrong. The course's purpose is to introduce the students to major ethical theories, guide them in making convincing arguments about the ethicality of IT-related acts, and make certain they know how different societies have viewed certain types of conduct. The course also introduces the major U.S. laws related to relevant issues, such as privacy, intellectual property, and free speech.

The pedagogy includes reading articles, news stories, and codes of ethics, as well as extensive class discussions in which the professor makes sure that the students are exposed to all facets of an issue, especially an analysis of the impact of each act on all stakeholders. The course opens with a review of the major ethical theories (e.g., those of Immanuel Kant and Jeremy Bentham), and proceeds to discuss the major IT-related ethical issues.

## Data Analysis

A paired- samples t-test was conducted for each of the six scenarios. The paired samples t-test is the appropriate statistic to employ when the same subjects respond on two occasions, and the difference between their two means is of interest (Howell, 1992). To enrich our understanding of the quantitative responses, we analyzed the written comments. In an effort to eliminate bias, we employed a research assistant with no prior information about the project to code the qualitative responses. The research assistant and one of the authors read and reread the comments, making margin notes of issues, themes, motifs and topics (Miles, 1983). The author and research assistant then compared the results and eliminated any redundancies. Next, we present the quantitative and qualitative results.

*Table 1. Paired samples statistics and t-test results*

| Issue | Mean Response | Standard Deviation | Paired Differences Mean | Paired Differences Standard Deviation | *p*-value |
|---|---|---|---|---|---|
| Violation of Intellectual Property | Pre 6.00<br>Post 6.12 | 3.378<br>3.439 | -.167 | 2.47 | .553 |
| Violation of Customer Privacy | Pre 1.50<br>Post 2.00 | 2.214<br>3.023 | -.500 | 3.26 | .179 |
| Violation of Employee Privacy | Pre 2.36<br>Post 2.22 | 2.962<br>2.840 | .141 | 3.16 | .695 |
| Spamming | Pre 5.94<br>Post 7.17 | 3.797<br>3.163 | -.123 | 3.37 | .002** |
| Violation of Free Speech | Pre 4.15<br>Post 3.46 | 3.497<br>3.297 | .692 | 3.15 | .056+ |
| Exaggeration of Information Systems Capabilities | Pre 2.26<br>Post 2.69 | 2.610<br>2.811 | -.436 | 2.48 | .125 |

## RESULTS

The results of the paired-samples t-test are presented in Table 1. Two of six t-tests results were significant in providing partial support for the alternative hypothesis that teaching ethics in information technology education changes students' ethical decisions. The pre-test means for spamming (5.94 to 7.17, $p<.01$) and violation of free speech (4.15 to 3.46, $p<.10$) were significantly different from the post-test means. In the case of spamming, the mean response increased, indicating that respondents were more willing to send spam e-mails at the end of the course then they were at the beginning of the course. The mean response decreased for the violation of free speech scenario, indicating that respondents were less likely to post potentially offensive content to a Web site at the end of the course, than they were willing to do at the beginning of the course.

While not significant, the means for violation of intellectual property rights (6.00 to 6.12, n.s.), violation of customer privacy (1.50 to 2.00, n.s.), and exaggeration of information systems capabilities (2.26 to 2.69, n.s.) increased at the end of the course, indicating that individuals were more likely to engage in the behavior in question at the end of the course than they were at the beginning of the course. Only the mean responses for violation of employee privacy (2.36 to 2.22, n.s.) were lower at the end of the term.

Thirty-one themes were garnered from the qualitative analysis. These themes are presented in Table 2. In an effort to understand the logic employed by the respondents in choosing whether to engage in the questionable behavior, we analyzed each scenario qualitatively. These results are presented in Tables 3a (pre-test) and 3b (post-test) and will be discussed separately for each scenario in the discussion section.

## DISCUSSION

The results of this study indicate that teaching ethics in information technology education has an effect on students' ethical decision-making behavior. In particular, the responses of the group changed significantly for two issues: spamming and violation of free speech. For spamming, the average response increased from 5.94 somewhat likely to spam for commercial gain to 7.17, a much greater willingness to spam. Perhaps the change came because the automatic reaction to spam is objection. However, when considering that many small businesses use spam as inexpensive advertising, the students became more sympathetic to spamming as a means of marketing, rather than to the spammed party. The comments suggest that individuals believe that being "desperate for a sale" will influence their decision to advertise through spam. In the pre-test comments some of the respondents indicated that a reason they would not spam is that they "hate it when it happens to me." This was not cited as a reason in the post-test. Furthermore, while "it's illegal" was given as a reason for some of the other scenarios, it was not cited as a reason to not send spam. The fact that spam is not illegal in the United States (only misrepresentation of the sending party is illegal in some states) may also influence respondents' decision to use spam as a marketing tool.

Students were significantly less likely to post a potentially offensive opinion to a company Web site after they took the class. The qualitative comments indicate that perhaps one reason for this change is that in the beginning of the term, "I have the authority" was cited as a reason to post the content, whereas at the end of the term, having the authority as webmaster does not appear to have played into the students' decisions. While we can only speculate, perhaps considering the ramifications of the posting to all stakeholder groups overrode the decision to support one's right to free speech. In terms of absolute mean, willingness to curb one's free speech on the Internet (4.15 to

*Table 2. Qualitative comments*

| |
|---|
| 1. It's illegal |
| 2. Company won't receive profit |
| 3. Promotes the game |
| 4. Software is overcharged/expensive |
| 5. It's a victimless crime |
| 6. Companies' responsibility to copy protect |
| 7. Depends on desire for game/need |
| 8. Only for personal use, would not distribute for profit |
| 9. Would support/respect people's privacy |
| 10. Depends on how desperate I am for a sale/business |
| 11. Depends on quality/credibility data/source |
| 12. Don't want to put my reputation/job at risk |
| 13. I hate it when it happens to me |
| 14. Would appreciate the potential response from customers |
| 15. Would rather build my business on honesty |
| 16. Would check for illegal activity |
| 17. Don't want to know other people's business/invasion of privacy |
| 18. No privacy on a company computer/protect company |
| 19. Only if topic is of concern/illegal/ harmful to others |
| 20. It is unethical/unprofessional/dishonest |
| 21. I have the authority |
| 22. Don't like/hate spam |
| 23. It's not illegal/no harm/would not get caught |
| 24. Would help my business/ I'd pay |
| 25. It's voluntary/no harm/they can delete/public info |
| 26. Legal ramifications are too high/risk/title VII issue |
| 27. Others may be offended/harmed |
| 28. Only if employee toned it down |
| 29. That's it's purpose/webmaster/employee choice |
| 30. Huge problem/don't lie/not good in long-term/potential business |
| 31. Need to be professional/that's my job/state facts/respect company |

3.46) was greater than willingness to commit the other acts described in the scenarios, but what constitutes or does not constitute free speech is subject to interpretation, which may explain the relatively high response means to this scenario.

The students did not change their attitudes significantly about the other four issues, although the qualitative responses offer some interesting insights. In the instance of violation of intellectual property rights, while individuals still indicated that they would likely engage in the behaviors at the end of the course, reasons such as "promotes the game" and "software is overcharged/expensive", and "it's a victimless crime" were not cited

*Table 3a. Qualitative comments by issue - pre-test*

| Issue | Comments | |
|---|---|---|
| Violation of Intellectual Property | 1. It's illegal<br>2. Company won't receive profit<br>3. Promotes the game<br>4. Software is overcharged/expensive<br>5. It's a victimless crime<br>6. Companies' responsibility to copy protect | 7. Depends on desire for game/need<br>8. Only for personal use, would not distribute for profit<br>18. No privacy on a company computer/ protect company<br>20. It is unethical/unprofessional/dishonest<br>23. It's not illegal/no harm/would not get caught |
| Violation of Customer Privacy | 1. It's illegal<br>6. Companies' responsibility to copy protect<br>7. Depends on desire for game/need<br>8. Only for personal use, would not distribute for profit<br>9. Would support/respect people's privacy<br>10. Depends on how desperate I am for a sale/business<br>11. Depends on quality/credibility data/ source<br>12. Don't want to put my reputation/job at risk | 13. I hate it when it happens to me<br>14. Would appreciate the potential response from customers<br>15. Would rather build my business on honesty<br>17. Don't want to know other people's business/invasion of privacy<br>20. It is unethical/unprofessional/dishonest<br>22. Don't like/hate spam<br>26. Legal ramifications are too high/risk/ title VII issue<br>27. Others may be offended/harmed |
| Violation of Employee Privacy | 1. It's illegal<br>9. Would support/respect people's privacy<br>11. Depends on quality/credibility data/ source<br>13. I hate it when it happens to me<br>16. Would check for illegal activity<br>17. Don't want to know other people's business/invasion of privacy<br>18. No privacy on a company computer/ protect company | 19. Only if topic is of concern/illegal/ harmful to others<br>20. It is unethical/unprofessional/dishonest<br>21. I have the authority<br>26. Legal ramifications are too high/risk/ title VII issue<br>27. Others may be offended/harmed<br>30. Huge problem/don't lie/not good in long-term/potential business<br>31. Need to be professional/that's my job/ state facts/respect company |
| Spamming | 10. Depends on how desperate I am for a sale/business<br>11. Depends on quality/credibility data/ source<br>12. Don't want to put my reputation/job at risk<br>13. I hate it when it happens to me<br>15. Would rather build my business on honesty<br>20. It is unethical/unprofessional/dishonest | 22. Don't like/hate spam<br>23. It's not illegal/no harm/would not get caught<br>24. Would help my business/ I'd pay<br>25. It's voluntary/no harm/they can delete/ public info<br>30. Huge problem/don't lie/not good in long-term/potential business |
| Violation of Free Speech | 11. Depends on quality/credibility data/ source<br>12. Don't want to put my reputation/job at risk<br>18. No privacy on a company computer/ protect company<br>19. Only if topic is of concern/illegal/ harmful to others<br>20. It is unethical/unprofessional/dishonest<br>21. I have the authority | 26. Legal ramifications are too high/risk/ title VII issue<br>27. Others may be offended/harmed<br>28. Only if employee toned it down<br>29. That's it's purpose/webmaster/employee choice<br>31. Need to be professional/that's my job/ state facts/respect company |
| Exaggeration of Information Systems Capabilities | 12. Don't want to put my reputation/job at risk<br>15. Would rather build my business on honesty<br>20. It is unethical/unprofessional/dishonest | 26. Legal ramifications are too high/risk/ title VII issue<br>30. Huge problem/don't lie/not good in long-term/potential business<br>31. Need to be professional/that's my job/ state facts/respect company |

*Table 3b. Qualitative comments by issue - post-test*

| Issue | Comments |
|---|---|
| Violation of Intellectual Property | 1. It's illegal<br>2. Company won't receive profit<br>6. Companies' responsibility to copy protect<br>7. Depends on desire for game/need<br>8. Only for personal use, would not distribute for profit<br>23. It's not illegal/no harm/would not get caught<br>24. Would help my business/ I'd pay |
| Violation of Customer Privacy | 8. Only for personal use, would not distribute for profit<br>9. Would support/respect people's privacy<br>12. Don't want to put my reputation/job at risk<br>13. I hate it when it happens to me<br>14. Would appreciate the potential response from customers<br>20. It is unethical/unprofessional/dishonest<br>26. Legal ramifications are too high/risk/title VII issue |
| Violation of Employee Privacy | 9. Would support/respect people's privacy<br>17. Don't want to know other people's business/invasion of privacy<br>18. No privacy on a company computer/protect company<br>19. Only if topic is of concern/illegal/ harmful to others<br>20. It is unethical/unprofessional/dishonest<br>21. I have the authority<br>25. It's voluntary/no harm/they can delete/public info<br>26. Legal ramifications are too high/risk/title VII issue<br>29. That's it's purpose/webmaster/employee choice |
| Spamming | 10. Depends on how desperate I am for a sale/business<br>11. Depends on quality/credibility data/source<br>12. Don't want to put my reputation/job at risk<br>17. Don't want to know other people's business/invasion of privacy<br>22. Don't like/hate spam<br>23. It's not illegal/no harm/would not get caught<br>24. Would help my business/ I'd pay<br>25. It's voluntary/no harm/they can delete/public info<br>30. Huge problem/don't lie/not good in long-term/potential business |
| Violation of Free Speech | 18. No privacy on a company computer/protect company<br>19. Only if topic is of concern/illegal/ harmful to others<br>20. It is unethical/unprofessional/dishonest<br>26. Legal ramifications are too high/risk/title VII issue<br>27. Others may be offended/harmed<br>28. Only if employee toned it down<br>29. That's it's purpose/webmaster/employee choice<br>31. Need to be professional/that's my job/state facts/respect company |
| Exaggeration of Information Systems Capabilities | 12. Don't want to put my reputation/job at risk<br>13. I hate it when it happens to me<br>14. Would appreciate the potential response from customers<br>15. Would rather build my business on honesty<br>20. It is unethical/unprofessional/dishonest<br>23. It's not illegal/no harm/would not get caught<br>30. Huge problem/don't lie/not good in long-term/potential business<br>31. Need to be professional/that's my job/state facts/respect company |

as justifications for the act at the end of the term. The students of this course started it and finished it with a relatively high likelihood that they would illegally copy copyrighted software. The response was 6.00 at the start of the course and 6.12 at the end of the course. This is disturbing to ethics educators who would like to see that their students graduate from such courses with higher moral attitudes. This is certainly so regarding acts that are not only unethical but also illegal.

The list of reasons cited as reasons to violate (or not violate) privacy in the beginning of the term was considerably longer than the reasons given at the end of the term. In terms of customer privacy, comments such as "Depends on how desperate I am for a sale/business" and "Depends on quality/credibility data/source" were not listed at the end of the term. The likelihood of violating employee privacy decreased slightly, and again, individuals seem to be making their decisions based upon different criteria at the end of the term. For example, students erroneously reasoned in the beginning of the term that it was "illegal" to key in on a subordinates Web-browsing habits. Even after taking the course and discovering that this is not an "illegal" practice, students were still less likely to engage in behavior that they viewed as a violation of employee privacy. We note further that willingness to violate privacy, both customer privacy (1.50 to 2.00) and employee privacy (2.36 to 2.22) was low at both the start and end of the course.

Willingness to exaggerate the capabilities of information systems for personal gain was low (2.26 to 2.69) both at the start and end of the course although doing so is not criminal. Students cite reasons such as "Would rather build my business on honesty" and "Need to be professional/respect company" at both the beginning and end of the term.

It is important to keep in mind that the purpose of this study was not to measure absolute moral attitude, but to see if ethics education in the form of a graduate course changes one's ethical decision-making approach. Based solely on the statistical analysis, we may conclude that such courses have a marginally significant impact on students' ethical decision-making. The quantitative data suggest that such courses may change initial attitudes if students are only partially familiar with an issue, or if an act is controversial but not illegal, such as spamming. These statistical results are disappointing at best.

Ethics educators looking for a silver lining may not find one by examining the qualitative responses in this study, to aid in understanding the cognitive processes taking place with respect to students' ethical IT decisions. The changes in reasoning behind the students' decisions over time may indicate that affective learning vis-à-vis internalization of values, beliefs, judgments, and practices (Bloom, 1956; Tomei, 2005) is taking place, however, that the values students are internalizing run counter to acceptable moral behavior.

The students started and ended the course with positive attitudes toward some acts that are immoral—let alone illegal—such as violation of intellectual property rights. This phenomenon may be explained by the concept of moral disengagement (Bandura, 1990; Bandura, Barbaranelli, Caprara, & Pastorelli 1996; Rogers, 1999), a type of cognitive distortion that may lead individuals to make decisions that are not aligned with their own internal moral principles. Individuals who know an act is probably unethical may justify it by convincing themselves that it meets a higher moral need or that it will rectify prior injustice. For example, individuals have justified unauthorized access to information systems (i.e., hacking) as helping organizations by drawing attention to weak security measures. Individuals who have made illegal copies of software have justified the act by claiming that software companies make exorbitant profits (Oz, 1990).

The findings of this study suggest that an important area for future research is the effect of common cognitive distortions on affective learning outcomes in ethics courses. Given that a positive relationship between moral disengagement and unethical decision-making behavior has previously been established (e.g., Detert, Treviño, & Baker, unpublished manuscript), it would be interesting to explore the role of moral disengagement as a potential moderator of the relationship between content and process outcomes associated with ethics courses.

## LIMITATIONS

While this study's strength lies in the matched pairs from two points of time, it is not without limitations. There are a number of factors that can come into play when individuals are taking self-report studies, not the least of which is the desire to respond in a way that society deems favorable. Further, it is not possible to directly assign course teachings as the cause of changes in students' ethical decisions. An interesting and potential future area of research might include administration of the vignettes at the beginning and end of an introductory course in information technology, and then to compare the means of those responses with the means of the responses from those in an ethics course later in the curriculum. This type of administration might serve the additional purpose of validating the hierarchical learning outcomes from literacy to tech-ology in the technology domain (Tomei, 2005).

Teaching any ethics course to college students—particularly graduate students—is challenging if the purpose is not simply to pontificate, "this is right, that is wrong." As mentioned above, the purpose of this course was to ensure that the students become familiar with major ethical theories and know how to use them for arguments, that the students know the current ethical and legal issues relating to IT, and understanding what society views—in general—as ethical or unethical regarding these issues. Had the purpose, and therefore the pedagogical approach been different, the results of this study might be different. Therefore, we must limit the validity of the findings to education that takes this pedagogical approach.

## CONCLUSION

The findings of this study provide partial support for the hypothesis that teaching ethics in information technology education changes students' ethical decision-making approach. While not always in the manner an ethics educator might predict, it is interesting to note that some attitudes did change over time. We conclude that IT ethics education can aid in students' affective learning, an important and necessary component in the overall learning process.

## REFERENCES

Bandura, A. (1990). Selective activation and disengagement of moral control. *Journal of Social Issues, 46*(1), 27-46.

Bandura, A., Barbaranelli, C., Caprara, G., & Pastorelli, C. (1996). Mechanisms of moral disengagement in the exercise of moral agency. *Journal of Personality and Social Psychology, 71*(2), 364-374.

Bloom, B. (1956). *Taxonomy of educational objectives.* New York: David McKay Company, Inc.

Calluzzo, V. & Cante, C. (2004). Ethics in information technology and software. *Journal of Business Ethics, 51*(3), 301-312.

Detert, J., Trevino, L., & Baker, V. Faulty cognitions and unethical decisions: The antecedents and outcomes of moral disengagement.

Etzioni, A. (2002). When it comes to ethics, B-schools get an F. *The Washington Post.* Retrieved from http://www.washingtonpost.com/ac2/wp-dyn/A38323-2002Aug2.

Haines, R. & Leonard, L. (2007). Situational influences on ethical decision-making in an IT context. *Information & Management, 44*(3), 313-320.

Howell, D. (1992). *Statistical methods for psychology*, 3rd ed. Belmont, CA: Duxbury Press.

Krathwohl, D. (2002). A revision of Bloom's taxonomy: An overview. *Theory into Practice, 41*(4), 212-218.

Krathwohl, D., Bloom, B., & Bertram, B. (1973). *Taxonomy of educational objectives, the classification of educational goals. Handbook II: Affective domain.* New York: David McKay Company, Inc.

Loch, K. & Conger, S. (1996). Evaluation ethical decision making and computer use. *Communications of the ACM, 39*(7), 74-83.

Logsdon, J., Thompson, J., & Reid, R. (1994). Software piracy: Is it related to level of moral judgment? *Journal of Business Ethics, 13*(11), 849-857.

Mangan, K. (2003). Accrediting board endorses stronger focus on ethics in business-school curriculum. *Chronicle of Higher Education* Retrieved from http://chronicle.com/prm/daily/2003/01/2003010805n.htm.

Miles, M. (Ed.). (1983). *Qualitative data as an attractive nuisance*. Beverly Hills, CA: Sage Publications.

Oz, E. (1990). The attitude of managers-to-be toward software piracy. *OR/MS Today, 17*(4), 24-26.

Peace, A., Galletta, D., & Thong, J. (2003). Software piracy in the workplace: A model and empirical Test. *Journal of Management Information Systems, 20*(1), 153-177.

Ramakrishna, H., Kini, R., & Vijayraman, B. (2001). Shaping of moral intensity regarding software piracy in university students: Immediate community effect. *Journal of Computer Information Systems, 41*(4), 47-51.

Reeves, M. (1990). An application of Bloom's taxonomy to the teaching of business ethics. *Journal of Business Ethics, 9*(7), 609-616.

Rogers, M. (1999). Psychology of hackers: Steps toward a new taxonomy. Retrieved from http://www.infowar.com.

Sims, R., Cheng, H., & Teegen, H. (1996). Toward a profile of student software pirates. *Journal of Business Ethics, 15*(8), 839-849.

Sims, R. (2002). *Teaching business ethics for effective learning.* Westport: Quorum Books.

Swanson, D. (2003). Business education puts corporate reputations at risk. *The Corporate Citizen, 3*, 34-35.

Swanson, D. (2004). The buck stops here: Why universities must reclaim business ethics education. *The Journal of Academic Ethics, 2*(1), 1-19.

Swanson, D. & Frederick, W. (2003). Are business schools silent partners in corporate crime?. *Journal of Corporate Citizenship, 9*, 24-27.

Tomei, L. (2005). *Taxonomy for the technology domain.* Hershey, PA: Information Science Publishing.

*This work was previously published in International Journal of Information and Communication Technology Education, Vol. 4, Issue 2, edited by L. Tomei, pp. 67-83, copyright 2008 by IGI Publishing (an imprint of IGI Global).*

# Chapter 20
# A Critical Discourse Analysis of Students' Anonymous Online Postings

**Dick Ng'ambi**
*University of Cape Town, South Africa*

## ABSTRACT

*It is difficult to understand students' social practices from artifacts of anonymous online postings. The analysis of text genres and discursive types of online postings has potential for enhancing teaching and learning experiences of students. This article focuses on analysis of students' anonymous online postings using Critical Discourse Analysis (CDA). The article argues that social practices reproduce during online interaction and artifacts embody such reproduction. A study involving more than 300 commerce students at a higher education institution (HEI) using a special purpose anonymous online consultation tool, the Dynamic Frequently Asked Questions (DFAQ), and social practices embodied in the artifacts is analyzed using CDA. The analysis used the three dimensions of CDA—description (text genres), interpretation (discursive type), and explanation (social practice)—and insights into students' social practices were inferred. The article concludes that CDA of anonymous postings provided insight into social practices of students and, in particular, highlighted the tension between perceptions of inflexibility of traditional teaching practices and student demands for flexible learning. Finally, CDA, as described in this article, could be useful in analyzing e-mail communications, short message service (SMS) interactions, Web blogs, and podcasts.*

## INTRODUCTION

Although Critical Discourse Analysis (CDA) has been used to provide social critique (Thompson, 2002; Willig, 1999), assist in developing appropriate social interventions (Willig, 1999), empower people (Panteli, 2003; Willig, 1999), and unravel "how language conspires to legitimate and perpetuate unequal power relations" (Willig, 1999), the potential of CDA for analyzing online artifacts has not been explored. The strengths of CDA lie in making connections between social and cultural structures and processes on the one hand, and properties of text on the other (Fairclough & Wodak, 1997:277). Other text analysis approaches, such as Exchange Structure Analysis (Pilkington, 1999) and text mining (Ng'ambi, 2002), do not link text to social and cultural structures.

Fairclough (1992) contends that every discourse instance has three dimensions: it is either spoken or written *text*; it is an *interaction* between people involving processes of producing and interpreting the text; or it is part of social *action*, and in some cases, virtually the whole of it. The activities on the right (see Figure 1) of the model represent the framework of analysis in which a piece of text is described, and then the discursive practices upon which it draws are identified and linked to the underlying power relations, which may be reproduced by the interaction (Thompson, 2004). The social interaction happens within the discursive practices, which produce text; through the analysis of text messages, evidence of social practices can be revealed or noted. Furthermore, the discursive practices are influenced by the situation or environment in which a participant is.

Atkins (2002) postulates three stages of understanding a discourse: (1) *social conditions of production and interpretation* (i.e., factors in society that led to the production of a text and how these factors affect interpretation); (2) the *process of production and interpretation of text* (i.e., how produced text affects interpretation); and (3) the product of the first two stages: the *text*.

The rest of the article is organized as follows: First a discourse theory is described, followed by a discussion on the research approach and analytical framework used. The case study is then discussed and an analysis of results explained. Finally, a conclusion is given.

*Figure 1. Discourse as text, interaction and context (Fairclough, 1989)*

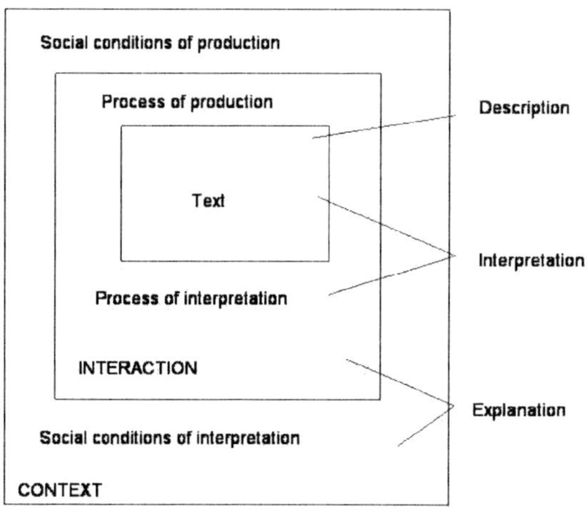

## DISCOURSE THEORY

CDA provides a way of thinking that analyzing text and discourse practices may give access to social identities and social relations. Phillips and Jorgensen (2002) observe:

*[D]iscourse practices—through which texts are produced (created) and consumed (received and interpreted)—are viewed as an important form of social practice which contributes to the constitution of the social world including social identities and social relations.* (p. 61)

The production of text draws its meanings from the social practice and *vice versa*. The discourse theory states that every word spoken draws its meaning from the social practices of which it is a part, or, recursively, from the sediment of prior practices (Burbules & Bruce, 2001). I infer from the discourse theory that the process of production and interpretation of online artifacts is not free from the social conditions of production and the social conditions of interpreting such text. Fairclough (1989) points out that discourse involves social conditions of production and social conditions of interpretation. Fairclough (1992) observes that the relationship between social action and text is understood in a context of interaction. It stands to reason that interaction is a product of social action, and the *traces* of its interactive processes left on the human mind reproduce social practices (interaction).

## RESEARCH APPROACH

In this section, a research process is described in terms of how communicative actions were set up, initiated, and controlled. It describes the notion of communication genres, scope of interactive audience, and the anonymous tool that mediated interaction.

### Planning

Planning a communicative event involves making decisions about the meeting time, venue, agenda, and so forth, and inviting participants to attend. A communicative event also can occur when a subject takes the initiative rather than wait for an invitation. Van Dijk (1996) gives two examples: a patient taking an initiative to talk to a doctor, and a student asking to talk to a professor. In either case, the doctor or the professor usually would decide about the setting. My view is that consciousness of powerful actors constrains interactions as it takes the focus away from the content of communication to the source of content. In this study, I was concerned with recursive social life in which a patient consults with a doctor, where the patient is also a doctor; a student consults an "expert," where a student is also an "expert." Such communication events do not need planning, and they are socially located. Students posted questions as need arose and were free to respond to any question. The study was located in an authentic context.

### Setting

Setting up a communicative event involves deciding who controls the interaction. As Van Dijk (1996) put it, "Who is allowed or obliged to participate, and in what role, may be decided by the chairperson or by other powerful participants who control the interaction" (p. 87). Van Dijk gives an example of the effect of positioning and the presence of props of power, such as the robes of a judge and the uniform of a police officer. My addition to Van Dijk's rationalization is that positioning and power are often unspoken or unwritten. Human actions are a product of consciousness of these unspoken or unwritten conditions. Mindful of the power and effect of powerful actors in a communication event, online interaction may enable or inhibit interactivity.

## Controlling Communicative Events

The power to regulate communicative events, as Van Dijk (1996) observes:

*consists of various dimensions of speech and talk; which mode of communication may/must be used; which language may/must be used by whom; which genres of discourse are allowed; which types of speech acts; or who may begin or interrupt turns at talk or discursive sequences.* (p. 88)

Van Dijk gives an example of defendants in court who may be required to speak the standard language, to answer questions only (and only when required to speak), to speak only about the topic being discussed, and to use a polite deferential style. In this study, there was no regulations of communicative events except to post a question or respond to one.

## Communication Genres

Paivarinta (2001) stresses, "A communication genre should be distinguished from the medium of communication; for instance, a fax or e-mail are not good examples of communication genres, whereas a hotel reservation or an invitation to a meeting, which can be mediated by fax or electronic mail, are" (p. 213). When someone mentions the word *e-mail*, they may be referring to e-mail as a medium of communication or e-mail as a message. It is because of this ambiguity that Paivarinta (2001) argues that e-mail was not a good example of communication genre.

## Scope and Audience Access

As Van Dijk (1996) observes, "When speakers are able to influence the mental models, knowledge, attitudes and eventually the ideologies of recipients, they may indirectly control their future actions" (p. 89). The influence on a student's mental model is in two ways: (a) exposure to other students' questions and responses, or (b) interpreting a response to a posted question. Access to a deluge of questions from other students mirrors understanding of a class to an individual and hence indirectly affects their questioning behavior.

## Anonymous Online Tool

The Dynamic Frequently Asked Questions (DFAQ) tool was designed and developed at the UCT as a special purpose question and consultation environment for students (Ng'ambi, 2003, 2004; Ng'ambi & Hardman, 2004). DFAQ provided a medium through which students consulted one another and the lecturer anonymously (Ng'ambi, 2003). Used as an educative, social, and communicative space, DFAQ dynamically created a knowledge resource from student consultations. DFAQ was available 24/7. Designed with a seamless Web/Short Messages Services interface, students had an anywhere-anytime anonymous consultation space. Time, place, and content of messages were self-regulatory and controlled by students.

# ANALYTICAL FRAMEWORK

The data analysis is carried on artifacts (text messages) from an anonymous knowledge-sharing environment using CDA, where certain generic specific genres and discursive types (Roode,

*Table 1. Text genres and discursive types (adapted from Roode et al., 2004)*

| Text Genre (TG) | Discursive Type (DT) |
|---|---|
| Confidence | Neutrality |
| Factual Information | Corporation |
| Humor | Technological optimism |
| Persuasion | Pragmatism |
| Uncertainty | Legitimacy |
| | Technocracy |

Speight, Pollock & Webber, 2004) are identified by examining issues of power and domination. There is a subjective judgment when identifying these text genres and discursive types (see Table 1) and applying them to sections of text (Roode et al., 2004).

In the context of this study, **neutrality discursive** type refers to discourses that are not taking sides on a topic of discussion. **Corporatism discursive** type refers to discourses that imply collaboration; **technological optimism** refers to discourses that acknowledge the technology's potentials. The **pragmatism discursive** type refers to discourse addressing practical issues. **Legitimacy discourse** discursive type refers to authoritative discourse, and technocracy discursive type refers to technocratic discourse. The text genres and discursive types are outcomes of the process of production and interpretation of text (see Figure 1). It follows that an iterative analysis (moving from text to social action) of CDA (i.e., Description, Interpretation, and Explanation) would help unravel social practices embodied in text.

## Description

Text is an outcome of an online interaction. Text is both a medium and an outcome of mental constructs (intentions). Thus, an author transforms intentionality into ostensive text messages. The reader of text interprets (attempts to deconstruct the author's intentions) the message and responds through another text (the cycle repeats). Thus, the focus of the description phase is to help identify text genres.

## Interpretation

While the description component focuses on text genre, interpretation focuses on understanding the production and interpretation process of text. Given that text is an outcome of an online interaction, the interpretation component involves an analysis of discursive types embodied in text.

## Explanation

The interpretation component connects interpretation (discursive types) and description (text genres). The explanation component links interaction to the social action (practices) or the modalities drawn upon during interaction. It is through the explanation that social practices are unraveled.

# CASE STUDY

The study was conducted at a medium-sized contact University in South Africa. More than 300 final year students registered for a degree in the Commerce Faculty participated in the study. The DFAQ tool was introduced to the students at the beginning of the semester in a six-month course. Most participants were full-time resident students. The class was well represented in terms of female and male students. For most of these participants, prior schooling did not prepare them for the critical demand of university courses, and although they were in their final year, the pressure of academic life was still a challenge. The objective of DFAQ was to provide an environment in which students would help one another, and the result of the interaction was a knowledge resource created by students for students with a subsidiary feedback benefit to faculty staff.

# ANALYSIS OF RESULTS

In this section, **unedited text** posted by students in the DFAQ is analyzed. For the sake of brevity, one posting [see Question 167] is analyzed. The posting attracted 16 responses (five are analyzed in this section). The question had 130 hits[1] (43% of the class read it) in two days. The analytical framework (see Table 1) is used to analyze both the posting and the responses.

*Text 1*

*[Question 167]: I have a general concern with the incredibely high volume of work thatwe are required to cover for the upcoming test. I know students always say this, however, i do not at all find it fair that we be expected to cover, literally 250 pages, some of us in a single night. All that ends up happening is students receiving low marks, not because of lack of knowledge of the content or whatever else, but aimply because the volume is too large to cover and make adequate sense. I for one have a problem with this upcoming test, as i am also a xyz 1009 student, have just written a test, have other equally demanding course to consider and yet am still expected to go home, get thru the 250 pages in time for the test tomorrow and get great marks! Please do something about this, even if it means breaking tests up so we have more class tests,thus less volumes to study for per test. Thanks and sorry to moan like this.*

The following comment was posted by a peer (a classmate).

*Text 2*

*[167-315]: I had the xyz in class assesment for monday, which I could not study for as I wrote a law test the next day which I needed better marks for, and then i now have 1 day to look through 250 pages of work for tomorrows test. I have decided to leave out 2 chapters for tomorrows test- If John (not his real name) tests us on chapters 1, or 5 then i will fail miserably, so here goes to spotting! but I completely agree with you!*

*Analysis 1.*

| Description (Text Analysis) | Interpretation (Discursive Type) | Explanation (Social Practice) |
|---|---|---|
| **Panic** (New TG) – the student is panicking because s/he has not studied for the test, and there is no time to understand the material.<br><br>**Factual Information** – is used such as 250 pages of a book, reference to a test just written in another course.<br><br>**Persuasion** – takes a form of an appeal to the lecturer to do something.<br><br>**Uncertainty** – is not sure what the options are for the lecturer but proceeds to suggest having many tests.<br><br>**Apologetic** (new TG) – is mindful that students always complain and that this is not one of such. Ends with an apology that s/he has "moaned." | **Pragmatism** => student is addressing practical issues: (i) volume of material to be covered before the test, (ii) lack of time to study, (iii) suggestion to break test into several tests.<br><br>**Legitimacy** => the student authoritatively demands that something be done about her/his complaint.<br><br>**Corporatism** => willing to accept alternatives if only test does not happen. | Generalizes that students do always complain but that this complaint is individualist and hence different.<br><br>Refers to a test just written (xyz 1009) "…I for one…" as though the test in question was not written by other students, and uses the phrase "I have other equally demanding course," as if it's the only student who takes these other courses.<br><br>Argues that despite this being an individualized complaint, failure to be listened to may lead many students to getting poor marks. |

*Analysis 2.*

| Description (Text Analysis) | Interpretation (Discursive Type) | Explanation (Social Practice) |
|---|---|---|
| **Humor** – it's not that this student would enjoy failing miserably, but s/he has taken a humorous approach.<br><br>**Factual information** – the student is honest in disclosing that the reason why s/he has no time to study for the test is because energies were directed at another course.<br><br>**Persuasion** – there is a gentle persuasion to go and study what you can and write the test. | **Neutrality** => rather than openly disagreeing with the posting and directly advising the fellow student, the student shares what s/he is going to do about it.<br><br>**Pragmatism** => the idea of leaving out two chapters is a practical solution, and hence pragmatic. | According to the statement, students receiving low marks is attributed to study strategies used. The student did not study for a class assignment in one course because s/he needed to focus on a test in another course. A student has disclosed her/his academic survival strategy. |

The following posting was from a lecturer.

*Text 3*

*[167-317]: The test timetables have been posted since the beginning of this semester/July. We make the information available to you in advance so that you will manage your time in a responsible manner.*

The following was a comment to the lecturer's comment:

*Text 4*

*[167-318]: I agree that there has been ample time to prepare. However, a lack of available textbooks has contributed to the inability of students to be prepared on time. Moreover, this has added to a general feeling of apathy which I have picked up from various students. Sad but true.*

The following statement is presumably from the student who asked the first question:

*Text 5*

*[167-326]: I totally agree with the last post. There's no point in stating that we've had since the beginning of semester, basically to prep for this test, thats just a (no offence) thick comment. Firstly, how do you study for this test from the beginning of semester when you haven't even been taught the stuff and secondly, some of has really do have other courses (try 3 other rather demanding courses) to think about. I wanna get out of this varsity come year end and all im saying is that giving students 250 pages to have to get through in less than 24 hours is asking them to basically fail the course, thus actually not graduate. It's not about time management anymore, there just aren't enough hours in a day to get through it all, sorry.*

The last posting was an interesting one from another student:

*Text 6*

*[167-333]: There's something very different with this semester's DFAQ... all people are doing is complaining! we are at ABC - one of the top universities in africa - and we're in third year, which means that we are expected to know how to deal with stress and deadlines. if you guys are struggling so much, then drop out, or something!*

*Analysis 3.*

| Description (Text Analysis) | Interpretation (Discursive Type) | Explanation (Social Practice) |
|---|---|---|
| **Factual information** – test timetables published at the beginning of the semester showed when tests would be taken. | **Legitimacy** – reference is made to the test timetables with which students are provided at the beginning of the semester. Adds that there is no excuse for studying for the test the night before. | The statements used the plural "we," suggesting a community of academics. And it seems to address all students—"so that you"—and not necessarily the author of the posting. It suggests that in posting test timetables in advance, faculty is blameless. |

*Analysis 4.*

| Description (Text Analysis) | Interpretation (Discursive Type) | Explanation (Social Practice) |
|---|---|---|
| **Confidence** – the response gives an impression of resolve.<br>**Factual information** – assuming the textbooks did not arrive in time, the student draws upon a fact that ought to be addressed. | **Pragmatism** => to refer to a textbook is to address a practical problem. | According to the posting, the community of students has not been able to prepare for the test on time because of the textbook problem. |

*Analysis 5.*

| Description (Text Analysis) | Interpretation (Discursive Type) | Explanation (Social Practice) |
|---|---|---|
| **Persuasion** – the student seeks to dismiss the fact that test timetable was published at the beginning of the semester, questioning the logic of such reasoning. | **Pragmatism** => refers to three other courses as demanding and concedes that there was no way to pass the test. | The student refers to self as having a single purpose of graduating at the end of that year but draws a community of students in the argument—"giving students…asking them…not graduate." |

*the work really isn't that difficult, and it's really not that much. by the way, i also wrote xyz 1009s yesterday, as well as the in-class assessment on monday, and i am still prepared for today's test - a little thing called time management. come on, people - stop complaining and manage your time! good luck for the test.*

## CONCLUSION

In this article, CDA was used to analyze anonymous student postings in an online environment. The article has shown how the text genres and discursive types serve as a vehicle for social critique. Two new text genres were observed: *panic* and *apologetic*. The article concludes as follows:

- Understanding the conditions of production and interpretation of online text provided insight into social practices of a community in which students were located. New text genres may emerge as conditions of production and interpretation change.
- Results of CDA would provide effective input in designing and developing interventions that affect communities of students.
- Use of CDA on anonymous online artifacts provided a way of understanding assumptions enshrined in the traditional practices (inflexibility) of institutions and the practicality as experienced by students (quest for flexible learning).

As mentioned in literature, identification of text genre and discursive types in CDA requires a subjective judgment. Thus, familiarity with the analytical instrument is required to get useful results. The application of CDA could be useful in the following: (i) analysis of e-mail correspondence; (ii) Short Message Services interaction; and (iii) analysis of Weblogs and podcasts.

*Analysis 6.*

| Description (Text Analysis) | Interpretation (Discursive Type) | Explanation (Social Practice) |
|---|---|---|
| **Confidences** - points out that s/he does the same courses, same time, but has done the assignment and is ready for the test that others are complaining about.<br><br>**Persuasion** – s/he persuades peers to stop complaining and start preparing for the test instead. **Humor** – the wishing of "good luck" to friends who are not ready for the test was meant to be humorous. | **Technological optimism** => refers to the posting in the online anonymous consultation tool (DFAQ) which students used for this engagement as having had many complaints that semester.<br><br>**Legitimacy** => attempts to calm the complainants to accept that it is never easy at a university. | Students' expectations of educational use of DFAQ differed from how fellow students used it (i.e., as a complaining space). The student calls on the community of students to "stop complaining and manage their time." The student accepts workload as part of university life and encourages fellow students to start working rather than complaining. |

## REFERENCES

Atkins. (2002). *Critical discourse analysis: A letter to expatriates from the Rt. Hon. Sir Norman Fowler MP*. Retrieved May 31, 2005, from http://www.cels.bham.ac.uk/resources/essays/Andrew%20Atkins%20Assignment%205.pdf

Burbules, C.N., & Bruce, C.B. (2001). *Theory and research on teaching as dialogue*. In V. Rochardson (Ed.), Handbook of research on teaching, 4th Ed. Washington, DC. Retrieved June 28, 2003, from http://faculty.ed.uiuc.edu/burbules/ncb/papers/dialogue.html

Fairclough, N. (1989). *Language and power*. London: Longman.

Fairclough, N. (1992). *Discourse and social change*. Cambridge: Polity Press.

Fairclough, N., & Wodak, R. (1997). Critical discourse analysis. In T.A. van Dijk (Ed.), *Discourse as social interaction* (pp. 259–284). London: SAGE Publications.

Ng'ambi, D. (2002). Pre-empting user questions through anticipation—Data mining FAQ lists. *Proceedings of the ACM International Conference on SAICSIT Enablement through Technology*, (pp. 101–109) Port Elizabeth, South Africa.

Ng'ambi, D. (2003). Students helping students learn—The case for a Web-based dynamic FAQ environment. *Proceedings of the International Association of Science and Technology for Development (IASTED) Conference on Information and Knowledge Sharing* (pp. 293–298). Scottsdale, Arizona.

Ng'ambi, D. (2004). Towards an organizational intelligent framework based on Giddens' structuration theory: A case of using an organizational memory system—DFAQ. *Proceedings of the IFIP World Congress Symposium on Professional Practice in Artificial Intelligence*, Toulouse, France.

Ng'ambi, D., & Hardman, J. (2004). Towards a knowledge-sharing scaffolding environment based on learners' questions. *British Journal of Educational Technology, 35*(2), 187–196.

Paivarinta, T. (2001). The concept of genre within the critical approach to information systems development. *Information and Organization, 11*, 207–234.

Panteli, N. (2003). *Discourse analysis in IS research: Constructing presence in virtual organizing*. Proceedings of the 11th European Conference on Information Systems, Naples.

Phillips, L., & Jorgensen, W.M. (2002). *Discourse analysis as theory and method*. London: Sage Publications.

Pilkington, M R. (1999). *Analysis educational discourse: The DISCOUNT scheme* [technical report No.99/2].

Rheingold, H. (2002). *Smart mobs: The next social revolution*. Cambridge: Perseus Publishing.

Roode, D., Speight, H., Pollock, M., & Webber, R. (2004). It's not the digital divide—It's the sociotechno divide! *Proceedings of 12th European Conference on Information Systems (ECIS2004)*, Turku, Finland.

Thompson, M. (2002). ICT, power, and development discourse: A critical analysis. In E. Wynn, A.E. Whitley, D.M. Myers, & D.M. DeGross, *Global and organizational discourse about information technology*.

Thompson, M.P.A. (2004). ICT, power, and developmental discourse: A critical analysis. *The Electronic Journal on Information Systems in Developing Countries, 20*(4), 1–25. Retrieved August 11, 2005, from http://www.ejisdc.org

Van Dijk, A.T. (1996). Discourse, power and access. In R.C. Caldas-Couthard, & M. Coulthard, *Texts and practices: Readings in critical discourse analysis* [pp. 85–104]. London: Routledge.

Viljoen, J-M., Preez, C., & Cook, A. (2005). The case for using SMS technologies to support distance education students in South Africa. *Perspectives in Education, 23*(4), 115–122.

Willig, C. (1999). Introduction: Making a difference. In C. Willig (Ed.), *Applied discourse analysis—Social and psychological interventions* (pp. 1–21), Buckingham: Open University Press.

## ENDNOTE

1. The number of times a posting was visited or opened to be read

*This work was previously published in International Journal of Information and Communication Technology Education, Vol. 4, Issue 3, edited by L. Tomei, pp. 31-39, copyright 2008 by IGI Publishing (an imprint of IGI Global).*

# Chapter 21
# Formative Assessment and Certification in Lifelong Learning with Cognitive and Metacognitive Measurements

**Edson P. Pimentel**
*Federal University of ABC, Santo André, SP, Brazil*

**Marcio Porto Feitosa**
*Mackenzie P. University, São Paulo, SP, Brazil*

**Nizam Omar**
*Mackenzie P. University, São Paulo, SP, Brazil*

## ABSTRACT

*Traditional forms of assessment used in face-to-face and distance learning education are insufficient to ascertain the increase of the knowledge acquired and the learners progress, therefore do not provide enough information to detect their learning gaps necessary to improve their competencies . Another point is that traditional assessment ways rarely involve the student in monitoring his own learning through his metacognitive abilities. Nowadays, professional skills to obtain a working position changes at the same velocity than the increase of knowledge and have to be considered by any professional and/or student to be qualified for a new job. This paper presents a model for formative assessment and certification in Lifelong Learning based on cognitive and metacognitive measurements that will make possible the identification of the professional learning gaps showing a roadmap to obtain educational and conceptual certification for his/her competence. Moreover, it presents the architecture of a computational environment for student knowledge mapping that will allow identifying more specifically the learning gaps in order to supply the educational system with qualitative information.*

DOI: 10.4018/978-1-60566-936-6.ch021

## INTRODUCTION

The rapid technological development and the growing changes in the profiles of professionals required to act in any area, in particular the area of IT, have taken people to seek ever more new capacities, by the other side Educational Institutions have sought to offer a lot of different types of courses and modes of learning for the maintenance and improvement of skill levels of these professionals called Longlife Learning.

Evaluate people, choosing training or its complement to obtain a good job placement involves many complexities. The combination of factors is very large, resulting in a number of personal profiles, almost infinite, and very difficult to compare.

The assessment process plays an important role in producing information that can help students and professionals, parents, teachers and educational administrators to know and deal better with the learning gaps. Teachers and the Intelligent Tutoring Systems (ITS) can use this information to adapt the instruction to the student's learning needs and difficulties and to work as guidelines to his/her formation.

The Assessment Reform Group (1999) based on their research stands that successful learning occurs when learners have ownership of their learning; when they understand the goals they are aiming for; when, crucially, they are motivated and have the skills to achieve success. Not only are these essential features of effective day-to-day learning in the classroom, they are key ingredients of successful Lifelong Learning.

Another important aspect in the learning process relates to the student's metacognitive abilities, *i.e.*, the process of reflecting about the own knowledge which Flavell (1979) called metacognition. Knowledge about knowledge itself is very important to the learning with quality.

Many teachers rely on a traditional, pre-test and pos-test design to document student progress as showed by Shepard (2001). Pretest results are used to establish each students' achievement level or location but are typically not used to gain insight into the nature of student's understanding, e.g., when a problem is missed, it is not known what partial knowledge or competing conception is at work. Moreover, to develop students' metacognitive knowledge about what helps in their own learning, there might be explicit discussion of both the facilitating and inhibiting effects of background knowledge.

The ongoing assessment that aims to diagnose and to improve the learning instead of merely classify the students is basic in distance learning education to increase the adaptability of the systems and the personalization of the education, increasing the motivation and reducing the evasion rate, besides increasing the quality and productivity of the learning. Moreover, it can help to minimize the problems of credibility lack on who effectively took the assessment, allowing monitoring the evolution of the learning instead of having only one measure at the end of the course. In distance learning education the majority of computational environments involves some kind of ongoing student assessment, in which observation is based on documentation of the student's interactions with the environment as showed by Silva and Vieira (2001).

This paper presents a model for formative assessment and certification in Longlife Learning based on cognitive and metacognitive measurements that will make possible the identification of the professional or student learning gaps. Moreover, it presents the architecture of a computational environment to implement the proposed model.

The model will support the monitoring and the development of metacognitive processes in order to allow the person to have control of his/her own learning through the process of self-regulation, which is self-monitoring, self-evaluation, and self-reinforcement. As a cognitive measurer this paper will propose the Knowledge Acquisition Level (KAL) obtained for each item of the knowledge domain, making possible the identification of the gaps of learning. KAL is not only a prompt measure, but its evolution can be tracked during the process, through continuous assessment.

The document is organized as follows. Next section discusses about continuous assessment and certification. Following a conceptual model for knowledge monitoring and certification is presented. In the sequence a computational learning environment, as well as, a certification environment are designed. Finally, additional considerations and some conclusions about this work and future work are made.

## CONTINUOUS ASSESSMENT AND CERTIFICATION

Management of Skills is a tendency for the market from the early 90s. The companies currently working to assess their internal talent and also to seek professionals in the market is based on a competency model to acquire more precise assessment in the process.

Perrenoud (2000) studied the relationship between training and skills, he explicitly declares that training is continuing to improve the development of skills, expanding the capacity of professionals to carry out their tasks and expect that this should also develop their professional skills. It is up to him include such features in their abilities and attitudes.

Competence to Perrenoud is a know-raise, this is not a technical or know one, but an ability to mobilize a number of resources, i.e., knowledge, know-how, the assessment schemes and actions, tools, attitudes -- in order to effectively deal with complex situations and unpublished. It is not enough, therefore, it is necessary enhance the range of resources availibles to teachers to automatically monitoring their learning. A virtual knowledge will continue until it is absorbed or mobilized as support to their ability to use it properly.

The same occurs in the process of formative assessment; we cannot say that any continued training alone leads to the construction of skills. In the process of continuous evaluation there is a recursive process, namely the development of skills that occurs in the conscience of apprentices, and possibly in a group of professionals. This takes the form of motivation and guides the continued training for skills. The completion of training in the working environment contributes to improve the competence of the student or trainee, not only because it is a collective training, but also because the training happens in the workplace. This reduces the distance between the absorption of a concept and its incorporation as competence.

Assessment and feedback are essential for helping people learn. An assessment process consistent with the learning principles should be continuous as part of the instruction and supply information about the student's learning level to teachers, parents and de own student: this is formative assessment (Bransford, 2003), (Perrenoud, 2000).

In a continuous learning assessment and accompaniment process, first of all, it is necessary to identify the purposes of assessment. Falchikov (2005) have classified these purposes into two main categories: summative and formative. In the first group the main purposes of assessment are restricted to selection, certification, accountability and effectiveness monitoring. Purposes in the latter group are more student-centered, and include diagnosis, motivation, feedback and learning improvement. Our focus in this work is on formative assessment.

### Lifelong Learning

Lifelong Learning is the process of acquiring knowledge or skills throughout life via education, training, work and general life experiences. This term recognizes that learning is not confined to childhood or the classroom, but takes place throughout life and in a range of situations (Aspin, 2007). During the last fifty years, constant scientific and technological innovation and change has had a profound effect on learning needs and styles. Learning can no longer be divided into a place and time to acquire knowledge, like school, and a place and time to apply the knowledge acquired: the workplace (Fisher, 2000).

Many aspects of effective teaching apply to all age groups. However, adults have had more life experiences and in many ways are differently motivated than children. Adults are more self-directed in their learning and have a greater need to know why they should learn something. Learning should be applicable to the learner's work or to other responsibilities valued by the learner. Thus, it is important that the instructor know the learner's needs and design learning activities that are relevant to those needs. The learner should be actively involved in learning, with the instructor acting as a facilitator (Collins, 2004).

To prevent skilled professionals from being phased out or forced into professions for which they are not talented, organized forms of lifelong learning are needed. Continuing professional development is an approach supporting Lifelong Learning (Aken, 2007).

## Continuous Assessment

A clear distinction should be made between **assessment of learning** for the purposes of grading and reporting, which has its own well-established procedures, and **assessment for learning** which calls for different priorities, new procedures and a new commitment.

Assessment of learning tends to be summative and is carried out periodically, e.g. at the end of a unit or year or key stage. Assessment for learning is the process of seeking and interpreting evidence for use by learners and their teachers to decide where the learners are in their learning, where they need to go and how best to get there. Assessment for learning is formative in nature and takes place all the time in the classroom (ARG, 1999).

The majority of distance learning computational environments involves some kind of student assessment; this is done by collecting the student's interactions with the environment. Silva and Vieira (2001) describe a method for the ongoing assessment of students in distance courses, based on the identification and structuring of relevant information regarding their interactions with the learning environment. These environments contain four ongoing assessment tools: tracking actions (log), redirection by test, records of messages from chats, and records of messages from discussions lists.

Learning on the Web (distance learning education) requires high self-regulatory skills. Virtanen et al (2003) stands that, in order to develop Web-based learning, we must pay more attention to learners' characteristics and help learners to be more aware of their learning processes and give guidance as to how to develop strategic learning. Therefore is still important to investigate effective ways of including metacognitive support in the design of natural and computer-based learning environments in order to improve the learning accompaniment.

## Knowledge Monitoring

In the past century, influential thinkers such as Dewey, Piaget and Vygotsky have argued that knowledge and control of one's own cognitive system play a key role in cognitive development as described by White et al (1999). One of the major conclusions from the research on cognition over the past 30 years is that students who monitor their learning are more effective learners than those who do not (Tobias and Everson, 2002).

The process of thinking about how we think, how we remember and how we learn was called metacognition by Flavell (1979). He claimed that through systematic training it's possible increasing the quantity and quality of children's self-monitoring skills as well as their metacognitive knowledge.

Many researchers have developed instruments and methods to measure metacognition as a whole or components of it. These methods range from self-questionnaires, where learners themselves rate their metacognitive skills and knowledge, to interviews or verbal-reports, in which the learners recall what they did and what they thought during a learning experience, as listed by Gama (2004).

Tobias and Everson (2002) have proposed a modular model of metacognition and it will be used as the theoretical foundation for this study. Their model assumes that the ability to differentiate between is known and unknown is a prerequisite for the effective self-regulation of learning. This skill is called knowledge monitoring and it supports the development of other metacognitive skills, such as comprehension monitoring, help seeking, planning, and revising. How much more students are aware of their thinking processes as they learn, much more they can control their own learning: self-awareness promotes self-regulation.

## Learning and Certification

The knowledge acquiring process has changed dramatically in the last years. The development and delivery of instructional material have been improved using the nine-stepped Robert Gagné (1970) model (Silveira et al., 2004), which could be organized as a set of learning objects through a multi-level structure that would allow information retrieval through ontology, for instance.

Though, in order to allow educators to advise this knowledge building process, an individualized profile for each student must be set up. Such an analysis could be performed by using computer-based tools which would support Adaptive Learning Environments (ALE), possibly through using Intelligent Tutoring Systems (ITS) techniques. Such environments would allow educators to build:

1. Competences and skills-based dynamic curricula, instead for static contents, which would require adaptive, coarse-grained Learning Objects.
2. Adequate representation mechanisms for competences and skills, which would allow continuous student tracking through cognitive grading obtained during learning process.
3. Structures for apprentices' initial evaluation, in order to settle a cognitive reference board for each student cognitive state, by taking into account their knowledge, skills and competences
4. Individualized learning plans which allow students to follow personalized learning paths from an initial landmark, established by an initial evaluation, to achieve a given certification.
5. Context-dependant, meaningful and interactive learning.
6. Learning certification strategies that take into account the process of knowledge building along with apprentice's evolutionary process while interacting with the system.

This proposal is linked to the following elements: knowledge evaluation (areas and concepts that are pre-requirements for certification process development), evaluation of previously obtained competences and skills, and what are those that must be developed in short and long-terms, and characterization of learning profiles, including learning styles, personal interests and motivations. By doing this, an ITS or ALE would be able to offer a detailed learning report, as well as to provide information that could be used to ascend student to upper hierarchical levels.

Certification represents the possibility of recognizing knowledge and abilities acquired along the life, as well as it helps to break barriers for job market and to continue and conclude studies. According to Rada (1999), certifying is "to attest as meeting a standard, and in educational arena, a certificate is a document that assures one has fulfilled some predefined requirements".

The certification process can be classified according to three different knowledge dimensions:

- **Conceptual Certification** is concerned about individual's capacity and the potential. A third-part emitted certificate is considered a test of truth on the acquired knowledge by any educational structures.
- **Professional Certification** is market-driven, since it is generally provided by corporations, according to the experience during a professional career. This is a sort of non-formal certification as the educational and conceptual certification are.
- **Educational Certification** is traditionally supplied by formal educational organization, like colleges and universities, which are responsible for evaluation and approval criteria, as well as the emission of the certificate or title.

According to CHEA (2005), accreditation is a process of external quality review used by higher education to scrutinize colleges, universities, and educational programs for quality assurance and quality improvement. In order to propose a model for knowledge certification it is necessary to structure and to organize the required knowledge for the area been certified.

For such, Omar (2005) shows that it is necessary to define the required contents and the structure of the knowledge, to establish the prerequisite for each item of the content, to identify the desired skills and abilities for each content and to structuralize an evaluation process that allows certifying this knowledge and competencies. These resources are showed in Figure 1.

*Figure 1. Basic architecture for certification (Omar, 2005)*

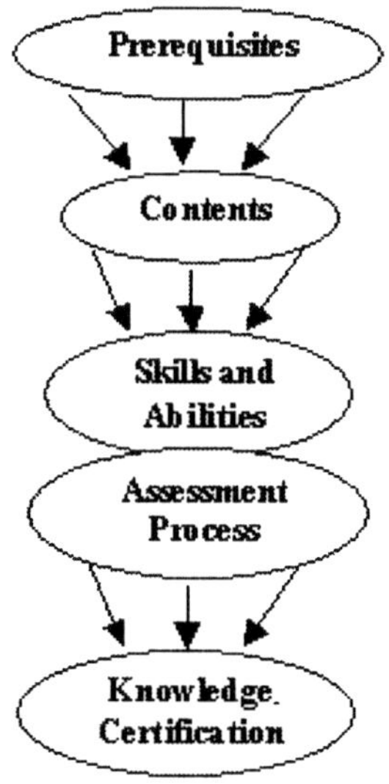


## A MODEL FOR KNOWLEDGE MONITORING AND CERTIFICATION

Working on the principle that assessment and feedback are essential for learning and improvement we have proposed a model for knowledge monitoring in distance learning education based on three measurers detailed as follow: KMA, KMB and KAL. The basic idea is to obtain these measurers from ongoing assessments on computer activities and to use them to monitoring to knowledge acquisition level of the student. The results obtained by learners could be used to the certification of skills and competences leading the professional to high position. In the sequence the description of the proposed metacognitive and cognitive measurers followed by the steps for certification.

### Metacognitive Measurers: KMA and KMB

Knowledge Monitoring Accuracy (KMA), created by Tobias & Everson's (2002) refers to how skilful a student is at predicting how he will perform on a learning task: it reflects his awareness of the knowledge he possess. The KMA resulted from the match between two pieces of information: first asking her confidence level in solving a problem and later asking him to solve the problem.

Knowledge Monitoring Bias (KMB) provides a statistical measure of any tendency or bias in the learner's knowledge monitoring ability. The KMB

*Table 1. KMA and KMB demonstration*

| ID | NP | CP | Errors | | | | KMA | | KMB | | Legend |
|---|---|---|---|---|---|---|---|---|---|---|---|
| | | | MO | MP | GO | GP | | | | | |
| 3 | 11 | 11 | 0 | 0 | 0 | 0 | 1.00 | High | 0.00 | Realistic | **ID:** Student Identifier |
| 61 | 8 | 5 | 3 | 0 | 0 | 0 | 0.44 | Medium | 0.19 | Random | **NP:** Number of Problems |
| 12 | 13 | 8 | 0 | 4 | 1 | 0 | 0.38 | Medium | -0.08 | Random | **CP:** Correct Student Preview |
| 67 | 18 | 9 | 1 | 7 | 1 | 0 | 0.22 | Medium | -0.11 | Random | **MO:** Medium Optimistic Errors |
| 14 | 9 | 4 | 1 | 4 | 0 | 0 | 0.17 | Medium | -0.17 | Random | **MP:** Medium Pessimistic Errors |
| 53 | 6 | 3 | 0 | 2 | 0 | 1 | 0.17 | Medium | -0.33 | Pessimistic | **GO:** Great Optimistic Errors |
| 34 | 13 | 5 | 2 | 5 | 0 | 1 | 0.04 | Medium | -0.19 | Random | **GP:** Great Pessimistic Errors |
| 50 | 10 | 3 | 2 | 2 | 0 | 3 | -0.20 | Medium | -0.30 | Pessimistic | |
| 44 | 4 | 0 | 0 | 4 | 0 | 0 | -0.50 | Low | -0.50 | Pessimistic | |

measure was created by Gama (2004) since the KMA does not provide a detailed account about the type of inaccuracies the student may show. The KMB takes into account the way student deviate from an accurate assessment of her knowledge monitoring. If there is no deviation, we say that the student is realistic about her assessment of her knowledge. On the contrary, the classification of student's current KMB state can be optimistic, pessimistic or random.

Table 1 shows some sample data obtained from empiric studies conducted by Pimentel (2006). KMA can assume High, Medium or Low values according to the preview and the student's assessment performance. For example, the student ID 12 is a Medium KMA level because he had realized 8 correct previews (CP) from 13 resolved problems. He had also performed 4 medium pessimistic errors (MP) and 1 great optimistic error (GO). Both, KMA and KMB ranges from -1.00 to 1.00. Concluding, he has a 0.38 KMA, classified as medium and a negative 0.08 KMB classified as Random, because sometimes he make right previews and sometimes he make mistakes. More details about calculations can saw at the referred paper.

## A Cognitive Measurer: KAL

The Knowledge Acquisition Level (KAL) indicates the learner's knowledge level in a specific subject of a knowledge domain. The zero value identifies total lack of knowledge. This measure can be obtained in several knowledge assessments units (AU) whose must associate the subjects or topics included in that UA. This will make possible to establish the student knowledge acquisition level in each topic of a domain.

Table 2 presents the simulation of one student's performance in four topics, during ten assessments activities. For example, Table 2 shows that "topic 1" in the T5 instant got the grade 0.8. The measure in T0-0 can be considered the student's initial mental state.

As shown in Figure 2, for "***topic 1***" and "***topic 2***", the KAL value can increase or decrease through ongoing assessments. This measure can be used for the accompaniment of the student's knowledge acquisition level.

It is known that a continuous assessment process will produce a large mass of data, demanding automatic or semi-automatic procedures for treatment and analysis. Advances in computer technology have made it possible store and to process a

*Table 2. The KAL evolution during 10 assessments*

| TOPICS | Assessments – Time Line (0 to 9) | | | | | | | | | |
|---|---|---|---|---|---|---|---|---|---|---|
| | T0 | T1 | T2 | T3 | T4 | T5 | T6 | T7 | T8 | T9 |
| **Topic 1** | 0.5 | 0.7 | 0.4 | 0.8 | 0.4 | **0.8** | 0.8 | 1.0 | 0.9 | 0.9 |
| Topic 2 | 0.2 | 0.5 | 0.5 | 0.4 | 0.5 | 0.5 | 0.7 | 0.6 | 0.7 | 0.8 |
| Topic 3 | 0.7 | 0.8 | 0.9 | 0.9 | 1.0 | 1.0 | 1.0 | 1.0 | 1.0 | 1.0 |
| Topic 4 | 0.2 | 0.0 | 0.3 | 0.1 | 0.3 | 0.2 | 0.2 | 0.4 | 0.3 | 0.4 |

*Figure 2. KAL evolution*

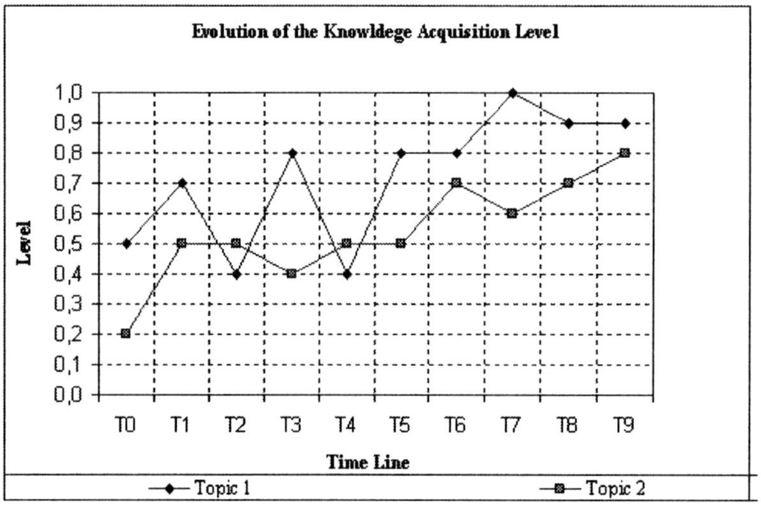


larger amount of data and new technologies have been developed to help extract information from these databases, emphasis laid on the Knowledge Discovery in Database (KDD) and the Data Mining (DM). KDD is the process of finding useful information and patterns in data. Data Mining is the use of algorithms to extract information and patterns (Fayyad, 2002).

## The Certification Process

The basis for the certification process is the definition of skills and abilities necessary to the performance of a specific function. At this point a person can then make a series of evaluations to determine their knowledge level which can be categorized as: basic, intermediate or advanced the Figure 3.

Table 3, gives an example for the Database Administrator function. The values in columns Basic, Intermediate and Advanced correspond to the minimum level needed for workers be classified at that level. The value zero indicates that the person don't´' need to have those skills, at that level. The value five indicates the maximum degree value.

If the function corresponds to a position in a company, to be framed in that position, the professional should obtain the minimum rates in all skills of it

In this proposal, each competence will be composed by "skills". Competence models have been defined by many consortiums: IMS standard

*Figure 3. Knowledge levels*

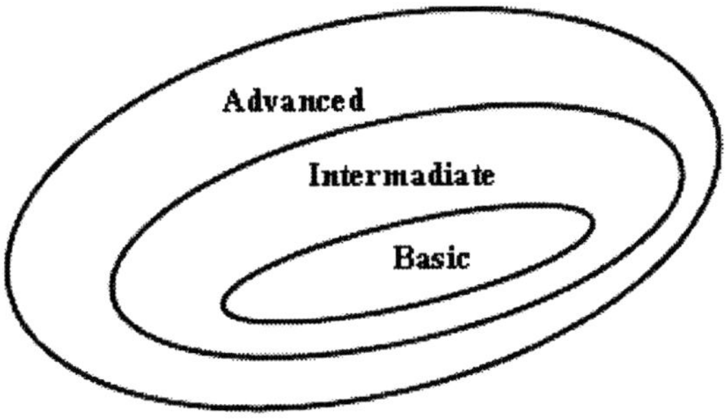


[2002], IEEE [2005] and HR-XML [2007] . For them the word competence serves to designate generically, knowledge, skills, attitudes and abilities. Here a competence will be used as a set of attributes and patterns needed to carry out a specific task. This patterns are associated to each job required by any employer by his own.

Another aspect to consider is the number of skills needed to compose a particular competence. For example, Bailey et. al. (2001) lists 85 skills needed for computer programmers working in the U.S. industrial sector.

In general, the standards above model as competency what was called skill in this study. The IMS (2002) and IEEE (2005), specify five compulsory elements, as showed in Figure 4:

1. **Identifier:** a unique and universal code enough to reference the skill any other system.;

*Figure 4. Outline of the standard competency model in IMS (2002)*

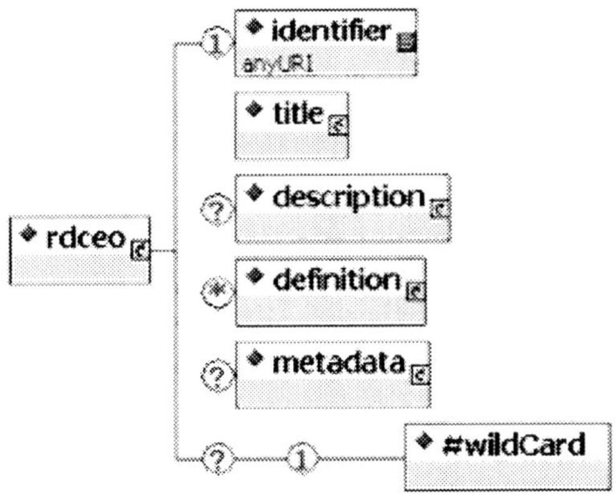


2. **Title:** a short name for the skill making sense to the human user; the title may be repeated in several idioms;
3. **Description:** a text, detailing and explaining about the skill, can also be repeated in several languages.;
4. **Definition:** describes the data types, default values, maximum and minimum;
5. **Metadata:** store the values measured in a particular skill instance.

The set of skills and abilities required for practicing a particular function also defines the knowledge (concepts) needed. Figure 5 presents a

*Table 3. Competences for the database administrators (DBA) function*

| Competences | Basic | Intermediate | Advanced |
|---|---|---|---|
| Logic and Physical Data Modelinga | 0 | 5 | 5 |
| Installation, Configuration and management of the. DBMS | 0 | 3 | 5 |
| Definition and Modification of the Data Schem | 3 | 5 | 5 |
| Maintenance and Data Retrieving. | 3 | 5 | 5 |
| Grant of Authorization for Data Access. | 0 | 3 | 5 |
| Monitoring and Performance Optimization | 0 | 3 | 5 |
| Definition of Strategies for Backup and Recovery Data. | 0 | 3 | 5 |

hierarchical scheme in which associated with each competence we have the knowledge (concepts).

It is from this structure of knowledge representation that assessments can be established for identifying the professional knowledge level as showed in the scheme of Table 4. First of all there are the competences (A) needed for a specific activity. After that, it is necessary to identify the set of concepts or knowledge (B) should be controlled to acquire each competence. Through continuous assessments (C) the person's (professional) metacognitive and cognitive profile is obtained. Finally, with this information is possible to prepare a learning plan to lead the person for certification.

## A LEARNING ENVIRONMENT ARCHITECTURE

Most of Computational Learning Environments (CLE) has incorporated some mechanisms of classroom assessment, classified as first generation of computer-based assessment based on objective tests (Ardigo, 2004).

This section presents the architecture of a computational environment for formative assessment in distance learning education based on the cognitive and metacognitive described in Section 3. A general view of this architecture is presented in Figure 6 that shows the relationships among the modules and sub-modules.

The use of a computational learning environment will favor the application of the formative assessment purposes and all its characteristics. Starting with a general setting capable of representing the knowledge to be reached or certified, the computational learning environment in its several modules will allow the diagnosis, learning monitoring, motivation, feedback, learning improvement, involvement and student awareness.

Moreover, intelligent tools will help to identify the cognitive and metacognitive student profile in order to assist the professor and the student himself in the result analysis and in setting the next learning stages.

There is a similarity between the proposed architecture and the traditional ITS architecture. The main difference is the new Assessment Module, which, by using information from the Student Module and the Domain Module will prompt the generation adaptive assessment according to the cognitive and metacognitive learner profile. Moreover, the learning gaps diagnosis will supply information for personalized learning plans, in such way that, once these gaps are filled, students may resume their learning, according to their expectations. In a traditional ITS, the assessment is generally concealed in the Tutoring Module.

It must be pointed out that the architecture proposed here does not present the Tutoring

*Figure 5. Knowledge levels*

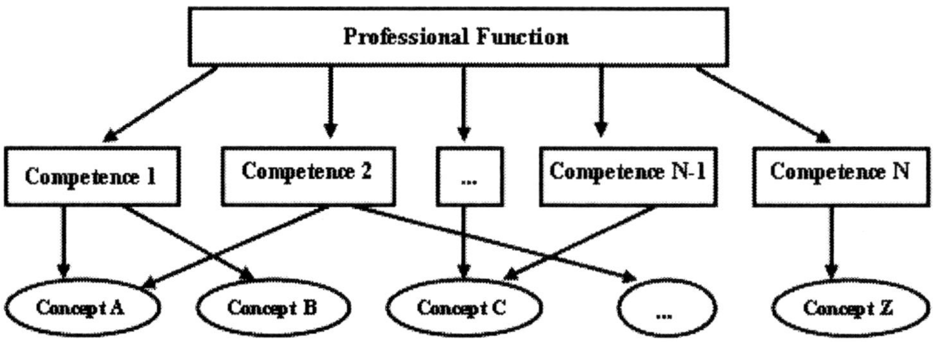

*Table 4. Competences for the database administrators (DBA) function*

| Certification Process for Function "A" Level "X" |
|---|
| **A) Competences** |
| • Competence 1<br>• Competence 2<br>• ...<br>• Competence N |
| **B) Contents** |
| • Concept A<br>• Concept B<br>• Concept C<br>• And so on |
| **C) Ongoing Assessment** |
| • Exercises that would allow verifying and certificating each competence in different knowledge levels (basic, intermediate and advanced)<br>• Each assessment must to associate the questions to the contents to make possible to establish knowledge measurer for each content |
| **D) Learning Planning** |
| • From the monitoring of cognitive and metaconitive knowledge sets up a study plan to achieve the desired level for certification |
| **E) Certification** |
| • Based on Knowledge measurers gotten from the assessments and minimum measurers established for each content of this learning object |

Module because the instructional process is not included in this work. However, this module can be normally connected to this architecture. The next subsection presents the description of the proposed environment modules as well as the relationships among them.

## Communication Module

Through this module, teacher and student will be able to interact with the environment in order to perform their respective tasks. The teacher will be able to directly interact with the management module and the accompaniment module and indirectly with the other three modules: student, assessment and knowledge.

The communication module (interface) will adapt to user-end (teacher or student) and to front-end device (personal computer, mobile device, palm-top, cell phone, etc).

## Management Module

The management module will be accessed by the teacher or anyone else responsible for the learning assessment and accompaniment process and will enable:

- To include data gathered from the pedagogical project of the course into knowledge module in order to reflect the learning objectives;
- To record assessment units associated to the learning objectives of each content, with the learning process involved equally specified;
- To create diagnostic, summative and formative assessment;
- To grade assessments performed by students when it was required.

## Knowledge Module

To allow the learning accompaniment, this module will structure the knowledge in learning hierarchies by using ontology that specifies a vocabulary relative to a set domain. This vocabulary defines the terms (classes, predicates, entities, properties and functions) and their relationships, representing a powerful tool to support the specification and the implementation even of complex computational systems (Guizzardi, 2000).

In order to establish the student cognitive state in each knowledge item, this module will record the knowledge representation and the concepts relationships. It could be said that this module will also contain the Learning Model. The objectives will be classified according to Blooms' revised taxonomy, enclosing the knowledge and the cognitive processes associated to each objective.

*Figure 6. General computational learning environment architecture*

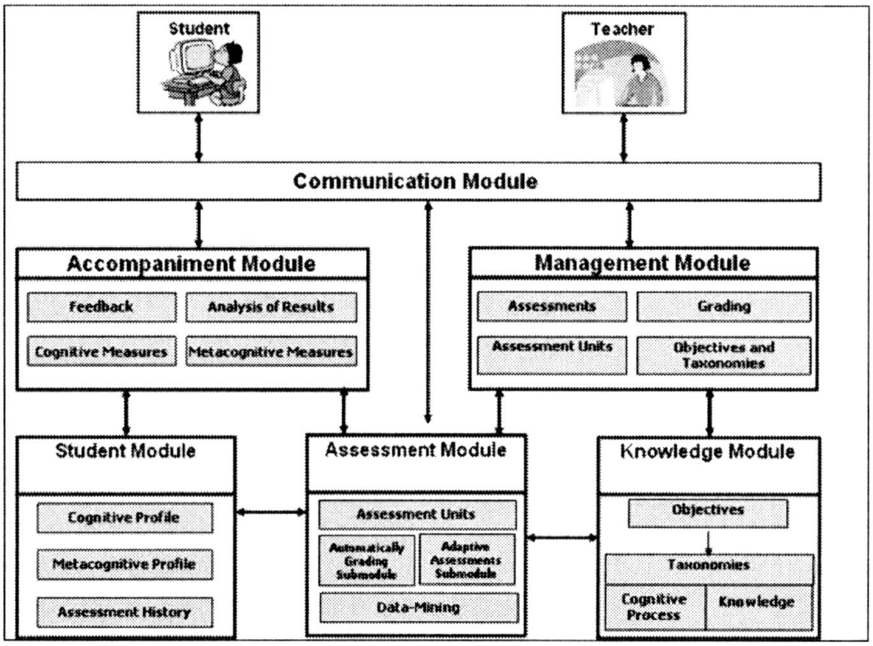


## Student Module

This module will store the cognitive and metacognitive student profile as well as the account of assessments.

Based on these data, it will be possible to report the student knowledge level immediately, select tasks or activities in that level or determine the learning needs to be filled to reach the objectives set.

## Assessment Module

This is the main module of this computer architecture and will store the assessment units that form the basis for the assessment settings. Each assessment unit will be associated with the objectives and consequently with the taxonomy.

In this paper, an assessment unit can assess even a simple concept, associated with either an indivisible content or some interrelated contents. For this reason, this paper does not use the term question found in traditional assessment methods. Hence, the attributing of a measure (right/wrong) to an assessment unit entails doing it in a detailed way, identifying the key failures.

Planning represents an important moment of assessment, since it must not only reflect the proposed objectives, but also record reliably the student's knowledge level. The elaboration of complex questions, what usually happens in the traditional system, without associating them with the addressed contents, hinders the identification of the learning gaps, that is, it is not favorable for a back-tracing in the sense of identifying accurately the contents that are blocking the student's learning improvement.

The assessment grading process (attribution of measures) could be automatic in objective questions or semi-automatic or manual when it comes to open questions. The automatic assessment grading is an open research field and deserves special attention since it consumes a long time of teachers' job. Moreover, assessment-grading makes room for certain subjectivity. However, this is not included among the objectives of this paper.

The adaptive assessment module should generate suitable assessments to the cognitive and metacognive student profile. The traditional educational system assesses all students in the same way, not considering the student's current knowledge level. Concerning formative assessment, it is necessary to take the student's current knowledge level into account and assess him accordingly, in order to contribute for his learning improvement.

To create adaptive assessments, the system will take into account the student's previous performance, stored in Student Module, besides attributes like: difficulty level of the assessment unit and contents relationship. According to the student's performance, the system will select suitable assessment units, with a lower or higher difficulty level, or will approach simpler or more complex contents.

As important as the data collection, in this module, is the treatment of these data in order to generate a set of simplified information that allows a human analysis or a computational analysis to support the decision-making. In this sense, the data-mining submodule should apply pattern-recognition algorithms to discover new knowledge in the assessment data. This submodule will supply information to the accompaniment module.

## Accompaniment Module

This module could be accessed by students and teachers in order to monitor the student's knowledge and learning improvements by means of cognitive and metacognitive measures gathered from assessments.

The feedback generated by the assessment grading process (automatic or manual) and the information provided by data-mining tasks will make it possible for teachers and students to accompany the learning process. Some decisions related to educational objectives could be taken from the information stored in this module, such as: to issue certificates and indicate the next instructional process stage based on the student cognitive and metacognitive level.

The cognitive measures will indicate the student's actual performance in the assessments. The possibility of selecting time periods will make it possible to track the student's evolution. The unstable student performance in some contents (high and low) could indicate that these contents are not adequately sedimented, while a continuous unsatisfactory student performance in some contents could indicate that this content is critical for the student to advance to the next step. In other words, the environment besides presenting results, will also allow for the Educational Users to verify the critical learning contents in individual level as well as in collective level.

Another important point in learning accompaniment in this work is continuity. Thus, having a unique or an interconnected knowledge database for several domains will make it possible to use it in several modules, courses or even in several education degrees.

The metacognitive measures are a differential in this work. The metacognitive accompaniment, mainly by the student, will offer him conditions to realize his actual level of knowledge. It is considered that metacognitive measures could be used in selecting the next instructional step and mainly in choosing the level of complexity of the next assessments. For example, the indication of a metacognitive pessimistic profile can suggest that the environment should select more complex contents for next assessments, besides keeping constant dialogue with the student to make him aware that his knowledge in fact is wider than he thinks.

The environment should not only show results to the student, but also compel him to hold a dialogue with it by selecting options, filling the gaps etc, in order to force him to read his results and in a certain way, to get his agreement.

## THE CERTIFICATION ENVIRONMENT ARCHITECTURE

In fundamental and high school people are prepared with a set of knowledge patterns, focusing on language, mathematics and science in general view. So, people choose a career and go to College or University. After that, in Lifelong Learning, usually when the person is already in a company, the market determines the required competences to take on a particular function. This section presents a general architecture to integrate a Learning Environment to a Certification Environment. As showed in Figure 7, the core is the required competences. Following, the Certification Environment (figure right side) is detailed.

### Competences Module

This module will hold the positions and functions in an organization. Associated with these positions will be a detailed set of competences (skills) represents as an ontology, indicating the depth degree needed for each competence in each function. It is the sum of the minimum competencies that will determine if a person can be or not certified in a particular function. Basically, this module will establish the minimum criteria for competence certification. At this module the system will incorporate informations about competence certificates acquired by the person formerly.

*Figure 7. General certification environment architecture*

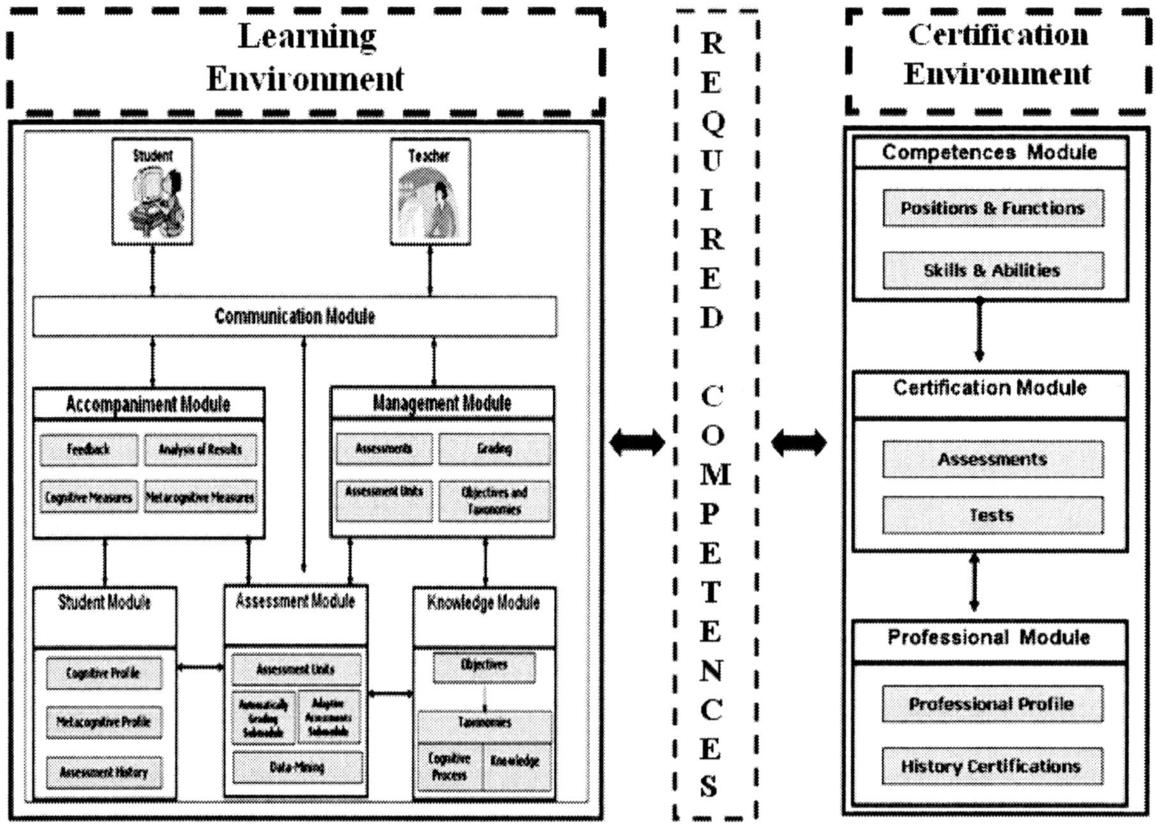

## Professional Module

This module will store the person's professional profile as well as his/her certifications history. This module is similar to the Student Model in the Learning Environment. The main difference is that the view here is connected to his/her competences. In the Student Model the focus is the Knowledge Acquisition Level. Based on these data, it will be possible to report immediately his/her certified competences and also his/her gaps. This will feed the learning plan for a desired certification and so on.

## Certification Module

This module is similar to the Assessment Model in the Learning Environment. The main difference is that here assessments are connected to required competences and not to objectives. This module will store assessments units and different kind of tests that form the basis for certification. It is important to note the connection between this module with the profile of vocational and skills required for each position / function.

# CONCLUSION

The Society of the Information has made possible that the people can learn in the most diverse places and moments. Many times this learning is full of gaps demanding a process of knowledge certification. Moreover, Management of Skills is a tendency and Companies are currently working to evaluate their internal talents and also to seek professionals in the market based on a competency model.

This paper has presented a model for formative assessment and certification in Lifelong Learning based on cognitive and metacognitive measurements that will make possible the identification of the professional learning gaps showing a roadmap to obtain educational and conceptual certification for his/her competence.

Most of Computational Learning Environments hides the assessment process and doesn't take into account the assessments inputs and outputs for the next step of a learning process. Formative assessment based continuous assessment can improve the learning in distance learning system education providing adaptive and personalization of the education, increasing the motivation and reducing the evasion rate. Besides, it can help to minimize the problems of credibility lack on who effectively took the assessment, allowing monitoring the evolution of the learning instead of having only one measure at the end of the course.

The environment architecture proposed in this article brings the assessment to the center in which assessment acts as a learning engine, gathering data that could identify precisely the student's knowledge level, and use this information to improve the learning process. We have proposed three measurers for monitoring the student knowledge acquisition level in distance learning education: KMA, KMB and KAL.

Concluding, the model integrates the Learning Environment with the Certification Environment, connecting the Educational System world, that deal with students, with the Organizations and Companies world, that deal with professionals an his/her required competences. This model is still under development and beyond the complete environment implementation case studies will be conducted for this proposal validation.

# REFERENCES

Aken, A., & Michalisin, M. I D.(2007). The impact of the skills gap on the recruitment of MIS graduates. *SIGMIS-CPR*. St. Louis, MI.

Aspin, D. N. (2007). *Lifelong learning concepts and conceptions' in philosophical perspectives on lifelong learning*. The Netherlands: Springer.

Assessment Reform Group – ARG. (1999). *Assessment for learning: Beyond the black box.* Cambridge: University of Cambridge, School of Education.

Bailey, J. t L., & Stefaniak, G. (2001). *Industry perceptions of the knowledge, skills, and abilities needed by computer programmers – SIGCPR 2001.* San Diego, CA USA.

Bransford, J., et al. (2003). *How people learn: brain, mind, experience, and school: Expanded edition.* Washington, D.C.: National Academy Press. National Research Council.

Collins, J. (2004). Education techniques for lifelong learning: principles of adult learning. *Radiographics, 24,* 1483–1489. Available at http://radiographics.rsnajnls.org/cgi/reprint/24/5/1483. doi:10.1148/rg.245045020

Council for Higher Education Accreditation (CHEA). (2005). *The fundamentals of accreditation.* Retrieved May 30, 2005, from http://www.chea.org/pdf/fund_accred_20ques_02.pdf

Falchikov, N. (2005). Improving assessment through student involvement: Practical solutions for aiding learning in higher and further education. New York: Routledge.

Fayyad, U., Grinstein, G. G., & Wierse, A. (2002). *Information visualization in data mining and knowledge discovery.* San Francisco, CA: Morgan Kaufmann Publishers Inc.

Fischer, G. (2000). Lifelong learning - More than training. *Journal of Interactive Learning Research, 7*(3), 265–294.

Flavell, J. H. (1979). Metacognition and cognitive monitoring: A new area of cognitive-developmental inquiry. *The American Psychologist, 34*(10), 906–911. doi:10.1037/0003-066X.34.10.906

Gagné, R. M. (1970). *The conditions of learning.* Longod, Holt, Rinehard and Winstn, Inc.

Gama, C. (2004). *Towards a model of metacognition instruction in interactive learning environments.* Doctoral Thesis. University of Sussex, England.

Guizzardi, G. (2000). *Uma Abordagem Metodológica de Desenvolvimento para e com Reuso, Baseada em Ontologias Formais de Domínio.* Dissertação (Mestrado). Universidade Federal do Espírito Santo, Vitória, ES, Brasil.

IEEE Competency Working Group. (2004). Draft Standard for Learning Technology- Reusable Competency Definitions. Retrieved from http://ltsc.ieee.org/wg20/files/ IEEE_1484.20.WD_01_rough.pdf

IMS Global Learning Consortium (2002). *IMS Reusable Definition of Competency or Educational Objective - Information Model. Version 1.0 Final Specification*

Omar, N., et al. (2005). *Towards a model for knowledge certification in engineering.* CEE- Congress on Engineering Education. Lisbon, Portugal

Perrenoud, P. (2000). *Dez Novas Competências para Ensinar.* Porto Alegre: Artmed, Pimentel, E.P. (2006). *Um Modelo para Avaliação e Acompanhamento Contínuo do Nível de Aquisição de Conhecimentos do Aprendiz.* Doutorado (Tese). Instituto Tecnológico da Aeronática (ITA), Brasil.

Pimentel, E. P., Omar, N., & França, V. F. (2005). *Monitoring the knowledge acquisition level in distance learning education.* Paper presented at the International Conference on Methods and Technologies for Learning, Palermo, Italy.

Rada, R. (1999). Sharing standards: IT skills standards. *Communications of the ACM, 42*(4), 21–26. doi:10.1145/299157.299162

Shepard, L. A. (2001). The role of classroom assessment in teaching and learning. In V. Richardson (Ed.), *The handbook of research on teaching* (4th ed.). Washington, DC: American Educational Research Association.

Silva, D. R., & Vieira, M. T. P. (2001). An ongoing assessment model in distance learning. In *Proceedings of Internet and Multimedia Systems and Applications*. Honolulu, USA.

Silveira, I. F., Mustaro, P. N., & Omar, N. (2004). Aprendizagem significativa baseada em uma arquitetura multicamadas de objetos de aprendizagem adaptativos. In *VII Congresso IberoAmericano de Informática Educativa, Monterrey. Anais do RIBIE.*

HRXML Specification (2007). Competencies (Measurable Characteristics) Recommendation April 2007.

Tobias, S., & Everson, H. T. (2002). *Knowing what you don't know: Further research on metacognitive knowledge monitoring.* College Board Research Report 2002-3, College Entrance Examination Board: New York.

Virtanen, P., Niemi, H., Nevgi, A., Raehalme, O., & Launonem, A. (2003). Towards strategic learning skills throught self-assessment and tutoring in web-based environment. In *Proceedings of ECER 2003 - European Conference on Educational Research*, Hamburg, Germany.

White, B. Y., Shimoda, T. A., & Frederiksen, J. R. (1999). Enabling students to construct theories of collaborative inquiry and reflective learning: Computer support for metacognitive development. *International Journal of Artificial Intelligence in Education, 10*, 151–182.

# Chapter 22
# Stories of Engagement with E-Learning:
## Revisiting the Taxonomy of Learning

**Geoffrey Lautenbach**
*University of Johannesburg, South Africa*

## ABSTRACT

*I argue that although university lecturers delve into the "shallow waters" of e-learning, they do not do so in sufficient depth and resign themselves to the perpetuation of cognitivist, behaviorist, and objectivist forms of knowledge without discovering more about the medium that could possibly liberate their restricted epistemologies. In this article, I explore possible reasons for varying engagement with e-learning, assuming that these reasons are located within the dimensions of the unit of analysis of the study; namely, lecturers' changing theories of knowledge and teaching in first encounters with e-learning. Using Lee Shulman's table of learning (Shulman, 2002) as a heuristic, I use excerpts from personal narratives to highlight the epistemological and pedagogical transformation of nine lecturers as they engage with educational technologies in their work.*

## INTRODUCTION

The concern of the larger study upon which this paper is based is the uptake and use of e-learning by lecturers in an education faculty at a university. Although e-learning forms only a part of the changing face of education, I identify the changing epistemologies and pedagogy of lecturers as a central issue in this process of transformation. According to Fullan and Stiegelbauer (1991), the successful implementation of learner-centered teaching depends to a large extent on development of the teacher (lecturers), which is not a top-down process but one in which the lecturer is very active; hence, the focus on the active participation of the selected lecturers in this study. To this end, nine lecturers of diverse technological ability consented to offer their stories about their engagement with e-learning in a number of narrative interviews.

One would assume that the incorporation of technology into higher education arguably should be a main priority for higher education practitioners, yet despite the cognizance of this necessity, e-learning uptake has been slow. Lecturers' limited engagements with e-learning were evident even in the early stages of the study. I now argue that the personal learning experiences and, to a degree, the teaching experiences of lecturers are directive indicators of their e-learning uptake. Moreover, I argue that these personal learning opportunities only become learning "events" for lecturers (and similarly also for the students) when they begin to fully engage with other lecturers, the larger community of lecturers using e-learning worldwide, the policies and strategies that guide them, and the divisions of labor that influence them as they engage with the tools of e-learning in order to ultimately change their inherent theories of knowledge and teaching (this is the activity system that is described in greater detail in the larger study). In this article, I see the lecturers' experiences within this activity system as the building blocks of their epistemological assumptions. I suggest, furthermore, that lecturers can only make meaning of their initial engagement with e-learning and the subsequent changes in their ways of teaching and thinking about teaching in general when they see the broader picture of how engagement with e-learning is not only on a physical level, but also strongly related to their geographical, historical, and cultural context.

I have also stated elsewhere that elements of lecturers' *resistance to* or *embracing of* technology in education are found in personal experiences, and it is in the "narrative situatedness" of lecturers' stories that I have found reasoning about their engagements with e-learning (Lautenbach & Van der Westhuizen, 2005a, 2005b; Lautenbach, Van der Westhuizen & Luca, 2006). Most faculty who embarked on a blending of online and face-to-face mediation and course presentation were doing so as novices at the time. By embarking on the study, I intended to capture temporally and spatially contextualized pictures of what happened in what has been, institutionally, a comprehensive adoption of blended learning. In telling their stories, lecturers exposed tensions within the activity system (Barab, Barnett, Yamagata-Lynch, Squire & Keating, 2002; Engeström, 1987, 1999; Kuutti, 1996; Leont'ev, 1978) that are critical to understanding what motivates specific actions within the system and, more generally, in understanding the dynamic nature (evolution) of the system in general (Barab et al., 2002). It is the description of this social interplay through lecturers' narratives that best illustrates the intricacies of emerging or "fossilized" epistemologies that, in many cases, have led to changes in the way these people teach using technology.

## THE UNIVERSITY AS A PLACE OF ENGAGEMENT WITH E-LEARNING

The uptake or adoption of e-learning is seen by university management and many lecturers as an essential component of what I call "professionalization of practice." Despite this, imperative adoption of e-learning is typically characterized by *nonuptake, adopt-and-abandon,* and *adopt-and-sustain.* I question the nature of the terms "adoption" and "uptake" and would rather, from this instant, refer to "engagement" as proposed by Lee Shulman (2002) in his table of learning. Rhem (2002) describes Shulman's interest in presenting a new taxonomy as something that more clearly reflects recent advances in understanding—"the world where people work"—and especially the place of "engagement." By using this new taxonomy, I resist the impulse to categorize or simplify the complex phenomenon of varying engagement with educational technologies, but at the same time, I make use of this heuristic to structure my findings.

Shulman's (2002) table of learning echoes the *taxonomy of educational objectives* devised by Benjamin Bloom (1956). In contrast, however, it posits that learning always involves engagement at some point. Shulman (2002) maintains that learning always begins with engagement, which in turn leads to knowledge and understanding. In Shulman's own words:

*Once someone understands, he or she becomes capable of performance or action. Critical reflection on one's practice and understanding leads to higher-order thinking in the form of a capacity to exercise judgment in the face of uncertainty and to create designs in the presence of constraints and unpredictability. Ultimately, the exercise of judgment makes possible the development of commitment. In commitment, we become capable of professing our understanding and our values, our faith and our love, our skepticism and our doubts, internalizing those attributes and making them integral to our identities. These commitments, in turn, make new engagements possible—and even necessary.* (Shulman, 2002)

This is a cyclic process where commitment and identity are followed by new engagements and motivations. The Shulman table of learning can be summarized briefly as follows:

- Engagement and motivation
- Knowledge and understanding
- Performance and action
- Reflection and critique
- Judgment and design
- Commitment and identity

In using this taxonomy, I am in no way proposing that things always happen in this sequence. It is used simply as a heuristic to structure university lecturers' stories of engagement with educational technologies or "pedagogical engagement." In this article, I see pedagogies of engagement as those that not only initially grab lecturers' interests in e-learning, but also those that *maintain this interest.* In other words, I see these as pedagogies that lead to what Rhem (2002) calls "deep learning." Engagement, therefore, cannot properly be understood as a means to an end; it is an end in itself. I believe that lecturers at the university, for example, do not try out these new technologies in order to merely increase their knowledge in the field of e-learning, but because they are engaged with what happens there. This is in line with the current focus on *active learning* and is based on the notion that people learn when engaged in worthwhile educational experiences.

I, therefore, highlight lecturers' stories of varying engagements with technology in their teaching, assuming that these stories are located within the dimensions of the unit of analysis of the study; namely, *lecturers' changing theories of knowledge and teaching in first encounters with e-learning.* In problematizing the notion of lecturers' engagements with educational technologies, I initially located the problem in three spheres: (1) lecturers' theories of knowledge and teaching; (2) the individual lecturer; and (3) the setting where e-learning must be implemented and sustained—the university and its related communities. Without knowing it at the time, I had stumbled across 3 major components that eventually formed part of the greater activity system described in the main study. In all three of these spheres mentioned previously, there were constraining factors, the most important perhaps being the unyielding epistemologies of lecturers.

Thus, I argue, that although lecturers delve into the shallow waters of e-learning, they do not do so in sufficient depth and resign themselves to the perpetuation of cognitivist, behaviorist, and objectivist forms of knowledge without discovering more about the medium that could possibly liberate their restricted epistemologies. They may not have learned to change their way of thinking about teaching using educational technologies or learned to perform certain required skills, and they may not always act in ways consistent with the norms, values, and conventions of the profession. Many lecturers, for example, still see e-learning as a way to access information and not as a process of distributed engagement and learning. These restricted epistemologies limit their pedagogic vision and influence the way they teach.

## METHODOLOGIES TO EXPOSE LECTURERS' VIEWS OF PERSONAL EPISTEMOLOGIES AND PEDAGOGIES OVER TIME

As a qualitative researcher, I am intrigued with the complexity of social interactions during engagement with e-learning, and the meanings that participants themselves attribute to these interactions. Within the natural setting of the university, I applied an interpretive and critical approach to the application of multiple methods of data collection and analysis. This research design was therefore pragmatic, both interpretive and critical, and grounded in the lived experiences of the participants. The specific genre of design for this inquiry can be seen as a triangular hybrid that included components of the ethnographic, the ethnomethodological, and the discursive tradition of qualitative inquiry (Alvesson & Sköldberg, 2000; Flick 1998; Henning, Van Rensburg & Smit, 2004). I contend that lecturers' changing theories of knowledge and teaching in first encounters with e-learning were adequately highlighted from a variety of perspectives using this combination of methodologies. It is also my conviction that the three methodologies were complementary and provided optimal richness and variety of data.

Conventional methods of interviewing did not provide me with enough viable options with which to record the multiple realities of lecturers and hence the incorporation of narrative interviews to supplement the ethnographic methods. All data analysis procedures in this inquiry searched for elements of the data that pertained specifically to narrative. For example, the ethnographic data reflected the life world of lecturers as identified from their observed behavior, practices, and activities, and also added to the development of the narrative portraits of the lecturers that showed how they constructed their preferred identities during their engagement with e-learning. Narrative analysis, therefore, featured as the common denominator in the process of data analysis. These narratives were analyzed performatively, as proposed by Langellier in 1989 (as cited by Riessman, 2002). I emphasize the performative approach because "a story involves story-telling, which is a reciprocal event between the story-teller and the interviewer" (Riessman, 2002). The storytellers' preferred identity is revealed in the stories they tell. The identity of the storyteller can be seen to be situated and accomplished in social interaction and in no way should be seen as inauthentic. Some lecturers experienced real epistemological change as a result of their initial engagement with e-learning; these changes are evident in the narrative excerpts that follow.

## NARRATIVE EXCERPTS REVEALING EPISTEMOLOGICAL AND PEDAGOGICAL CHANGE

This section deals with the notion of lecturers' changing theories of knowledge and teaching as depicted in their narratives. The way in which these lecturers made these changes (or not) is evident in the issues they describe in their stories in the quest for the development of *a functional online pedagogy*. Pseudonyms have been used for the nine lecturers, and the subtitles give the reader some idea of the type of story from which the extracts come.

### David's Story: The Conquering Crusader Who Lived to Tell the Tale

David admits to a limited theory of teaching with e-learning in his early career, but credits his general knowledge of teaching to his early years of experience as a school teacher. Initially, David began to identify problems with face-to-face teaching and recognized colleagues perpetuating the type of teaching he had become accustomed to at the school level. At this early stage, he had already identified differences between online and face-to-face modes of teaching and his changing focus from technology to methodology. He soon progressed to integrate the two modes of teaching where he discovered the importance of sound theoretical and subject knowledge. From that moment on, David substantiated all of his activities in his online teaching with theory. He currently shows awareness of the latest available literature and most up-to-date research in the field of e-learning and repeatedly stresses the importance of theoretical knowledge in his narrative. He had not only engaged with the tools of e-learning but had *reflected on his actions*. This is congruent with the notion that *action without reflection is unlikely to produce learning*. A further development of this theme is that David highlights the contextualization of theory and the adaptation of theory to the unique local situation. This is a concrete sign of David's exercise of *judgment and design*. In the future, David aims to continue trying to find the best methodologies to improve his teaching, and his hands-on approach should lead to experimentation with various methodologies. He now sees learning activities as the main focus of his teaching with technology, and he professes to choose a specific pedagogy based on teaching goals for every activity in his courses. *Commitment* is evidenced in the internalization of sound educational values; his well-developed e-learning persona; and his willingness to commit himself to the larger e-learning community at the university, worldwide, and within the educational profession. In other words, he has taken the values and principles of the greater e-learning community seriously enough to make them his own.

## Susan's Story: The Chameleon Who Learned to Blend In

Concerning her own emerging epistemology, Susan reveals her field of expertise to be human learning. She also professes not to have changed her teaching methods much over the past few years, and with regard to her pedagogy, she is still using the "same old principles" for teaching online. This seems to work for her because of her notion that "teaching must be seen as dialogue" and that it is a "process of collective inquiry in which students and the teacher explore together." She ascribes this to the fact that the tools within the course management system support this co-inquiry. To this end, she claims to be "doing what she has always done" when teaching with technology, indicating that a well-developed personal epistemology and pedagogy is perhaps the secret to success when teaching online. She claims to *understand* the notions of teaching and learning with technology and bases her actions (her performance or practice) on this understanding. Although she seems to have reflected on her actions, Susan's judgment and design is limited. Commitment is also not evident in this story.

## Brian's Story: The Man Who Found a New Lease on Life

Regarding his personal theories of knowledge and teaching, Brian goes right back to his roots as a physical science teacher at the school level and elaborates on how his teaching has not changed much over the length of his career. He claims to have always been creative. He ascribes his general knowledge of teaching to his early career as a teacher and admits to having tried various ways of teaching in the past. He even admits to boredom with the way in which he taught certain concepts in the past and adds that he had not, until recently, even considered teaching using the tools of e-learning. At present, Brian tells of being able to find new ways to teach and of how he has improved his own subject knowledge by reflecting on the design of learning activities. He addresses the similarities between face-to-face teaching and teaching with technology, indicating a practical working knowledge of the basic theory behind each mode, but also expresses the desire to "do more advanced things in the future." Brian already envisages using more interaction and more complex animations in his future teaching and speaks of complex online tutorials, the simulation of real-life activities, and online assessment in future courses. He does not speak of the implications for the design and development of such activities at present and does not elaborate on his role in this process. He seems to realize his limitations but has enough enthusiasm and theoretical knowledge of teaching with technology to dream of these activities becoming a reality.

## Mark's Story: The Traveler Who Lost the Urge to Explore

With regard to his own theories of knowledge and teaching, it is important to note that Mark sees e-learning as a field on its own and even describes it as "someone else's field." Mark claims to have been initially uninformed about the use of e-learning in teaching and admits that his initial focus was on applying the technology. Based on a brief period where he was actively involved in online modules, he is now able to expose and discuss a number of topical issues related to e-learning and teaching in general. Even so, Mark expresses uncertainty about e-learning potential usefulness at present based on his observation of other lecturers. As an educator, he is aware that this is an example of "petrified pedagogies" or pedagogies that have not changed to meet the demands of a new medium. Mark's awareness of the need to adapt his pedagogy to teaching online is evident from his discourse, but it is also clear that he has not yet personally attempted to do so. He does, however, indicate that he will seriously consider using e-learning in his teaching in the future if it can offer something that normal face-to-face teaching at the university cannot. He speaks of this as "value added." Commitment, as the highest attainment an educated person can achieve, has not been realized in this story.

## Ellen's Story: Finding the Foot to Fit the Glass Slipper

Ellen narrates the story of how she initially wanted to directly transfer content from an existing course into an online version of the same course, and how she came to discover that this was not an effective and practical way of teaching online. She ascribes the demise of her first attempts to good course design but poor use of the technology. At present, Ellen is comfortable with the idea of teaching online and admits to having theoretical knowledge of good practice in this field. She admits to changing her whole way of thinking about teaching with technology and couples this with a change in her way of thinking about education in general. She is aware of her personal theories of knowledge and teaching, and uses the terms "epistemology" and "pedagogy" freely in her story, indicating that she is comfortable with these aspects in her career as an educator. She expresses knowledge of her own personal epistemology and admits to not having to change her epistemology when teaching online. Ellen stresses her advanced knowledge of teaching based on many years in education and tells how she is still exploring her e-learning pedagogy. The one thing Ellen professes to have gotten out of this process from a pedagogical point of view is the way in which she now can combine face-to-face teaching with online components of her courses. She clearly states that this is something that cannot be learned from books and that her online pedagogy is dependant on her practical experience in the field. She implies in her story that her current experience in e-learning has led her to change her pedagogy, which seems to impact everything else.

## Hester's Story: The Lonely Path of the Long Distance Runner

Hester is comfortable with what she does and is obviously well read in the field of education. As an educational psychologist, she shows her need for interaction and the formation of relationships online and also proposes infusing the theory of teaching into e-learning. She does not, however, speak about changing her pedagogy when teaching online. It appears in Susan's narrative that a well-developed personal pedagogy is a major contributor to success online. Even though Hester and Susan do not say so directly, it may be easy to adapt existing theories of teaching for online teaching, whereas not having a well-developed pedagogy does not afford one this luxury in the first place. Hester ascribes her success online to keeping up to date with developments in the field of e-learning, using the tools available to her differently each time in order to see what works for her, and trying out different tools and activities all the time. She continually comes up with new uses of the Web that can be implemented in her teaching in the future.

## Irma's Story: The Professional Woman Keeping One Step Ahead

Regarding her theories of knowledge and teaching, Irma tells the story of being comfortable with her own epistemology and repeatedly brings up the issue of striving to be the subject and knowledge expert. By reducing or even removing her technological concerns, Irma is confident that she then will be able to concentrate on remaining the expert in her field in the future. With regard to teaching, Irma speaks of projecting the self into the technology, indicating that she is consciously trying to address the issue of distance when teaching online as well as other issues like "giving heart to the technology." She narrates the story of being conscious of her personality and teaching style being projected through the online activities and of the student "being with her," "hearing her voice," and "interacting with her and not the technology." Irma claims that using e-learning in teaching does not take much intelligence (i.e., IQ) but rather demands emotional intelligence (EQ), which simply involves a change in attitude toward teaching this way.

## Walter's Story: Moses Seeing the Promised Land for the First Time

Concerning his own emerging epistemology and pedagogy, Walter admits to adapting the way he teaches for implementation on the Web, and clearly sees the link between teaching and technology. In the past, he expressed e-learning as the cutting edge of education and saw great potential in pursuing e-learning-related teaching projects. At that time, he only saw e-learning as an aid to teaching, but now he sees it in all aspects of his personal and professional life. He tells how he has included e-learning as part of his everyday thinking. By seeing the link between teaching and technology, Walter proposes to blend e-learning into his teaching style in the future (and not the opposite).

## Rose's Story: Conflict on the Playground

Rose demonstrates an example of a *frozen and unyielding epistemology*. She professes to be comfortable with what works for her at present and does not see herself changing her teaching in the near future. She claims that eye contact and face-to-face teaching should be the first priority at the university without considering the many benefits of using e-learning at all. With regard to her pedagogy, Rose tells of a sense of losing control when using e-learning in her teaching. She expresses feelings that she is no longer the expert when teaching with technology, and the technology becomes the focus instead of the teaching activities. For this reason, she sees e-learning as cold and impersonal. She is unable to take multiple factors about educational technologies into consideration and compare them to values and standards of education that themselves are shifting. In other words, she may understand technological and educational issues, but she is unable to go beyond understanding in order to foster judgment and design. Her reflection about teaching with technology is that in her opinion, most lecturers are currently using technology without sufficient knowledge of the pedagogy that is needed in order to do so effectively.

## DISCUSSION

These accounts, summarized from individual narratives, indicate that learning about *teaching using technology* will be more effective if lecturers are committed. Successfully committed people are more disposed to engage (Shulman, 2002). Commitment engenders new engagements, which in turn engender new understandings, and so on. Engagement is inherently collaborative in nature, and commitment involves the development of and involvement in healthy communities. For this reason, if lecturers are presented with educational technologies as a meaningful whole within a healthy community of practice, I argue that this will help them accept the value of the knowledge before they shift their focus to the appropriation and the ability to use the knowledge. This is in line with Vygotsky's concept of developmental teaching where knowledge is only seen as useful when it "moves ahead of development" (Vygotsky, 1978).

Although some lecturers still see e-learning as *access to information* and not as a *process of distributed engagement and learning* (compare Henning 2003), I have discovered from the narratives that some lecturers have changed their pedagogy based on a flexible and evolving epistemology. In so doing, they have begun to exemplify what Nardi and O'Day (1999) refer to as "keystone species" in the establishment of learning and information ecologies in their workplace. These are like key components of a natural ecosystem that are essential to the survival and existence of the system. Furthermore, I suggest that this pedagogical change has led to, if only emergently, some form of learning and information ecology within the work environment. The formation of this "ecology" seems to be closely related to interactions with lecturers and other key members of the "ecosystem" highlighting the potential of a healthy and vibrant "community of practice" for these lecturers. The fact that there is as yet no coherent set of established pedagogies for e-learning begs for continued questioning of what may be adopted or even fast-tracked as pedagogies in a rapidly evolving medium of learning and teaching. With no formal curriculum for lecturers to follow in this process and no formal pedagogies to follow, each one has entered the system with a life history (and thus a lived experience) that has ultimately played a role in how they engaged with technology, and due to the uniqueness of each life history, each of their stories differs. They have all landed in the system in a different way and their *varying levels of engagement with the tools of e-learning can be related very much to how they position themselves in their stories.*

Excerpts from the nine narratives originating from this inquiry, rendered as brief portraits of the larger picture, address a wide spectrum of themes that have in the interim proved to be the experiences of many other lecturers at the university. These narratives and the ideas that have emerged in this article using a simple taxonomy as a thinking tool, must not be seen as trivial, as they offer a coherent way to think about how university lecturers use educational technologies in their teaching. This article merely highlights some thoughts and hopefully sheds more light on the topical issue of lecturers' engagements with e-learning at institutions of higher learning.

## REFERENCES


Alvesson, M., & Sköldberg, K. (2000). *Reflexive methodology: New vistas for qualitative research*. London: Sage.

Barab, S.A., Barnett, M., Yamagata-Lynch, L., Squire, K., & Keating, T. (2002). Using activity theory to understand the systemic tensions characterising a technology-rich introductory astronomy course. *Mind, Culture, and Activity, 9*(2), 76–107.

Bloom, B.S. (1956). *The taxonomy of educational objectives: Cognitive domain*. New York: David McKay.

Engeström, Y. (1987). *Learning by expanding: An activity-theoretical approach to developmental research*. Helsinki: Orienta-Konsultit.

Engeström, Y. (1999). Activity theory and individual and social transformation. In Y. Engeström, R. Miettinen, & R. Punamäki, (Eds.), 1999. *Perspectives on activity theory*. Cambridge: Cambridge University Press, 19–38.

Flick, U. (1998). *An introduction to qualitative research*. London: Sage Publications.

Fullan, M.G., & Stiegelbauer, S. (1991). *The new meaning of educational change*. New York: Teachers College Press.

Henning, E. (2003). I click, therefore I am (not): Is cognition "distributed" or is it "contained" in borderless e-learning programmes. *International Journal of Training and Development*, 7(4)._

Henning, E., Van Rensburg, W., & Smit, B. (2004). *Finding your way in qualitative research*. Pretoria: Van Schaik Publishers.

Kuutti, K. (1996). Activity theory as potential framework for human-computer interaction research. In B. Nardi (Ed.), 1996. *Context and consciousness*. Cambridge, MA: The MIT Press, 17–44.

Lautenbach, G.V., & Van der Westhuizen, D. (2005a). *Investigating lecturers changing epistemologies and pedagogies during engagement with e-learning in an education faculty*. Proceedings of the ED-MEDIA Conference, Montreal, Canada, AACE, 2880–2887.

Lautenbach, G.V., & Van der Westhuizen, D. (2005b). Hybridising methods towards narratives in an educational ICT inquiry. *Education as Change. Special Issue: ICT in Education, 9*(2), 46–73.

Lautenbach, G.V., Van der Westhuizen, D., & Luca, J. (2006). *Rags to riches and conflict on the playground: Contrasting narratives of e-learning in an education faculty*. Proceedings of the ED-MEDIA Conference, Orlando, Florida, AACE, 1700–1707.

Leont'ev, A. (1978). *Activity, consciousness and personality*. Englewood Cliffs, NJ: Prentice-Hall.

Nardi, B.A., & O'Day, V.L. (1999). *Information ecologies. Using technology with heart*. Cambridge, MA: The MIT Press.

Riessman, C.K. (2002). *Analysis of personal narratives*. In J..F. Gubrium, & J.A. Holstein (Eds.), 2002. *Handbook of interview research: Context and method*. Thousand Oaks, CA: Sage Publications, 695–709.

Rhem, J. (2002). Of diagrams and models: Learning as a game of pinball. *The National Teaching and Learning Forum, 11*(4). Retrieved February 2, 2007, from http://lihini.sjp.ac.lk/careers/edreform/n_amer/carnegie/diagrams.htm

Shulman, L. (2002). Making differences: A table of learning. *Change, 34*(6), 36–44.

Vygotsky, L. (1978). *Mind in society. The development of higher psychological processes*. In M. Cole, V.J. Steiner, S. Scribner, & E. Souberman (Eds.). Cambridge, MA: Harvard University Press.


*This work was previously published in International Journal of Information and Communication Technology Education, Vol. 4, Issue 3, edited by L. Tomei, pp. 11-19, copyright 2008 by IGI Publishing (an imprint of IGI Global).*

# Special Section
# A Focused Discussion on Educational Technologies

*This section revisits a 2007 special issue of the International Journal of Information and Communication Technology Education, which was edited by Dr. Bruce Howard. Following an introduction by Dr. Howard and Dr. Lawrence Tomei, editor of this collection, the next six chapters explore six key elements of the authors' examination of the classroom of the future. Particular emphasis on implications for science education is provided.*

# Chapter 23

# Emerging Educational Technologies and Science Education:

## A Multifaceted Research Approach

**Bruce C. Howard**
*Center for Educational Technologies® Wheeling Jesuit University, USA*

**Lawrence Tomei**
*Robert Morris University, USA*

## INTRODUCTION

*"The truth is that the want of common education with us is not from our poverty, but from the want of an orderly system. More money is now paid for the education of a part than would be paid for that of the whole if systematically arranged."*

*--Thomas Jefferson to Joseph C. Cabell, 1820.*

When discussing emerging educational technologies, the complaint around the globe is common enough: we may be outfitting schools with classrooms of the future, but teaching methods remain mired in the past. In the six articles that follow we describe our research on choosing and applying emerging educational technologies in the light of what we know about best practice teaching methods. Whereas many well-respected experts have addressed the need for new methodologies, we chose to focus on the *process of choosing* the technologies themselves. We set out to determine how to evaluate the individual promise an educational technology may hold and to provide guidelines to those who choose and use the technologies for teaching and learning.

The research was conducted by a team in the United States from the NASA-sponsored Classroom of the Future at the Center for Educational Technologies® in Wheeling, West Virginia. Among our team of researchers and instructional designers, the process was dubbed "benchmarking." A multi-

DOI: 10.4018/978-1-60566-936-6.ch023

faceted, two-phase approach was developed that blended classic research methodology with those used in market research studies. We gathered data and expertise from a variety of sources, including academic research articles, industry reports, interviews with leaders and national trend-setters, and the experiences of our own veteran staff.

Among the international community, NASA is well respected. Materials created to achieve NASA's educational goals are typically highly regarded as well. In our experience, NASA curriculum developers strive to incorporate innovative, effective uses of a broad range of educational technologies into their program offerings. This process involves a great deal of experimentation that is time-consuming, risky, and costly. As developers of educational websites, CD-ROMs, informal education programs, and teacher professional development experiences, the NASA-sponsored Classroom of the Future™ often faces the question of how to make the best use of educational technologies to inspire, engage, and educate.

To effectively integrate technology into NASA educational offerings, instructional designers should begin with an examination of the capabilities and limitation of various technologies—how particular ones could best support their curricular goals and how to use them for maximum impact (Bromley, 1997, 1998; Bruce & Hogan, 1998; Summerville & Reid-Griffin, 2008). Moreover, research has found that designers and developers need to be aware of the contextual factors, or *enabling conditions,* of the technology they plan to use (Downing & Holtz, 2008; Zhao, Byers, Pugh, & Sheldon, 2001). A list that succinctly identifies which educational technologies are better and why would become an indispensable tool for classroom teachers. In the past, coupling a constantly evolving field of education with the highly dynamic nature of technology development has made such a task nearly impossible. As soon as a list is generated, it becomes outdated.

Given how rapidly educational technologies change, this study sought to create a means by which decisions about capabilities and limitations, and effective use of classroom technologies could be made in a just-in-time fashion. In addition to the traditional conclusions and implications for future work found in academic research, pragmatic recommendations were also posed for instructional designers, developers, and classroom users of educational technologies. Practical principles and metrics were derived from exemplars with thoughts that these principles would be durable across range types of technologies, over multiple generations of products, and in most every country.

## Phase One

The initial phase of the project involved the following activities and is covered in more detail in subsequent articles in this issue.

- Milestones and seminal works conducted on the topic of benchmarking educational technologies were examined over 15 years worth of research studies. Investigations sought criteria for effective educational technologies; specifically, implementations and applications that resulted in design principles, decision-making principles, or measurables for gauging effectiveness.
- An 18-month investigation examined US National Science Foundation trends, determining the degree to which program solicitations included educational technologies.
- A cross-section of *pacesetters* in the realm of educational technologies were considered, including futurists, authors, journal editors, government officials, and leaders of professional organizations. Subjects were asked how they gauge the effectiveness of educational technologies and which technologies hold promise for improving the teaching and learning process.
- Lists of promising educational technologies, tools, websites, resources, software,

and hardware were reviewed in an attempt to extract common design features and develop practical lists of design principles.

- Finally, an in-depth commentary was solicited from a scientist-turned-educator to provide insight into emerging educational technologies such as the use of real-time data, science instrumentation, caves and immersive virtual reality, simulations, videogames, 3-D object creation, eBooks, and science television.

### Phase Two

In a subsequent investigation, Phase One findings were applied to a new development project. The result was the Educational Technology Collaborative website that uses Web 2.0 technologies to bring the curriculum development community together with classroom users, exploring issues related to design principles, metrics for choosing and using technologies, and methodologies for sharing best practices. This project is described in detail in the final article.

## CONTEXT

### The NASA-Sponsored Classroom of the Future

Since 1992 the NASA-sponsored Classroom of the Future has developed innovative educational products, teacher training activities, and theoretical and applied research that have benefited hundreds of thousands of students and teachers worldwide. The unique relationship with NASA allows the Classroom of the Future to develop and test high-quality materials that are technology intensive. In 2006 NASA put forth three major education goals:

1. Strengthen NASA and the future workforce of the United States—*NASA will continue contributing to the development of the U.S. science, technology, engineering, mathematics, and geography (STEM-G) workforce of the future through a diverse portfolio of education initiatives that target students at all levels.*
2. Attract and retain students in STEM-G disciplines—*To compete effectively for the minds, imaginations, and career ambitions of young people, NASA will focus on engaging and retaining U.S. students in STEM-G education programs to encourage their pursuit of educational disciplines critical to NASA's future engineering, scientific, and technical missions.*
3. Engage the U.S. citizenry in NASA's mission—*NASA will build strategic partnerships and linkages between STEM-G formal and informal education providers. Through hands-on, interactive, educational activities, NASA will engage students, educators, families, the general public, and all agency stakeholders to increase the science and technology literacy in the United States.*

NASA Education has embarked on the sizeable task of inspiring the next generation of explorers. Doing so will entail millions of hours of labor, hundreds of millions of dollars, and a shared vision for how to accomplish that goal. It is clear that during this effort, a growing emphasis will be placed upon the distribution of digital materials, the Internet as a dissemination vehicle, and other forms of educational technologies.

### Purpose

The Classroom of the Future is funded by NASA to provide leadership and guidance about educational technologies to the NASA-related community. In particular, there are many who develop educational products and programs, including subject matter experts, instructional designers, educational researchers, writers, and video producers. Another audience, which is not mutually exclusive, is those who use NASA products and programs in formal classrooms and informal venues. This includes teachers, school administrators, museum staff, and other outreach professionals.

## The Taxonomy for the Technology Domain

In our experience both the developers and users represent a continuum of understanding of and fluency with technology. Tomei (2003) has put forth a *Taxonomy for the Technology Domain* that helps to encapsulate the purposes of this project succinctly, by describing a continuum of six hierarchical levels:

- Technology for Literacy (Level 1.0) is the minimum degree of competency with respect to technology, computers, educational programs, office productivity software, the Internet, and their synergistic effectiveness as a learning strategy.
- Technology for Collaboration (Level 2.0) is the ability to employ technology for effective interpersonal interaction, such as written and aural communication, the professional exchange of information, and interpersonal collaboration.
- Technology for Decision Making (Level 3.0) refers to the ability to use technology in new and concrete situations to analyze, assess, and judge. Level 3.0 assumes mastery of the concepts and skills from the previous two levels in applied learning situations.
- Technology for Infusion (Level 4.0) recognizes technology as a powerful strategy for discovering and exploring academic content, including the identification, harvesting, and application of existing technology to unique learning situations.
- Technology for Integration (Level 5.0) is the ability to create new technology-based materials, combining otherwise disparate technologies to develop new, previously nonexistent, instructional materials to enhance learner understanding.
- Tech-ology (Level 6.0) refers to the ability to judge the universal impact, shared values, and social implications of technology use and its influence on teaching and learning. Tech-ology is a combination of "tech" (technology) and "ology" (the study of); therefore, the final stage of the taxonomy addresses the study of technology.

Within the next six chapters, which originally appeared as a Special Issue of the International Journal of Information Communication and Technology Education (IJICTE), we seek to provide the necessary informational resources as a professional development mechanism. Research is conducted at all levels of the Taxonomy with aspirations to move educational content developers, teachers and students toward a Level 6.0 perspective and the highest pinnacle of the hierarchical pyramid. As technologies are discussed and explored in this issue, references will be made to the Taxonomy and the appropriate levels of technology being demonstrated.

## The Research Context

Recently, a convergence of three major movements within education has been noted; specifically, the standards movement, the educational technology movement, and the teacher quality movement (Wenglinsky, 2005). Because of this, the need to justify technology expenditures and how classroom time is spent has increased dramatically as policymakers and the public demand evidence of results and accountability from teachers; think *No Child Left Behind*. Simultaneously, we are in the midst of two additional movements not in education, but in technology use in general: the use of open-source technologies and the broad scale use of "Web 2.0" user-generated content for

*Figure 1. Taxonomy for the technology domain*

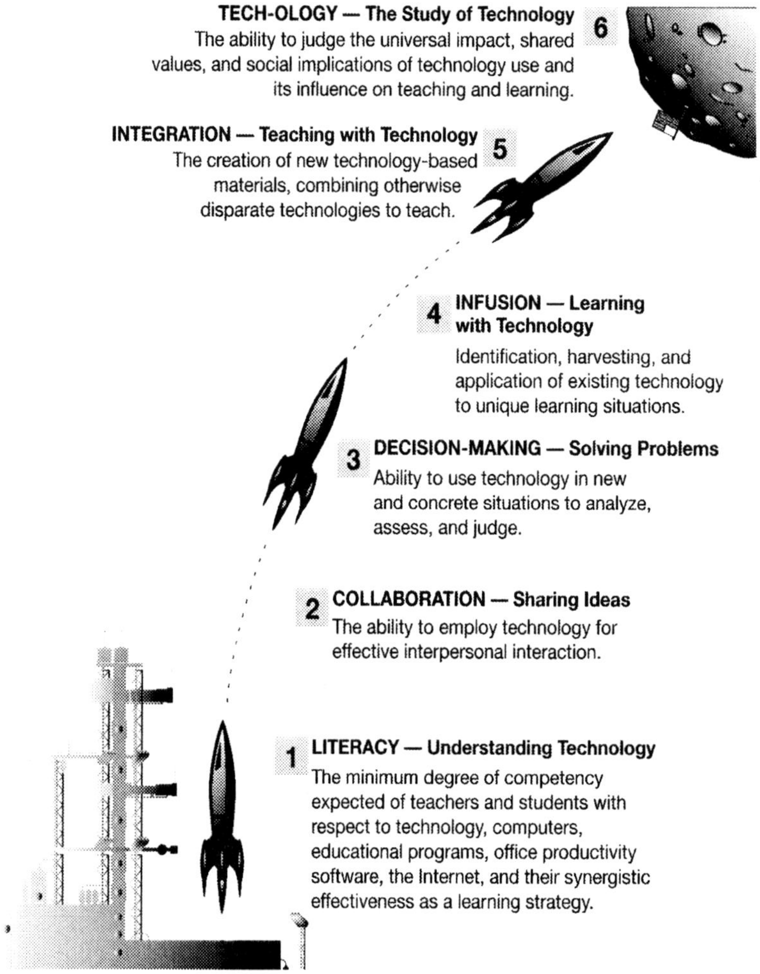


social and professional networking. It is unclear as yet the manner in which the dynamics inherent in these various movements will converge and reach equilibrium for the good of education.

Research on educational technologies is not keeping up with these rapid transformations. It is a question of scale. Traditionally, researchers have concentrated on designing and developing new technology tools and conducting small-to-medium scale evaluations to measure their impact. Meanwhile, the macroscopic question of whether or not technology makes a difference seems to dominate the discourse. Lawmakers and district administrators want large-scale studies to know where to put their resources (Lawless & Pelligrino, 2007).

What influence do small-scale studies have on changing theory and practice in educational technology? Some would say such contributions are limited (e.g., Zhao, Byers, Pugh, & Sheldon, 2001). Zhao and colleagues cite three reasons for this conclusion: Effects-based studies have a limited time of actual usefulness; the effect of technology in such studies is often misunderstood and misrepresented—deemphasizing the methodology and content; and, decision makers in the debate rarely base their decisions to use such technologies on research.

There are a growing number of venues in which these issues are being debated. In parallel there are growing calls for a national research agenda in educational technology research (e.g., Baker, 2001; Bull & Bell, 2006; Bull, Knesek, Roblyer, Schrum, & Thompson, 2005; Lawless & Pellegrino, 2007; Means & Haertel, 2003; Means, Wagner, Haertel, Javitz, 2003; Milken Exchange, 1997; No Child Left Behind Act, P.L. 107-110; Perez, Cherniavsky, & Hamilton, 2006; Presidents Committee of Advisors on Science and Technology, Panel on Educational Technology, 1997). In their appeal for a large-scale research study, Baker and O'Neil state, "Despite best efforts, technology-based innovations seem to have persistently avoided significant, innovative evaluation" (2006, p.3).

In its most succinct account, the context of the present work can be described as follows: a growing frustration with the limits of educational technology studies, a lack of innovation in evaluation, the need to justify precious resources spent on educational technology, a lack of information about what actually works, and ever-changing technology tools.

In this context as researchers, we need to learn to respond more rapidly with robust evaluations of what works and why. For this work, a multimodal, multifaceted pragmatic approach to evaluation was employed; but even that was not fast enough. Even in the months since beginning the work until publishing these articles, some of the imperatives presented herein may be less important now given new trends such as interoperability, mobility, or anywhere-anytime learning. As the work continues, we seek to test and verify our methodology, blending classic research methods with those used in market research, while maintaining professional academic standards.

## Data Sources

This research study made use of several data sources:

- Resources freely available on various websites, such as those of professional organizations, nonprofits, and government entities.
- Scholarly publications and books, such as the *Journal of Computers in Teacher Education (JCTE)*, *Journal of Technology and Teacher Education (JTATE)*, *Contemporary Issues in Technology & Teacher Education (CITE Journal)*, *Journal of Research in Technology and Education,* and *Educational Technology, Research and Development (ETR & D).*
- Other periodicals and trade magazines focusing on educational technologies, such as *Technological Horizons in Education (T.H.E.), Converge, Technology and Learning, Learning and Leading with Technology, Innovate, Edutopia, Campus Technology, eLearn Magazine,* and *eSchool News.*
- Experts and leaders in educational technology.
- Market research reports.
- Federal program announcements and requests for proposals.

# CONCLUSION

## Setting the Stage

Employing the Taxonomy as a guide, the following chapters explore six key elements of our examination of the classroom of the future. As we do so, we provide particular emphasis on implications for science education.

***Evaluating Educational Technologies: Historical Milestones*** reviews 15 years of published results from which a list of the characteristics of effective educational uses of technology was harvested. The studies presented in this paper considered technical, administrative, and learning features as well as recent investigations that emphasized administrative characteristics necessary to support No Child Left Behind, Summary of Criteria for Effective Educational Technologies, in particular, extends the Taxonomy for the Technology Domain by examining a series of technical, administrative, and learning features that add important considerations to the classification of technologies for teaching and learning.

Chen, Calinger, Howard, and Oskorus combine the perspectives of various sources to create a set of recommended design principles for technology-enhanced learning environments. In ***Emerging EdTech: Expert Perspectives and Design Principles***, the contributors create a new set of design principles to guide the evaluation of how educational technologies can be used in the classroom or help instructional designers in creating exemplary ways to implement technologies. In their search for answers as to what constitutes good practice, the reader is encouraged to involve the use of the Taxonomy as a guide for designing new, technology-based instructional content.

***The Best EdTech of 2007: Promising Features and Design Models***, by the assistant director for the Center for Educational Technologies, shares examples from six categories of promising educational technologies, tools, websites, resources, software, and hardware. The categories include knowledge and comprehension tools, interactive technologies and problem-solving tools, product-creation tools, efficiency and productivity tools, communication and collaboration tools, and technology tutors and are now part of a larger project for the NASA-sponsored Classroom of the Future. The six categories also mirror the six levels of the Taxonomy for Technology and bear your consideration as you read this important article.

In ***Setting Trends for Educational Technologies within the National Science Foundation***, the project team reviewed National Science Foundation (NSF) program announcements and awards to discern the amount and type of emphasis placed upon educational technologies with an emphasis on cognitive tutors/intelligent agents, distance learning, and online communities. The review of NSF's K-16 educational efforts in science, technology, engineering, mathematics, and geography (STEM-G) incorporates instructional technologies at all levels of the taxonomy.

Charles Wood, director of the NASA-sponsored Classroom of the Future, works with real data, uses authentic scientific instruments, explores immersive simulations and acts as scientists in ***Science for Everyone: Visions for Near-Future Educational Technology***. The capabilities discussed in this manuscript raise questions about the role of schools and the effectiveness of directed learning traditionally supplied by teachers and their ability to transform society into a culture of learning. Readers should consider using the taxonomy to identify where *Seismometers, Telescopes, Robots, Satellite Image Observatories, Image Processors,* Virtual reality videogames, Simulations, and eBooks (to name a few of the technologies discussed in this article) fall within the six levels of the new classification system.

Our final chapter for this section is ***Instructional Design, Web 2.0 Style.*** The author examines how we move ahead with the implementations of technology while confronted with delays waiting for research to defend and substantiate our efforts. Howard introduces a website called the *EdTech Collaborative* whose purpose is to provide an information-rich resource around which various professional communities may communicate and collaborate. It's also a great place for readers of these articles to discuss them by creating entries, offering constructive comments, editing articles, and reading reviews at one of many different levels.

## REFERENCES

Baker, E. L. (2001). Technology: How do we know it works? In W. F. Heineke, L. Blasi, & J. Willis (Eds.), *Methods of evaluating educational technology* (pp. 77-84). Greenwich, CT: Information Age.

Baker, E. L., & O'Neil, H. F. (2006). Evaluating web-based learning environments. In H. F. O'Neil & R. S. Perez (eds.), *Web-based learning: Theory, research, and practice* (pp. 3-20). Mahwah, NJ: Erlbaum.

Bromley, H. (1997). Thinking about computers and schools: A skeptical view. In P. E. Agre & D. Schuler (Eds.), *Reinventing technology, rediscovering community: Critical explorations of computing as a social practice* (pp. 107-126). Greenwich, CT: Ablex.

Bromley, H. (1998). Introduction: Data-driven democracy? Social assessment of educational computing. In H. Bromley & M. Apple (Eds.), *Education, technology, power* (pp. 1-28). Albany, NY: SUNY Press.

Bruce, B. C., & Hogan, M. P. (1998). The disappearance of technology: Toward an ecological model of literacy. In D. Reinking, M. C. McKenna, L. D. Labbo, & R. D. Kieffer (Eds.), *Handbook of literacy and technology: Transformations in a post-typographic world* (pp. 281). Mahwah, NJ: Erlbaum.

Bull, G., Knezek, G., Roblyer, M. D., Schrum, L., & Thompson, A. (2005). A proactive approach to a research agenda for educational technology. *Journal of Research on Technology in Education, 37*(3), 217–220.

Bull, G. L., & Bell, L. (2006). The 2006 national technology leadership summit. [Online serial]. *Contemporary Issues in Technology & Teacher Education, 6*(3).

Downing, K. F., & Holtz, J. K. (2008). *Online science learning: Best practices and technologies*. Hershey, PA: IGI Global.

Hoffer, E. (1982). *Between the devil and the dragon: The best essays and aphorisms of Eric Hoffer.* New York: Harper & Row.

Januszewski, A. (2005). Stasis and change in the definition of educational technology: The rationale and decision making process. *TechTrends, 49*(1), 45–46.

Lawless, K. A., & Pellegrino, J. W. (2007). Professional Development in Integrating Technology into Teaching and Learning: Knowns, Unknowns, and Ways to Pursue better Questions and Answers. *Review of Educational Research, 77*(4), 575–614. doi:10.3102/0034654307309921

Means, B., & Haertel, G. (2003). Cross-cutting themes in research designs for technology. In G. D. Haertel & B. Means (Eds.), *Evaluating educational technology: Effective research designs for improving learning* (pp. 257-269). New York: Teachers College Press.

Means, B., Wagner, M., Haertel, G. D., & Javitz, H. S. (2003). Studying the cumulative impacts of educational technology. In G. D. Haertel & B. Means (Eds.), *Evaluating educational technology* (pp. 230-256). New York: Teachers College Press.

Milken Family Foundation. (1997). *A call for a national research agenda.* Retrieved January 9, 2009, from http://www.mff.org/edtech/article.taf?_function=detail&Content_uid1=100

Nardi, B. A., & O'Day, V. L. (1999). *Information ecologies: Using technology with heart.* Cambridge, MA: MIT Press.

Perez, R. S., Cherniavsky, J., & Hamilton, E. R. (2006). Summary and conclusions. In H. F. O'Neil & R. S. Perez (Eds.), *Web-based learning* (pp. 393-404). Mahwah, NJ: Erlbaum.

President's Committee of Advisors on Science and Technology. (1997). *Report to the president on the use of technology to strengthen K-12 education in the United States, March 1997.* Retrieved January 9, 2009, from http://www.ostp.gov/cs/pcast/documents_reports/archive

Riddle, W. C., et al. (2007) *No Child Left Behind Act* (pp. 107-110). Washington, DC: Congressional Research Service.

Russell, M. (2001). Framing technology program evaluations. In W. Heineke, L. Blasi, & J. Willis (Eds.), *Methods of evaluating educational technology* (pp. 149-162). Greenwich, CT: Information Age.

Summerville, J., & Reid-Griffin, A. (2008). Technology Integration and Instructional Design. *TechTrends*, *52*(5), 45–51. doi:10.1007/s11528-008-0196-z

Tomei, L. A. (2005). *Taxonomy for the technology domain*. Hershey, PA: Information Science Reference.

Wenglinsky, H. (2005). *Using technology wisely: The keys to success in schools*. New York: Teachers College Press.

Zhao, Y., Byers, J., Pugh, K., & Sheldon, S. (2001). What's worth looking for? Issues in educational technology research. In W. F. Heineke, L. Blasi, & J. Willis (Eds.), *Methods of evaluating educational technology* (p. 260). Greenwich, CT: Information Age.

## KEY TERMS AND DEFINITIONS

**Educational Technology:** A device or system that makes use of digital media to enhance the teaching and learning process.

**Benchmarking:** The process of identifying exemplary educational technologies and their use.

**Design Principles:** A descriptor or characteristic which can be used by educators and designers to structure the content and features of an educational technology and its implementation into curriculum.

**Pacesetter:** A person who is a leading influence in his or her field or study or work.

**Exemplar:** A model representing excellence as a whole.

**Metric:** An objective standard for measurement for content, structure, or performance.

**Instructional Design:** A systematic approach to the design and development of instructional materials and products using objectives, teaching strategies and evaluation to meet learning needs.

**Benchmarking:** The process of identifying exemplary educational technologies and their use.

**Educational Technology:** A device or system that makes use of digital media to enhance the teaching and learning process.

# Chapter 24
# Evaluating Educational Technologies:
## Historical Milestones

**Laurie Ruberg**
*Center for Educational Technologies®, Wheeling Jesuit University, USA*

**Manetta Calinger**
*Center for Educational Technologies®, Wheeling Jesuit University, USA*

**Bruce C. Howard**
*Center for Educational Technologies®, Wheeling Jesuit University, USA*

## ABSTRACT

*Our team of researchers reviewed published results from the last 15 years to compile a list of the characteristics of effective educational uses of technology. All the studies considered technical, administrative, and learning features, while more recent investigations emphasized administrative characteristics necessary to support No Child Left Behind reporting. Recommended characteristics have evolved over time as expectations for technology integration have shifted from a focus on technology skill development to integrated use of technology as part of effective teaching and learning practices. Technology literacy is now considered as an integrated component of curriculum support and professional development. A timeline of relevant historical milestones in the evaluation of educational technologies illustrates how the understanding of and expectations for effective use of educational technologies has progressed to keep pace with advances in technological affordances.*

## INTRODUCTION

In an effort to improve student learning, educators, administrators, and researchers over the years have sought to identify the disparities across policy, curriculum, teacher professional training, student learning needs, and classroom environments. What followed were their recommendations for closing the gaps through the promise of new emerging technologies. In the last 15 years nearly every school district has completed several phases of major computer purchases, upgraded their bandwidth, and conducted professional training on the use of technology.

In this chapter we review the most relevant educational technology reports and studies of the last 15

DOI: 10.4018/978-1-60566-936-6.ch024

years to gain perspective on their results and the evolution of the educational technology environment. The studies from 1993 through 2002 share a common feature—they include, either implicitly or explicitly, specific criteria for evaluating the effectiveness of technology or for identifying the technology features necessary for use in formal or informal education environments. For example, such criteria might include cost, ease of use, or impact on student learning. More recent studies reflect a shift in focus away from evaluation of the technology (ISTE, 2002) to a perspective that examines how integrated use of a given technology (or system of technologies) improves teaching and learning processes (ISTE, 2008).

Our analysis includes only reports and studies that issued multiple recommendations. Each study represents a major effort or relevant perspective to our current objective of establishing design principles and metrics for choosing and using educational technologies. We briefly describe the context of each report and list its recommendations. Additionally, we provide a much longer timeline of relevant historical milestones in the evaluation of educational technologies. See Figure 1.

## REVIEW OF MAJOR EFFORTS IN EVALUATING EDUCATIONAL TECHNOLOGIES

### CEO Forum on Education and Technology (1996-2001)

The CEO Forum on Education and Technology was founded in 1996 and committed to a five-year partnership between business and education leaders to assess and monitor the progress toward integrating technology in America's schools. The forum published annual *School Technology and Readiness Reports*. A few of the reports focused on topics such as teacher preparation programs and professional development, but evaluating and assessing the role of technology in education were the focus of several of the annual reports.

The last report, *Key Building Blocks for Student Achievement in the 21st Century* (2001), culminated a five-year study that assessed varying aspects of assimilating technology into U.S. classrooms. This report identified four critical elements of effective technology implementation in American schools: connectivity, hardware, content (to allow integration of technology into the curriculum), and professional development.

The investigation reported several other key findings, most importantly that educational technology can improve student achievement. Assessments of scores in basic skill areas showed increases in proficiency in all areas. Eleven percent of the gain directly correlated to the technology implementation in the basic skills and computer education classes begun 10 years earlier. Interestingly, the study also found that achievement tests were based primarily on accessed state standards, and these standards generally emphasized lower order skills. The study recommended that standards be updated to reflect the need for preparing students with the technology skills needed for the 21st century environment.

### The Milken Family Foundation (1997, 1998)

The Milken Family Foundation is a private organization whose goal is "to discover and advance inventive and effective ways of helping people help themselves to…lead productive and satisfying lives." The foundation works toward this objective primarily thorough education and medical research initiatives.

Its report, *A Call for a National Research Agenda (1997),* is brief and, at first glance, seemingly outdated now more than a decade later. However, the essay implored the United States to outline a national agenda researching the benefits of educational technology in America's classrooms. The agenda had four tasks: (a) catalog

*Figure 1. Timeline of relevant historical milestones in the evaluation of educational technologies*

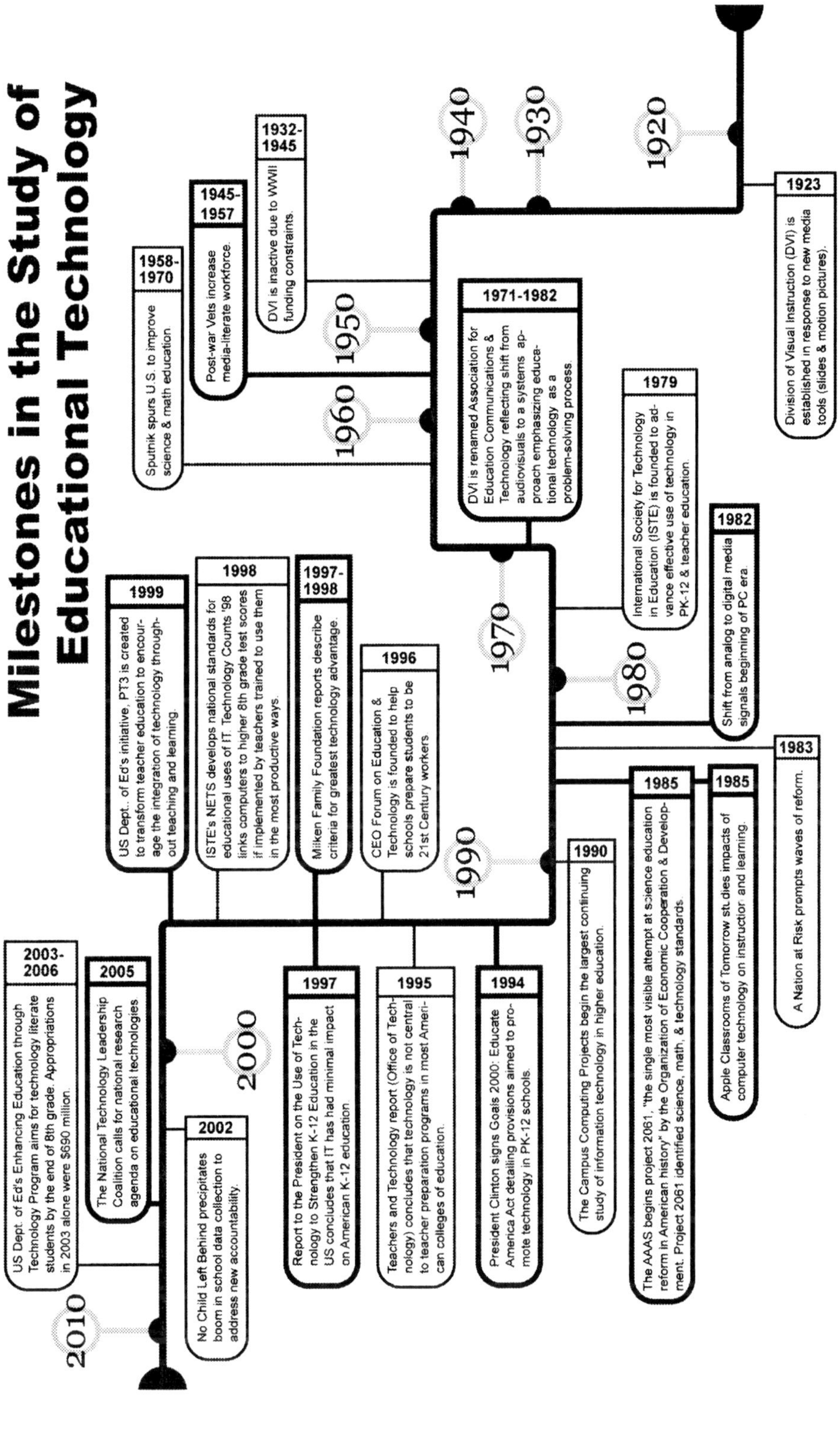

what is known and identify significant gaps in knowledge, (b) formulate and prioritize appropriate research questions, (c) mobilize resources, and (d) disseminate results.

The significance of this report is its call for a national technology plan *almost a decade ago.* The four tasks identified are still relevant, although the substance and work of the plan would be greatly different in light of today's complex technology world.

The Milken Family Foundation's 1998 report, *Technology in American Schools: Seven Dimensions for Gauging Progress*, described learning outcomes necessary for technology to be used to its greatest educational advantage in a classroom. Effective technology should:

1. Accelerate, enrich, and deepen basic skills.
2. Motivate and engage students in learning.
3. Relate academics to practices of today's workforce.
4. Increase economic viability of tomorrow's workers.
5. Strengthen teaching.
6. Contribute to positive change in schools.
7. Connect schools to the world.

The publication focused on systemic change. It acknowledged that the transformation necessary to effectively integrate technology into the student learning process is complex, requiring new ways of thinking, teaching, and learning. Other essential criteria for effective educational technology included core technology fluency for students, curriculum and learning assessment tools, increased relevancy of technology used, potential to increase student motivation, professional competency, system capacity, community connections, technology capacity, and accountability.

Within two years after the 1997 report, 36 states had established technology standards, and 9 others were in the process of doing so; 22 of the 36 had already integrated them into their overall standards for the basic subject areas. The Milken Exchange recognized the importance of assessing the impact of technology and commissioned a study in West Virginia that subsequently reported that 11 percent of the academic gains in mathematics and reading for fifth-graders in 1995 were directly attributable to technology interventions. A later analysis in 1999 found that technology was a cost-effective way to improve student learning when compared to other strategies, such as reducing class size.

The West Virginia basic skills/computer education (BS/CE) program was evaluated because its comprehensive teacher training and its 10-year history made it the longest-running U.S. program for the implementation of technology in education. In addition to the findings cited above, the Milken Exchange study found the BS/CE program to be highly successful in providing equal opportunities for low income and rural students and for females, widely considered to be at a disadvantage in learning technology programs.

An analysis of factors responsible for the success of the program reinforced the criteria identified in the previously discussed study. Integration of technology into the instructional curriculum was a significant factor in the program's success. The report also concluded that computers inside the classrooms were more effective than centralized computer labs, and that comprehensive teacher training was also a key factor in West Virginia's success.

## President's Committee of Advisors on Science and Technology (1997)

The President's Committee of Advisors on Science and Technology (PCAST) created a bridge between the private sector and the academic community for technology, scientific research, and math and science education. Reports primarily involved technology impacts on economics, the environment, health concerns, and sustainable development with educational technology research constituting a minor area of study. In 1995 PCAST addressed this lack of educational focus with the formation of the Panel on Educational Technology, which commissioned the *Report to the President on the Use of Technology to Strengthen K-12 Education in the United States* (1997) to advise the president on the development and application of technology to K-12 education in the U.S.

The report made a number of significant recommendations for using technology in K-12 education, such as focusing on learning outcomes (not the technology itself), emphasizing content and pedagogy (not just the hardware), providing teacher professional development, and significantly increasing spending for technology-related improvements. The panel also recommended a "large-scale program of rigorous, systematic research on education in general, and educational technology in particular."

## National Governors Association (2001, 2003)

In the United States the National Governors Association develops and implements solutions to public policy challenges. As part of its education initiative and its commitment to advancing the use of technology in the classroom, the association commissioned a series of reports and studies to investigate the status of technology implementation and its effectiveness in achieving educational objectives in K-16 education. The focus of these efforts was to identify potential solutions to improve state economies. The governors reasoned that if more citizens are better educated with 21st century skills, the economy would thrive. Therefore, they argued that supporting education supports the economy and justifies the investment in technology.

One report, *State of e-learning in the States* (2001), focused on identifying the states that are promoting access to e-learning, defined as "instructional content or learning experiences by electronic technology." The quality of technology infrastructure, the use of financial incentives, and the application of technology to increase opportunities to economically disadvantaged students were criteria for investigation.

The study found that most states had begun to implement new e-learning opportunities. Significant challenges, however, were recognized. Recommendations included the need for developing content that optimally used technology, the improvement of the technology infrastructure, and the need to reduce the digitally underserved by providing technology to schools serving socioeconomically disadvantaged populations. The report also recommended that investments in technology education must meet rigorous criteria for demonstrating significant positive impact on student achievement via improved test scores. The report also emphasized that workplace skills should be emphasized in technology curricula.

The association's 2003 report, *Higher Expectations I,* was a collection of essays that focused on the state of higher education. One essay, *Technology: Creating New Models in Higher Education,* proposed that technology would increase access to higher education and would radically change educational delivery. The essay also projected that portability, another important criterion of exemplary educational technology, would be largely responsible for this outcome.

## North Central Regional Technology in Education Consortium (1997-2005)

The North Central Regional Educational Laboratory (NCREL) was funded by the U.S. Department of Education until its contract expired in September 2005. Many of the educational studies published by NCREL are still available on its websites.

In 1997 the North Central Regional Teaching in Education Consortium (NCRTEC) and NCREL began developing a suite of tools to assist in the planning of integration of technology in education. The *Learning with Technology Profile Tool* was a computer program to help educators assess their instructional practices in areas of learning and technology. The program contained indicators of engaged learning and technology implementation, such as alignment, relevance, range of use, assessment strategies, connectivity, support, facility resources, and provision for equity. The tool allowed teachers to identify and categorize their instructional practices regarding technology integration by answering questions concerning their teaching habits. Their answers were analyzed, and strengths and weaknesses of current instructional practices given. The report concluded that evaluations of the effectiveness of educational technologies need to include a thorough assessment of implementation practices.

## Consortium for School Networking: Digital Learning Spaces 2010 (2005)

The mission of the Consortium for School Networking (CoSN) is to "serve as the national organization for K-12 technology leaders who use technology strategically to ultimately improve teaching and learning." CoSN's 2005 annual report, *Digital Learning Spaces 2010*, listed the important characteristics of the most effective emerging educational technology as affordable speed and power, wireless (portable) technology, and individualized student learning.

CoSN also published *Hot Technologies for K-12 Schools* in 2004. It introduced five needs that should be considered when evaluating the "learning space" of emerging technologies. Technology metrics could be directly written from the consortium's explicit recommendations:

1. Does this technology promote authentic learning and galvanize the instructional process?
2. Does this technology improve assessment and evaluation at all levels?
3. Does this technology address diverse learning styles and needs of students?
4. Does this technology build community and foster communication?
5. Does this technology improve the efficiency of school administration?

In this report CoSN committee members evaluated only emerging technologies. They used three criteria for effectiveness: The technology must address one of the five key educational needs, fundamentally change schools and learning, and meet standards for feasibility within schools and systems.

## Australian Capital Territory Department of Education: Emerging Technologies: A Framework for Thinking (2005)

Commissioned by the Australian Capital Territory Department of Education, *Emerging Technologies: A Framework for Thinking* is a comprehensive report on educational technologies, their impact on Australian education, and the necessary reforms needed to give students skills for an ever-increasing technological world.

Recognizing the difficulty in predicting types of future technology, the report identifies desired characteristics that should typify effective emerging technologies. These characteristics included mobility, interoperability, convergence, divergence, integration, richness of content, security, creativity, interactivity and collaboration, and utilization of open source software.

Interestingly, this report stressed the critical importance of policy and environmental factors, which affect implementing changes in classroom instruction using technology. These factors included considerations such as the political, legal, social, and cultural impacts on educational systemic change. Further recommendations included the need for educators to build new literacies, engage in strategic professional development opportunities, embrace visionary leadership, and incorporate communication technologies into curriculum, instructional delivery, assessment, and reporting practices.

## Education Week: Technology Counts (2006)

*Education Week* published *Technology Counts 2006,* a comprehensive report on technology in U.S. schools (2006). Education Week surveyed state technology officials in all 50 states plus the District of Columbia. The survey analysis made use of 14 indicators in three areas of technology policy and practice: access to technology, use of technology, and capacity of use. The analysis used specific criteria to assign points, which were then totaled and averaged, and a letter grade assigned.

Vetting procedures were strictly followed throughout the research survey. The initial analyses were sent to state officials for comments and supporting documentation. Researchers from *Editorial Projects in Education* corroborated the documentary evidence through phone calls and e-mails. Any proposed changes to procedure required appropriate documentation as evidence of need.

One of the more relevant findings concerned the need for computerized systems that can reliably handle large amounts of educational data. The U.S. *No Child Left Behind Act* precipitated a sudden increase in school data collection to address new accountability demands. Recommendations included increasing both computer availability and literacy in American classrooms; providing teachers, administrators, and technology officials with statistical and analytic tools needed to improve administrative and reporting efficiency; and implementing training opportunities on the effective use of data in classroom instruction.

Access to downloadable data files that allow for school data to be more effectively analyzed both for student test performance and for school characteristics was reported as an important criterion for evaluating a state's educational technology readiness.

The research also surveyed state and district practices for using data to improve teaching and learning. The findings stated that technology has the potential to greatly assist in the complex administrative tasks involved in the ever-increasing requirements for tracking and reporting student progress. However, key problems, such as the lack of resources and professional development opportunities, prohibit full realization of the technology potential.

## National Educational Technology Standards for Teachers: Preparing Teachers to Use Technology (2002, 2008)

The International Society for Technology in Education (ISTE) received a Preparing Tomorrow's Teachers to Use Technology ($PT^3$) grant from the U.S. Department of Education to bring together a national committee representing experts in preschool through senior high school and postsecondary teacher education to develop national standards for educational uses of technology (Kelly, 2002). The standards proposed by this ISTE initiative were designed to bring faculty, administration, school district technology support personnel, and postsecondary teacher educators together to plan technology use as a component of school curriculum and goal setting for all grade-level teaching and learning.

This ISTE report (Kelly, 2002) divided technology standards into six broad categories for teacher professional growth and capabilities. Briefly summarized, these standards for teachers address:

1. **Basic Operations and Concepts:** knowledge, skills, and understanding of technology concepts and skills as well as a plan for continued growth in these areas.
2. **Planning and Designing Learning Environments and Experiences:** design of

technology-rich learning opportunities, application of technology research, evaluation of suitable technology resources, plans for management of these technologies, and plans for how teachers will manage student learning in this technology-rich environment.

3. **Teaching, Learning, and the Curriculum:** methods and strategies for applying technology to maximize student learning.
4. **Assessment and Evaluation:** the application of technology to facilitate effective assessment and evaluation strategies.
5. **Productivity and Professional Practice:** use of technology to enhance teaching productivity and professional practice.
6. **Social, Ethical, Legal, and Human Issues:** understanding of how these issues are embedded in the use of technology and application of that understanding in practice.

Recognizing that today's teaching and learning environment requires educators to be even more sophisticated technology users, ISTE revised its Educational Technology Standards for Teachers in 2008. These new standards recognize teachers' positions as "less about *staying ahead* and more about *moving ahead* as members of dynamic learning communities" (ISTE, 2008). The focus of the 2008 technology standards for teachers reflects a shift from trying to benchmark teacher and student performance and capabilities to goals that emphasize abilities to orchestrate technology integration and promote technology abilities that are forward focused and global in context.

This brief summary of the 2008 standards illustrates how they differ from the previous ISTE standards. The 2008 National Educational Technology Standards for Teachers (NETS-T) emphasize teacher abilities to (1) facilitate and inspire student learning and creativity, (2) design digital-age learning experiences and assessments, (3) model digital-age work and learning, (4) promote and model digital citizenship and responsibility, and (5) engage in professional growth and leadership. Three themes behind the revisions to the NETS-T reflect the impact of ever-changing technologies, related changes in teaching expectations, and changes in our culture and society toward a more global economy and knowledge structure (ISTE, 2008).

The increasing global nature of our society requires that teachers and students become adept at what is referred to as "21st century skills" (Partnership for 21st Century Skills, 2008). However, teachers and education policy authors should be aware of areas where the goals for technology evaluation for global businesses and educators clash. Educators ultimately are interested in providing technology instruction that offers enduring knowledge and generic process skills whereas global system economies desire a workforce prepared for immediate career placement (Ruberg, 2008; Spring, 2008). The tensions between these conflicting interests demonstrate how shifts toward global systems, knowledge economies, lifelong learning, and multicultural factors impact the evaluation of educational technologies and attempts to identify disparities across policy, curriculum, teacher professional training, student learning needs, and classroom environments.

## SUMMARY

Each of these seminal works provides a perspective on choosing and using educational technologies. We summarize the implicit and explicit criteria used in these works in *Table 1*. Across the top the works are arranged from oldest (left) to most recent (right). Each row represents a desired characteristic related to effective educational technology. The criteria fall into three groups: administrative features, technical features, and learning features.

The table shows there has been little temporal change in emphasis over the last 15 years. In terms of the learning features, our study included only

*Table 1. Summary of criteria for effective educational technologies*

| **Report/ Study >** | **CEO Forum:** *Key Building Blocks* | **Milken Foundation:** *Seven Dimensions* | **PCAST:** *Report to the President* | **National Governors Association:** *New Models* | **National Governors Association:** *State of E-Learning* | **NCRTEC:** *Learning with Technology* | **CoSN:** *Digital Learning Spaces 2010* | **CoSN:** *Hot Technologies for K-12 Schools* | **Australian Capital Territory:** *Emerging Technologies* | **Education Week:** *Technology Counts 2006* |
|---|---|---|---|---|---|---|---|---|---|---|
| **Criteria for Effective Educational Technologies** | | | | | | | | | | |
| Administrative Features | | | | | | | | | | |
| Allows feasibility in implementation | | | | | | | | | | • |
| Improves administrative efficiency | | | • | | | | | • | | • |
| Enhances speed | | | | | | | • | | | |
| Provides more equitable access | | | | | • | • | | | | |
| Assists professional development | | • | • | | | • | | | • | |
| **Technical Features** | | | | | | | | | | |
| Includes wireless features | | | | • | | • | • | | • | |
| Uses standards for interoperability | | | | | | • | | | • | |
| Provides for more open architecture | | | | | | • | | | | |
| Provides effective helps | | | | | | • | | | | • |
| Makes use more educator friendly | | | • | | | | | | | • |
| Facilitates training and support | | • | | | | | | | | • |
| Adds functionality | | | | • | | • | | | | |
| Includes diverse tools | | | | | | • | | | • | |
| Accommodates limited access | • | | • | | • | | | | | • |
| Supports production | | | | • | | | | | | |
| Provides uncomplicated updates | | | | | | | | • | | |
| Improves technical performance | | • | | | | | | | | |
| Accommodates security needs | | | | | | | | | • | |
| Utilizes open source | | | | | | | | | • | |
| Accommodates limited connectivity | • | | • | | • | | | | | |
| **Learning Features** | | | | | | | | | | |
| Uses a constructivist framework | | | • | | | • | | | | |
| Promotes authentic learning | | | | | | • | | • | | |

*Table 1. continued*

| **Report/ Study >** | **CEO Forum:** *Key Building Blocks* | **Milken Foundation:** *Seven Dimensions* | **PCAST:** *Report to the President* | **National Governors Association:** *New Models* | **National Governors Association:** *State of E-Learning* | **NCRTEC:** *Learning with Technology* | **CoSN:** *Digital Learning Spaces 2010* | **CoSN:** *Hot Technologies for K-12 Schools* | **Australian Capital Territory:** *Emerging Technologies* | **Education Week:** *Technology Counts 2006* |
|---|---|---|---|---|---|---|---|---|---|---|
| Supports inquiry & analysis | | | • | | | | | | | |
| Uses appropriate content & pedagogy | | | • | | | • | | | | |
| Provides an interactive experience | | | | | | • | | | • | |
| Encourages learning by doing | | | | | | • | | | | |
| Engages learners | | • | | | | • | | | | |
| Supports project design skills | | | | | | • | | | | |
| Promotes programming & authoring | | | | | | • | | | | |
| Supports creativity | | | • | | | | | | • | |
| Allows for user contributions | | | | | | • | | | | |
| Builds community | | • | | | | | | • | | |
| Encourages collaboration | | | • | • | | • | | | • | |
| Increases knowledge and skills | | • | | | • | | | • | | |
| Provides rich content | | | | • | | | | | • | |
| Relates academics to workplace skills | | • | | • | • | | | | | |
| Increases technology skills | | • | • | | • | | | | | |
| Promotes use of challenging tools | | | | | | • | | | | |
| Accommodates diverse learning styles | | | | | | | | • | | |
| Allows adaptability & scalability | | | | | | • | | • | | |
| Can be individualized | | | | | | | • | | | |
| Aligns with standards | | | • | | • | | | | | |
| Integrates into curriculum | • | • | | | • | | | | • | |
| Improves assessment/evaluation | | • | | | | | | • | | • |
| Allows asynchronous, aspatial communication | | | | | | | | • | | |
| Fosters changes to the teaching process | | • | • | • | | | | | | |

those reports that provided the most notable perspectives on criteria for educational technologies that lead to effective teaching and learning, so we expected to see constancy there. The exception is the criterion that an effective educational technology should *increase technology skills*. It appears that this emphasis from the 1990s is no longer of concern. In terms of the technical and administrative features, the summary figure shows that there is no less interest in these issues now as compared to the 1990s. It appears over the years that even as educational technology solutions are provided, managers change their expectations to focus on the next "problem." For example, in the 1990s educators emphasized having an adequate number of computers for students, while in recent years the emphasis is wireless capabilities for classroom sets of laptops.

The criteria most cited within the administrative features are *improving school administrative efficiency* and *improving teacher professional development*. At least half of the reports also cited professional development in their recommendations for better use of technologies. The criteria most cited within the technical features area are *improved connectivity, increased access to computers, and increased wireless capability*. Fundamentally, these are nearly the same: Any evaluation of the effectiveness of an educational technology should consider whether students and teachers will be able to have adequate access to it and access to the resources it provides, such as the Internet. The criteria most cited within the learning features are *encourages collaboration, increases knowledge and skills, relates academics to workplace skills, integrates into the curriculum, improves assessment/evaluation*, and *changes the teaching process*. These are the same concerns for any new developments in education.

## CONCLUSION

Our review of 10 major studies identifies some consistent criteria that have emerged as important capabilities of effective educational technologies. These include improving professional development, encouraging collaboration, and integrating technology into curricula. In addition to these, there are a large number of other recommendations—46 from 10 studies—suggesting that there is widespread hope that technology will improve many aspects of teaching and learning. Perhaps expectations are too high, but the recommendations from these studies provide pointers for developers of future educational technology materials of what users want.

The historical milestones referenced here illustrate the evolutionary changes in recommended approaches to evaluating educational technologies. As our society has changed its expectations for teaching and learning outcomes, the intentional use of and approaches to evaluating educational technologies reflect these shifts in expectations and goals. The current ISTE standards for educational technologies reflect our social and information shift to a more global, rather than national setting for learning. Teacher professional development today focuses on how technology tools can help educators support knowledge, social, and infrastructure goals that will prepare students with 21st century technology skills and abilities. Lawless and Pellegrino (2007) suggest a systematic framework for studying technology integration in schools, teacher adoption of technologies, and long-term impacts of technology integration on teaching and learning. We suggest that a future article about the evaluation of educational technologies examine this subject by using the 2008 ISTE NETS-T standards and current school and district technology plan documents as a framework for categorizing technology evaluation research.

## REFERENCES


Australian Capital Territory Department of Education. (2005). *Emerging technologies: A framework for thinking*. Retrieved Sept. 25, 2006, from http://www.det.act.gov.au/publicat/pdf/emergingtechnologies.pdf

CEO Forum on Education and Technology. (1997, October). *School technology and readiness report: From pillars to progress*. Retrieved January 14, 2009, from http://www.eric.ed.gov/ERICDocs/data/ericdocs2sql/content_storage_01/0000019b/80/15/40/09.pdf

CEO Forum on Education and Technology. (2001, June). *School technology and readiness report: Key building blocks for student achievement in the 21st century.* Retrieved January 14, 2009, from http://www.ceoforum.org/downloads/report4.pdf

CoSN. (2005) *Digital learning spaces 2010.* Retrieved January 14, 2009, from http://cosn.org/resources/emerging_technologies/learningspaces.cfm

CoSN. (2006) *Hot technologies for K-12 schools.* Retrieved January 14, 2009, from http://cosn.org/resources/emerging_technologies/hot.cfm

Education Week. (2006), *Technology counts 2006.* Retrieved January 14, 2009, from http://www.edweek.org/ew/toc/2006/05/04/index.html

International Society for Technology in Education (ISTE). (2008) *National Educational Technology Standards for Teachers (2nd Ed.).* Retrieved January 14, 2009, from http://www.iste.org/nets

Jones, B. F., Valdez, G., Nowakowski, J., & Rasmussen, C. (1995). *Plugging in: Choosing and using educational technology.* Washington, DC: Council for Educational Development and Research; Oak Brook, IL: North Central Regional Educational Laboratory.

Kelly, M. G., & McAnear, A. (Eds.). (2002) International Society for Technology in Education (ISTE): Eugene, OR.

Lawless, K. A., & Pellegrino, J. W. (2007). Professional development in integrating technology into teaching and learning: Knowns, Unknowns, and ways to pursue better questions and answers. *Review of Educational Research, 77*(4), 575–614. doi:10.3102/0034654307309921

Lemke, C., & Coughin, E. (1998, July 10). *Technology in American schools: Seven dimensions for gauging progress*. Retrieved Sept. 25, 2006, from http://www.mff.org.publications

Mendenhall, R. W. (2003). *Technology: Creating new models in higher education*: National Governors Association. Retrieved Jan. 14, 2009, from http://www.nga.org/Files/pdf/HIGHEREDTECH.pdf

National Governors Association. (2001, June 6). *The state of e-learning in the states*. Retrieved January 14, 2009, from http://www.nga.org/Files/pdf/060601ELEARNING.pdf

North Central Regional Technology in Education Consortium. (1997). *Learning with technology profile tool.* Retrieved January 14, 2009, from http://www.ncrtec.org/capacity/profile/profwww.htm

Partnership for 21st Century Skills. (Winter 2008). *News 2.1.* Retrieved January 14, 2009, from http://www.21stcenturyskills.org/newsletters/P21-winter2008/

President's Committee of Advisors on Science and Technology. (1997). *Report to the president on the use of technology to strengthen K-12 education in the United States, March 1997*. Retrieved Sept. 25, 2006, from http://www.ostp.gov/PCAST/k-12ed.html#2.1

Ruberg, L. (April, 2009). Blending globalization and curriculum analysis to study technology-focused teacher training. Paper will be presented at the annual meeting of American Educational Research Association, San Diego.

Spring, J. (2008). Research on Globalization and Education. *Review of Educational Research*, 78.

The Milken Family Foundation. (1997). *A call for a national research agenda.* Retrieved January 14, 2009, from http://www.mff.org/edtech/article.taf?_function=detail&Content_uid1=100

U.S. Department of Education. (2006). *No Child Left Behind Act*. Retrieved Sept. 25, 2006, from http://www.ed.gov/nclb/landing.html

## KEY TERMS AND DEFINITIONS

**Educational Technology:** The use of technology to improve teaching, learning, and the school environment.

**Design Principles:** A descriptor or characteristic that can be used by educators and designers to structure the content and features of an educational technology and its implementation in curriculum.

**Learning Outcome:** A statement that defines the skills or understanding that a learner will be

**Constructivist Framework:** The opinion that learners construct new information from current and past knowledge within an instructional situation.

**Authentic Learning:** Learning in a context that is embedded in real-life perspectives, making the learning more meaningful and more motivating to the learner.

**Learning Environment:** The place and setting where learning occurs, it is not limited to a physical classroom and includes virtual settings in which learning takes place.

**Criteria:** The standards by which an evaluation is made.

# Chapter 25
# Emerging Edtech:
## Expert Perspectives and Design Principles

**Ching-Huei Chen**
*Center for Educational Technologies®, Wheeling Jesuit University, USA*

**Manetta Calinger**
*Center for Educational Technologies®, Wheeling Jesuit University, USA*

**Bruce C. Howard**
*Center for Educational Technologies®, Wheeling Jesuit University, USA*

**Anna Oskorus**
*TiER 1 Performance Solutions, USA*

## ABSTRACT

*Design principles are universal and may be translated onto the newest trends and emergent technologies. In this research study, the authors combined the perspectives provided by two sources to create a set of recommended design principles for technology-enhanced learning environments. One source was the How People Learn framework (Bransford, Brown, & Cocking, 2000). The second source was a series of interviews conducted with pacesetters in the field of educational technologies. With the knowledge gained from these two sources, the authors created our own set of design principles. These principles may be used to guide evaluation, instructional design efforts, or best practice models for exemplary use of educational technologies in the classroom.*

## INTRODUCTION

Hundreds of millions of dollars have been spent in recent years to improve and maintain technology infrastructure for schools. Now policymakers and the public want to know what impact this technology has had on student learning. To answer that question, states and school districts need parameters for evaluating their technology-related activities and using the data to guide their decision making. However, researchers have cautioned against drawing inappropriate cause-and-effect conclusions based on experimental studies (Olson & Wisher, 2002; Russell, 2001). What scientifically based evidence is available on the impact of educational technology often is focused on the degree to which a particular technology leads to changes in learning or teaching (Lawless & Pellegrino, 2007). In fact, a better

DOI: 10.4018/978-1-60566-936-6.ch025

way of judging the impact of new technologies is to examine how they are used and the context in which the use occurs (Schifter, 2008; Zhao, Byers, Pugh, & Sheldon, 2001).

Instructional designers and researchers have stressed the need for robust design principles to guide the production of products and programs (e.g., Kali, 2006; Kali, Spitulnik, & Linn, 2004; Underwood et al., 2005). In this research we also chose to emphasize design principles because they are universal, and they translate to the newest trends and emergent technologies. To do so, we set out to combine the latest research perspectives with the most current leadership perspectives. We began by summarizing key perspectives of the How People Learn (HPL) framework for learning environments. The HPL framework is widely respected and provides recommendations that can be applied to the design of technology-enhanced learning environments. We interviewed pacesetters in educational technologies and reported emerging themes based on the thoughts of those pacesetters. These sources provided the foundation for creating our own set of recommended design principles for technology-enhanced learning environments. This approach combined the best of the past with fresh perspectives from the present.

## IMPORTANT PRINCIPLES ABOUT LEARNING AND TEACHING

The National Academy of Sciences *How People Learn* book synthesized decades of research on how people learn to develop a framework for understanding the connections between cognition and instruction (Bransford, Brown, & Cocking, 2000). This report is widely embraced as a seminal work for educators and researchers alike. In fact, How People Learn (HPL) is becoming widely accepted as a theoretical framework, and that is how we use it here. That work provided the theoretical foundation for designing and conducting the interview study of the pacesetters.

Although the HPL framework provides many important teaching and learning implications, we highlight four of the principles that have particular importance in the design of technology-enhanced learning environments. Each has a solid research base as well as important implications for how teachers teach. Each principle also helps designers think about technology's role in the design and delivery of effective learning environments.

One important principle about the way people learn is that "students come to the classroom with preconceptions about how the world works" (Bransford et al., 2000, p. 14), which include beliefs and prior knowledge acquired through various experiences (e.g., Lin, 2001; Pressley et al., 1992). This learning principle suggests that students start to make sense of the world at a very young age. In many cases students already hold multiple conflicting views before learning new information, as a result, they create their repertoire of views without reflecting on their existing knowledge. This principle implies that designers of effective technology-enhanced learning or instruction should build on students' preconceptions and learning styles, allow decision making, and foster students' multiple intelligences.

Another HPL principle is that "to develop competence in an area of inquiry, students must have a deep foundation of factual knowledge, understand facts and ideas in the context of a conceptual framework, and organize knowledge in ways that facilitate retrieval and application" (Bransford et al., 2000, p. 16). Numerous studies comparing performance by experts and novices have shown that experts not only obtain richly structured knowledge bases that allow them to plan a task, notice patterns, generate reasonable arguments, and draw analogies to other problems, but they also exhibit more organized conceptual frameworks that allow for greater transfer.

This learning principle suggests effective learning environments are knowledge centered and based on developing richly structured information foundations. Instructional designers should center the learning environment on what is taught, why it is taught, and what competence or mastery looks like.

A third principle from the HPL framework is that "a metacognitive approach to instruction can help students take control of their own learning by defining goals and monitoring their progress in achieving them" (Bransford et al., 2000, p. 18). Most educators agree that metacognition is one of the most important cognitive skills among intelligent human activities. Good teachers provide the time, space, and materials necessary to promote metacognitive skills such as self-regulated learning. Reciprocal teaching, for example, is a technique designed to improve students' reading comprehension by helping them explicate, elaborate, and monitor their understanding as they read (Palinscar & Brown, 1984). This learning principle suggests designers must incorporate metacognitive activities into the subject matter that students are learning in order to improve understanding and help students transfer learning to new settings and events.

The fourth principle of the HPL framework is that "learning is influenced in fundamental ways by the context in which it takes place. A community-centered approach requires the development of norms for the classroom and school, as well as connections to the outside world, that support core learning values" (Bransford et al., 2000, p. 25). Learning is a collaborative process in which students actively construct and receive explanations and ideas and negotiate meanings. Such collaboration can help students engage in deeper cognitive processing, such as clarifying thinking, reorganizing information, correcting misconceptions, and developing new understanding. This principle suggests that designers must incorporate a collaborative mechanism in the design process for students to experience cognitive conflicts, hear different perspectives, and ultimately accomplish the learning tasks with the help of their peers or experts.

## METHOD

### Participants

For this study we identified key pacesetters in educational technologies from a multitude of specialties. These pacesetters were also directly or indirectly at the forefront of the newest initiatives. We sought representation from the following categories at the highest level of management or peer reputation: the U.S. Department of Education Educational Technology Office, program officers for federal educational technology initiatives, grant awardees, state government technology directors, educational technology professional organizations, educational technology futurists, gaming and simulation experts, educational technology journal editors (both peer-reviewed and trade journals), and university professors. We wanted a widely representative group to eliminate single-field bias. The pacesetters' diversity added depth to the information and insights we collected from them. Their experiences, from both their personal use of various technologies and the individual circumstances of their experience, informed their perspective.

### Procedure

We began our research by recruiting three educational technology experts as advisors and developing a list of possible interview participants. These advisors have been involved in large educational technology initiatives and represent broad perspectives: national initiatives, curriculum development, teacher professional development, professional organizations, and the academic world. The research team teleconferenced with these advisors multiple times. We sought consensus on the study method and interview items.

We also conducted interviews with three people not on the research team to test the item wording and length of interview. We then made minor revisions to the protocol. The group consensus

*Table 1. Interview questions*

| |
|---|
| *Question 1. What makes an educational technology or its use exemplary?* This question sought the criteria that experts consciously or subconsciously use to rate educational technologies. We phrased the question to show that the choice of technology and the way it is used matter. |
| *Question 2. Based on your criteria from question 1, identify which technologies on the list below you would consider exemplary. Please note that this list is not exhaustive.* Here we wanted to prompt the interviewee to provide perspective on a wide range of technologies. We were concerned that without the list interviewees might not think broadly or that many of them would focus too narrowly on topics that were highly salient at the time, such as videogames or the latest social networking site. By the same token, we also were concerned about skewing their perspective by providing a list. To address this concern, we eschewed a categorization scheme for now and listed the items in random order. |
| *Question 3. Now that you have identified which technologies from the list that you feel are exemplary, please select at least three of those technologies to discuss with us. We would like two types of information about each technology: Identify the features you like about that technology, and give the advantages and disadvantages of using that technology.* We wanted to focus the discussion on the design features that make a technology or its use particularly effective. From this information we hoped to generate a list of *exemplars,* which are discussed elsewhere in this special issue. |
| *Question 4. In your opinion, which technologies show the most future promise for educational use?* We wanted to focus on the idea of "promise" here. Perhaps there are existing technologies that are underutilized or could be ported over to education. Perhaps there are some that have had an impact in other areas, such as business, which should be considered for educational uses. Perhaps some are being used in extraordinary ways somewhere but have received little publicity. |

and initial interview process were themselves informative to our overall research study. Table 1 shows the final four questions with a rationale for each.

Recruitment consisted of calling or e-mailing participants to request interviews. When a candidate agreed to participate, we sent a follow-up e-mail to confirm participation, scheduled an interview, and mailed or e-mailed a cover letter, consent form, and the interview protocol. We interviewed 18 pacesetters; however, two declined to consent. Therefore, this paper features the perspectives and insights of 16 pacesetters. We conducted the interviews by phone; the calls typically lasted about 40 minutes. One team member took notes, while another guided the interview and asked the questions. Interviewees also had opportunities to comment about topics or technologies not included in the interview protocol. For each interview the notes were typed into summary format, and both team members revised them until it was agreed that the summary fairly represented the conversation. In many cases the interviewee either e-mailed or faxed follow-up details.

## Data Analyses

The interviews consisted of four questions. In the present article we report on the findings for questions 1 and 4. The results for questions 2 and 3 are described elsewhere in this special issue. From the interview data we conducted a content analysis to identify thematic patterns for each question. We first did this vertically within individual cases before moving across cases, using the constant comparative technique outlined by Miles and Huberman (1994). Then we refined analytic categories and tested for reliability between coders and used the categories to examine data from all interviews.

## RESULTS

Pacesetters generally provided detailed, specific comments about what makes an educational technology or its use exemplary. They typically reiterated their initial comments from various angles throughout the interview. Here we present the seven major themes that arose from the interviews along with pertinent comments from pacesetters.

## Technology Theme 1: Create a Learner-Centered Environment

Technology itself is not an answer in education. The technology must have a purpose, such as to enhance teaching or learning or to promote problem solving. Pacesetters considered a technology exemplary if it resulted in substantial gains for the learner. To that end, technology that emphasizes a learner-centered environment and allows learner control has a significant potential to result in more substantial learner gains, making the technology exemplary. Pacesetters also recommended that the focus be on the learning, not the technology. Furthermore, pacesetters said technology that employs experiential or situated learning has significant impact on increasing student learning, motivation, and retention of content. Additionally, they considered technology that supports individualized learning and fits students' needs as highly effective. Here are some comments from pacesetters:

*To get at what helps people learn, use four broad areas: Who is the learner, how do they learn, what are their intellectual capabilities, and what are their strengths and weaknesses? If we're going to envision a future of technology and learning, we should use this framework for thinking.*

*Technology is exemplary if you can tie it to student achievement. There are a lot of tools thrown out into the classroom, but if you can't tie it to student objectives, it's meaningless. A tool without a task has no context using technology. Until we can help schools with methodology, it is ineffective.*

*Good educational technologies engage learners (scaffold learners) toward meaningful problem solving. Exemplary technology use is a function of the alignment between the specific problem-solving/reasoning and the technology tool used to assist that reasoning. Exemplary lessons with technology would be those which allow students to make discoveries for themselves, such as a scientist would.*

*When you talk about exemplary learning, you aren't really talking about the technology at all... It's when the technologies are applied to certain circumstances that makes all the difference. For instance, in some cases the laptop is effective; in some cases it's not. It depends on many factors, such as the school districts and principals. We should be looking at what is going on in the classroom—get away from the discussion on technology, and focus on teaching and learning.*

*Experimental educational technologies are the new paradigm in learning, as they are interactive. The outdated 20th century paradigm involves passive technologies where you click on something and read information.*

## Technology Theme 2: Support Engagement, Interactivity, and Motivation

Educational technology must be engaging, interactive, and motivating. Pacesetters said engagement was extremely important for getting students interested in further learning. Thanks to videogames and communication tools, technology already is prevalent in many students' lives, and can be highly motivating and conducive to student learning. Likewise, interactive technology that allows students to exchange work, ideas, and data and that also allows for multiple platforms provides exemplary educational experiences, pacesetters said. Students who use technology to see and analyze real-time data are much more engaged and interested in what they are learning. Here's what pacesetters said:

*Education technology moves away from text-based to engaging game and simulation education.*

*I think technologies that allow kids to experiment in a dynamic environment are exemplary. Kids can "push" in one place and "see what happens" in another place. It's not static. Google Maps are an exemplary technology because they are not static and you can experiment.*

*Technology provides firsthand experience rather than reading (the material) from a textbook, and therefore becomes a more memorable and exciting learning experience. Students are more likely to remember and retain this information. When students see real-time information and real events happening, they are much more engaged and interested in what they are learning about. More learning takes place, students remember more, and dig deeper. Students are actually learning more in the situated learning perspective.*

## Technology Theme 3: Promote Ease of Use and Cost Effectiveness

Pacesetters said exemplary technology is easy to use and versatile. Several reasoned that a technology that works consistently and is simple to implement would naturally reach a larger audience. Cost is also a consideration. Cost includes production, training, use, maintenance, and the need for updates. Several pacesetters said technology must be "lower end" enough to work in the school and within the school's resources. One pacesetter said the iPod was an exemplary educational technology:

*The iPod is ...a technology that has gained success because of pervasive use and ... it could be easily adapted to the classroom...Ease of use is an important feature for exemplary technology because it can make a big impact when it is used on a widespread basis.*

## Technology Theme 4: Provide Assessment

Exemplary technology features automated ongoing assessment and formative feedback embedded in the technology for quicker and more sophisticated reinforcement. Learners should be able to receive feedback and assessment in all contexts or lessons, pacesetters said. An exemplary technology works for teachers too, the pacesetters noted. If teachers can integrate a technology effectively into their teaching—even if that means a realistic amount of professional development—the technology ultimately has a better chance of impacting student achievement. Add to that teachers' ability to combine a technology with another in their teaching, and you've got an exemplary technology, pacesetters said. Here's what one pacesetter—a technology expert representing a state department of education—said:

*...feedback needs to be quicker and can be quicker via use of technology and to also get more sophisticated feedback. If we're going to envision a future of technology and learning, we should use this framework for thinking. Add up everything, but if it doesn't help someone learn, even if it's great, it doesn't really make someone better or help them learn. So we need quick assessment and feedback.*

## Technology Theme 5: Support Social and Community Building

An educational technology that incorporates a high degree of social and community-building opportunities is effective, pacesetters said. This type of tool lets students access current data and information and build a social network around their learning task. The pacesetters predict technologies that effectively use communication and collaboration will have a significant impact on education. Communication technologies that enable students to access information and resources from global intelligence pools will quickly revolutionize how students work through academic courses, obtain information, and collect their own data. Students will communicate with mentors and peers in other countries, unconstrained by geography. Social networking technologies will provide students with invaluable expertise and support as they take advantage of greatly increased options for attending classes and as they make better use of distance human interactions and artificial intelligence. Here are some insights from pacesetters:

*The vast number of interactions made possible through technology, including application sharing, tools for chat rooms, bringing people together for sharing models, and roles, and addressing the different aspects of community coaching are exemplary features of educational technologies.*

*The social and community building aspect is an overall advantage of technology. Social network is crucial to learning. Technology that connects people is what's needed to get learning communities going.*

## Technology Theme 6: Promote Scalability, Utility, Dissemination, and Portability

Pacesetters said scalability, utility, dissemination, and portability are important criteria for exemplary educational technology. A technology should be able to handle a large number of students and reach a large audience, but it should also adapt easily to other contexts and across different platforms. Future educational technology must be portable too, pacesetter said, so students can take a tool with them outside of school to learn anywhere, any time. Here are some other pacesetter comments:

*Some technologies have one primary use, and the user can't get beyond that one use. An example of this is using a graphing calculator to graph a mathematical equation. This technology doesn't apply to other contexts; if there are a dozen ways in which the technology can be used, there will be more educational opportunities.*

*The content delivery platform will always change; there will always be new software and new hardware. The educational assets should not go obsolete when you go from a tablet to a handheld. Ideally, what you have in your educational initiative is a set of learning experiences that cuts across different technologies.*

## Technology Theme 7: Foster Knowledge Construction/Integration

Embedding components of knowledge construction within a technology has potential for helping students develop a deeper conceptual understanding of the material they're learning. Pacesetters said knowledge construction tools could promote immersive environments in which students not only create their own environments or genres, but share them with other students. Here are some comments from pacesetters:

*Argumentation tools and virtual reality technologies are potentially effective approaches to developing a deeper conceptual understanding.*

*Technologies that have potential as concept builders and problem-solving scaffolders would be future exemplary technologies. Immersive environments have this potential as well as knowledge construction programs such as STELLA or Wikis.*

*The technology is enabling multiple paths to exploring information. Computer offers you interesting ways to do research- structure information.*

*StorySpace is a hypermedia construction program. This is a non-linear and promotes student construction of their understanding. If the technology is non-linear, students can make more links to the concepts involved.*

## DESIGN PRINCIPLES FOR EXEMPLARY EDUCATIONAL TECHNOLOGIES

In this section, we describe six design principles that resulted from clustering learning principles from HPL with the technology themes from pacesetters. In Table 2 we illustrate the connection between HPL, pacesetters, and design principles. Examples for each design principle are also included.

### The Technology-Based Instruction Should Promote Learner-Centered Experiences

According to Callahan and Switzer (2001), technology plays an important role in facilitating quality education. They also wrote that the very first dimension of facilitating a quality education is that students should be at the center of their own learning. Pacesetters agreed in theme 1. Thus, designers should pay close to attention to the knowledge, skills, and attitudes that students bring to their understanding of knowledge. How much the technology impacts students and their learning depends on the instructional strategies that accompany its use (Russell, 2001; Summerville & Reid-Griffin, 2008). To create a learner-centered environment, designers must integrate effective instructional strategies and recognize students' prior knowledge, skills, and attitudes and a variety of approaches to their own learning (Trinidad, 2003).

### The Technology-Based Instruction Should Scaffold Knowledge Construction

According to the theory of HPL, learning occurs when a learner actively builds meaningful cognitive representations and mental models. An educational technology should help students make sense of new materials. For example, a technology-supported argumentation tool engages students in constructing scientific knowledge claims, evaluating the claims constructed, and establishing the objectivity of scientific explanation (Duschl, Ellenboger, & Erduran, 1999).

### The Technology-Based Instruction Should Support Problem-Solving and Metacognitive Skills

Technology can provide numerous representational tools that expand and strengthen human

*Table 2. Illustrative connections between HPL, pacesetters, and 21st century educational technology design principles*

| Important Principles from HPL | Pacesetter Perspectives | 21st Century Educational Technology | |
|---|---|---|---|
| | | Design Principles | Examples of Learning Environment Features |
| Build on student preconceptions and learning styles: Students come to the classroom with preconceptions about how the world works. | Emphasize a learner-centered environment: Technology should allow students to investigate and make discoveries for themselves. | The technology-based instruction should promote learner-centered experiences. | Support learners' preconceptions, prior knowledge, beliefs, and experiences.<br>Offer learners choices.<br>Foster learners' multiple abilities.<br>Reinforce multiple distributed intelligences.<br>Support both visual and verbal learners.<br>Encourage learners to investigate personally relevant problems. |
| Support development of richly structured knowledge foundations: Students must have a deep foundation of factual knowledge, understand facts and ideas in the context of a conceptual framework, and organize knowledge in ways that facilitate retrieval and application. | Foster knowledge construction/ integration: Technology should provide multiple paths for students to explore information. | The technology-based instruction should scaffold knowledge construction. | Enhance organized content knowledge.<br>Provide knowledge construction or organizational tools.<br>Create a cognitive conflict that contradicts learners' intuition.<br>Encourage learners to reconsider their repertoire of ideas.<br>Enable manipulation of factors in models and simulation. |
| Provide metacognitive activities: Students should take control of their own learning by defining goals and monitoring their progress in achieving them. | Support engagement, interactivity, and motivation: Technology should use real-time data and live events to encourage students to be more engaged and interested in what they are learning. | The technology-based instruction should support problem-solving and metacognitive skills. | Promote self-regulated learning.<br>Help learners engage in expert-like thinking processes.<br>Provide ample times for decision making.<br>Provide learners with opportunities to practice real-world situations.<br>Present options for learning skills in a virtual world for application to real-world situations.<br>Allow the transfer of pragmatic learning changes to real-world environments.<br>Support learners in articulating complex scientific ideas. |

cognitive capacities such as problem solving. The problem-based learning framework uses a real-world problem to create a cognitive connection to complex, real-world problem solving. To that end, instructional designers should incorporate meaningful problems with appropriate goals central to subject matter concepts.

## The Technology-Based Instruction Should Support Assessment

Technology that facilitates cycles of ongoing assessment is considered exemplary. An online threaded discussion forum, for example, lets students post their work where they can get feedback from a wider range of critics, reflect

*Table 2. continued*

| Important Principles from HPL | Pacesetter Perspectives | 21st Century Educational Technology | |
|---|---|---|---|
| | | Design Principles | Examples of Learning Environment Features |
| Encourage cognitive conflicts and hear different perspectives: Learning is influenced in fundamental ways by the context in which it takes place. | Support social and community building: Technology should allow students to build a social network around their learning task. | The technology-based instruction should provide a community of learning. | Support social/community building.<br>Facilitate communication and collaboration among learners or with experts.<br>Allow multiple platforms of communication.<br>Employ multiple social activity structures.<br>Enable multiple ways to participate in online discussions.<br>Provide opportunities for learners to serve as instructors of their peers.<br>Scaffold the development of classroom norms. |
| n/a | Promote scalability, utility, dissemination, and portability: Technology should be able to handle a large number of students, reach a large audience, and adapt easily to other contexts and platforms.<br>Promote ease of use and cost effectiveness: Technology should be ease to use and affordable for the general public. | The technology-based instruction should support versatility. | Support utility.<br>Allow dissemination.<br>Allow learners to learn everywhere.<br>Incorporate features for ease of use.<br>Reach a large audience.<br>Provide multiple opportunities for use.<br>Enable virtual navigation for exploring complex physical systems. |
| n/a | Provide assessment: Technology should offer ongoing feedback and automated assessment to enhance student achievement. | The technology-based instruction should support assessment. | Support assessment.<br>Provide feedback.<br>Provide subsequent review.<br>Enable just-in-time evaluation.<br>Remedy weaknesses or skill deficits.<br>Enable learners to evaluate the work of their peers. |

on what they've learned, and compare their work with peers' products. Such technology supports ongoing assessment, has the potential to promote meaningful learning, and contributes significantly to teaching and learning for understanding (Wiske, 2006).

## The Technology-Based Instruction Should Provide a Community of Learning

Communication and collaboration are two important components that an exemplary educational technology offers students. For instance, students use digital media and environments to communicate and work collaboratively, including from a distance, to support their own learning and contribute to the learning of others. Such processes also encourage students to develop cultural understanding and global awareness when they communicate and work with students from other cultures.

### The Technology-Based Instruction Should Support Versatility

Although HPL does not explicitly address versatility, several pacesetters said educational technologies should be widespread, easy to use, and portable.

## DISCUSSION

In conducting this study, we were faced with the familiar question of how to define educational technology and the relative importance of a technology versus its use. The pacesetters provided us valuable perspectives. In future research we must grapple with how educational technology research is actually used for decision making. One end of the spectrum proposes it is really not about the technology, it is solely about the context. Studies from this perspective are fundamentally so context bound that you can't generalize their findings. The other end of the spectrum portrays a tidy world in which an input equals a measurable output and how dollars are spent can predict with certainty the outcomes achieved. Although the *No Child Left Behind*-like need for accountability prefers the latter, reality may be closer to the former.

The interviews yielded many intriguing themes, the most often cited being the importance of engagement and interactivity in a technology. Implementation variables such as cost, ease of use, and teacher receptivity and adoption were also important. Pacesetters gave input on how exemplary technologies enhance teaching and learning, with problem solving and collaboration being notable. Moreover, community networking and social networking technologies were most often cited for future education impact. As some sociologists accuse technology of isolating students and decreasing social skills, our findings make it increasingly apparent that students (and teachers, mentors, and educational technology leaders) recognize that technology has the power to bring people together and that collaboration via technology can greatly enhance learning.

## CONCLUSION

Wager (1992) claimed that "the educational technology that can make the biggest difference to school and students is not the hardware, but the process of designing effective instruction" (p. 454). Our pacesetters delivered a similar message: It is not technology *per se* that improved student outcomes, but rather how the technology was used and integrated into instructional processes. The pacesetter interviews support the importance of the classroom context and the role of teachers rather than the technology itself.

In this study we searched for answers as to what constitutes good practice in the use of educational technology. Continued research will refine those answers. We recommend conducting ongoing dialog among the experts who see students and teachers on a day-to-day basis. We think their unique perspectives would further the knowledge base we've laid out. Future studies could also substitute an online survey for the live interviews. Regardless of the approach, we think the study of best practices in educational technology must be ongoing. After all, in the time it took you to read this chapter, a new technology has likely been developed or used in a different way by an educator.

## REFERENCES

Bransford, J. D., Brown, A. L., & Cocking, R. R. (2000). *How people learn*. Washington, DC: National Academy Press.

Callahan, W. P., & Switzer, T. J. (2001). Technology as facilitator of quality education: A model. In W. Heineke & L. Blasi (Eds.), *Methods of evaluating educational technologies* (pp. 215-235). Greenwich, CT: Information Age Publishing.

Kali, Y. (2006). Collaborative knowledge-building using a design principles database. *International Journal of Computer-Supported Collaborative Learning*, *1*(2), 187–201. doi:10.1007/s11412-006-8993-x

Kali, Y., Spitulnik, M., & Linn, M. (2004). Building community using the design principles database. In P. Gerjets, P. A. Kirschner, J. Elen & R. Joiner (Eds.), *Instructional design for effective and enjoyable computer-supported learning* (pp. 294-305). Tuebingen: Knowledge Media Research Center.

Lawless, K. A., & Pellegrino, J. W. (2007). Professional Development in Integrating Technology into Teaching and Learning: Knowns, Unknowns, and Ways to Pursue better Questions and Answers. *Review of Educational Research*, *77*(4), 575–614. doi:10.3102/0034654307309921

Lin, X. (2001). Designing metacognitive activities. *Educational Technology Research and Development*, *49*(2), 23–40. doi:10.1007/BF02504926

Miles, M. B., & Huberman, A. M. (1994). *Qualitative data analysis* (2nd ed.). Thousand Oaks, CA: Sage.

Olson, T., & Wisher, R. A. (2002). The effectiveness of Web-based instruction: An initial inquiry. *International Review of Research in Open and Distance Learning*, *3*(2), 1–17.

Palinscar, A. S., & Brown, A. (1984). Reciprocal teaching of comprehension-fostering and comprehension-monitoring activities. *Cognition and Instruction*, *1*(2), 117–175. doi:10.1207/s1532690xci0102_1

Pressley, M., Wood, E., Woloshyn, V. E., Martin, V., King, A., & Menke, D. (1992). Encouraging mindful use of prior knowledge: Attempting to construct explanatory answers facilitates learning. *Educational Psychologist*, *27*(1), 91–109. doi:10.1207/s15326985ep2701_7

Russell, M. (2001). Framing technology program evaluations. In W. Heineke & L. Blasi (Eds.), *Methods of evaluating educational technology* (pp. 149-162). Greenwich, CT: Information Age Publishing.

Schifter, C. C. (2008). *Infusing Technology into the classroom: Continuous practice improvement.* Hershey, PA: IGI Global.

Summerville, J., & Reid-Griffin, A. (2008). Technology Integration and Instructional Design. *TechTrends*, *52*(5), 45–51. doi:10.1007/s11528-008-0196-z

Trinidad, S. (2003). Working with technology-riched learning environments: Strategies for success. In M. S. Khine & D. Fisher (Eds.), *Technology-rich learning environments: A future perspective*. Hackensack, NJ: World Scientific.

Underwood, J. S., Hoadley, C., Lee, H. S., Holcbrands, K., DiGiano, C., & Renninger, K. A. (2005). IDEA: Identifying design principles in educational applets. *Educational Technology Research and Development*, *53*(2), 99–112. doi:10.1007/BF02504868

Wager, W. (1992). Educational technology: A broader vision. *Education and Urban Society*, *24*(4), 454–465. doi:10.1177/0013124592024004003

Wiske, M. S. (2006). Teaching for meaningful learning with new technologies. In E. A. Ashburn & R. E. Floden (Eds.), *Meaningful learning using technology* (pp. 26-44). New York: Teachers College Press.

Zhao, Y., Byers, J., Pugh, K., & Sheldon, S. (2001). What is worth looking for? Issues in educational technology research. In W. Heineke & L. Blasi (Eds.), *Methods of evaluating educational technology* (pp. 163-179). Greenwich, CT: Information Age Publishing.

## KEY TERMS AND DEFINITIONS

**Design Principle:** A descriptor or characteristic which can be used by educators and designers to structure the content and features of an education technology and its implementation into curriculum.

**Pacesetter:** A person who is a leading influence in his or her field of study or work.

**Educational Technology:** The use of technology to improve teaching, learning and the school environment.

**Learning Environment:** Place or setting where learning occurs; not limitied to a physical classroom and includes virtual settings in which learning takes place.

**Instructional Design:** A systematic approach to the design and development of instructional materials and products using objectives, teaching strategies and evaluation to meet learning needs.

**Metacognition:** The study of thinking and learning.

**Criterion:** The standard by which an evaluation is made

**How People Learn:** A book by the National Academies Press which describes a framework for understanding the learning sciences.

# Chapter 26
# The Best Edtech of 2007: Promising Features and Design Models

**Bruce C. Howard**
*Center for Educational Technologies®, Wheeling Jesuit University, USA*

## ABSTRACT

*As part of a larger project for the NASA-sponsored Classroom of the Future to benchmark the effectiveness of educational technologies, researchers used multiple data sources to develop a list of exemplars and delineate common design features. The exemplars included promising educational technologies, tools, websites, resources, software, and hardware. Each exemplar was placed into one of six categories: knowledge and comprehension tools, interactive technologies and problem-solving tools, product-creation tools, efficiency and productivity tools, communication and collaboration tools, and technology tutors. The features of each exemplar were described, and a set of common design principles for that category was developed.*

## INTRODUCTION

Emerging technologies present enormous potential for improved teaching and learning as schools and universities worldwide embrace and use the new media. The growth of multimedia and digital information and communication technologies has revolutionized the opportunity to learn. We are on the cusp of a new era that will see the ability for information, audio, and video to be accessed nearly anytime and anywhere in the world. Users expect their media to be cross-platform and available for consumption on a variety of devices. Social networks, especially among young people, are expansive, immediate, and important. Information is consumed in nuggets, much like fast food, and shared virally—sometimes reaching millions of consumers within minutes of release.

As educators we must look forward—testing promising developments—while holding on to the established best practices of the past. We must integrate the new and the old using a thoughtful, principle-driven approach. The continuum of opinion on how to do so is broad. At one end is the desire

DOI: 10.4018/978-1-60566-936-6.ch026

of the early adopter to "try it out." The advantage to this approach is that the novelty alone increases student motivation (Clark & Sugrue, 1991). The disadvantage is the inefficient use of class time during the trial and error process and the need to maintain motivation through novelty. At the other end of the continuum is the desire, sometimes imposed by the administration, to use only research-based tools, techniques, and approaches. The advantage here is in using proven methodologies. The disadvantage is that such methodologies might be disconnected by more than a few years from the most current educational environments and constraints.

In our approach to this study, we accommodated both ends of the continuum. With a futuristic perspective we developed a list of exemplary educational technologies from 2007 that had promise for improved teaching and learning in the coming five to ten years. With an eye toward integrating existing best practice, we derived the common *design principles* from these exemplars.

## METHOD

### Data Sources

We derived the items on the list of exemplars from three sources: comments from interviews of educational technology leaders, articles from trade journals, and expertise derived in house. The interview process with the educational technology leaders is described elsewhere in this special issue. These leaders, or *pacesetters*, were chosen from a multitude of regions in the United States and represented the forefront of the newest national or regional initiatives in educational technology. Pacesetters included staff from the U.S. Department of Education, program officers for federal and state educational technology initiatives, grant awardees, professional organizations, futurists, gaming and simulation experts, journal editors, and university professors. As research subjects, their identities are not disclosed here. The trade journals were from the United States and included *Campus Technology Magazine, Converge, Education World®, Edutopia, eLearn Magazine, eSchool News, Innovate, T.H.E. (Technological Horizons in Education),* and *Technology and Learning*. Our in-house expertise from the NASA-sponsored Classroom of the Future™ included a team of educators, researchers, instructional designers, programmers, multimedia producers, technology specialists and subject matter experts.

### Procedure

As in other studies in this special issue, we defined educational technology as *a device or system that makes use of digital media to enhance the teaching and learning process*. Here is the procedure our team used, with a detailed description following:

1. Draft an initial list of exemplary technologies.
2. Elicit comments from pacesetters on the list. Revise list.
3. Examine trade journals from 2007. Revise list.
4. Finalize the list.
5. Categorize exemplars into six categories.
6. Conduct background research on each exemplar.
7. Examine features and derive design principles for each category.

#### Initial List of Exemplary Technologies

Our first step was to use the combined expertise of more than a dozen Classroom of the Future™ team members to generate an initial list of wide-ranging educational technologies. The list was to be used in the pacesetter interviews. Members were asked to identify assorted technologies that they have seen, experienced, or heard about that they considered powerful educational tools or have potential to be effective in education. We generally stayed away from hardware, unless it was specifically for use in education such as probeware or electronic white boards. We also stayed away from technologies used for those with disabilities. This was a topic outside the scope of the project. Items were added to and deleted from the list until group consensus was formed. The list focused primarily on types of technologies, but we used specific instances when the product was well known. For instance, simulations and games were on the list without mentioning any specific titles, but PowerPoint® and Inspiration® were listed specifically.

The purpose of the initial list was to provide interviewees a catalyst for discussion on a variety of exemplars. We were concerned that without the list, interviewees might not think broadly or that many of them would focus too narrowly on topics that were highly salient at the time, such as videogames or the latest social networking site. By the same token, we were concerned about skewing their perspective by providing a list. To accommodate this concern, we eschewed a categorization

*Table 1. List of exemplary educational technologies, Version 1*

| | |
|---|---|
| PowerPoint® | Tapped In® (web-based learning environment for teacher professional development) |
| Discussion boards | Simulations |
| Instant messaging | Planners (Outlook Express®, Project®) |
| Video games | Virtual reality |
| HyperStudio® | Online games, console games, CD-ROM games |
| Authoring tools (Dreamweaver®, FrontPage®, Director®) | Interactive theaters |
| SMART Board™ (interactive electronic whiteboard) | World Wind, Google™ Earth |
| Kid Pix® | Modeling, visualization tools |
| Digital video editors (Premiere®) | Online instruments, labs |
| Photo editing software (Photoshop®) | Emerging multimedia technologies |
| Learning objects/Java Applets (used to provide interactive features to web applications that cannot be provided by HTML) | Video, audio streaming |
| Argumentation tools (COTF DiSC tool) | Podcasting |
| Journaling software (Greymatter, LifeJournal™) | Screencasting (a digital recording of computer screen output, often containing audio narration) |
| Blogs | Narrowcasting (display of content on a digital signage network for a narrow audience, not the general public) |
| Inspiration®, Kidspiration® | Internet protocol television |
| Videoconferencing | WebCT®, Blackboard® |
| Webcasts | Virtual schools |
| Online chats | Cell phones used in education |
| Social networking, collaborative technologies (including MySpace.com®) | Palm® Treo™ |
| | Bluetooth® |
| | Handheld computers, PDAs |
| | Tablet PCs |
| | Closed captioning and compliancy technologies |
| | Other____________ |

scheme for now and listed the items in random order. Table 1 shows our initial list.

## Interviews with Pacesetters

As described in detail elsewhere in this special issue, we conducted an interview study of pacesetters in the field of educational technologies. Questions two and three of the four interview questions were aimed at identifying exemplars and their features:

- **Question 2.** Based on your criteria from question 1, identify which technologies on the list below you would consider exemplary. Please note that this list is not exhaustive.
- **Question 3.** Now that you have identified which technologies from the list that you feel are exemplary, please select at least three of those technologies to discuss with us. We would like two types of information about each technology: Identify the features you like about that technology, and give the advantages and disadvantages of using that technology.

## Examination of Trade Journals

Trade journals offer a wealth of information and resources to track developments in the field of educational technology. Their contributors scan widely to find stories of innovative and interesting uses of technology and emerging trends. Their editors sort through lists of stories and separate the mediocre from the noteworthy. Although their perspective may perhaps be biased by the need to increase readership and generate ad revenue, trade journals provide a professional and commercial perspective, adding more data points in the desire to triangulate our conclusions. To eliminate some of this bias, we focused on articles that provided case studies or reviews of promising educational technologies, such as those offering recommendations, lists of "top picks," editors' choices, or the like.

## Finalizing the List

With the interviews and the review of the trade journals completed, our team conducted a holistic evaluation to finalize the list of exemplars. We did this twice, once near the conclusion of the study and again about eight months later just before submitting the research study for publication, using additional researchers. The second round of holistic evaluation was a double-checking procedure. It also allowed us to see if any new exemplars had emerged recently. Those technologies that generally met the following criteria were included on the list of exemplars:

- **Inspiring, engaging, educational:** Could be used in curricular materials to inspire, engage, or educate; promotes authentic learning; increases knowledge or skills; well-designed and implemented.
- **Likely to be used:** Would and could easily be adopted by educators; affordable; flexible; versatile; easy to use; easy to deliver and maintain.
- **Popularity:** Has been widely embraced; bugs have been worked out.
- **Promise/ Innovation:** Could be used in education for high impact or educational reform.

With each step in our procedure our research team revised the previous list by considering the recommendation of the source and either adding the exemplar to the list or occasionally removing one. For example, after the interviews, we created a second version of the list of exemplars by integrating the pacesetters' responses into the first list. The responses they gave highlighting features and advantages/ disadvantages were used later in deriving design principles. For example, one pacesetter mentioned several "mind tools" one of which was called STELLA®. He stated, "This exemplary software of the future will let people create things more fluidly, and it turns drawings into quick models placed geographically in a virtual world." We added this to the list of exemplars, and noted its main feature as an easy-to-use visualization tool. As a contrary example, after examining the trade journals and reviewing the pacesetter interviews we could find few instances of the use of Tablet PCs as a promising educational tool, so we removed it from the list.

## Categorizing the Exemplars

We sorted the list of technologies into meaningful categories for ease of understanding and discussion. We used six categories, loosely based on Russell (2001). In this scheme Russell uses three categories for technologies, according to the role they play: (a) defining aspects of the curriculum, (b) providing instructional tools, and (c) supporting productivity and communications. In the first the technologies are used as part of a curriculum. For example, in the 1990s this was exemplified by the push to include computer-skills teaching on top of the existing subjects, such as teaching word processing, spreadsheet use, or the conduct of an Internet search. Today, teachers are encouraged to think of the computer as a tool that should act invisibly in all curriculum areas, much like the pocket calculator is commonly used in a mathematics classroom. In Russell's second category are those technology tools that assist with the teaching and learning process. These include tools that help students and teachers acquire and organize information and practice skills. It also includes those that help them to create products through which they may acquire or demonstrate knowledge and skill. In the third category are those technology tools that are used by teachers and students to support classroom productivity and student-to-student or teacher-to-student communication.

Our categorization scheme is similar, but uses more categories:

1. **Knowledge and Comprehension Tools:** These technologies help learners to acquire and organize content and concepts. Roughly corresponds to the low end of Bloom's Taxonomy (Bloom, Englehart, Furst, Hill, & Krathwohl, 1956).
2. **Interactive Technologies and Problem-solving Tools:** Using these technologies, learners make use of information to apply it to a situation, make decisions, analyze, or synthesize to solve problems. Requires a response on the part of the learner. Roughly corresponds to the high end of Bloom's Taxonomy.
3. **Product-creation Tools:** Students use these technologies to develop artifacts that represent their learning.
4. **Efficiency and Productivity Tools:** Technologies in this category assist in increasing the efficiency of the teaching and learning process, helping learners and educators to be organized and productive.
5. **Communication and Collaboration Tools:** These technologies support an educational, industrious exchange of information among people.
6. **Technology Tutors:** These technologies provide scaffolds for learning skills related to technology use, such as typing, using spreadsheets, or word processing. These are the types of skills addressed by the National Educational Technology Standards.

## Features and Design Principles

Each exemplar was examined for its features, and these features were aligned with design principles. In the results section we list each exemplar and its noteworthy features, sorted by technology category. Looking across all the exemplars within that technology category, we were able to derive thematic design principles. In moving from the results of the study to the implications of the study, deriving the design principles is an important step. Design principles provide a more universal and generalizable approach to choosing and using educational technologies. As the technologies change, the design principles provide a means for developing or using instructional units that make the most effective use of the technologies.

## RESULTS

We present the results of this research in two lists. The List of Exemplars provides a description of each exemplar, sorted into six categories (a) Knowledge and Comprehension Tools, (B) Interactive Technologies and Problem-Solving Tools, (C) Product-Creation Tools, (D) Efficiency and Productivity Tools, (E) Communication and Collaboration Tools, and (F) Technology Tutors. Quotes about the advantages and disadvantages from the pacesetters are included. The second list of Features and Design Principles highlights the key features of each exemplar with a summary of the relevant design principles.

### List of Exemplars

### 1. Knowledge and Comprehension Tools

These technologies help learners to acquire and organize content and concepts. They roughly correspond to the low end of Bloom's Taxonomy.

- **Instructables®:** (www.instructables.com) With Instructables, users create and share instructions for making almost anything from food to fire to paper. Popular example of user-generated content with educational value. Other users comment and add their advice. Also provides categories, search abilities, and communities.
- **Ology:** (www.amnh.org/ology/) This American Museum of Natural History site allows children to explore, contact scientists to ask questions, and play games. With an engaging interface, it is ideal for an after-school program and may be customized for topic and grade level.
- **Lucas Learning Programs®:** (www.lucaslearning.com) *Star Wars* creator George Lucas has lent his name to this collection of visually stunning software products, games, and K-12 classroom instructional materials. Topics include math problems, energy, forces, simple machine, light, and magnetism.
- **Science Toys You Can Make with Your Kids:** (www.kk.org/cooltools/archives/000156.php) Parents and educators make learning-rich toys and gadgets alongside their kids. The components of each device are inexpensive and can be assembled quickly. The website instructions and visuals are simple, which facilitates use and understanding.
- **NOAA's Ocean Challenge Puzzle:** (www8.nos.noaa.gov/oequizx/) Features a real-time, multiplayer, online game that teaches facts related to ocean history. As players collaborate from sites around the world, their proficiency reveals a secret marine image.
- **KidBiz3000®:** (www.kidbiz3000.com) This is a highly individualized reading and writing program for students in grades 2-8. The program is sensitive to a student's reading level and customizes accordingly. Software delivers assignments to the entire class but tailors the lesson to student ability. Student growth is assessed, monitored, and reported. Students also receive feedback by e-mail.

- **NutshellMath®:** (www.nutshellmath.com) This website talks students through math problems with animated, audio lessons. There are separate sections for parents, teachers and students. Nutshell covers math from pre-algebra to algebra II, giving parents checkups and reports along the way.
- **LeapFrog®:** (www.leapfrog.com) The LeapFrog® suite of engaging games, materials and products is immensely popular and assists parents and teachers in helping children grow to love learning.
- **CXOnline.Bridges.com—Choices Explorer®:** (www.bridges.com/us/) With this program K-16 students can create career-oriented portfolios online. It provides resources for students, teachers, and guidance counselors. Contains more than 900 articles on different jobs and fields, more than 200 videos, interviews, career information by school subject, a student blog site, and materials for college test preparation.
- **SAS inSchool: Curriculum Pathways®:** (www.sasinschool.com) An educational resource that targets students in grades 8-12. Interactive virtual classrooms customize material for all subject disciplines and provide innovative integration of technology and curriculum. Provides learner-centered activities with measurable assessments mapped to state and national standards and an emphasis on higher-order thinking skills.
- **United Streaming™:** (streaming.discoveryeducation.com) Holds the largest and most current K-12 video digital library available today. Videos are standards based and have been shown to increase student achievement. Additional features are assignment and quiz builders, sites for teacher postings that students can access, and a writing prompt generator to help students create papers on site material.
- **Safari Montage®:** (www.safarimontage.com) This is a resource for educational videos from leading educational video publishers. The database contains videos broken down into segments for smaller topic use and also offers lesson plans and quizzes. This device is provided to schools and public libraries with digital access. Users can customize segments from existing videos and create playlists.
- **First in Math®:** (www.firstinmath.com) Helps students achieve mathematical and critical thinking skills through online skill set games, math practice and test preparation activities.
- **WriteBrain:** (www.sunburst.com/writebrain) This standards-based online writing environment actively engages students in grades 2-8 to learn and improve their writing and grammar skills.
- **Cosmeo™:** (www.cosmeo.com) Students explore and discover through the video clips, and read related articles to gain in-depth knowledge about a wide variety of topics.
- **BrainPOP®:** (www.brainpop.com) Offers more than 600 animated standards-based movies for K-12 students. These movies provide explanations to questions that arise from seven categories: science, math, english, social studies, health, arts and music, and technology. Each movie is supported by activities, such as experiments, quizzes, and printable worksheets.
- **AceReader Pro Deluxe Plus:** (www.acereader.com) Includes many features to improve reading level and speed such as interactive flash cards and text to speech activities.

**Podcasting/Screencasting**

How are podcasts different from what has gone before, like taped lectures? Primarily it's all about the accessibility. The files sizes are reasonable, the hardware is ubiquitous, and teenagers think it's cool. The files may be automatically downloaded every time the audioplayer is synchronized with the computer. This is known as RSS, or really simple syndication, and it acts like a subscription to digital content delivered right to the audioplayer. This technology is here to stay:

*You don't have to have an iPod® to access podcasts. Kids are digital natives. There is a learning curve for adults with this technology, but not for kids. That's where kids are, using mobile devices. We should be looking for ways to move education outside of the school. How do we turn them on to science using a different medium? Podcasts provide opportunities to listen and share. We are so text/verbal based traditionally. You can do anything on mobile devices. We need to look at these mediums.*

**Visualization Tools**

Allow users to see either real things, like the planet, or create items that have real properties to better facilitate understanding of how the object looks and works.

*All visualization tools, like Inspiration™, Kidspiration®, etc., are important, especially the modeling tools. Model building tools, such as Madonna®, STELLA®, or iThink® and VinSim® (out of MIT) need to be made more accessible. STELLA is geared toward K-12 learning. This exemplary software of the future will let people create things more fluidly, and it turns drawings into quick models placed geographically in a virtual world.*

*Visualization tools enable students to see trends in data that normally only experts might see. They provide visualization of real-time data like ocean water temperature or air quality. There are disadvantages too. Any organization that is providing the visualization tools of real data needs sophisticated programming and modeling techniques to really compile this data and provide an accurate visualization of it. The more data you have, the better the model. There are not as many organizations that could provide this type of visualization because of the sophistication—NASA and the EPA could, but not a small college.*

*These tools have powerful potential. There are good chemistry tools which help to visualize molecules, isotopes, etc. Chemistry is so abstract that tools like this can help to visualize the topics and help learning.*

- **Google Earth™:** (earth.google.com) Google Earth™ is especially useful for applying learning to student's everyday lives: they can examine the geomorphology of their own neighborhood, search for various real-life structures, or create maps. Interactive features include importing images like a topographic map and layering them. Lessons plans for Google Earth™, Google Moon™ and Google Mars™ are available from Scholastic (teacher.scholastic.com/lessonplans/exploreyourearth/) . Google SketchUp® 3-D modeling program enables easy creation of 3-D models of houses, gizmo-building projects, and spaceships. These can be published for others to see.

Google Earth and Worldwind® provide dynamic, experiential learning relating to geography and place-based phenomena. Educators need to be well-trained to use these technologies effectively. In a mapping program like GLOBE, the teacher must understand geography to guide students through activities.

- **NCES Create-A-Graph:** (www.nces.ed.gov/nceskids/createagraph/) This is a free graphing tool for children. It includes a tutor feature and explains how to create five types of graphs and charts from actual educational data.
- **Fathom Dynamic Data Software™:** (www.keypress.com/fathom/) Powerful, concept-building software used to explore the relationship between data and what the data represents. Users move from abstract concepts to concrete examples in the form of graphs, charts, build-it-yourself simulations, and statistical processing. More than 300 datasets are included. Importing additional data is also a possibility for customization.

### Modeling

Visualization and modeling are very similar. There are many types of modeling programs. Some allow the user to specify the relationships among nodes and build a conceptual map. Others allow the mathematical relationships between nodes to be specified or derived.

Much like visualization, modeling allows learners to build a conceptual map.

Modeling tools Inspiration, Kidspiration, etc. are ideal for education because of the high level of interactivity they provide. STELLA, for instance, is geared toward K-12 learning. This exemplary software of the future will let students create more fluidly. It turns drawings into quick models placed geographically in a virtual world.

Exemplary concept mapping software promotes thinking, which is often nonlinear, so students are encouraged to construct their own understanding. Modeling software that is basically a graphics library is constraining and puts students at risk for developing flawed conceptual understandings.

- **Inspiration®/KidSpiration®:** (www.inspiration.com) Inspiration®/ KidSpiration® allows students to organize thoughts, comprehend language, and communicate verbally. It features visual, textual, and audio modes. Users learn to communicate more precisely and effectively by drawing their thoughts and adding words to accompany the picture images. InspireData™ gives students a way to look at and understand data and data analysis by allowing them to see different graph types or an animation of the data over time.
- **Berkeley Madonna™:** (www.berkeleymadonna.com) Berkeley Madonna™ modeling software provides modeling and analysis of dynamic systems using differential equations.
- **Model-It™:** (www.goknow.com/Products/Model-It/) This is a powerful scaffolding tool, helping students create models of relationships among variables without knowing calculus.
- **ModelKinetix™:** (www.modelkinetix.com) This is commercial software that creates on-screen models for any system within all fields of sciences.
- **STELLA™:** (www.iseesystems.com/softwares/Education/STELLASoftware.aspx) This is designed for teaching and learning. It allows for modeling of complex systems along with tools for presenting the models.

## 2. Interactive Technologies and Problem-Solving Tools

Interactive technologies require a response on the part of the learner. These technologies include simulations, video games, and virtual reality.

- **BioInteractive's Virtual Labs:** (www.hhmi.org/biointeractive/vlabs/) Hosts computerized simulations for interactive high school biology labs. Students see a visual demonstration of major biological principles. Examples include Transgenic Fly Virtual Lab, Bacterial Identification Lab, Cardiology Lab, and Neurophysiology Lab.

- **BlueZones®:** (www.quest.bluezones.com/quests/) BlueZones® revolves around the question of why people live longer than is average worldwide in four parts of the world called "Blue Zones." Users embark on an expedition to discover possible answers to the question, interacting with others along the way. Users vote to direct the mission.
- **NOAA's Ocean Explorations:** (www.explore.noaa.gov) Offers engaging live broadcasts of marine life, underwater archeology, and underwater geology.
- **Journey North:** (www.learner.org/jnorth/) K-12 students and the general public can track migration patterns of butterflies, bald eagles, robins, and a variety of other birds and mammals and other natural events.
- **Bugscope:** (www.bugscope.beckman.uiuc.edu) Remotely operate a scanning electron microscope to image "bugs" at high magnifications. Design an experiment and receive images in real time.
- **Logger Pro 3:** (www.vernier.com/soft/lp.html) Allows students to collect and analyze data with real-time graphing. Activities are completed in three steps: measure (data collection), analyze, and learn. The software can analyze data using integrals, linear progressions, tangents, and histograms. It can test hypotheses and has video analysis and synchronization functions. There are also photo analysis functions.
- **Stellarium:** (www.stellarium.org) Stellarium gives students a firsthand look at real time planetary movement, star formations, and a collection of cosmological information.
- **Telescopes in Education:** (www.telescopesineducation.com) K-12 students use telescopes remotely to view galaxies, stars and nebulae among other activities and experiments. It can be used at no cost and users can save images.

*[In TIE] students manipulate or provide the coordinates in the direction of the telescope and within 24 hours they will see photos of the area they specified. Since telescopes are expensive, time-consuming to share use and set up, and hazardous for younger students to use, these online instruments permit students to participate in an activity that might otherwise be impractical for classroom use.*

### Simulations

*Simulations and games are on a continuum. They're in the same family. They're highly engaging.*

*Simulations are a huge, untapped resource. They're really important for the future. In certain communities, such as in Somalia, where it is extremely hot, simulated cadavers are extremely beneficial for teaching medicine because the extreme heat causes accelerated decay of bodies.*

*Simulations involve behaviors and engagement, depending on what and where it is, and are completely immersive. Students created a real experience, in general, scripted and storyboarded it, and were able to write and engage with technology. They had to draw (arts) and animate. All tools were embedded in one environment, using technology to go from text to multimedia. This would be exemplary in that it facilitates something you could not have done before.*

- **Virtual History: Ancient Egypt, Settling America:** (www.knowledgematters.com) This suite of social studies simulations is designed to teach students about history from a different perspective. The program allows the student to take on the role of a person in the community. He or she must make decisions depending on arising situations which may determine the outcome of the community.
- **Architectural Studio 3D:** (www.architectstudio3d.org) Study, design, and build a virtual home. This free online tool lets students pair up with a mock client to build a home based on the client's personal needs and lifestyle preferences.
- **Making History: The Calm and the Storm:** (www.muzzylane.com) Allows students to make decisions about key moments in history with this interactive educational game. Encourages critical thinking, discussion, and debate skills using history and international relations as a platform.

**Games**

*Video games are both incredibly relevant and incredibly distracting. The notion, the philosophy that video games bring to bear with complex interfaces is all hugely guidable and revolutionary.*

*Most if not all computer games with the exception of the Sims are locked into educational uselessness. Games are a challenge and opportunity but can't be looked at too much as a role model. The challenge is to build the game around education and not entertainment.*

*Virtual Worlds I like the most, such as SecondLife®. It's excellent for engaging someone; it's a completely immersive environment. The advantage is you can manipulate and feel like you're actually in it. It's completely immersive. Virtual worlds do not require as much expertise…Another advantage is that there's real people on the other end. You're not limited to the number of people, and content is available.*

*Online games, not so much console-based ones, create a community and also because of the assessment by experts within the online game community, they can be rapidly updated and won't become obsolete because of these continuing updates, unlike consoles.*

*Some games are simulations, like SimCity™. That is exemplary. It's a dynamic environment where kids can experiment. The new game coming out from Maxis Score, Will Rider™, builds on the idea of the other sim games. On the humanity side, Age of Empire™ is about different groups of people in the renaissance, such as the saracens. You have to build a technology tree—for example, build a better plow to feed your people. It's a good education of the history of the Middle Ages. When a kid is at the end of the game, he knows what a trebuchet is and other artifacts of the Middle Ages that otherwise you would not get until graduate school. The game builder has created a scaffold so that kids understand the context of information.*

*There's a right way and wrong way to use virtual environments. The wrong way is to create an environment you can float through and click on a few things, but you can't do anything. The experiences are incredibly shallow and expensive. Kids get tired of them really quickly…They can pick up an object. It's got properties, and if it interacts with a second thing, a third thing will happen. Objects with properties without predetermined actions are engaging. They have the ability to ask what if I did this, what would change?*

*...the more interactive a virtual environment the more exemplar it is going to be. Virtual realities where users simply use the mouse to go from one activity to the next are not employing the full interactive power of a simulation. They have to be able to see that their decisions in the game affect its destiny in a unique way; it cannot just be a simply formula or multiple-choice like turnout. Objects with properties without predetermined actions are engaging. The end result of the game must be very flexible and customizable.*

- **Quest Atlantis®:** (atlantis.crlt.indiana.edu/) This learning and teaching project aimed at children 9 to 12 years old uses an online role-playing game to teach younger children about learning and motivation. Users travel through virtual villages and locate and complete quests.

*[Quest Atlantis] is a good example of an immersive environment. Immersion into the environment and the problems within the environment result in an incredibly high level of engagement. I would say there are two criteria for motivation: Does it engage the learner, and does the learner persist in engagement because of the technology? Gaming elements can be embedded in the tasks, and this makes it very effective*

- **GameShow Pro™:** (www.learningware.com) This is game show software that allows educators to customize interactive learning experiences and create and play Hollywood-style game shows around any content. Features are easy to use and users can choose to set up teams, use the timers, add multimedia items, and enter questions and answers.
- **Oregon Trail®:** (www.learningcompany.com) Although this is not a new game it still survives in the world of online educational gaming. The information is dense but so integral in the fun of the game that students absorb it seamlessly.
- **Virtual Leader©:** (www.simulearn.net) Teaches leadership skills. This interactive virtual leadership tool consists of practiceware, a workbook, professional on-demand coaching, and facilitation and enhances interpersonal skills.

**Videoconferencing**

- **Videoconferencing with the Center for Educational Technologies' e-Missions™:** (www.e-missions.net) Provides simulations that immerse students in mock emergencies. Students use critical thinking and problem-solving skills to solve crises related to natural disasters.
- **Smithsonian Environmental Research Center Electronic Field Trips®: (**www.serc.si.edu/education/dl/eft/) Allows K-12 students to interact with a Smithsonian scientist on a live satellite broadcast using videoconferencing equipment.
- **COSI® Live Knee Replacement Surgery®** (www.edheads.org/activities/knee/) The operation is targeted to grades 7-12 and allows students to perform virtual knee replacement surgery. Students can also view real knee replacement surgery and talk to the surgeon or surgical personnel while the operation is taking place.

**Probeware**

Should be standard equipment in science labs. Couples probes and other scientific measuring devices with computers or other data collection devices for classroom experimentation and data analysis. Notables include:

- **Pitsco:** (www.pitsco.com)
- **Pasco:** (www.pasco.com/products/probeware/passport)
- **Vernier Dynamics Sensor System®:** (www.vernier.com/labequipment/)

## 3. Product-Creation Tools

Students use these technologies to develop artifacts which represent their learning.

- **Stagecast Creator®:** (www.stagecast.com) Users create their own interactive games, stories, and simulations with visual programming, logic, and critical thinking.
- **Docs:** (docs.google.com) Formerly known as Writely® this web word processor facilitates document sharing and real-time collaboration. Includes access and versioning controls.
- **The Internet Radio Project®:** (www.projectkir.org/irp/public/cms/category/10) This gives youth a free forum to develop internet radio broadcasts. Radio station mentors worldwide provide advice and media literacy guidance to the groups of youth producers.
- **Camtasia®:** (www.techsmith.com/camtasia.asp) Allows the recording of screen events and actions such as narrate, add emphasis, create quizzes, and add titles. They are then distributed as video clips.

*Use the Camtasia® software with the tablet PCs to record lectures and make them available for download on demand. You can use this method to build a bridge from school to home. For example, (when you are teaching fractions), say 50 percent get it, and 50 percent don't. You can take this technology, and deliver instruction anywhere for remediation, and reinforcement. Students can be riding home from school on the bus and can download the teacher's instructions to a mobile device to reinforce the lesson.*

- **Project Poster®:** (poster.4teachers.org) This is technology's answer to in-class poster presentations. Students can create streamlined posters online, presenting one image and four links.
- **ThinkQuest:** (www.thinkquest.org) Lets students build their own educational websites which are added to the public ThinkQuest Library. Students build research, writing, teamwork, and technology skills and compete for prizes.
- **Nvu®:** (www.nvu.com) This is a free web page authoring software. Similar to Adobe® Dreamweaver® it uses friendly page layout WYSIWYG and works on all major operating systems. Nvu allows ftp, site management, and use of templates.
- **Pixie:** (www.tech4learning.com) Like Photoshop® for elementary school students. Students can take art to a new level as they use the software to express themselves creatively and create slideshows to show off their work. This program also helps students learn the basics of digital design.
- **PowerPoint®:** (www.office.microsoft.com/en-us/powerpoint/) The gold standard presentation tool. Helps students to clarify their understandings and present them to others.

*PowerPoint® represents a lot of different things to a lot of different people. For many people it represents a philosophy of bullet points. If PowerPoint is used well as an authoring tool, it's very powerful, but if you have a bad philosophy, then it's not powerful. It's death by PowerPoint.*

*I've done bad presentations where people emerge bludgeoned and battered.*

*Advantages: It can be used for lecture notes, it can be interactive, it is multimedia, it's recognizable, it has a standardized template, it can be sent to people to get feedback or to share, you can build a template, you can collaborate with others, it's low-end creativity.*

*Disadvantages: It loses some of the multimedia effect, it lacks interactivity, and it can promote more lecturing.*

*Disadvantage: It's used in a linear fashion, but it doesn't have to be. A linear presentation doesn't reflect the way people think or learn.*

## 4. Efficiency and Productivity Tools

Technologies in this category assist in increasing the efficiency of the teaching and learning process, helping learners and educators to be organized and productive.

- **PowerMediaPlus.com:** (www.powermediaplus.com) A learning management system with multimedia resources. It combines a large library of media like audio files, video clips, clip art, and images with tools for simple lesson plans, development and podcasting.
- **TeacherSource™:** (www.pbs.org/teachers/) Presents online lesson plans and other resources to reinforce PBS broadcasts and provide further educational opportunities.
- **Rubistar™:** (rubistar.4teachers.org) Helps teachers create rubrics for project-based learning activities. It provides common categories to help develop a format for each rubric and teachers can also search for pre-made rubrics.
- **ThinkTank and NoteStar:** (thinktank.4teachers.org) These two help students develop research paper topics and assists in the setup of topics and subtopics, collect and organize notes, and prepare a bibliography.
- **netTrekker.com:** (www.nettrekker.com/di/) This website contains educator-selected online resources. They are organized by reading level and aligned with state standards. This site has a variety of materials for diverse groups of students, including ELL/ESL, students with special needs, and more.
- **KnowledgeBox:** (www.uk.knowledgebox.com) This resource is geared to K-6 age students and has entertaining videos and electronic media and sites for teachers to share lessons, ideas, and resources.
- **Answers.com™:** (www.answers.com) This is a combination of an encyclopedia, a dictionary, and an almanac.
- **CompassLearning®:**(www.compasslearning.com) A fully online student assessment program. Each state's standards are used for assessment and learning.
- **BackPack™:** (www.backpackit.com) A free, online organizer with to-do lists, calendars, and open sharing features. With an open and flexible system allowing for liberal customization; some may use it for homework notes, others could use it to plan a vacation.
- **Grokker:** (www.grokker.com) This research tool provides data visualization and organization of online searches.
- **Pageflakes:** (www.pageflakes.com) Using Pageflakes, educators arrange "Flakes"--small, movable versions of popular web sites, interactive research tools, and education-specific applications--on a customized web page. Pageflakes has worked with

educators to develop multi-user Flakes specifically for the classroom environment, including a Grade Tracker, Class Schedule, To-Do-List, Message Board, Class Blog, and Class Calendar, in addition to popular online reference tools such as Wikipedia and a dictionary.

- **TeacherTube®:** (www.teachertube.com) TeacherTube creates a way to incorporate watch and upload videos which promote authentic learning on a wide assortment of topics such integrating technology and applying skills to everyday life.
- **Center for Interactive Learning and Collaboration:** (www.cilc.org) Help schools utilize videoconferencing in order to connect with the world around them. Allows access to videoconferencing activities, online lessons, workshops, and virtual field trips.

·**Google**™: (www.google.com) This has developed into more than a search engine and now hosts many educational features. New applications and functionality are continually added to Google™. A number of high quality products are available at no cost from Google™:

- Alerts™, lets a user know when new content such as web sites, group postings, and news feeds are found, as per the user's search terms.
- 'Apps for Your Domain' are a set of Google applications that help administrate web sites for small businesses and educational institutions.
- Blog Search™ allows users to search blogs, filtering the results by date.
- 'Book Search™' is a search engine for full text books.
- 'Blogger™' is a weblog publishing tool that the everyday person can use.
- Code™ is a web API programming interface that helps programmers use the information cached at Google for other applications.
- Groups™ is a searchable usenet. It leads users to topic-specific groups, discussions and communities. Searchers can join and contribute or just search and use the information added by others.
- Hello™ allows users to send images to their blogs.
- Images™ is a search engine for images that uses the name of the file and the text around it to meet the search requirements.
- Notebook™ allows users to clip information while researching online.
- Orkut™ is a social networking program.
- Reader™ allows users to subscribe to RSS.
- Scholar™ is a search engine for full-text scholarly publications.
- 'Sets™' can be searched for information on sets of data.

**Learning Management Systems**

Learning management systems are used for the management and delivery of learning materials and resources. Users can track progress completely in multiple modes of learning. These systems are also useful for constructing course content.

*[LMSs] can be especially helpful in educational environments with limited access to professional development. It has significantly changed the nature of coursework in distance learning situations, where previously education was conducted through telephone.*

- **OpenAcademic®:** (www.openacademic.org) Provides a collection of tools that allow the management of all aspects of an institution's learning environment. Running on open source tools, users are able to contribute to the site's educational tools. OpenAcademic® combines several different programs, including Elgg, Drupal, Moodle, and MediaWiki.
- **ClassLink2000.com®:** (www.classlink2000.com) Teachers can view and grade work, make comments, and create lesson plans with this versatile application. Provides curriculum, rubrics for assessments, and standards that map to student work. Administrators can improve communication with school personnel and parents. It automatically sends assignments to staff and students and generates feedback to students.
- **Desire2Learn.com®:** (www.desire2learn.com) Boasts more robust functionality than standard LMSs. It has a blog feature and software for creating individual learning paths for learners. Stores, tags, searches, and reuses learning objects for users. Delivers synchronous chat, whiteboard, and presentation features for real-time collaboration.
- **Moodle:** (www.moodle.org) Moodle is a free open-source course management system with hundreds of thousands of registered users. This program specializes in creating strong online communities of learners and features forums, content management, quizzes with multiple question types, blogs, wikis, surveys, chats, glossaries, peer assessments, and support in 60 languages.

**Teacher Resource Exchanges**
Community exchange websites have become abundant as educators become more immersed in the digital world. They are a powerful connecting force between educators and typically include lesson plans, media, ideas, resources, and discussion boards. At sitesforteachers.com there is a database of these sites. Some noteworthy ones include:

- **Apple Learning Interchange®:** (www.ali.apple.com) This is a teacher community site that has, in addition to all the usual resources, a system where educators connect through simple searching, messaging, iChat, and collaborative publication tools.
- **Discovery Educator Network®:** (www.discoveryeducatornetwork.com) Includes a rating system for uploaded lesson plans. It also has a blog for teachers to exchange stories and ideas.
- **teAch-nology.com®:** (www.teach-nology.com) Provides basic resources for no cost. Yearly membership is available, and the site includes games and rubric makers.

**Electronic White Boards (SmartBoards ™)**
These allow for computer content to be displayed on a whiteboard-like screen. The content can be dynamically altered with digital pens and saved for later use.

*Electronic White Boards are used in the classroom for interactive lessons. The content can be manipulated according to classroom needs and saved for review and future use, printed, or sent to each student electronically. These can be valuable tools when utilized in a way that changes classroom dynamics from lectures to interactive learning.*

This type of learning appeals to a variety of learning styles. Many younger students especially appreciate the kinesthetic nature of whiteboards. An LCD is required to use this tool. A laptop and LCD creates a whole system. The large screen size also facilitates classroom activities.

## 5. Communication and Collaboration Tools

These technologies support an educational, industrious exchange of information among people.

Collaborative technologies include online chats and blogs. These do not have to be in real time, facilitating their occurrence. Students can also collaborate with others worldwide, enriching their educational experience. If assigned to analyze water samples in their home town, students can connect a digital camera to a microscope and share their information through a blog, web site, or e-mail, inquiring about the nature of other students' water samples from around the globe.

- **Tapped In®:** (www.tappedin.org) A career-long online "home" that transforms teacher professional development. "Tenants" encounter development experiences that are not offered in their school or district. Educators, administrators, and consultants from all over the world share ideas, strategies, and content. The Tapped In interface metaphor is a campus with buildings, offices, and group rooms, each equipped with text chat, threaded discussion, file and link storage. Other features include a calendaring system, branded pages, a job bank, student accounts, conversation transcripts, room sticky notes, and online courses.
- **Elgg:** (www.elgg.net) Join other educators passionate about new approaches to teaching and learning! Users learn by creating online learning profiles and interacting within both formal and ad-hoc communities.
- **Global Schoolhouse®:** (www.gsn.org) The original resource for global project-based learning, problem-based learning, and online collaborative learning.
- **SchoolNotes.com®:** (www.schoolnotes.com) Gives teachers and parents a way to ensure that education comes home with children. Teachers leave notes about homework on this web site for students and their parents.

**Blogs**

Blogs are variously understood as discussion forums, venues for opinion pieces, or webpages where the user may update content of a personal nature easily.

*Blogs are exemplary by nature because they facilitate communication. Even when not specialized for a course's content, social science teachers will find them useful. Often teachers set up blogs for the classroom where they can communicate with students in other parts of the world.*

*Because they have the potential to be invaluable to learning blogs are becoming popular in education. Academically, each blog holds a twofold purpose. They are a 'first draft' of history; an account of life (preferably focused on particular subject matter). There is learning value in reading others' blogs. However this learning is often unstructured and unfocused. As a second use blogs can be manipulated into a formal learning environment. Academic journals are being used successfully for education.*

*Blogs permit students to use creativity as a channel to express what they know, and the creative quality of a blog lures in reluctant readers.*

*With 'kudos' and other response features, students can give and receive feedback on their thoughts.*

**Wikis**

Wikis are information-based websites with user-generated content. Users add and edit content. The advantage to this method is the power of so many minds when working together to make the information as accurate and articulate as possible.

*Wikis also have potential for social construction. They enable people to build and participate in a community voice and be the expert in a subject. Teaching others is also an effective way for students to learn. Wikis typically work better with large communities. Recently devised wikis are easier technologies to use.*

*It can be difficult for educators to track and monitor a collaborative environment. Since there is a risk of sabotage or misinformation, guidance is necessary. The educational paradigm must change for wikis to be embraced. Unless specified at the creation of a wiki, there could be confusion and debate over authority, both intellectual and social.*

- **Wikispaces for Educators:** (www.wikispaces.com/site/for/teachers) Educators can use this to start their own class wiki free of charge.
- **MediaWiki:** (www.mediawiki.org) This website is free for others to use and improve. MediaWiki is the software engine developed for and used by Wikipedia among many others.

### 6. Technology Tutors

These technologies provide scaffolds for learning skills related to technology use, such as typing, using spreadsheets, or word processing. These are the types of skills addressed by the National Educational Technology Standards. Their purpose is to teach technology skills by bringing material to the students in an organized and time-efficient fashion.

- **Atomic Learning**™: (www.atomiclearning.com) Provides web-based software training for more than 100 applications. The short and straightforward tutorials can be used in technology classes or for professional development.
- **KidzOnline:** (www.nnkol.org) These educational videos teach children and teens about skills such as digital audio and video editing, Microsoft PowerPoint presentations, website creation, and 3-D animation.

## Features and Design Principles: The Key Features of Each Exemplar with a Summary of the Relevant Design Principles

### 1. Knowledge and Comprehension Tools

- Instructables
  - Involves parents
  - Easy to use
  - Teaches skills
  - Allows user-generated content
- Ology: Earth; Cosmeo; BrainPOP
  - Customizable for topic and grade level
  - Organized
  - Uses multimedia capabilities
  - Visually appealing
- Lucas Learning Programs
  - Visually appealing
  - Promotes enthusiasm
- Science Toys You Can Make with Your Kids
  - Involves parents
  - Easy to use
  - Encourages hands-on activities
  - Uses toys to teach science
- NOAA's Ocean Challenge Puzzle
  - Real-time
  - Collaboration

    - Multiplayer
- KidBiz 3000
    - Customizable
    - Provides user with feedback
- NutshellMath
    - Provides multimedia examples
    - Uses audio
    - Involves parents
- LeapFrog
    - Uses the mass appeal of toys to educate
    - Portable
- CXOnline.Bridges.com—Choices Explorer; SAS inSchool: Curriculum Pathways
    - Valuable resources rarely found elsewhere
    - Age-appropriate style
    - Assessment integrated throughout
- United Streaming; Safari Montage
    - Ease of use has led to mass appeal
    - Standards-aligned
    - Uses learning objects
- First in Math; WriteBrain
    - Uses multimedia to teach skills
    - Interactive
    - Standards-aligned
- AceReader Pro Deluxe Plus
    - Promotes skill building
- Google Earth
    - Uses visualization for learning
    - Easy to use
    - Captivating
    - Promotes curiosity
- NCES Create-A-Graph; Fathom Dynamic Data Software
    - Enables manipulation of factors
    - Allows use of real datasets
    - Students develop authentic science skills
    - Allows visualization of data
    - Promotes model-building
- Inspiration/KidSpiration
    - Assists mental organization
    - Easy to use
- Berkeley Madonna; Model-It; ModelKinetix; STELLA
    - Enhances thought process
    - Assists mental organization
    - Promotes model-building
    - Allows visualization/ testing of conceptual relationships

**Design Principles**

Knowledge and comprehension tools should:

- Offer learners choices.
- Allow customize for topic and grade level.
- Promote motivation.
- Provide opportunity for learners to be interactive with feedback and user control.
- Engage learners in a variety of inquiry processes.
- Allow learners to make discoveries by themselves.
- Enhance organized content knowledge.
- Provide meaningful patterns of information.
- Facilitate communication and collaboration among learners or experts.
- Allow learners to learn everywhere.
- Provide access to important aspects of knowledge.
- Pique learners' curiosity.
- Reach a large audience.

## 2. Interactive Technologies and Problem-solving Tools

- BioInteractive's Virtual Labs; Bugscope; Stellarium
    - Allows scientific thinking
    - Builds science inquiry skills
    - Authentic-feeling
    - Interactive
    - Provides virtual representations
- BlueZones
    - Promotes collaboration

    - Encourages decision making
    - Uses critical thinking
    - Builds science inquiry skills
- NOAA's Ocean Explorations
    - Live
    - Interaction with scientists
- Journey North
    - Promotes collaboration
    - Builds science inquiry skills
    - Real-time
    - Authentic science data
- Vernier's Logger Pro 3
    - Enables manipulation of factors
    - Students develop authentic science skills
    - Allows visualization of data
- Telescopes in Education
    - Builds science inquiry skills
    - Real-time
    - Authentic science data
- Virtual History: Ancient Egypt, Settling America
    - Teaches problem solving
    - Promotes collaboration
    - Encourages decision making
    - Uses critical thinking
- Architectural Studio 3D
    - Allows for authentic tasks
- Making History: The Calm and the Storm; Quest Atlantis
    - Interactive
    - Appealing and engaging
    - Encourages decision making
    - Uses critical thinking
- GameShow Pro
    - Uses a game within a game to teach concepts
- Oregon Trail
    - Encourages decision making
    - Uses simulation to teach history
- Virtual Leader
    - Builds thinking skills
- Center for Educational Technologies' e-Missions
    - Teaches problem solving
    - Promotes collaboration
    - Encourages decision making
    - Uses critical thinking
- Smithsonian Environmental Research Center Electronic Field Trips: COSI Live Knee Replacement Surgery
    - Real-time events

**Design Principles**

Interactive technologies and problem-solving tools should:

- Provide ample times for decision making.
- Present options for learning skills in a virtual world for application to real-world situations.
- Help learners understand and internalize the structure of argumentation.
- Help learners making thinking visible.
- Scaffold the understanding and use of scientific explanations of the natural world in context of problem-based learning.
- Encourage inquiry and understanding.
- Employ multiple social activity structures.
- Allow the transfer of pragmatic learning changes to real-world environments.
- Allow experiential learning.
- Reach a large audience.
- Allow learners to control, explore, and ask questions.
- Allow learners control of real-time data or events.
- Support creativity.
- Promote critical thinking.
- Present options for learning skills in a virtual world for application to real-world situations.
- Promote productive interactions.
- Enhance learners' problem solving skills.

## 3. Product-Creation Tools

- Stagecast Creator
  - Inspires creativity
  - Develops problem-solving skills
  - Promotes logic skills
- Docs
  - Promotes collaboration
- Internet Radio Project
  - Uses student-generated content
  - Uses live events to motivate
- Camtasia
  - Promotes portability of products
- Project Poster
  - Uses student-generated content
  - Easy to use
- ThinkQuest
  - Uses student-generated content
  - Uses competition to motivate
  - Team-based learning
- Nvu; KidPix; Pixie
  - Student-generated content
  - Inspires creativity
  - Younger versions of professional tools
- PowerPoint
  - Organizes thinking
  - Expandable
  - Universal

**Design Principles**
Product-creation tools should:

- Enable learners to evaluate their own.
- Allow learners to contribute the knowledge base.
- Allow access to live events and real data.
- Promote productive interactions.
- Provide learners with opportunities to practice real-world situations.
- Support creativity.
- Support for team-based learning.
- Motivate learners to generate content knowledge.

## 4. Efficiency and Productivity Tools

- PowerMediaPlus.com
  - Makes use of multimedia
- TeacherSource
  - Well-organized
  - Leverages PBS resources
- Rubistar
  - Provides professional development
- ThinkTank and NoteStar
  - Helps organize
- netTrekker.com
  - Customizable
  - Standards-aligned
  - Peer-reviewed resources
- KnowledgeBox
  - Uses multimedia
- Answers.com
  - Many resources
  - Well organized
- CompassLearning
  - Assists assessment
  - Provides professional development
  - Customizable
  - Standards-aligned
- BackPack
  - Helps organize
  - Customizable
- Grokker
  - Helps organize online resources
- Pageflakes
  - Customizable
  - Organization
- TeacherTube
  - Teacher resource
  - User-generated content
- Google: Alerts, Apps for Your Domain, Blog Search, Book Search, Code, Groups, Hello, Images, Notebook, Orkut, Reader, Scholar, Sets
  - Easy to use
- OpenAcademic; ClassLink2000.com; Desire2Learn.com; Moodle
  - Multimodal

  - Helps organize
  - Encourages parent participation
  - Standards-aligned
  - Customizable by administrators
- Apple Learning Interchange; Discovery Educator Network; Teach-nology.com
  - Collaborative
  - User-generated content
  - Peer-reviewed

**Design Principles**
Efficiency and productivity tools should:

- Facilitate development of organizational skills.
- Customize for topic and grade level.
- Motivate learners to learn.
- Provide multiple opportunities for use.
- Encourage social learning.
- Promote productive interactions.
- Provide opportunities for learners to serve as instructors of their peers.
- Facilitate communication and collaboration among learners or with experts.
- Promote discovery and curiosity.
- Allow experiential learning.
- Provide dynamic visual aids for the perception of 3-D phenomena.

## 5. Communication and Collaboration Tools

- Tapped In
  - Multimodal
  - Allows peer-review
  - Encourages information sharing
  - Helps professional development
- Elgg; Global Schoolhouse
  - User-generated content
  - Encourages information sharing
- SchoolNotes.com
  - Encourages parent involvement
- Blogs
  - User-generated content
  - Easy to use
- Wikis
  - User-generated content
  - Customizable

**Design Principles**
Communication and collaboration tools should:

- Encourage social learning.
- Encourage inquiry and understanding.
- Support social/community building.
- Facilitate communication and collaboration among learners or with experts.
- Allow multiple platforms of communication.
- Connect learners to facilitate the formation of a learning community.
- Enable community building.
- Employ multiple social activity structures.
- Enable multiple ways to participate in online discussions.
- Allow learners to inspect the work of other learners.
- Provide opportunities for learners to serve as instructors of their peers.
- Promote productive interactions.
- Scaffold the development of classroom norms.
- Customize for topic and grade level.

## 6. Technology Tutors

- Atomic Learning
  - Trains technology skills
  - Short multimedia clips
- Nortel Network's KidzOnline
  - Trains technology skills
  - Appealing and engaging

**Design Principles**
Technology tutors should:

- Engage learners in a variety of inquiry processes.

- Provide learners with several learning sequences to cope with the vast amount of possibilities.
- Promote motivation.
- Allow learners to control, explore, and ask questions.
- Allow learners to learn everywhere.
- Incorporate features for ease of use.
- Reach a large audience.

## CONCLUSION

This study was conducted as part of a larger research project to establish design principles and metrics for choosing and using emerging educational technologies. In this study we developed a list of promising educational technologies, tools, websites, resources, software, and hardware, extracted the common design features, and developed lists of design principles.

The list of exemplars is overwhelming. It is hard to identify which among these notables are truly effective for education. In the future there is a need for a systemic evaluation on how their features lead to outcomes of interest. From our examination of exemplars, we noted convergence around educational technologies making use of these features:

- Learning management systems
- Collaboration and communication
- Mobility & interoperability
- User-generated content (blogs, wikis, educator exchanges)
- Games and simulations
- Visualization and modeling
- Real-time live events
- Commercial careers-focused resources
- Creation of student products
- Peer-reviewed, juried educator resources

Using this list of exemplars we derived the common design principles. These principles are durable across types of technologies and multiple generations of products. In future research we recommend developing metrics for effectiveness. The metrics could be validated by applying them to this list of exemplars and examining the discriminant and convergent validity. In the concluding article for this special issue we highlight how such metrics could be used in a Web 2.0 collaborative application to foster information-resource sharing among instructional developers and users of emerging technologies.

## REFERENCES

Bloom, B. S., Englehart, M. B., Furst, E. J., Hill, W. H., & Krathwohl, O. R. (1956). *Taxonomy of educational objectives: The classification of educational goals. Handbook 1: The cognitive domain.* New York: Longman.

Clark, R. E., & Sugrue, B. M. (1991). Research on instructional media, 1978-1988. In G. J. Anglin (Ed.), *Instructional technology: Past, present, and future* (pp.327-343). Englewood, CO: Libraries Unlimited.

Russell, M. (2001). Framing technology program evaluations. In W. Heineke, L. Blasi, & J. Willis (Eds.), *Methods of evaluating educational technology* (pp. 149-162). Greenwich, CT: Information Age.

## KEY TERMS AND DEFINITIONS

**Educational Technology:** Device or system that makes use of digital media to enhance the teaching and learning process.

**Exemplar:** A model for representing excellence as a whole.

**Design Principle:** A descriptor or characteristic which can be sued y educators and designers to structure the content and features of an educational technology and its implementation into curriculum.

**Digital Information:** The electronic technology that generates, stores, and processes information in terms of the numbers 0 and 1.

**Communication Technology:** Technology that transmits data; are either cable (land) or wireless (radio, microwave, or satellite).

**Pacesetter:** A person who is a leading influence in his or her field of study or work.

**Instructional Design:** A systematic approach to the design and development of instructional materials and products using objectives, teaching strategies and evaluation to meet learning needs.

# Chapter 27
# Setting Trends for Educational Technologies within the National Science Foundation

**Bruce C. Howard**
*Center for Educational Technologies®, Wheeling Jesuit University, USA*

**Laura J. Curtis**
*Center for Educational Technologies®, Wheeling Jesuit University, USA*

## ABSTRACT

*Our research team evaluated 18 months of National Science Foundation (NSF) program announcements and awarded programs to discern the amount and type of emphasis placed upon educational technologies. NSF issued 65 solicitations for proposals with 53.8 percent calling for educational technology components. A sampling of 366 of the 1,180 funded projects, showed that 34.7 percent included educational technology. Twenty-five percent of the projects were in biology and cognitive science, with another 40% in general science, computer science, technical education, engineering, and math. Many types of educational technologies were funded, with an emphasis on cognitive tutors/intelligent agents, distance learning, and online communities.*

## INTRODUCTION

In the United States the National Science Foundation (NSF), with its $5.5 billion annual budget, funds approximately 20 percent of all federally supported basic research, and it is a major funding source for new educational initiatives as well. An essential element of NSF's mission is integrating this research with education to help ensure a skilled workforce and plenty of capable teachers.

DOI: 10.4018/978-1-60566-936-6.ch027

In 2004 the NASA Learning Technologies group presented a benchmarking study in which it examined the funding trends for K-16 education taking place within the NSF's Division of Elementary, Secondary, and Informal Education (ESIE). The survey found that three programs in particular have a large number of awards and significant overall funding: the Informal Science Education Program, the Instructional Materials Development Program, and the Teacher Enhancement Program. Additionally, the report cited significant support for funding the Centers for Learning and Teaching Program and the Urban Systemic Program. Extrapolating from these findings, the report concluded that the overall trends for NSF's K-16 educational efforts in science, technology, engineering, mathematics, and geography (STEM-G) were (a) development of tools/applications for STEM-G education outside of the classroom; (b) development of instructional classroom material that is compelling, hands on, and standards based that will advance science, math, and technology education in schools; (c) teacher professional development; (d) the research of teaching methods in STEM-G; and (e) improvement of STEM-G education in urban schools.

We set out to update and extend the 2004 study by conducting an analysis across multiple divisions within NSF, focusing particularly on educational technologies. The National Science Foundation includes an education component in all of its program announcements/solicitations (also commonly known as requests for proposals or requests for applications). We presumed an analysis of how this money was allocated over the past 18 months would yield conclusions about emerging trends in STEM-G education. In particular, we examined the trends from two perspectives. First, we wanted to determine the degree to which program solicitations prescribed or recommended the use of educational technologies. This would yield an indicator of the federal agency's emphasis on educational technology. Second, we sought to examine the emphasis placed on educational technologies by those who received the awards. Presumably this would yield a broader range indicator of the emphasis placed on educational technologies by proposal reviewers and the practitioners themselves across the country.

## METHOD

### Data Source

There are 11 program areas and one cross-cutting area within NSF generally arranged by science discipline. Education is included as a program area called Education and Human Resources (EHR). All EHR programs are in support of research conducted in the other 10 program areas. NSF program announcements/solicitations (hereafter, "solicitations") for EHR programs were accessed from an online database (http://nsf.gov/funding/). There were 506 solicitations in the database, 65 of which fell into the 18-month period of interest.

We conducted three stages of analysis, using two types of data: the text from *solicitations* for each of the 65 programs and *abstracts* from awarded proposals. In stage one we wanted to develop an operational definition of educational technology to be used in the subsequent stages. We derived the definition from the data to control for the varying perspectives of the investigators. We randomly selected 100 abstracts from awarded proposals from among the 65 programs. Four investigators read the abstracts and wrote down the terms indicating an educational technology was included. The group reached consensus around the terms and then used them to create the operational definition. The results section explains this process in more depth.

In stage two we used the definition to examine the text of the 65 solicitations to determine the degree to which they included educational technologies. For the subset of solicitations that did include educational technologies, we developed a dataset of abstracts from the awarded proposals. In stage three we analyzed the text of the abstracts to identify specifics of the proposed educational technologies. Figure 1 provides an illustration of the process.

## RESULTS

### Stage One: Defining Educational Technologies

A review of 100 abstracts from EHR programs provided the context for creating an operational

*Figure 1. The three stages of the research process*

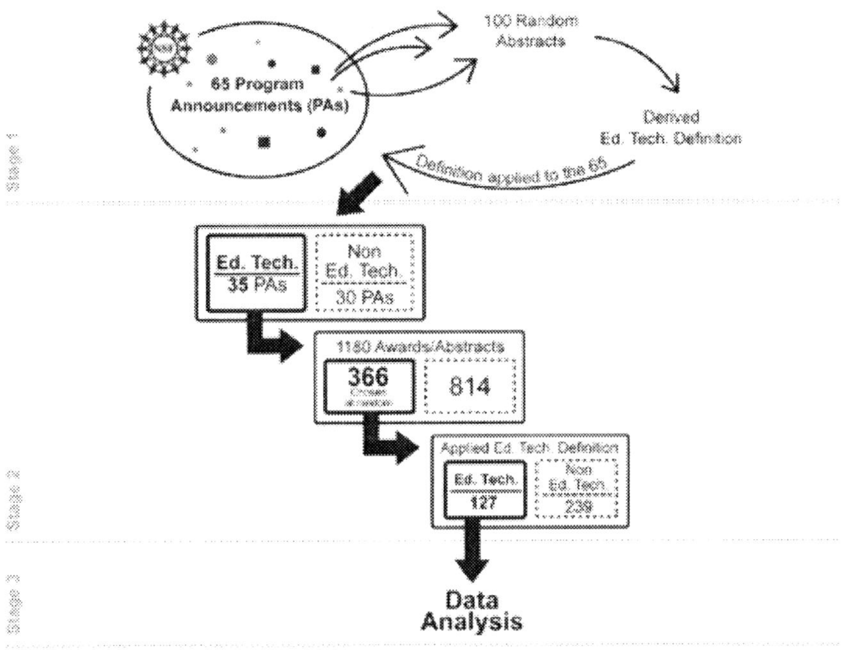


definition of educational technologies. The four investigators agreed on this definition: "The integration of electronic or digital products and systems with knowledge and theories from one or more of the following domains—cognitive science, social science, research on learning, and research on education, and/or classic sciences, such as biology, physics, or chemistry—specifically to increase the capacity of the user to learn content and/or cognitive processing skills."

The definition is by necessity strictly defined, and some may disagree with it. Many definitions have been proposed over the years by various professional societies, agencies, textbook writers, and researchers (e.g., Januszewski, Molenda & Harris, 2007). In this case the definition emerged from the data itself and was written by group consensus. We do not expect that this definition would apply outside of this study.

The definition is comprised of three primary parts: *type of technology*, *the field of application*, and *educational purpose*. Knowing each of these was pertinent to make judgments about particular proposals. For instance, suppose a proposal suggests using a toaster for training undergraduates about mechanical engineering. In this case the *type of technology* is not electronic or digital and so fails the test. While a toaster might be a technology, it's not an educational technology by our definition. A simulated toaster, however, might pass the test. Also, we would need to know if the use of the toaster is based on a scientific *field of application*. If the toaster (or even a simulated toaster) was there merely to brown bread, it is not educational. If the toaster served as a teaching tool for users to take apart and study how resistance is used to generate heat, thereby solving the problem of how to toast, then it is educational. Finally, the technology must be used for an *educational purpose* to learn content or skills. If the toaster provided a wonderful means by which to collect data on resistance, but students were not taught about resistance or inquiry skills, then it is not educational.

## Stage Two: Analysis of Solicitations

Using the operational definition, we identified 35 of the 65 solicitations (53.8 percent) as including educational technologies. Two investigators for each solicitation conducted the review, and interrater reliability was checked. There was 91.7 percent consistency, and every inconsistency was resolved by group consensus among all four investigators.

*Table 1* lists the 35 solicitations along with the number of awards for each and total funding. Overall there were 1,180 awards, representing a total of $565,123,893.

## Stage Three: Analysis of Abstracts

From the 1,180 awards we randomly sampled 366 (31%) of the abstracts. *Table 2* shows the distribution of sampled abstracts across the 35 programs and the funding allotted to those programs. Using the operational definition of educational technologies, and the same review process used in Stage Two, we determined which proposals included educational technologies. The table shows the number of abstracts sampled and how many included educational technologies. The final two columns show the total funding for that sample and the percentage of that funding for educational technology projects. Overall, 127 (34.7 percent) of the funded proposals were educational technology related. From a funding perspective, 27% of the sampled projects included educational technologies ($61,721,078 of $228,040,647).

Using the text from the abstracts, we conducted thematic analyses to examine three areas of interest: (a) discipline represented, (b) type of technology represented, and (c) the product or website resulting from the award.

**Disciplines Represented.** *Figure 2* shows 16 science disciplines represented by the sampled abstracts. Some abstracts were placed in multiple categories because their projects represented multiple disciplines so the total is larger than 100 percent.

**Type of Technology Represented.** We examined the range of educational technologies represented among the sampled abstracts. *Figure 3* lists the range of technologies. Note that some awarded programs employed more than one technology. The research team used the same set of definitions for various technologies to improve interrater reliability.

**Types of Products.** In many cases the awarded projects proposed the creation of a product. Many products incorporate intelligent agents or cognitive tutors. The *Simulated Student* is a programming project to develop an open source cognitive tutor that enables efficacy for teacher authoring. *INVISSIBLE* is a web-based platform that allows students to create authentic multimedia scenarios. *TYCHO* is a physics-related cognitive tutor, and *EPOCH* enables chemistry tutoring. Online communities were a central component of 13 funded projects. Grid computing was integrated into eight of the programs. Seven of the programs are part of NSF's *Cyberinfrastructure Initiative* (CI team), which enables K-16 access to high-performance computing, networking, data collection, and visualization tools and resources. The *Net-SEAL* computer science project encompasses a series of networking experiments and animations based on simulating small, large, and global networks.

One videogame was identified among the abstracts. The online *Fantasy Basketball Game* teaches decision-making strategies and scaffolds metacognitive abilities to help players acquire statistical knowledge. Six of the abstracts funded digital library initiatives. *CalWomenTech* provides strategies for increasing recruitment of women into predominantly male-oriented careers, such as technology and law enforcement. The website offers a digital library of resources, such as multimedia CDs, online training, PowerPoint presentations, workshops, and consultation. Four

*Table 1. NSF EHR programs that include educational technologies*

| Program | Number of Awards | Total Funding |
|---|---|---|
| Active Nanostructures and Nanosystems (ANN) | 154 | $83,711,680 |
| Advanced Learning Technologies (ALT) | 10 | $4,009,937 |
| Advanced Technological Education (ATE) | 107 | $45,610,930 |
| Alliances for Broadening Participation in STEM (ABP) | N/A[1] | N/A[1] |
| Centers for Ocean Science Education Excellence (COSEE) | 9 | $5,835,424 |
| Centers of Research Excellence in Science and Technology (CREST) | 15 | $11,541,940 |
| Chemistry Research Instrumentation and Facilities: Cyberinfrastructure and Research Facilities (CRIF:CRF) | N/A[1] | N/A[1] |
| Communicating Research to Public Audiences | N/A[1] | N/A[1] |
| Course, Curriculum, and Laboratory Improvement (CCLI) | 300 | $39,358,554 |
| Cyberinfrastructure Training, Education, Advancement, and Mentoring for Our 21st Century Workforce (CI-TEAM) | 16 | $5,835,313 |
| Engineering Research Centers (ERC) | 25 | $31,949,024 |
| Federal Cyber Service: Scholarship for Service (SFS) | 19 | $6,089,512 |
| George E. Brown, Jr. Network for Earthquake Engineering Simulation Research (NEESR) | 24 | $11,794,126 |
| Historically Black Colleges and Universities Undergraduate Program (HBCU-UP) | 23 | $14,570,643 |
| Informal Science Education (ISE) | 65 | $51,652,202 |
| Information Technology Experiences for Students and Teachers (ITEST) | 21 | $21,787,228 |
| Instructional Materials Development (IMD) | 21 | $13,266,950 |
| International Polar Year (IPY) | N/A[1] | N/A[1] |
| Materials Use: Science, Engineering, and Society (MUSES) | 8 | $19,743,456 |
| Math and Science Partnership (MSP) | 2 | $7,682,375 |
| Nanotechnology Undergraduate Education (NUE) | 17 | $11,136,067 |
| National Science, Tech., Engineering, & Mathematics Education Digital Library (NSDL) | 22 | $12,889,674 |
| Next Generation Cyberinfrastructure Tools | N/A[1] | N/A[1] |
| NSF Academies for Young Scientists (NSFAYS) | N/A[1] | N/A[1] |
| NSF Director's Award for Distinguished Teaching Scholars (DTS) | 9 | $2,285,376 |
| Partnerships for Innovation (PFI) | 25 | $8,833,140 |
| Research and Evaluation on Education in Science and Engineering (REESE) | 82 | $40,046,103 |
| Research Experiences for Teachers (RET) | 24 | $6,692,189 |
| Research Experiences for Undergraduates (REU) | 13 | $3,936,440 |
| Research in Disabilities Education (RDE) | 13 | $3,549,298 |
| Research on Gender in Science and Engineering FY 2006 (GSE) | 28 | $8,459,912 |
| Science of Learning Centers (SLC) | 8 | $5,180,839 |
| Science, Technology, Engineering, and Mathematics Talent Expansion Program (STEP) | 25 | $25,515,860 |
| Teacher Professional Continuum (TPC) | 98 | $57,111,300 |
| Tribal Colleges and Universities Program (TCUP) | 10 | $8,984,841 |
| Total: | 1180 | $565,123,893 |

[1] No Abstracts Available

*Table 2. Awarded proposals: Number sampled, funding levels, ed tech emphasis*

| Program | Number of Abstracts Sampled | Subset Including Ed Tech | Total Funding (for Sample) | Percent of Funding for Ed. Tech. |
|---|---|---|---|---|
| Active Nanostructures and Nanosystems (ANN) | 10 | 0 | $3,565,729 | 0% |
| Advanced Learning Technologies (ALT) | 10 | 10 | $4,009,937 | 100% |
| Advanced Technological Education (ATE) | 7 | 3 | $2,606,357 | 38% |
| Alliances for Broadening Participation in STEM (ABP) | N/A[1] | N/A[1] | N/A[1] | N/A[1] |
| Centers for Ocean Science Education Excellence (COSEE) | 10 | 8 | $6,488,560 | 76% |
| Centers of Research Excellence in Science and Technology (CREST) | 13 | 2 | $10,235,940 | 18% |
| Chemistry Research Instrumentation and Facilities: Cyberinfrastructure and Research Facilities (CRIF:CRF) | N/A[1] | N/A[1] | N/A[1] | N/A[1] |
| Communicating Research to Public Audiences | N/A[1] | N/A[1] | N/A[1] | N/A[1] |
| Course, Curriculum, and Laboratory Improvement (CCLI) | 27 | 24 | $3,583,101 | 87% |
| Cyberinfrastructure Training, Education, Advancement, and Mentoring for Our 21st Century Workforce (CI-TEAM) | 16 | 9 | $4,585,313 | 73% |
| Engineering Research Centers (ERC) | 15 | 0 | $23,218,636 | 0% |
| Federal Cyber Service: Scholarship for Service (SFS) | 19 | 0 | $6,089,512 | 0% |
| George E. Brown, Jr. Network for Earthquake Engineering Simulation Research (NEESR) | 10 | 0 | $4,355,150 | 0% |
| Historically Black Colleges and Universities Undergraduate Program (HBCU-UP) | 10 | 2 | $6,307,934 | 18% |
| Informal Science Education (ISE) | 43 | 9 | $46,126,913 | 25% |
| Information Technology Experiences for Students and Teachers (IT-EST) | 7 | 7 | $6,770,558 | 100% |
| Instructional Materials Development (IMD) | 12 | 6 | $6,018,050 | 52% |
| International Polar Year (IPY) | N/A[1] | N/A[1] | N/A[1] | N/A[1] |
| Materials Use: Science, Engineering, and Society (MUSES) | 8 | 0 | $19,743,456 | 0% |
| Math and Science Partnership (MSP) | 2 | 0 | $7,682,375 | 0% |
| Nanotechnology Undergraduate Education (NUE) | 7 | 1 | $1,078,516 | 19% |
| National Science, Tech., Engineering, & Mathematics Education Digital Library (NSDL) | 10 | 9 | $6,020,789 | 89% |
| Next Generation Cyberinfrastructure Tools | N/A[1] | N/A[1] | N/A[1] | N/A[1] |
| NSF Academies for Young Scientists (NSFAYS) | N/A[1] | N/A[1] | N/A[1] | N/A[1] |
| NSF Director's Award for Distinguished Teaching Scholars (DTS) | 4 | 2 | $882,191 | 54% |
| Partnerships for Innovation (PFI) | 10 | 0 | $3,194,575 | 0% |
| Research and Evaluation on Education in Science and Engineering (REESE) | 39 | 5 | $19,477,659 | 11% |
| Research Experiences for Teachers (RET) | 13 | 3 | $3,887,803 | 29% |
| Research Experiences for Undergraduates (REU) | 5 | 1 | $1,030,321 | 18% |
| Research in Disabilities Education (RDE) | 8 | 3 | $1,968,340 | 45% |
| Research on Gender in Science and Engineering FY 2006 (GSE) | 11 | 6 | $3,098,798 | 48% |
| Science of Learning Centers (SLC) | 7 | 5 | $1,280,000 | 73% |

*continued on the following page*

*Table 2. continued*

| Program | Number of Abstracts Sampled | Subset Including Ed Tech | Total Funding (for Sample) | Percent of Funding for Ed. Tech. |
|---|---|---|---|---|
| Science, Technology, Engineering, and Mathematics Talent Expansion Program (STEP) | 8 | 2 | $6,472,771 | 29% |
| Teacher Professional Continuum (TPC) | 10 | 4 | $5,547,180 | 30% |
| Tribal Colleges and Universities Program (TCUP) | 15 | 6 | $12,714,183 | 37% |
| Total: | 366 | 127 | $228,040,647 | 27% |

[1] No Abstracts Available

*Figure 2. Frequency of science disciplines in awarded projects*

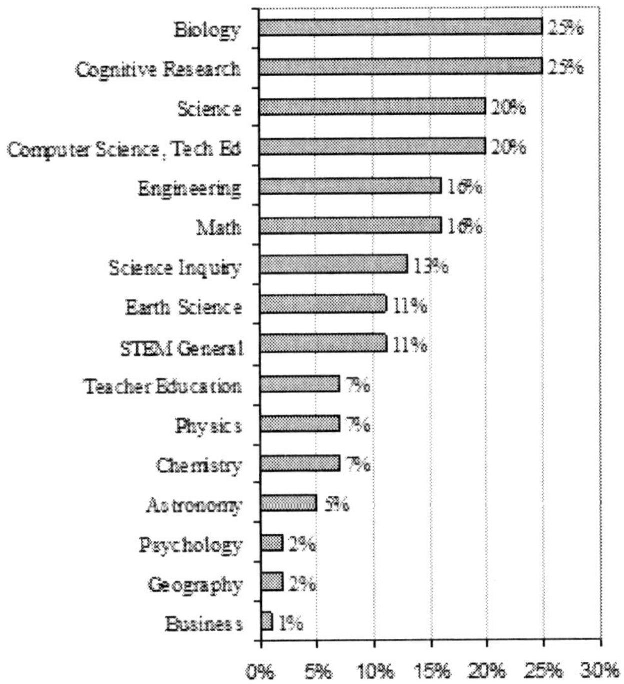


funded projects are a component of the *NEEDS* initiative. *NEEDS* (National Engineering Education Delivery System) is a digital library of learning resources for engineering education. Ten projects involved virtual laboratories. The projects, such as *IMAPS, VISTA, C-NERVE,* and *Powerlab,* provide computer-aided data collection and interactive tools. One distance learning effort, the *Algebra Project,* is an online collection of narratives related to Hurricane Katrina.

*Ecobeaker's* introduces teachers and students to simple computer modeling, applications of simulations in teaching and in science, and GIS data manipulation. Ten of the identified projects represented nationally disseminated online courses. *Cheminformatics* is an awarded SBIR phase II project to develop online tools in chemistry topics. *Webwork* is a community forum for open source software developers. The website *bcurbanecology.com* is part of an urban ecology course that

*Figure 3. Frequency of types of technologies in awarded projects*

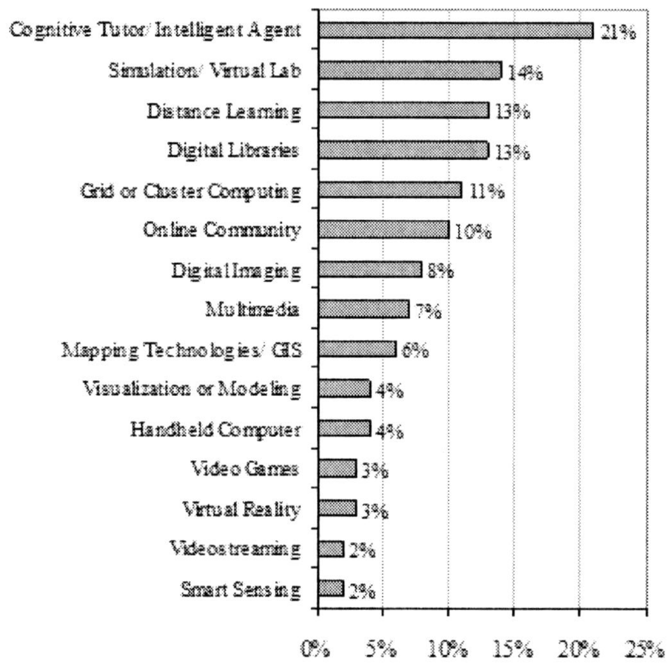


involves student collection of ecological data via GIS handheld technologies.

## CONCLUSION

NSF issued 65 solicitations for proposals over an 18-month period with 53.8 percent calling for educational technology components. Based on our sampling of 366 of the 1,180 funded projects, 34.7 percent included educational technology components. Further extrapolation suggests that about $150 million was spent on educational technology, or about $100 million per year.

A quarter of the funded education technology projects were in biology and cognitive science; adding general science, computer science, technical education, engineering, and math brings the total to about 65 percent. Major STEM-G disciplines, such as physics, chemistry, astronomy, and geography, were each less than 5 percent of the funded projects, and teacher preparation was only 6 percent. Clearly the pace of experimentation in developing educational technology varies greatly among STEM-G disciplines, and the most challenging at the high school level—physics and chemistry—lag behind biology. This may be a selection effect since there are many more biologists than physicists and chemists in universities that submit NSF proposals. However, the educational technology need is strong for physics and chemistry.

Many types of educational technologies were funded, with the single largest number being cognitive tutors/intelligent agents (CI/IA). These types of technologies rarely appear on lists, suggesting that NSF has recognized a technological intervention that most pundits and experts have overlooked. Distance learning and online communities make up the same percentage as CI/IA, showing that this more commonly acknowledged technology is being strongly supported by NSF and the authors of winning proposals. Videogames, virtual reality, and smart sensing are at the bottom of the list—perhaps they are still on the emerging frontier. But the other bottom dweller, video streaming, may be there because it is now a widely used educational technology that no longer needs NSF development support.

The process of science inquiry and collaboration is relying more on information technologies in order to store and analyze vast amounts of data. The trend of cyber-based inquiry is evidenced in NSF's call for building an information technology infrastructure to support complex data storage and analysis tools. NSF's CyberInfrastructure Vision for the 21st century states that "today's scientists and engineers need access to new information technology capabilities, such as distributed wired and wireless observing network complexes, and sophisticated simulation tools that permit exploration of phenomena that can never be observed or replicated by experiment. Computation offers new models of behavior and modes of scientific discovery that greatly extend the limited range of modes that can be produced with mathematics alone, for example, chaotic behavior." In the present study NSF funding integrated cyberinfrastructure (grid computing) across biology, engineering, computer science, astronomy, and teacher education.

## REFERENCES

Januszewski, A., Molenda, M., & Harris, P. (2007). *Educational technology: A definition with commentary*. Mahweh, NJ: Lawrence Erlbaum.

Learning Technologies, N. A. S. A. (2004). *Advanced technology applications for education benchmark study*. Retrieved January 9, 2009, from http://learn.arc.nasa.gov/benchmark/

## KEY TERMS AND DEFINITIONS

**Cognitive Tutor:** A computer program which allows a cognitive model of a student to be developed through his or her interaction with the program. Individualized instruction is them provided based on this model.

**Cyberinfrastructure:** An infrastructure of application-specific software, tools, and databases. cognitive science: the study of thinking and learning.

**Educational Technology:** (as defined in this article) the integration of electronic or digital products and theories from one or more of the following domains—cognitive science, social science, research on learning, and research on education, and/or classic sciences, such as biology, physics, or chemistry—specifically to increase the capacity of the user to learn content and/or cognitive processing skills.

**Intelligent Agent:** Software that performs an automated information gathering task and reports to the user at programmed intervals.

**Solicitation:** The request for applications for financial assistance which describes program objectives, eligibility requirements, performance activity, evaluation criteria, award conditions and other significant information about the award opportunity.

# Chapter 28
# Science for Everyone: Visions for Near-Future Educational Technology

**Charles A. Wood**
*Center for Educational Technologies®, Wheeling Jesuit University, USA*

## ABSTRACT

*Recent and emerging technologies offer many opportunities for exploration and learning. These technologies allow learners (of any age) to work with real data, use authentic scientific instruments, explore immersive simulations and act as scientists. The capabilities soon to be available raise questions about the role of schools and do rely on directed learning traditionally supplied by teachers. The prevalence of new tools and data streams can transform society, not just kids, into a culture of learning.*

## FRAMEWORK

Predicting the future of educational technology is difficult. New ideas, products and capabilities spring into existence and are developed within months, making it nearly impossible to predict the exciting new opportunities even a few years from now. But we can say that technology will be increasingly incorporated in most aspects of formal, informal, and casual education, and that it will build upon today's capabilities. In this review I discuss some current and emerging technologies and suggest how they might be used to increase learning in science, technology, engineering, mathematics, and geography (STEM-G). I don't describe well-entrenched tools, nor administrative or teacher management applications, and am not limited to classroom uses. These are personal choices of tools with high opportunities for engaging learning.

The foundation of most educational technologies is the World Wide Web and similar networks (e.g., wireless cell phone nets) that are already widely available in the Western World and many parts of Asia (though issues of financial access still loom everywhere). These networks are becoming more pervasive (even invasive), ever faster, and ultimately everyone in developed nations will be connected. The educational value of evolving networks is that learners will be able to connect to almost every conceivable kind of learning opportunity, anytime,

DOI: 10.4018/978-1-60566-936-6.ch028

from almost any place. Once online there are already a great variety of interactive learning activities, including control of real and simulated scientific instruments, expeditions to tag along with, courses to take, simulations and games to play, and literally billions of content-rich web pages to peruse. And increasingly these activities are not done alone, but rather within communities. *Second Life, Halo* and a growing number of other synthetic worlds bring teams together to talk, solve problems, share experiences, and collaborate. This is immersive, shared learning that young people seem to do naturally.

The value of educational technology is often unvalidated through formal assessments, yet is widely considered important because it makes learning more lively and more participatory, plus develops skill in using technology, itself a learning goal. If a learner becomes engaged in the tasks, it is assumed that there is a higher likelihood that the experience will be productive. The uses of technology described here are generally ones that require involvement and interactions—observing, collecting, displaying, and interpreting data; making decisions that have learning consequences; and using instruments normally beyond typical educational experiences. And the learning opportunities typically focus on important problems worthy of a learner's time and effort. Trivial labs and make-work exercises with non-real data are intrinsically boring, but activities based on real data and socially significant STEM-G issues capture attention. The Internet and online tech tools bring the world live into our learning environments.

## REAL TIME DATA

Many organizations place near real-time data online, providing opportunities for classes and individuals to experience authentic data analysis, often using professional tools. One extraordinarily successful example is the discovery of comets in the daily solar images obtained by the SOHO spacecraft and placed online. As of July, 2008 (http://home.earthlink.net/~tonyhoffman/SOHOleaderboard.htm), 1500 comets have been discovered by 67 amateur astronomers from 17 countries. Most of these comets would not have been found without the amateurs because professional astronomers do not have the time to search the daily flood of data. Other examples of discoveries from online astronomy data are numerous, including the discovery of asteroids, variable stars, and supernovae. School kids have even discovered proto-planetary objects out beyond Pluto. With the imminent arrival of massive surveys that map the entire sky every three nights, there will be more near real-time astronomical data online than all the astronomers in the world can review. With the creation of appropriate tools, there can be an explosion of science discoveries and explorations by adults and youth everywhere. Science is becoming an activity for everyone.

The GLOBE Project (http://www.globe.gov/fsl/html/aboutglobe.cgi?intro&lang=en&nav=1) is another successful example of youth making and analyzing observations, this time of their local environments, which contribute to global scientific understanding. More than 40,000 teachers have been trained to use GLOBE in their classrooms, and 19 million measurements have been contributed by more than one million students in 110 countries. GLOBE must be one of the largest international data collection/education programs ever.

Other types of near real-time data allow students to share the excitement of current geophysical activity. For example, near real-time seismic data are displayed on interactive maps by the U.S. Geologic Survey (http://earthquake.usgs.gov/eqcenter/) and the Incorporated Research Institutions for Seismology (http://www.iris.edu/seismon/). With creation of easily mastered online tools, learners could determine where earthquakes are centered, and estimate magnitudes and potential damage. Follow-up on CNN and other news sources would provide ground truth for major events.

The most widely available real-time information is weather data. Through the Weather Channel and every news channel on TV, near real-time weather images and radar data are visible nearly constantly. And the Internet brings even more satellite images of various types from around the Earth (and sometimes other planets). What is commonly overlooked by nearly all educators is that these continually updated streams of data can be used in the study of various aspects of science, geography, and even social studies. What needs to be developed are educational modules that easily import current data and guide directed investigations, perhaps using an artificial intelligence interface.

Using real-time data is doing real science with potentially important but unpredictable discoveries. It gives participants actual experience as scientists, hopefully turning them on to the excitement of intellectual exploration. All of these technology-enabled explorations are already being used, often by individuals with their home computers—the problem is that too few teachers or science center staff are proficient enough to help their learners get involved. Such authentic learning opportunities will become more widespread as technologies and learning guides come into existence as turnkey systems.

## REAL SCIENCE INSTRUMENTS IN YOUR HANDS

School is largely pretend; students learn content and skills just to learn them. Using real scientific instruments to collect real data transforms learning into an activity with a purpose. Working with real instruments that have to be properly used to make real measurements that in turn need to be carefully analyzed and interpreted is exciting. Compare sitting in a planetarium to being in an observatory with a real telescope. Schools rarely have real instruments (other than microscopes), but a number of capable scientific instruments are now available at costs that are achievable through PTAs, bake sales, and even federal and state technology dollars. But like classroom laptop programs (Zucker and Light, 2009), equipment purchases are typically only about 20% of the total cost of use, with training, service and support being more than 50%.

### Seismometers

A remarkably sensitive seismometer kit (http://jclahr.com/science/psn/as1/index.html) can be bought for $600 that allows students to record global earthquakes as they occur, and free software is available to calculate earthquake magnitudes and distances. Success in setting up and using the instrument requires training and perseverance by the educator, and support groups exist. All science centers should have a working seismometer on display, but apparently no turnkey solution is available yet.

You may already have access to a seismometer, because one is built into every Macintosh laptop computer. All modern laptops have sudden motion sensors (accelerometers) to detect rapid movement (falling) as an input to protect the hard drive. Macintosh makes access to that data available, and free software (SeisMac; http://www.suitable.com/tools/seismac.html) transforms Macs into portable seismometers. There appear to be no materials to support the educational use of SeisMac – an opportunity for someone.

### Telescopes

Telescopes and associated electronic cameras are becoming more common in schools, but their greatest applicability is outside school hours, so access is usually restricted. Of course, the Sun can be observed during school hours (as often can be the Moon), and new technology binoculars and small solar telescopes offer dramatic views of the solar corona and sunspots. Small astronomical telescopes typically costing $200 to $1,000 can be used in early evenings to study the moon and planets, leading to immersive exploration of geologic processes and the role of gravity and temperature in explaining the existence of atmospheres and water. The same small telescope also can be used for science projects by high schoolers, igniting a passion for science. The visual excitement of putting an eyeball in front of an eyepiece should not be underestimated—it is often an inspiring moment.

Radio telescopes do work during the day, and NASA has long had a radio telescope educational program. Project Jove (http://radiojove.gsfc.nasa.gov/) is an innovative $300 radio telescope specifically for studying changing radiation from Jupiter and the Sun. This educational program, which has been used by 850 schools and individuals, gives the real experience of doing science. More capable radio telescopes with moveable dish antennas capable of studying the galaxy are $7,000 (http://www.cassicorp.com/), which makes them more suitable as virtual instruments. The University of Florida (http://ufro1.astro.ufl.edu/liveaccess.htm) periodically streams live data from their radio telescope observatory.

## Robots

Robots were a science fantasy for much of the last 100 years but now are pervasive in industry and are entering personal living spaces and educational venues. Robots are intrinsically intriguing because, like a dog, they have some human traits and theoretically can be made to do what you want. And because robotic rovers have been used on Mars for more than a decade, there is an immediate tie to NASA science. Lego® Mindstorms® robots (http://mindstorms.lego.com/) are available from $250 and have extensive associated lesson plans and activities for formal and informal settings; unfortunately, too few learners have opportunities to use them. There are some good educational training materials, but more are needed, especially in using robots to improve teaching of science and math. Cheaper robots (coming from China) will make it easier for more educators to use robots in different learning environments.

## Satellite Image Observatories

The continuing and often unexpected natural changes of the Earth is a driver for learning why change happens. One of the easiest tools to bring change into a classroom or science center is a receiver that captures images of Earth directly from NOAA's passing weather satellites. These receivers capture about a dozen color images a day with 4 km resolution. Satellite capture cards and antennas (http://www.hamtronics.com/r139.htm) are available for $450 and connect to an existing PC for saving and displaying images. Satellite images allow tracking of weather systems, including hurricanes, but also detection of ocean temperatures and currents, and monitoring of agriculture growth, forest fires, floods, blizzards, and many other dynamic phenomena. Learning opportunities abound, and some teacher training materials exist, but there need to be national champions to promote the excitement of this opportunity for education rather than amateur radio operators.

## Image Processors

Computers provide the mathematical power to quickly manipulate vast matrices of numbers—and that is all a digital image is. Nearly all of the instruments described here produce digital displays of data that need to be enhanced and visualized. With inexpensive or free software such as Photoshop Elements, the ancient NIH image and its online version ImageJ (http://rsb.info.nih.gov/ij/), the information within data can be explored. Seeing remarkable images on NASA web sites is not as engaging as enhancing your own image, which creates ownership, pride and understanding.

## ONLINE SCIENCE INSTRUMENTS: REAL AND SIMULATED

Scientists use instruments to make observations and conduct experiments. Most students don't have access to quality scientific instruments and miss the experience of collecting real data that requires planning and analysis. Through the Internet there is now easy access to actual and simulated science instruments, especially online telescopes. The best known program is Telescopes in Education (TIE; http://www.telescopesineducation.com/), which has been used by students in hundreds of classrooms to plan, collect, and analyze real astronomical data. The decade-old program allows students to actually control a telescope and digital camera to acquire images. Other online telescopes, such as the oldest, the Bradford Robotic Telescope (http://www.telescope.org/) – now in the Canary Islands - allow users to submit observing requests and later receive the images by e-mail. The excitement of receiving "your" image is there, but the more complex involvement in operating a large scientific instrument is completely lacking. Slooh (http://www.slooh.com/) is a commercial online observatory that necessarily has an easy interface and sells telescope time for about $10/hr. Imagine if a school sold candy bars to raise money for observing with a real telescope!

A few other types of online instruments are available today. Students mail their own insects to the University of Illinois' Bugscope project (http://bugscope.beckman.uiuc.edu/) and actually control an electron microscope to image them. Nearly all other virtual electron microscopes (e.g. http://www.vcbio.science.ru.nl/en/fesem/intro/) are simply viewers that allow users to simulate control of focus, contrast, brightness, and magnification for pre-imaged specimens. They are good for training but not inspiration.

One of the most famous groups of simulated scientific instruments is produced by Project CLEA—Contemporary Laboratory Experiences in Astronomy (http://www3.gettysburg.edu/~marschal/clea/CLEAhome.html). Thirteen simulated astronomy labs have been developed that present real data and simulated tools to analyze them. Learners can measure the heights of mountains on the moon, the rotation rate of the sun, the expansion of the universe, and many other interesting projects. Afterwards students can use the same analysis tools with data collected with their own telescopes.

### Caves and Immersive Virtual Reality

In the 1980s virtual reality was the next big thing for research and education. It still is, and maybe someday will be. NSF has funded various organizations to experiment with virtual reality. Researchers at Brown University built an "eight-foot cubicle in which high-resolution stereo graphics are projected onto three walls and the floor to create an immersive virtual reality experience." I have entered the Brown cave (http://graphics.cs.brown.edu/research/cave/home.html) to explore canyons on the surface of Mars. The experience is captivating (actually dizzyingly real at first) and allows researchers to get inside data, to explore visualization in three dimensions. But caves are expensive and difficult to create and maintain; they may be excellent educational activities for research centers, but probably not for classrooms. Three-dimensional goggles are a lower cost option for classes and personal virtual reality exploration. Goggles are basically two tiny TV monitors embedded in a pair of spectacles with associated software to display the virtual images. But immersive data and guided exploration materials are needed for this technology to get beyond experimenters' labs.

Second Life requires neither a cave nor special glasses. It is a free 3-D virtual reality simulation (http://secondlife.com/) displayed on your monitor. You customize an avatar and explore and interact with 3-D environments and other avatars. Second Life currently has 40,000 to 50,000 people online at any time, and there are many science education areas to explore. Many simply display material that can be seen equally well on normal websites, but because your avatar can meet and speak with others a strong social component of learning is possible. A key component of Second Life is that anyone can build a virtual environment with provided tools that are simple to get started with. Currently Second Life is limited to people over the age of 18, and younger people are restricted to a Teen Grid; this will change.

## Simulations

Real scientists discover by reading what their predecessors and colleagues have done, conducting experiments, making observations, and exploring simulations of physical or biological processes. In classrooms, students read textbooks and conduct labs but rarely investigate simulations. Although hands-on experimentation is a critical learning experience, computers and the Internet offer opportunities for another kind of experiential learning where a virtual reality simulation is manipulated and its responses noted.

STELLA is a famous simulation program that can model many different types of systems and processes, but STELLA (http://www.iseesystems.com/softwares/Education/STELLASoftware.aspx) is not graphically interesting and has a steep learning curve. Fortunately, there are many other types of less intimidating simulations.

Virtual frog dissections were one of the first software labs to simulate experiments with higher quality graphics. There is now a vibrant cottage industry of transposing standard school lab experiments to computer interactions. The best—some of the physics simulators—offer a chance to learn by exploring parameter space. Most test the processes of experimentation in that the correct sequence of steps has to be performed to achieve success but don't give much feel for the underlying science. A study has demonstrated that at least one simulation of an electric circuits lab is more effective at teaching concepts than actually conducting the lab with lights, wire and batteries (Finkelstein et al, 2005).

A number of planetary science sims have been created to help students understand geologic processes by playing with them. Splat! (http://volcano.und.nodak.edu/vwdocs/crater_sim/intersect.html) was a Java applet that allowed the learner to vary the size distribution and number of impact craters formed on a planetary surface and watch both a view of the changing surface and a crater frequency graph as used by scientists. Comparison of Splats! visual maps of craters with real photos of the moon quickly reveals the different impact conditions of older and younger lunar surfaces. A companion applet (http://volcano.oregonstate.edu/kids/fun/volcano/volcano.html) models the distribution of ash from an explosive volcanic eruption. Users can vary the energy of the eruption, particle sizes, wind speed, and even gravity to explore what eruptions would be like under many different conditions—including on other planets.

Simulations such as these fit nicely into in-school learning because they require only tens of minutes to play and can embed a personal understanding of complex processes. Easy sim-building software would allow teachers to construct just the sims they need, but an approach more likely to be successful is to create libraries of sims for downloading.

## Serious Games

Serious games—a phrase that avoids the negative connotations of "educational games"—is a concept struggling to become a reality. In the 1980s there were two acclaimed educational computer games, *Oregon Trail* and *Where in the World is Carmen Sandiego*? Many other efforts, most famously *Math Blasters*, were really just drill and kill exercises that provided a video treat as a reward for completing numbing problems. The success of *OT* and *WITWICS* stemmed from their interesting story lines, rather than their graphics or interactivities. Originally *Oregon Trail* was very effective as a text game, and *Carmen Sandiego* was a hunt for an elusive criminal that was relatively engaging.

The best examples of existing serious games are simulations that teach procedures or sequential steps to achieve goals. For example, business games require maximizing profit in management simulations. Commercially successful games that mimic a business model are construction sims such as *Roller Coaster Tycoon* and *Sim-Earth*.

Video games offer a natural, motivating way of learning (Gee, 2003). Video games require solving a seemingly unending stream of increasingly complex challenges. Players constantly make decisions based on what they have just learned experientially. Unlike school where a test is the end of learning on a particular topic, no matter what the grade, game players keep playing until they master the knowledge and skills needed for success. Video games immerse players in a realistic, almost authentic environment, which players must absorb to be successful. In video games the medium can be the (educational) message.

Because of success that the U.S. Army has had in utilizing a game (*America's Army*) for training and recruiting, and because of the failure of many schools to excite students for learning, serious video games are now being rapidly discussed and even developed, with funding from NSF, NASA, and other US agencies. No one is yet certain if serious games can be created that successfully meld engaging game play with desired domain learning outcomes.

Serious video games typically aren't for classroom use because most take tens of hours to play, and there just isn't that much time available during the school day. Can an educationally meaningful game be built that requires only tens of minutes to play? Most serious games are for personal, self-directed learning, perhaps as homework reinforcement of what was taught in formal environments. Consider *Viral Escape*, a game to learn about fighting a cold. The scene is within a body, with movement possible along arteries, through tissues, and into bones, with scale changes available. To fight a cold, you (as dissolving chemicals from a cold relief pill) stream through the body looking for white blood cells to destroy and red ones to clone; T-cells must be fought off. Players of *Viral Escape* will become intimately familiar with the layout of the human body, its components, and what a cold really is. The student who plays this game at home will come to biology class the next day with a visceral familiarity of things that had only been words the day before.

Generally, serious games with STEM-G subjects have not been successful, and there have been few of them. Most successful serious (or semi-serious) games are simulations that explore ideas in history and warfare. There need to be more serious games based on the puzzles and clues (real and false) of scientists at work.

## 3-D Object Creation

Sometimes new technical capabilities for industry create new opportunities for education. One of the recent examples pregnant with possibilities is 3-D printers (http://www.zcorp.com/products/printersdetail.asp?ID=1). These were invented to provide a rapid and cheap way for industrial developers to prototype new products. These amazing devices "print" thin layers of plastic that fuse together, gradually building up a solid and strong three-dimensional structure. Printers that build black and white physical structures about 4"x4"x4" cost $20,000, and color constructors (my new word for 3-D printers) that make things up to 24" wide are about $60,000. If normal trends apply, these costs should be 10-20 times lower within five years. The materials to construct a fully-colored, fist-sided object currently cost about $10. One of the first public uses of 3-D printing technology is constructing and selling 3-D representations of user-created avatars for Will Wright's new videogame *Spore*.

3-D printers work with existing 3-D CAD software (http://www.solidworks.com/) and new 3-D scanners ($2,500; https://www.nextengine.com/indexSecure.htm) that allow duplication and modification of even intricate objects, such as chains. There will be many educational uses of this integrated scan-CAD manipulate-print 3-D technology. An obvious use will be creating 3-D models of chemical and biological structures, e.g. a DNA strand or the crystal structure of a mineral. More interesting constructions might include a human femur, which could be transformed with the CAD to a horse's femur (illustrating different strength needs) or more exotically a femur for a mammal on a planet of different gravity. Models of planetary landforms, such as impact craters, volcanoes, and stream channels, could be created from planetary digital terrain maps. Other possible examples include 3-D models of complex mathematical surfaces (Palais, 2006) or heads of unwrapped mummies revealed by computed tomography (Cheng et al., 2006). One researcher is experimenting with 3-D printers that might be able to reproduce themselves—this is probably not in the business plan of the 3-D printer manufacturers! 3-D constructors will provide real-life visualization, especially valuable for tactile learners and the seeing-impaired. And 3-D constructions take visualization to the next level—tactilization.

## eBooks—Content and Readers

The idea of electronic books that fit within a compact and ergonomic reader has been around for more than a decade; various companies promoting them have sprung into existence with fanfare and disappeared quietly months later. There are many advantages of an etext, including low cost, easy updating, and bundling with teacher notes and activities. The current lack of success of ebooks is because of their high cost (a good reader is $300), lack of a standard format that permits any book to be read on any reader, not quite good enough legibility, and the lack of a strong core market. Excellent progress is being made on the legibility, and the price should fall as sales volume builds. A format standard is being promoted, but probably none will fully emerge until a single vendor—and their format—becomes dominant. Perhaps it will be Amazon's Kindle (http://tinyurl.com/6n34al).

I propose that the killer app for ebooks is, in fact, textbooks, which, in their traditional form, is a billion dollar annual market. The state of Texas has discussed requiring all schools to buy etexts and a few districts have experimented with them. If a large state like Texas replaces physical books with etexts, a strong market will be created overnight, a format standard will likely emerge, and reader cost should decline.

eText readers will presumably be paperback size with very high text resolution, color for illustrations, and the standard features of keyword searches and electronic bookmarking and highlighting. To be more useful as general eTutors, eBook readers will evolve to increase their computing capability so that they can run simulations, access the Internet, and e-mail for interacting with teachers and fellow students. These eTutors will essentially be paperback-size computers that wirelessly connect anywhere (like cell phones) to the entire world of digital learning. And they must cost less than $100/yr to be compatible with current school spending on textbooks. The proposed ideal etext reader could really be just a larger screen iPhone, which already has reader software and all the other requested features, except for the price.

It might be thought that this is the same as using a laptop computer, but eBook readers will be as different from a laptop as a cell phone is. Laptops are too big and too expensive to carry all the time. Netbook computers are a significant step in the right direction, but still too large. A paperback size eBook reader provides a large enough screen size to be useful for reading (unlike a cell phone or PDA) but small enough to be put in a purse or a large pocket.

A smaller market that can develop for eBooks is specialized libraries for particular disciplines. This would be of value to college students and professional learners and to advanced hobbyists. Examples might be eMed, which could be an exhaustive collection of reference materials for pre-med and medical students, including live subscriptions to relevant journals and news. eSky could be a library for amateur astronomers, containing the entire 60 year run of *Sky & Telescope* magazine, a collection of observing guidebooks, and live updating from NASA and other astronomy web sites.

## iExplorer

iPhones are the coolest and most successful portable technology ever (displacing iPods). The ease of operation, powerful software, huge capacity, portability and Zen design make iPhones an ideal platform for a new advanced mobile education technology – the iExplorer. With a built in camera that easily uploads images to a website, a GPS receiver to pin-point location, and a web browser, the iPhone is a powerful platform to transform into a multi-purpose learning tool. For example, students conducting a botany field project could capture an image of a plant, record an audio description, tag the lat/long/elevation/time, and post it all online. Or they could use an online identification tool to compare what they see with images of typical specimens. As a roving tutor, specially written iPhone applications could lead learners on field trips, guiding them where to go, when to stop and observe specific features, and provide tools to create their own record of the experience to share.

At least one standalone mobile education module already exists. The Celestron SkyScout (http://www.celestron.com/skyscout/) is a $200 handheld, zero power telescope with a built in GPS unit and sky catalog. Point the SkyScout at any naked eye natural object in the sky and it will display and say what the object is and give information on its nature.

## Cool Science TV

Motivating youth toward careers in science requires overcoming a common cultural stereotype (certainly common in the US) that science is boring and uncool. We know how to overcome such stereotypes. The *CSI* (*Crime Scene Investigation*) television programs have portrayed forensic science as an exciting career, and schools all across the country are scrambling to develop courses and programs to satisfy the strong educational demand that has resulted.

Commercial TV is where effective educational/motivational efforts belong, but traditionally, educationally enlightening TV has been on public TV. In the US, *Nova* and other award-winning TV science shows are seen by an older, wealthy, white PBS audience, not kids from various backgrounds whom we need to transfix with self-efficacy visions. Additionally, traditional science programs are not a continuing fictional drama, but rather a one-off factual report. We need a continuing story with STEM-G characters we gradually build positive relations with. Here is a sample storyline that might work. A team of engineers and scientists is designing the Orion spacecraft that will carry new adventurers to the Moon. They have to discover the best design features and build a spacecraft, incorporating (or repulsing) inputs and demands from lunar scientists, administrators, contractors and politicians, and with rivalries and uncertainty within their own extended team. This storyline is in the context of the growing recognition of a generational competition with China, India, Japan, Europe, and maybe Russia for space exploitation, not just exploration. There would be many opportunities for character development and interaction as well as audience exposure to actual technical issues that are finally, just barely sometimes, overcome each episode.

A TV program might not appear to be a significant innovative educational technology, but it actually can be because a program creates a strong public appreciation of anything positively depicted. If program producers can be cajoled into creating TV series that portray scientists as real-life explorers of exciting ideas and unravelers of significant mysteries, such series may be more effective at turning kids on to STEM-G careers than all other efforts combined.

## MAKING TECHNOLOGY WORK

Critical to almost every idea here is that teachers (or science center staff or learners themselves) will need to learn something new, how to operate new equipment or software, and how to tie it to meaningful (standards-driven and test-driven) learning goals. There will always be a few teachers who undertake such projects, and these are the ones who win awards and speak at national conferences. The real question for federal agencies that promote education improvement is how do we make these technology-centric learning tools easy enough for widespread adoption so that millions of kids experience them? Piecemeal solutions with occasional workshops and massive online collections of learning activities aren't the answers. We have had these for years, but youth achievement in STEM-G learning has not improved.

Just as the clocks on home VCRs have been blinking 12:00 for years, much modern educational technology is too hard for easy use. We don't need more ed tech innovations, we need more adoptions of programs and tech tools that have been proven to work. I propose that a new organization (a National Center to Make Technology Easy) is needed not just to promote best practices, but to transform proven ideas and tools into near-turnkey solutions with appropriate training and curricular materials and ties to standards. Developers and early adapters of technologies often get caught up in the minutiae of the technology and leave behind the educator who simply wants to use the tool. The NCMTE organization would develop and market not kits nor do-it-yourself instructions, but ready-to-use tech tools (like seismometers, satellite receivers, etc.), with simple instructions on how to incorporate them into lessons. A key to success would be the development of a community of users for mutual support. Tech tools have to be reduced to the simplicity of a toaster so that they will be widely adopted. The goal is not mastery of technogizmos, but exploration of concepts and knowledge, and falling in love with learning.

## INTERNET LEARNING: THE END OF SCHOOLS

Schools are failing. The evidence is everywhere from test scores to shootings to the growing number of alternatives such as online academies and home schools. The products and ideas described here provide guidance for a coming educational technology revolution that will transform the idea of learning. Just saying *learning* opens the mind to new ways of learning that are obscured by the word *school*. Learning does not have to occur in certain buildings at certain hours in groups of 24 under the direction of people trained in only one discipline (education). Learning is becoming a 24/7 activity with tools and information available anywhere through mobile connections to everything on and off the planet.

Learning outside of a highly structured school building may not be appropriate for everyone at all times, and a culture will have to evolve to allow it to augment if not ultimately replace traditional school practices. Guided learning and being in contact with the world offer an infinite number of learning styles and directions. Guidance is critical, as is assessment—but by achievement rather than by testing. The traditional idea of educating a well-rounded citizen is one to strive for—which we are failing at in many schools—and the proposed internet learning will result in people being differently learned, some having great expertise in some topics and less awareness of others. But each person can find the path that works for her or him, and in aggregate, for society.

## Creating a Science-Centered Society

The activities described above are largely aimed at the youth whom society need to convince to become the scientists and engineers necessary for our modern economies to continue. But science could become a dominant social element worldwide for children and adults. With the coming easy access to immense amounts of near real-time data and the tools to analyze and explore it, understanding our planet and the universe beyond could become a global pastime. Instead of being a passive receptor of whatever science is occasionally selected for the nightly news, any curious person with Internet access will be able to monitor many dynamic characteristics of the Earth, the sun and the night sky. I envision large numbers of WorldWatchers tracking developing hurricanes, retreating polar caps, pollution from nearby smokestacks, the migration of fall foliage, mining on the moon, a supernova in a distant galaxy, and the scream (with x-ray frequencies remapped as audio) of gases being sucked into the black hole at the center of our galaxy. *Nova* will become the most watched (or podcast) TV program, and competition between comet hunting teams will be featured on Monday night TV. Soon federal agencies will be pleading for students to study business and political science.

## REFERENCES

Cheng, R., Brown, W. P., Fahrig, R., & Reinhart, C. (2006). An Egyptian child mummy. *Science.* Retrieved Aug. 3, 2006, from http://www.sciencemag.org/cgi/content/full/313/5794/1730

Gee, J. P. (2003). *What video games have to teach us about learning and literacy.* Palgrave Macmillan.

Palais, R., & Benard, L. (2006). Still life: Five glass surfaces on a tabletop. *Science.* Retrieved Aug. 3, 2006, from http://www.sciencemag.org/cgi/content/full/313/5794/1730

Zucker, A. A., & Light, D. (2009). Laptop programs for students. *Science, 323*, 82–85. doi:10.1126/science.1167705

# Chapter 29
# Instructional Design, Web 2.0 Style

**Bruce C. Howard**
*Center for Educational Technologies®, Wheeling Jesuit University, USA*

## CONCLUSIONS

In the previous articles, we reported on the results of a multifaceted research study on how to benchmark and use emerging educational technologies. Our approach blended classic research methods with those used in market research studies. We gathered data and expertise from a variety of sources, including academic research articles, industry reports, interviews with leaders and national pacesetters, and the experiences of our own veteran staff. Our objective was to create a means by which decisions about affordances, constraints, and effective use could be made in a just-in-time fashion. We have only scratched the surface.

To summarize, we conducted a variety of studies across two phases. We looked back at important reports and studies from the United States to identify criteria for effectiveness. To date, there have been few research studies of effective technologies and their use that have resulted in design principles, decision-making principles, or metrics for gauging effectiveness. Of those we reviewed, we found that they recommended the following learning features: encourages collaboration, increases knowledge and skills, relates academics to workplace skills, integrates into the curriculum, improves assessment/evaluation, and changes the teaching process.

We interviewed pacesetters. Some emphasized that the classroom context and the role of the teacher are the most important objects of study, not the technology. Engagement and interactivity were important themes. Pacesetters cited teacher development and implementation variables. Collaboration via technology was the most often cited technology for future education impact.

We developed a list of exemplars from multiple sources and uncovered their commonalities and design principles. The list included notable trends, such as learning management systems; collaboration and communication; mobility and interoperability; engagement; user-generated content; games and simulations; visualization and modeling; real-time, live events; commercial careers-focused resources; creation of student products; and peer-reviewed, juried educator resources. The large number of teacher resource exchange communities indicates that many educators need and seek support for more

DOI: 10.4018/978-1-60566-936-6.ch029

effective integration of technology. The programs represented by this list will serve as an additional data source for future phases of this research.

We examined funding trends from the National Science Foundation, determining the degree to which program solicitations included educational technologies. We also examined the emphasis placed on educational technologies by those who received the awards. Many types of educational technology projects were solicited and funded, with cognitive tutors/intelligent agents and distance learning/online communities the most common initiatives.

In forecasting the future, the executive director of the NASA-sponsored Classroom of the Future program wrote about the potential of real time data, real science instruments in learner's hands, online science instruments, caves and immersive virtual reality, simulations, videogames, 3-D object creation, e-books, and science television, and he ended with a vision for a society fully engaged in science.

In the second phase of this project we took the findings from Phase One and applied them to a new development project. The result was the *EdTech Collaborative* website, which uses web 2.0 technologies to bring the curriculum development community together with classroom users, exploring issues related to design principles, metrics for choosing and using technologies, and the sharing of best practices. In the next section we report on the progress of this development project.

## NEXT STEPS

We know from the research literature and conventional wisdom that educational technologies hold a great deal of promise to move inspired learners to the next level, to fully engage them, and to help them to learn rich content knowledge and skills. We also know from the research literature and our own experiences that all too often emerging technologies can be engaging without really being educational. How do we move ahead if we are delayed waiting for the research to catch up? How do we make reasoned choices when the technologies keep changing? We have to adapt, get organized, and collaborate with divergent communities.

Take podcasting, for instance. Even before any research has come out on podcasting's efficacy as an educational tool, universities have embraced it widely. To create podcasts, a user merely needs something to say, a way to record those thoughts (microphone and/or live screen capture utility), and a means to disseminate the final product. University professors are teaching themselves how to do this. Within NASA, educational video producers and instructional designers are teaching themselves too. NASA podcasts may be downloaded for free from iTunes®, along with lectures on every assorted subject from colleges worldwide. We met with a team from a K-12 private school in Baltimore, Maryland, who taught its science teachers how to create their own video podcasts. To reinforce each day's lesson, these teachers uploaded the podcasts to the school learning management system for students to download. While this may appear to be a revolution, it is perhaps an unfortunate one. That is, while time and energy are being spent on learning the particulars of *how* to do it, these early adopters might be missing opportunities to learn *how best* to do it. Mistakes are replicated every day. Learners are watching or listening to *bad* podcasts among the good. Little is known about the pedagogy, and little will be published for a few years.

Downing and Holtz (2008) phrased the problem like this: "...much of the world is restricted from emerging best practices simply because of a lack of collaboration between educators and information/ communication technology developers... Developers often know what is possible, but lack an application, while educators know what students need, but are unaware of the extent to which technology can meet those needs" (page 10).

We have a solution. In one small corner of this

revolution, we are adapting, organizing, and collaborating. We have developed a website called the *EdTech Collaborative.* Its purpose is to provide an information-rich resource around which various professional communities may communicate and collaborate. The site targets two communities: instructional designers and education practitioners. By designers we mean anyone who finds themselves in a position of integrating educational technologies into the design or development of curricular products or programs, such as master teachers, college professors, lesson plan creators, professional curriculum developers, or informal education providers. Practitioners are those people who make use of these products and programs with the target audience.

The site's content is created by our staff and the users. The programming allows for multiple levels of user input, from creating entries (like a blog), making comments (like a discussion board), editing articles (like Wikipedia), and completing or reading reviews (like Epinions.com™). After registering, users may contribute at many levels, from creating new strands to reviewing existing entries, commenting, or rating. All entries are posted immediately, but site moderators may pull them or edit them. As the site grows with additional content, we hope to recruit more and more moderators with various types of expertise.

For the designer/developer we feature three areas: *Design, Develop,* and *Research.* For the educator/practitioner we feature three parallel areas: *Choose, Use,* and *Test.* To view the information, a user first selects a strand and then clicks on one of the six tabs corresponding to each of these main areas. For example, in Figure 1 we show the "podcasting" strand.

Within the strand under *Design*, the user sees design principles that should be considered before the instructional event is developed. Each principle is explained in the context of that strand. For instance, in the podcasting strand one design principle is, "Use research questions to pique learners' curiosity." The explanation reads, "In the script for the podcast, the speaker could incorporate a description of how the present topic addresses a fundamental research question. The research could be posed in such a way as to pique learners' curiosity and spark the listeners' ideas about the answer."

Under *Develop* we provide a forum for discussion about lessons learned during the development process itself. For instance, with podcasting producers might provide examples of methods

*Figure 1. Screenshot from the EdTech Collaborative website showing an example for podcasting and the six featured areas: Design, Develop, Research, Choose, Use, Test*

that worked well or did not work so well in the production process, such as what editing settings to use or how to maximize exposure on *iTunes*. College professors might recommend inexpensive microphones to buy.

Under *Research* users read summaries of seminal academic research articles related to the strand topic. For example, the podcasting strand includes an article on how to design multimedia for maximum cognitive impact.

Educators and practitioners might be more interested in the *Choose, Use,* and *Test* areas. Under *Choose* users see a series of metrics, organized into categories. The metrics are provided to help users choose or apply the topic of the strand to their own personal situation. For instance with podcasting, in the *Retention* category, are these metrics: "The technology enables the content to become more memorable to the learner," "The technology models complex thinking processes," and "The technology provides the mechanisms to help the learner recognize what has been learned or experienced." Next to each metric is a graphic showing other users' ratings on a scale from 1 to 5 stars. Users may also rate the technology on each metric and provide comments.

In the *Use* area educators have a forum for dialog on best practices related to the strand topic. These best practices have to do with the actual use of the technology in an integrated lesson in the classroom. In the *Test* area users may see summaries of market research, statistics, or nonacademic articles, which provide information related to choosing and using the technology. For instance, under podcasting one entry reads, "ComScore study reveals that males between the ages of 18-24 are more likely to download podcasts via *iTunes*." It is presumed that as technologies emerge, the marketing information will precede associated academic research. Think of how valuable this information resource would have been a few years ago when the academicians were conducting classroom research on handheld PDAs (personal digital assistants), and the market share of such devices was plummeting. As a consequence, it is questionable how valuable those research studies really were with such a scarcity of the technology itself in present classrooms.

In addition to these six main areas for the strand topics, the website features an exploratory area for discussion of informal experimentation with emerging technologies. The idea behind this area is to stay up to the minute. As users experiment, they have the opportunity to learn from each other. As the group of experimenters starts to converge on similar opinions, then it would be time to turn the topic of the experimentation into a more formal strand. For instance, a present topic of experimentation is the educational applications of *SecondLife*, an immersive first-person virtual world that has been the topic of many speculative articles in educational technology users' circles. Within a year or two as consensus builds about its educational merits and as it is tried out in classrooms, one of the website's contributors will write a strand entry on it, including design principles, lessons learned, metrics, best practices, and any market research or academic research available.

In summary, the *EdTech Collaborative* is adaptable, organized, and collaborative. As a website it is more responsive to changing classroom conditions than a printed resource, and it is expandable and editable. Organizationally, it brings together information from a variety of sources, making it worthy of frequent visits by the community of users. In terms of collaboration, the website seeks balance and emphasizes communication among divergent communities of practice. The academic researchers have opportunity to read or provide summaries of relevant papers. The commercial industry may view and contribute market studies and reports. Developers can hear from classroom users more directly. Each of these communities can inform each other in meaningful ways.

As users and developers we strive to incorporate innovative and more effective uses of educational technologies into our program and product offerings. By specifying the design principles, metrics, and best practices in the use of learning technologies today, we can better guide the development of more effective learning tools. We must explore different applications of new technologies to identify the most innovative and effective uses, and we should strive to make those applications more adaptable, organized, and collaborative.

## REFERENCES

Downing, K. F., & Holtz, J. K. (2008). *Online Science Learning: Best Practices and Technologies*. Hershey, PA: IGI Global.

## KEY TERMS AND DEFINITIONS

**Educational Technology:** The use of technology to improve teaching, learning, and the school environment.

**Pacesetters:** A person who is a leading influence in his or her filed of study or work.

**Exemplar:** A model for representing excellence as a whole

**Design Principle:** A descriptor or characteristic which can be used by educators and designers to structure the content and features of an educational technology and its implementation into curriculum.

**Instructional Design:** A systematic approach to the design and development of instructional materials and products using objectives, teaching strategies and evaluation to meet learning needs.

**Educational Practitioner:** One who uses the products and programs developed by instructional designers in a learning environment with a specific target audience.

**Metric:** An objective standard for measurement for content, structure, or performance.

# Compilation of References


3DS Max. Retrieved from http://www4.discreet.com/3dsmax/

Abbott.,L., Austin, R., Mulkeen, A., & Metcalf, N. (2004). The global classroom: Advancing cultural awareness through collaborative work using ICT. *European Journal of Special Needs Education, 19*(2).

Accounting Education Change Commission (AECC). (1990). Objectives of education for accountants: Position statement number one. *Issues in Accounting Education, 5*(2), 307-312.

Accounting Education Change Commission (AECC). (1992). Objectives of education for accountants: Position statement number two. *Issues in Accounting Education, 7*(2), 249-251.

Adams, D. A., Nelson, R. R., & Todd, P. A. (1992). Perceived usefulness, ease of use, and usage of information technology: a replication. *MIS Quarterly, 16*(2), 227–247. doi:10.2307/249577

Adams, E. C., & Freeman, C. (2000). Communities of practice: Bridging technology and knowledge assessment. *Journal of Knowledge Management, 4*(1), 38–44. doi:10.1108/13673270010315939

Agarwal, R., & Karahanna, E. (2000). Time flies when you're having fun: Cognitive absorption and beliefs about information technology usage. *MIS Quarterly, 24*(4), 665–694. doi:10.2307/3250951

Agarwal, R., & Prasad, J. (1999). Are individual differences germane to the acceptance of new information technologies? *Decision Sciences Journal, 30*(2), 361–391. doi:10.1111/j.1540-5915.1999.tb01614.x

Ajzen, I. (1988). *Attitudes, personality, and behavior.* The Dorsey Press, Chicago.

Ajzen, I. (1991). The theory of planned behavior. *Organizational Behavior and Human Decision Processes, 50*(2), 179–211. doi:10.1016/0749-5978(91)90020-T

Ajzen, I., & Fishbein, M. (1980). *Understanding attitudes and predicting social behavior.* Englewood Cliffs, NJ: Prentice-Hall.

Ajzen, I., & Fishbein, M. (1980). *Understanding attitudes and predicting social behavior.* Englewood Cliffs, NJ: Prentice-Hall.

Aken, A., & Michalisin, M. I D.(2007). The impact of the skills gap on the recruitment of MIS graduates. *SIGMIS-CPR.* St. Louis, MI.

Albrecht, W. & Sack, R. (2000). Accounting education: Charting the course through a perilous future. *American Accounting Association, Accounting Education Series, 16.* Retrieved from http://aaahq.org/pibs/AESv16/toc.htm.

AlgoViz. Retrieved from http://www.cgv.tugraz.at/people/ullrich/AlgoViz;internal&action=_setlanguage.action

Allen, I., & Seaman, J. (2005). *Growing by degrees: Online education in the United States, 2005.* The Sloan Consortium.

Allen, I.E., & Seaman, J. (2008). Staying the course: Online education in the United States. *The Sloan Consortium*, 1-28.

Allen, M., Bourhis, J., Burrell, N., & Mabray, E. (2002). Comparing student satisfaction with distance education to traditional classrooms in higher education: A meta-analysis. American Journal of Distance Education, 16(2), 83–97. doi:10.1207/S15389286AJDE1602_3doi:10.1207/S15389286AJDE1602_3

Ally, M. (2004). Foundations of educational theory for online learning. In T. Anderson & F. Elloumi (Eds.), *Theory and practice of online learning* (p.5). Athabasca University, Alberta: Creative Commons.

Althaus, S. L. (1997). Computer-mediated communication in the university classroom: An experiment with online discussions. *Communication Education, 46*, 158–174. doi:10.1080/03634529709379088

Alvesson, M., & Sköldberg, K. (2000). *Reflexive methodology: New vistas for qualitative res*earch. London: Sage.

American Institute of Certified Public Accountants (AICPA). (1998). CPA vision project: Focus on the horizon. *Journal of Accountancy, 3*(6), 186.

American Institute of Certified Public Accountants (AICPA). (1999). AICPA core competency framework for entry into the accounting profession. Retrieved from http://www.aicpa.org/edu/corecomp.htm.

Anderson, J., & McCormick, R. (2005). *Pedagogic quality in policy and innovation in education: Quality criteria*, Brussels: EUN Schoolnet

Anderson, J., & McCormick, R. (2006). Pedagogic quality–supporting the next UK generation of e-learning, Section 26. In U-D. Ehlers & J.M. Pawlowski (Eds.), *Handbook on quality and the standardisation in e-learning.* Springer, Berlin.

Anderson, T. (2004). Toward a theory of online learning. In T. Anderson & F. Elloumi (Eds.), *Theory and practice of online learning.* Athabasca University, Alberta: Creative Commons.

Anderson, T., Rourke, L., Garrison, D. R., & Archer, W. (2002). Assessing teaching presence in a computer conferencing context. *Journal of Asynchronous Learning Networks, 5*,(2). Retrieved April 20, 2003 from, http://sloan-c.org/publications/jaln/v5n2/v5n2_anderson.asp

Anderson, W. (2008). Online degrees: Are they right for you? *Certifications Magazine*, *10*(6), 18–23.

Angel Learning. (2009). Retrieved January 2009 from, http://www.angellearning.com/

Answers.com. (2007). *Folksonomy.* Retrieved from http://www.answers.com/ topic/folksonomy?cat=technology

Arbaugh, J. B., & Duray, R. (2002). Technological and structural characteristics, student learning and satisfaction with Web-based courses. An exploratory study of two on-line MBA programs. *Management Learning, 33*(3), 331–147. doi:10.1177/1350507602333003

Arbaugh, J. B., & Hiltz, S. (2005). Improving quantitative research on aln effectiveness. In S. Hiltz & R. Goldman (Eds.), *Learning together online: Research on asynchronous learning networks* (pp. 81-102). London: Lawrence Erlbaum Associates.

Argyris, C. (1999). *On organizational learning* (2nd Edition). Malden, MA: Blackwell Publishers Inc.

Aspin, D. N. (2007). *Lifelong learning concepts and conceptions' in philosophical perspectives on lifelong learning.* The Netherlands: Springer.

Assessment Reform Group – ARG. (1999). *Assessment for learning: Beyond the black box.* Cambridge: University of Cambridge, School of Education.

Atkins. (2002). *Critical discourse analysis: A letter to expatriates from the Rt. Hon. Sir Norman Fowler MP.* Retrieved May 31, 2005, from http://www.cels.bham.ac.uk/resources/essays/Andrew%20Atkins%20Assignment%205.pdf

Atkins-Sayre, W., Hopkins, S., Mohundro, S., & Sayre, W. (1998). *Rewards and liabilities of presentation software as an ancillary tool: Prison or paradise?* ERIC Document Reproduction Service; ED 430260.

Austin, R. (2006). The role of ICT in bridge-building and social inclusion: Theory, policy and practice issues. *European Journal of Teacher Education, 29*(2), May, 145-161.

Austin, R., & Anderson, J. (2008). *E-schooling: global messages from a small island.* Oxford: Routledge.

Austin, R., Abbot, L., Mulkeen, A., & Metcalfe, N. (2003). Dissolving boundaries: Cross-national co-operation through technology in education. *The Curriculum Journal, 14*(1), 55-84.

Austin,R. (2007). Reconecting young people in Northern Ireland. In Loader, B. (Ed.) *Young Citizens in the Digital Age*, Routledge.

Australian Capital Territory Department of Education. (2005). *Emerging technologies: A framework for thinking.* Retrieved Sept. 25, 2006, from http://www.det.act.gov.au/publicat/pdf/emergingtechnologies.pdf

Bailey, J. t L., & Stefaniak, G. (2001). *Industry perceptions of the knowledge, skills, and abilities needed by computer programmers – SIGCPR 2001.* San Diego, CA USA.

Baker, E. L. (2001). Technology: How do we know it works? In W. F. Heineke, L. Blasi, & J. Willis (Eds.), *Methods of evaluating educational technology* (pp. 77-84). Greenwich, CT: Information Age.

Baker, E. L., & O'Neil, H. F. (2006). Evaluating web-based learning environments. In H. F. O'Neil & R. S. Perez (eds.), *Web-based learning: Theory, research, and practice* (pp. 3-20). Mahwah, NJ: Erlbaum.

Banas, E. J., & Emory, W. F. (1998). History and issues of distance learning. *Public Administration Quarterly, 22*(3), 365–383.

Bandura, A. (1990). Selective activation and disengagement of moral control. *Journal of Social Issues, 46*(1), 27-46.

Bandura, A., Barbaranelli, C., Caprara, G., & Pastorelli, C. (1996). Mechanisms of moral disengagement in the exercise of moral agency. *Journal of Personality and Social Psychology, 71*(2), 364-374.

Barab, S.A., Barnett, M., Yamagata-Lynch, L., Squire, K., & Keating, T. (2002). Using activity theory to understand the systemic tensions characterising a technology-rich introductory astronomy course. *Mind, Culture, and Activity, 9*(2), 76–107.

Bartsch, R., & Cobern, K. (2003). Effectiveness of PowerPoint presentations in lectures. *Computers & Education, 41*, 77-86.

Bassilli, J. N., & Joordens, S. (2008). Media player tool use, satisfaction with online lectures and examination performance. Journal of Distance Education, 22(2), 93–108.

Bates, A. (1995). *Technology, open learning and distance education.* London: Routledge.

Bates, R., & Khasawneh, S. (2007). Self-efficacy and college students' perceptions and use of online learning systems. *Computers in Human Behavior, 23*, 175–191. doi:10.1016/j.chb.2004.04.004

Baumann, M. (2006). Caught in the web 2.0. *Information Today, 22*(8), 38.

Beaubien, J. (2002). Harnessing the power of complexity in an online learning environment. In K.E. Rudestam, & R. Schoenholtz(Eds.). *Handbook of oOnline learning: Innovations in higher education and corporate training.* Thousand Oaks: Sage Publications, 221-236.

Becta (British Educational Communications and Technology Agency). (2005). *An introduction to learning platform.* Retrieved from http://www.becta.org.uk/leas/display.cfm?section=1_2

Bee, R., & Usip, E. (1998). Differing attitudes of economic students about web-based instruction. *College Student Journal, 32*, 258–269.

Bellamy, K., & McLean, D. (2003). The mechanics of PowerPoint (Electronic Version). *Journal of Audiovisual Media in Medicine, 26*( 2), 74-78.

Benbunan-Fich, R., Hiltz, S., & Harasim, L. (2005). The online interaction learning model: An integrated theoretical framework for learning networks. In S. Hiltz & R. Goldman (Eds.), *Learning together online: Research on asynchronous learning networks* (pp. 19-37). London: Lawrence Erlbaum Associates.

Benefits & Compensation Digest (2008). Tools and techniques, *45*(8), 10.

Benenson, I., Torrens, P.M. (2004). Geosimulation: object-based modeling of urban phenomena: Computers, Environment and Urban Systems. *Computer, Environment and Urban System*, *28*, 1-8.

Berge, Z. (1995). Facilitating computer conferencing: Recommendations for the field. *Educational Technology*, *35*(1), 22–30.

Bernard, R. M., Abrami, P. C., Lou, Y., Borokhovski, E., Wade, A., Wozney, L., et al. (2004). How does distance education compare to classroom instruction? A meta-analysis of the empirical literature. Review of Educational Research, 74(3), 379–439. doi:10.3102/00346543074003 379doi:10.3102/00346543074003379

Berson, I. (2003). Grooming cybervictims: The Psychosocial effects of online exploitation for youth. *Journal of School Violence*, *2*(1), 5–18. doi:10.1300/J202v02n01_02

Bhattacherjee, A. (2001). Understanding information systems continuance: An expectation-confirmation model. *MIS Quarterly*, *25*(3), 351–370. doi:10.2307/3250921

Billari, C.F., Prskawetz, A. (2003). *Agent-based computational demography: Using simulation to improve our understanding of demographic behaviour.* Germany: Phisica-Verlag.

Blackboard. (2009). Retrieved January 2009 from, http://www.blackboard.com/

Blender. Retrieved from http://www.blender.org

Bloom, B. (1956). *Taxonomy of educational objectives.* New York: David McKay Company, Inc.

Bloom, B. S., Englehart, M. B., Furst, E. J., Hill, W. H., & Krathwohl, O. R. (1956). *Taxonomy of educational objectives: The classification of educational goals. Handbook 1: The cognitive domain.* New York: Longman.

Bloom, B.S. (1956). *The taxonomy of educational objectives: Cognitive domain.* New York: David McKay.

Bocchi, J., Eastman, J. K., & Swift, C. O. (2004). Retaining the online learner: Profile of students in an online MVA program and implications for teaching them. *Journal of Education for Business*, *79*(4), 245–253. doi:10.3200/JOEB.79.4.245-253

Bødker, S. (1989). A human activity approach to user interfaces. *Human-Computer Interaction*, *4*, 171–195. doi:10.1207/s15327051hci0403_1

Bonk, C. J. *(2002). Collaborative tools for e-learning.* Chief Learning Officer Magazine, November. *Retrieved-January2009 from,* http://www.clomedia.com/

Bonk, C. J., & Graham, C. R. (Eds.). (2007, in press). *Handbook of blended learning: Global perspectives, local designs.* San Francisco, CA: Pfeiffer Publishing.

Bower, B. L. (2001). Distance education: Facing the faculty challenge. *Online Journal of Distance Learning Administration,4*(2). Retrieved March 22, 2006 from http://www.westga.edu/%7Edistance/ojdla/summer42/summer42.html

Braksick, L. W. (2000). *Unlock behavior, unleash profits.* New York: McGraw-Hill.

Bransford, J. D., Brown, A. L., & Cocking, R. R. (2000). *How people learn.* Washington, DC: National Academy Press.

Bransford, J., et al. (2003). *How people learn: brain, mind, experience, and school: Expanded edition.* Washington, D.C.: National Academy Press. National Research Council.

Braun, T. (2008). Making a choice: The perceptions and attitudes of online graduate students. Journal of Technology and Teacher Education, 16(1), 63–92.

Breeding, M. (2007). We need to go beyond web 2.0. *Computers in Libraries, 27*(5), 22-25.

Bromley, H. (1997). Thinking about computers and schools: A skeptical view. In P. E. Agre & D. Schuler (Eds.), *Reinventing technology, rediscovering community: Critical explorations of computing as a social practice* (pp. 107-126). Greenwich, CT: Ablex.

Bromley, H. (1998). Introduction: Data-driven democracy? Social assessment of educational computing. In H. Bromley & M. Apple (Eds.), *Education, technology, power* (pp. 1-28). Albany, NY: SUNY Press.

Brookfield, S. D. (1999). *Discussing a way of teaching.* Jossey-Bass: San Francisco.

Brooks, B. (1999). *Professional development for teaching technology across the curriculum*. Minister of Education, Edmonton, Alberta, Canada. Retrieved June, 23, 2005 from, http://ednet.edc.gov.ab.ca/technology

Brown, D. (2001). Developing Strategies for collecting and presenting grooming evidence in a high tech world. *National Center For Prosecution of Child abuse. Update, 14*(11).

Bruce, B. C., & Hogan, M. P. (1998). The disappearance of technology: Toward an ecological model of literacy. In D. Reinking, M. C. McKenna, L. D. Labbo, & R. D. Kieffer (Eds.), *Handbook of literacy and technology: Transformations in a post-typographic world* (pp. 281). Mahwah, NJ: Erlbaum.

Bruner, J. (1990). *Acts of meaning.* Cambridge, MA: Harvard University Press.

Buchanan, E. A., Myers, S. E., & Hardin, S. L. (2005). Holding your hand from a distance: Online mentoring and the graduate library and information science student. *The Journal of Educators Online, 2*(2), 1-18.

Buendia, F., Diaz, P., & Benlloch, J. V. (2002). A framework for the instructional design of multi-structured educational applications. In P. Kommers & G. Richards (Eds.), *Proceedings of world conference on educational multimedia, hypermedia & telecommunications 2002,(*pp. 210-215). Chesapeake, VA: AACE.

Bull, G. L., & Bell, L. (2006). The 2006 national technology leadership summit. [Online serial]. *Contemporary Issues in Technology & Teacher Education, 6*(3).

Bull, G., Knezek, G., Roblyer, M. D., Schrum, L., & Thompson, A. (2005). A proactive approach to a research agenda for educational technology. *Journal of Research on Technology in Education, 37*(3), 217–220.

Burbules, C.N., & Bruce, C.B. (2001). *Theory and research on teaching as dialogue*. In V. Rochardson (Ed.), Handbook of research on teaching, 4th Ed. Washington, DC. Retrieved June 28, 2003, from http://faculty.ed.uiuc.edu/burbules/ncb/papers/dialogue.html

Burgess, L. A. (2003). WebCT as an e-learning tool: A study of technology students' perception. *Journal of Technology Education, 15*(1), 6–15.

Callahan, W. P., & Switzer, T. J. (2001). Technology as facilitator of quality education: A model. In W. Heineke & L. Blasi (Eds.), *Methods of evaluating educational technologies* (pp. 215-235). Greenwich, CT: Information Age Publishing.

Calluzzo, V. & Cante, C. (2004). Ethics in information technology and software. *Journal of Business Ethics, 51*(3), 301-312.

Campbell, M. (2005). Cyber bullying: An old problem in a new guise. *Australian Journal of Guidance & Counselling, 15*(1), 68–76. doi:10.1375/ajgc.15.1.68

Campus Computing Survey. (1995). Retrieved May 9, 2006 from, http://www.campuscomputing.net/

Cardinaels, K., Meire, M., & Duval, E. (2005). Automating metadata generation: The simple indexing interface. In *Proceedings International Conference on World Wide Web* (pp. 1-8).

Carman, J. (2002). *Blended learning design: Five key ingredients.* Retrieved January 26, 2007 from, http://www.agilantlearning.com/instructionaldesign.html

Carnwell, R. (1999). Distance education and the need for dialogue. *Open Learning, 16(2), 58-60.*

Carrington, V. (2008). 'I'm Dylan and I'm not going to say my last name': some thoughts on childhood, text and new technologies. *British Educational Research Journal, 34*(2), 151–188. doi:10.1080/01411920701492027

Carrol, J. M., Rosson, M., Convertino, G., & Ganoe, G. (2006). Awareness and teamwork in computer-supported collaboration. *Interacting with Computers, 18*, 21–46. doi:10.1016/j.intcom.2005.05.005

Cavanaugh, J. (2006). Comparing online time to offline time: The shocking truth. *Distance Education Report, 10*(9), 1–6.

Cebeci, Z., & Erdogan, Y. (2005). Tree view editing learning object metadata. *Interdisciplinary Journal of Knowledge and Learning Objects, 1*, 99–108.

Center for Teaching and Learning. (1990). Improving Multiple Choice Questions. Retrieved from http://ctl.unc.edu/fyc8.html.

CEO Forum on Education and Technology. (1997, October). *School technology and readiness report: From pillars to progress*. Retrieved January 14, 2009, from http://www.eric.ed.gov/ERICDocs/data/ericdocs2sql/content_storage_01/0000019b/80/15/40/09.pdf

CEO Forum on Education and Technology. (2001, June). *School technology and readiness report: Key building blocks for student achievement in the 21st century*. Retrieved January 14, 2009, from http://www.ceoforum.org/downloads/report4.pdf

CGEMS. CG Educational Materials Source. Retrieved from http://cgems.inesc-id.pt/

CGT. Computer graphics tutorials. Retrieved from http://web.sbu.edu/cs/cgtutorials/

Chang, Y. (2005). *The use of PowerPoint as a learning aid for international students*. Research Project Report Presented to the School of Education Indiana University, South Bend.

Changchit, C., Cutshall, R., & Elwood, S. (2006). Students' perceptions of the laptop program: What factors should be considered before implementing the program? *International Journal of Information and Communication Technology Education*, *2*(2), 53–61.

Chau, P. Y. K., & Hu, P. J.-H. (2001). Information technology acceptance by individual professionals: A model comparison approach. *Decision Sciences*, *32*(4), 699–719. doi:10.1111/j.1540-5915.2001.tb00978.x

Chen, B., & Cheng, H.H. (2005). Interpretive OpenGL for computer graphics. *Computers & Graphics*, *29*(3), 331–339.

Chen, Y. J. (2001). Dimensions of transactional distance in the world wide Web learning environment: A factor analysis. *British Journal of Educational Technology*, *32*, 459-470.

Cheng, C. Y. Y., & Yen, J. (1998). Virtual Learning Environment (VLE): A Web-based collaborative learning system. In *Proceedings of the Thirty-First Annual Hawaii International Conference on System Sciences, January, 6-9, 1998, Hawaii* (pp. 480-491). IEEE.

Cheng, R., Brown, W. P., Fahrig, R., & Reinhart, C. (2006). An Egyptian child mummy. *Science*. Retrieved Aug. 3, 2006, from http://www.sciencemag.org/cgi/content/full/313/5794/1730

Chickering, A. & Gamson, Z. (2001). Implementing the seven principles of good practice in undergraduate education: Technology as lever. *Accounting Education News, spring*, 9-10.

Chorianopoulos, K., Spinellis, D. (2006). User interface evaluation of interactive TV: A media studies perspective. *Univiversity Access Information Society*, *5*, 209-218.

Chu, H. C., & Hinton, B. E. (2001). *Factors affecting student completion in a distance*

Chute, E. (2005, May 08). Cyber schools spring up in state. *The Pittsburgh Post-Gazette*.

Cicciarelli, M. (2006). *Telementoring and computer mediated discussions: A description of online instructors' support*. Doctoral Dissertation, Duquesne University of Pittsburgh, Pennsylvania, 2004. (UMI N0. 3238573)

Cicciarelli, M. (2007). Behavioral, cognitive, and humanistic theories that guide online course design. *The International Journal of Information and Communication Technology Education, 3*(4).

Clark, J. (2000). Collaboration tools in online learning environments. *ALN Magazine*, *4*, 1.

Clark, R. E., & Sugrue, B. M. (1991). Research on instructional media, 1978-1988. In G. J. Anglin (Ed.), *Instructional technology: Past, present, and future* (pp.327-343). Englewood, CO: Libraries Unlimited.

Cochran, D., Conklin, J., & Modin, S. (YEAR). A new bloom: Transforming learning. *Learning and Leading with Technology, 34*(5), 22-25.

Collins, J. (2004). Education techniques for lifelong learning: principles of adult learning. *Radiographics*, *24*, 1483–1489. Available at http://radiographics.rsnajnls.org/cgi/reprint/24/5/1483. doi:10.1148/rg.245045020

Collins, M. (2000). Comparing web correspondence and lecture versions of second-year non-major biology course. *British Journal of Educational Technology*, *31*(1), 21–27. doi:10.1111/1467-8535.00132

Collins, M., & Berge, Z. (1996). *Facilitating interaction in computer mediated online courses.* Paper presented at the FSU/AECT Distance Education Conference, June. Retrieved June 29, 2005 from, http://www.emoderators.com/moderators/flcc.html

Collins, R., E. M., Berry S., Kanouse D., Kunkel D, Hunter S., and Miu A. (2004). Watching sex on television predicts adolescent initiation of sexual behavior. *Pediatrics, 114*, 280–289. doi:10.1542/peds.2003-1065-L

Cook, J. (2002). The Role of dialogue in computer-based learning and observing learning: An evolutionary approach to theory. *Journal of Interactive Media in Education.*

Coombs, K. (2007). Building a library Web site on the pillars of web 2.0. *Computers in Libraries, 27*(1), 16-19.

Cooper, L. (1999). Anatomy of an online course. *T.H.E. Journal, 26*(6), 49–51.

CoSN. (2005) *Digital learning spaces 2010.* Retrieved January 14, 2009, from http://cosn.org/resources/emerging_technologies/learningspaces.cfm

CoSN. (2006) *Hot technologies for K-12 schools.* Retrieved January 14, 2009, from http://cosn.org/resources/emerging_technologies/hot.cfm

Costello Report. (2004). *Future post-primary arrangements in Northern Ireland.*, ¶5.14, 3 http://www.deni.gov.uk/pprb/costello_report.htm (See also the DE pages about the Costello report in http://www.deni.gov.uk/pprb/index.htm)

Council for Higher Education Accreditation (CHEA). (2005). *The fundamentals of accreditation.* Retrieved May 30, 2005, from http://www.chea.org/pdf/fund_accred_20ques_02.pdf

Course Genie. (2009). Retrieved January 2009 from, http://www.wimba.com/products/wimba_classroom/

Course-management systems. (2005). *Library Technology Reports, 41*(3), 7-11.

Cox, J. (2005). Higher ed learns notebook PC lessons. *Network World,* Apr 11, *22*(14), 25-26.

Crawford, S. Z., Greenwell, C., & Andrew, D. (2007). Exploring the relationship between perceptions of quality in basic instruction programs and repeat participation. Physical Educator, 64(2), 64–71.

Cunningham, S. (1999). Reports of the graphics and visualization education 99 workshop. Retrieved from http://www.eg.org/WorkingGroups/GVE/GVE99

Cunningham, S. (2000). Re-inventing the introductory computer graphics course: Providing tools for a wider audience. *Computers & Graphics, 24*(12), 293–296.

Cunningham, S. (2001). Lessons from scene graphs: Using scene graphs to teach hierarchical modeling. *Computers & Graphics, 4.*

Dagdilelis, V., Evangelidis, G., Satratzemi, M., Efopoulos, V., & Zagouras, C. (2003). DELYS: A novel microworld-based educational software for teaching computer science subjects. *Computers & Education, 40,* 307–325. doi:10.1016/S0360-1315(02)00133-1

Dahl, D. (2008). *Integration of learning objects on metadata level – A Web 2.0-coined approach (German),* Akademische Verlagsgesellschaft AKA, Heidelberg.

Dahl, D., & Vossen, G. (2007). share.loc: Transparenz und Integration von Lernobjekten an der Universität Münster. *Working Paper E-Learning Competence Center Münster.*

Dahl, D., Lechtenbörger, J., Sieberg, J., & Vossen, G. (2008). Learnr – Web 2.0-Driven Learning. In *Proceedings International Conference Web Based Education* (pp. 136-141).

Dahl, D., Vossen, G., & Westerkamp, P. (2006). share.loc – A Mulit-Tiered Interoperable E-Learning Metadata Repository. In *Proceedings IEEE International Conference on Advanced Learning Technologies (ICALT)*( pp. 891-895).

Daniel, B., Schwier, R. A., & McCalla, G. (2003). Social capital in virtual learning communities and distributed communities of practice. *Canadian Journal of Learning and Technology, 29*(3).

Daniels, A. C. (2000). *Bringing out the best in people: How to apply the astonishing* power of positive reinforcement (2nd ed.). New York: McGraw-Hill.

Davidson, J., & Martellozzo, E. (2005, Sept. 8-10). *Policing the Internet and protecting children from sex offending online: When strangers become 'virtual friends'.* Paper presented at the Cybersafety Conference, Oxford, England.

Davies, J., & Graff, M. (2005). Performance in e-learning: Online participation and student grades. British Journal of Educational Technology, 36(4), 657–663. d oi:10.1111/j.1467-8535.2005.00542.xdoi:10.1111/j.1467-8535.2005.00542.x

Davis, F. (1989). Perceived usefulness, perceived ease of use and user acceptance of information technology, *MIS Quarterly*, *13*(3), 319-340.

Davis, F. D. (1989). Perceived usefulness, perceived ease of use, and user acceptance of information technology. *MIS Quarterly*, *13*(3), 319–340. doi:10.2307/249008

Davis, F. D. (1989). Perceived usefulness, perceived ease of use, and user acceptance of information technology. *MIS Quarterly*, *13*(3), 319–340. doi:10.2307/249008

Davis, F. D., Bagozzi, R. P., & Warshaw, P. R. (1989). User acceptance of computer technology: A comparison of two theoretical models. *Management Science*, *35*(8), 982–1003. doi:10.1287/mnsc.35.8.982

Davis, N., Cho, M.O., & Hagenson, L. (2006). Intercultural competence and the role of technology in teacher education. *Contemporary Issues in Technology and Teacher Education,* *4*(4).Retrieved May 7, 2006 from, http://www.citejournal.org/vol4/iss4/maintoc.cfm

DeBourgh, G. A. Simply elegance: Course management systems as a pedagogical infrastructure to enhance science learning. *The Technology Source Archives.* Retreived May 26, 2006 from, http://www.technologysource.org/article/simple_elegance/

Demb, A., Erickson, D., & Hawkins-Wilding, S. (2004). The laptop alternative: student reactions and strategic implications. *Computers & Education*, *43*(4), 383–401.

Detert, J., Trevino, L., & Baker, V. Faulty cognitions and unethical decisions: The antecedents and outcomes of moral disengagement.

DirectX. Retrieved from http://www.microsoft.com/windows/directx/default.mspx

Downes, S. (2001). Learning objects: Resources for distance education worldwide. *International Review of Research in Open and Distance Learning, 2*(1), 29.

Downing, K. F., & Holtz, J. K. (2008). *Online Science Learning: Best Practices and Technologies.* Hershey, PA: IGI Global.

Downing, K. F., & Holtz, J. K. (2008). *Online science learning: Best practices and technologies.* Hershey, PA: IGI Global.

Drago, W., & Peltier, J. (2004). The effects of class size on effectiveness of online courses. *Management Research News*, *27*(10), 27. doi:10.1108/01409170410784310

Drennan, J., & Kennedy, J. (2005). Factors affecting student attitudes toward flexible online learning in management education. *The Journal of Educational Research*, *98*(6), 331–338. doi:10.3200/JOER.98.6.331-338

Driscoll, K. (2007). Collaboration in today's classrooms: New Web tools change the game. *MultiMedia & Internet@ Schools, 14*(3), 9-12.

Duval, E., Hodgens, W., Sutton, S., & Weibel, S.-L. (2002). Metadata principles and practicalities. *D-Lib Magazine*, *8*(4). doi:10.1045/april2002-weibel

Dykman, C., & Davis, C. (2008). Online education forum - part three a quality online educational experience. *Journal of Information Systems Education*, *19*(3), 281–289.

Eastman, J., & Swift, C. (2001). New horizons in distance education: The online learner-centered marketing class, *Journal of Marketing Education*, *23*(1) 25-34.

Education Week. (2006), *Technology counts 2006.* Retrieved January 14, 2009, from http://www.edweek.org/ew/toc/2006/05/04/index.html

Engeström, Y. (1987). *Learning by expanding: An activity-theoretical approach to developmental research.* Helsinki: Orienta-Konsultit.

Engeström, Y. (1999). Activity theory and individual and social transformation. In Y. Engeström, R. Miettinen, & R.

Punamäki, (Eds.), 1999. *Perspectives on activity theory*. Cambridge: Cambridge University Press, 19–38.

Erstad, O. (2003). Electracy as empowerment: Student activities in learning environments using technology. *Young: Nordic Journal of Youth Research, 11*(1), 11–28.

Etzioni, A. (2002). When it comes to ethics, B-schools get an F. *The Washington Post.* Retrieved from http://www.washingtonpost.com/ac2/wp-dyn/A38323-2002Aug2.

Excess Voice. *RSS Explained.* Retrieved from http://www.excessvoice.com/rssexplanation.htm

Exploratories. Retrieved from http://www.cs.brown.edu/exploratories/home.html

Fabos, B., & Young, M.D. (1999). Telecommunication in the classroom: Rhetoric versus reality. *Review of Educational Research, 68*(3), 322-349.

Fairclough, N. (1989). *Language and power.* London: Longman.

Fairclough, N. (1992). *Discourse and social change*. Cambridge: Polity Press.

Fairclough, N., & Wodak, R. (1997). Critical discourse analysis. In T.A. van Dijk (Ed.), *Discourse as social interaction* (pp. 259–284). London: SAGE Publications.

Falchikov, N. (2005). Improving assessment through student involvement: Practical solutions for aiding learning in higher and further education. New York: Routledge.

Fauske, J., & Wade, S. E. (2003-2004). Research to practice online: Conditions that foster democracy, community, and critical thinking in computer-mediated discussions. *Journal of Research on Technology n Education, 36*(2), 137-153.

Fayyad, U., Grinstein, G. G., & Wierse, A. (2002). *Information visualization in data mining and knowledge discovery.* San Francisco, CA: Morgan Kaufmann Publishers Inc.

Felix, U. (2001). A multivariate analysis of students' experience of web-based learning. *Australian Journal of Educational Technology, 17*(1), 21–36.

Finn, S., & Inman, J. (2004) Digital unity and digital divide: Surveying alumni to study effects of a campus laptop initiative. *Journal of Research on Technology in Education, 36*(3), 297-317.

Fischer, G. (2000). Lifelong learning - More than training. *Journal of Interactive Learning Research, 7*(3), 265–294.

Fjermestad, H. S. & Zhang, Y. (2005). Effectiveness for students: Comparisons of "in-seat" and aln courses. In S.Hiltz & R.Goldman (Eds.), *Learning together online: Research on asynchronous learning networks* (pp. 39-79). London: Lawrence Erlbaum Associates.

Flavell, J. H. (1979). Metacognition and cognitive monitoring: A new area of cognitive-developmental inquiry. *The American Psychologist, 34*(10), 906–911. doi:10.1037/0003-066X.34.10.906

Fleming, M., & Levie, W. (1993). *Instructional message design: Principles from the behavioral and cognitive sciences* (2nd ed.). New Jersey: Educational Technology Publications.

Flick, U. (1998). *An introduction to qualitative research.* London: Sage Publications.

Fogg, B.J. (2003). *Persuasive technology.* Morgan Kaufmann.

Fornell, C., & Larcker, D. F. (1981). Evaluating structural equation models with unobservable variables and measurement error. *JMR, Journal of Marketing Research, 18*, 39–50. doi:10.2307/3151312

Fowler, D. (March 2005). Are on-site courses as effective as online? *Online Cl@ssroom.*

Frantz, P. L., & Wilson, A. H. (2004). Student performance in the legal environment course: Determinants and comparisons. Journal of Legal Studies Education, 21(2), 225. doi:10.1111/j.1744-1722.2004.tb00318.xdoi:10.1111/j.1744-1722.2004.tb00318.x

Frey, B., & Kirby, D. (2002). Student perceptions on the value of PowerPoint. In D.G. Brown(Ed.), *Faculty development programs that work.* Bolton, MA: Anker Publishing Inc.

Freyer, W. (2006). To blog or not to blog? You decide. *Interactive Educator, 2*(2), 28-33.

Friesen, N. (2001). What are educational objects? *Interactive Learning Environments, 9*(3).

Friesen, N. (2003). Three objections to learning objects and e-learning standards. In R. E. McGreal (Ed.), *Online education using learning objects,*pp. 59-70. London: Routledge.

Friesen, N. (2004). *International LOM Survey: Final Report.* Retrieved January 1, 2009 from http://jtc1sc36.org/doc/36N0871.pdf

FSU. (2009). Retrieved January 2009 from, http://learningforlife.fsu.edu/ctl/catalog/index.cfm /

Fullan, M.G., & Stiegelbauer, S. (1991). *The new meaning of educational change.* New York: Teachers College Press.

Fung, Y. Y. H. (2004). Collaborative online learning: Interaction patterns and limiting factors. *Open Learning, 19*(2), 135–149. doi:10.1080/0268051042000224743

Furtado ,V., & Vasconcelos E. (2007) Geosimulation in education: A system for teaching police resource allocation. *International Journal of Artificial Intelligence in Education, 17.*

Furtado, E., Furtado V., & Vasconcelos E. (2007). A conceptual framework for the design and evaluation of affective usability in educational geosimulation systems. *Proceedings of Interact.* Rio de Janeiro.

Gagné, R. M. (1970). *The conditions of learning.* Longod, Holt, Rinehard and Winstn, Inc.

Gagne, R. M., Wager, W. W., Golas, K. C., & Keller, J. M. (2005). *Principles of instructional design* ($5^{th}$ ed.). Belmont, CA: Wadsworth/Thomson Learning.

Galvão, J. R., Martins, P. G.,& Gomes, M. R. (2000). Modeling reality with simulation games for a cooperative learning. In *Proceedings of the Winter Simulation Conference*, (pp. 1692-1698). Orlando, Florida: Society for Computer Simulation International.

Gama, C. (2004). *Towards a model of metacognition instruction in interactive learning environments.* Doctoral Thesis. University of Sussex, England.

Garrison, D. R. (1992). Critical thinking and self-directed learning in adult education: An analysis of responsibility and control issues. *Adult Education Quarterly, 42*(3), 136–148.

Garrison, D. R. (1993). A cognitive constructivist view of distance education: An analysis of teaching-learning assumptions. *Distance Education, 14*(2), 199–211. doi:10.1080/0158791930140204

Garrison, D. R. (2002). Cognitive presence for effective asynchronous online learning: The role of reflective inquiry, self-direction and metacognition. *Fourth Annual Sloan ALN Workshop.*

Garrison, D.R., & Archer, W. (2000). *A transactional perspective on teaching and learning: A framework for adult and higher education.* Oxford, United Kingdom: Pergamon.

Gates, K. (1998) Should colleges and universities require students to own their own computers? *CAUSE/EFFECT Journal, 21*(3).

Geary, G. (1991). *Electronic performance support system* (1st ed.). Tolland, MA: Gery Associates.

Gee, J. P. (2003). *What video games have to teach us about learning and literacy.* Palgrave Macmillan.

Gefen, D., & Straub, D. (2000). The relative importance of perceived ease of use in IS adoption: A study of e-commerce adoption. *Journal of the Association for Information Systems, 1*(8), 1–28.

Gfeller, M. K., Niess, M. L., & Lederman, N. G. (1999). Preservice teachers' use of multiple representations in solving arithmetic mean problems. *School Science and Mathematics, 99*, 250-257.

Ghani, J. A., Supnick, R., & Rooney, P. (1991). The experience of flow in computer-mediated and in face-to-face groups. In *Proceedings of the Twelfth International Conference on Information Systems.* New York, NY.

Gibb, J. (1995). *Tribes.* Sausalito, CA: Center Source Systems.

Gibbons, A. S., Lawless, K. A., Anderson, T. A., & Duffin, J. (2001). The web and model-centered instruction. In

B.H. Khan (Ed.), *Web-based training*. Englewood Cliffs, NJ: Educational Technology Publications, 137-146.

Gibson, J., & Herrera, J. (1999). How to go from classroom based to online delivery in eighteen months or less: A case study in online program development. *T.H.E. Journal*, *26*(6), 57–60.

Gill, T., Gilliland-Swetland, A., & Baca, M. (1998). *Introduction to metadata: Pathways to digital information*. Getty Research Institute.

Gillis, A., Winston, J., Braid, A., MacDonald, P., & MacQuarrie, M. A. (2000). The learning needs and experiences of women using print-based and CD-ROM technology in nursing distance education. *Journal of Distance Education, 15(1)*,1-20.

GL4Java. Retrieved from http://gl4java.sourceforge.net

Godwin-Jones, R. (2004). Learning objects: Scorn or SCORM? *Language Learning & Technology, 8*(2), 7-12.

Gordon-Murnane, L. (2006). Social bookmarking, folksonomies, and web 2.0 tools. *Searcher: The Magazine for Database Professionals, 14*(6), 26-38.

Grabe, M., & Christopherson, K. (2005). Evaluating the advantages and disadvantages of providing lecture notes: The role of internet technology as a delivery system and research tool. *The Internet and Higher Education*, *8*(4), 291–298.

Grant, L. (2006) Using Wikis in schools: A Case Study. *Futurelab*. Bristol. Retrieved from www.futurelab.org.uk

Green, K. (2004). Campus computing 2004:Fifteenth national survey of computing and information technology in US higher education. *The Campus Computing Project,* University System of Georgia.

Greenberg, J. (2001). A quantitative categorical analysis of metadata elements in image-applicable metadata schemas. *Journal of the American Society for Information Science and Technology*, 917–924. doi:10.1002/asi.1170

Greenberg, J., Patuelli, M. C., Parsia, B., & Davenport Robertson, W. (2001). Author-generated Dublin core metadata for web resources: A baseline study in an organization.In *Proceedings International Conference on Dublin Core and Metadata Applications* (pp. 38-46).

Greenlaw, S. A., & DeLoach, S. B. (2003). Teaching critical thinking with electronic discussion. *Journal of Economic Education, 34*(1), Winter, 36-52.

Guizzardi, G. (2000). *Uma Abordagem Metodológica de Desenvolvimento para e com Reuso, Baseada em Ontologias Formais de Domínio*. Dissertação (Mestrado). Universidade Federal do Espírito Santo, Vitória, ES, Brasil.

Haase, A. (2005). Trends in online learning communities. *ACM SIGGROUP Bulletin. Special Issue on Online Learning Communities*, *25*, 2–6.

Hagstrom, F., & Wertsch, J. V. (2004). The social construction of identity: The clinical challenge. *Topics in Language Disorders*, *24*(3), 162–173.

Haines, R. & Leonard, L. (2007). Situational influences on ethical decision-making in an IT context. *Information & Management, 44*(3), 313-320.

Hair, J., Jr., Anderson, R. E., Tatham, R. L., & Black, W. C. (1992). *Multivariate Data Analysis with Readings*. New York: MacMillan Publishing Company.

Hall, T. (1999). Intelligence community collaboration baseline study, final report. Retrieved January, 2006 from http://collaboration.mitre.org.

Hammoud, L., Love, S., & Brinkman, W. P. (2008). The affect of lecturers' attitude on students' use of an online learning environment. In *Proceedings of the 15th European conference on Cognitive ergonomics: the ergonomics of cool interaction ECCE, September 15-19, 2008, Funchal, Portugal.* (ACM International Conference Proceeding Series; Vol. 369)

Hangzhou, C. (2006). Lecture notes in computer science. Berlin: Springer-Verlag.

Hanisch, F., & Straßer, W. (1996-2003). GRIS/ILO interactive learning objects. WSI/GRIS University of Tuebingen, Germany. Retrieved from http://www.gris.uni-tuebingen.de/projects/ilo

Hanisch, F., & Straßer, W. (2002). *Highly interactive Web-based courseware*. Proceedings of the CGE02 - Eurographics/SIGGRAPH Workshop on Computer Graphics Education, 31–35.

Hankin, J. (1999) Alice, the college teacher and the rottweiler in wonderland: The prospect and problems of distance education. *Executive Speeches*, 1492: 18-21.

Hara, N., & Kling, R. (2000). Students' frustrations with a web-based distance education course. *First Monday*, *4*(12), 6–24.

Harasim, L. (2000). Shift happens: Online education as a new paradigm in learning. The Internet and Higher Education, 3, 41–61. doi:10.1016/S1096-7516(00)00032-4doi:10.1016/S1096-7516(00)00032-4

Harding, G. (2005). The students' feedback: A survey of three approached to teaching with modern PowerPoint presentations. *American Society for Engineering Education*. Retrieved April 1-2, 2005 from,.

Harknett, R.J., & Cobane, C.T. (1997). Introducing instructional technology to international relations. *Political Science & Politics, 30*, 496-500.

Hartley, J. (2007). Teaching, learning and the new technology: A review for teachers. *British Journal of Educational Technology, 38*(1), 42-62.

Hasan, B. (2003). The influence of specific computer experiences on computer self-efficacy beliefs. *Computers in Human Behavior, 19*(4), 443–450. doi:10.1016/S0747-5632(02)00079-1

Hathorn, L. G., & Ingram, A. L. (2002). Cooperation and collaboration using computer-mediated communication. *Journal of Educational Computing Research, 26*, 325-347.

Hauser, J. (2007). Media specialists can learn web 2.0 tools to make schools more cool. *Computers in Libraries, 27*(2), 6-8, 47-48.

Havelka, D. (2003). Predicting software self-efficacy among business students: A preliminary assessment. *Journal of Information Systems Education, 14*(2), 145.

Hazari, S. (2004). Strategy for assessment of online course discussion. *Journal of Information Systems Education, 15*(4), 349–355.

Heckman, R., & Annabi, H. (2003). A content analytic comparison of FTF and ALN case-study discussions. In *Proceedings of the 36th Hawaii International Conference on System Sciences (HICSS'03)*. Honolulu, Hawaii, USA.

Heckman, R., & Annabi, H. (2006). cultivating voluntary online learning communities in blended environments. *Journal of Asynchronous Learning Networks*, 10.

Henning, E. (2003). I click, therefore I am (not): Is cognition "distributed" or is it "contained" in borderless e-learning programmes. *International Journal of Training and Development*, 7(4)..

Henning, E., Van Rensburg, W., & Smit, B. (2004). *Finding your way in qualitative research*. Pretoria: Van Schaik Publishers.

Henry, J. W., & Stone, R. W. (2001). The role of computer self-efficacy, outcome expectancy, and attribution theory in impacting computer system use. *Journal of International Information Management, 10*(1), 1–16.

Hickman, L. A. (2001). *Philosophical tools for technological culture: Putting pragmatism to work*. Bloomington: Indiana University Press.

Hisham, N., Campton, P., & FitzGerald, D. (2004). A tale of two cities: A study on the satisfaction of asynchronous e-learning system in two Australian universities. In *Proceedings of the 21st ASCILITE Conference, December 5-8, 2004, Perth, Western Australia* (pp. 395-402).

Hitchner, L.E., & Sowizral, H.E. (2000). Adapting computer graphics curricula to changes in graphics. *Computers & Graphics, 24*(2), 283–288.

Ho, C.M.L. (2000). Developing intercultural awareness and writing skills through email exchange. *The Internet TSL Journal, VI*(12). Retrieved September 29, 2005 from, http://iteslj.org/Articles/Ho-Email.html

Hodas, S. (1993). Technology refusal and the organizational culture of schools. *Education Policy Analysis Archives, 1*(10).

Hoffer, E. (1982). *Between the devil and the dragon: The best essays and aphorisms of Eric Hoffer.* New York: Harper & Row.

Holmberg, B. (1989). *Theory and practice of distance education.* London: Routledge.

Holmberg, B. (1999). The conversational approach to distance education. *Open Learning, 14(3)*, 58-60.

Holmberg, B. (2003). A theory of distance education based on empathy. In M.G. Moore, & W.G. Anderson (Eds.), *Handbook of distance education.*London: Lawrence Erlbaum Associates, 79-86.

Hong, K. (2002). Relationships between students' and instructional variables with satisfaction learning from a Web-based course. *The Internet and Higher Education, 5*(3), 267–281. doi:10.1016/S1096-7516(02)00105-7

Hong, K. S., Lai, K. W., & Holton, D. (2003). Students' satisfaction and perceived learning with a web-based course. *Educational Technology and Society, 6*(1), 116–124.

Höök, K. (2004). User-centred design and evaluation of affective interfaces. In Zs. Ruttkay, & C. Pelachaud (Eds.), From brows till trust: Evaluating embodied conversational agents. Kluwer. http://www.sics.se/~kia/papers/hook1.pdf

Hoskins, S. L., & Hooff, J. C. V. (2005). Motivation and ability: Which students use online learning and what influence does it have on their achievement? *British Journal of Educational Technology, 36*(2), 177–192. doi:10.1111/j.1467-8535.2005.00451.x

Howell, D. (1992). *Statistical methods for psychology*, 3rd ed. Belmont, CA: Duxbury Press.

Howland, J.L., & Moore, J. L. (2002). Student perceptions as distance learners in Internet-based courses. *Distance Education, 23*(2), 183-195.

HRXML Specification (2007). Competencies (Measurable Characteristics) Recommendation April 2007.

Huang, H. M. (2002). Student perceptions in an online mediated environment. *International Journal of Instructional Media, 29,* 405-422.

Huang, H. M., & Liaw, S. S. (2004). Guiding distance educators in building Web-based instructions. *International Journal of Instructional Media, 31*(2), 125-137.

Huffmann, K. (2006). Web 2.0: Beyond the concept-practical ways to implement RSS, podcasts, and wikis. *Education Libraries, 29*(1), 12-19.

Hughes, S. C. (2002). Overcoming social and psychological barriers to effective online collaboration. *Educational Technology, 5*(1), 86–92.

Huifen, L., Xin, H., & Xinyu, Z. (2002). Human communication and interaction in web-based learning: a case study of the digital media Web course. In *Proceedings of the International Conference on Computers in Education (ICCE'02), December 3-6, 2002, Auckland, New Zealand.* (pp. 217-221).

Huysman, M., & Wulf, V. (2005). Information technology in building and sustaining the relational base of communities. *The Information Society, 21,* 81–89. doi:10.1080/01972240590925285

HyperGraph. Retrieved from http://www.siggraph.org/education/materials/HyperGraph/hypergraph.htm

ICG. Introduction to computer graphics. Retrieved from http://www.cgv.tugraz.at/CGTutorial/CGTutorial.html

IEEE Competency Working Group. (2004). Draft Standard for Learning Technology- Reusable Competency Definitions. Retrieved from http://ltsc.ieee.org/wg20/files/ IEEE_1484.20.WD_01_rough.pdf

Igbaria, M., & Greenhaus, J. H. (1992). Determinants of MIS employee's turnover intentions: A structural equation model. *Communications of the ACM, 35,* 35–49. doi:10.1145/129630.129631

Igbaria, M., Zinatelli, N., Cragg, P., & Cavaye, A. L. M. (1997). Personal computing acceptance factors in small firms: A structural equation model. *MIS Quarterly, 21*(3). doi:10.2307/249498

Iivari, J. (2005). An empirical test of the DeLone-McLean model of information system success. *The Data Base for Advances in Information Systems, 36*(2), 8–27.

ILO. Graphical-interactive systems, interactive learning objects. Retrieved from http://www.gris.uni-tuebingen.de/projects/ilo/index.html

Im, Y., & Lee, O. (2003-2004). Pedagogical implications of online discussion for preservice teacher training. *Journal of Instructional Media, 31*(2), 125-137.

IMS Global Learning Consortium (2002). *IMS Reusable Definition of Competency or Educational Objective - Information Model. Version 1.0 Final Specification*

International Society for Technology in Education (ISTE). (2008) *National Educational Technology Standards for Teachers (2nd Ed.).* Retrieved January 14, 2009, from http://www.iste.org/nets

Issroff, K., & Scanlon, E. (2001). Case studies revisited-What can activity theory offer? In *Proceedings of First Euro-CSCL Conference, Maastricht.*

Jang, K. S., Hwang, S. Y., Park, S. J., Kim, Y. M., & Kim, M. J. (2005). Effects of a Web-based teaching method on undergraduate nursing students' learning of electrocardiography. The Journal of Nursing Education, 44(1), 35–39. PubMed

Januszewski, A. (2005). Stasis and change in the definition of educational technology: The rationale and decision making process. *TechTrends, 49*(1), 45–46.

Januszewski, A., Molenda, M., & Harris, P. (2007). *Educational technology: A definition with commentary.* Mahweh, NJ: Lawrence Erlbaum.

Jenlink, P., & Carr, A. A. (1996). Conversation as a medium for change in education. *Educational Technology,* 31–38.

Jiang, M., & Ting, E. (2000). A study of factors influencing students' perceived learning in a web-based course environment. *International Journal of Educational Telecommunications, 6*(4), 317–338.

Johnson, C. M. (2001). A survey of current research on online communities of practice. *The Internet and Higher Education, 4,* 45–60. doi:10.1016/S1096-7516(01)00047-1

Jonassen, D. (2003). Using cognitive tools to represent problems. *Journal of Research on Technology in Education, 35,* 362-381.

Jones, A., & Kemper, D. (1994). Approaches to learning and student acceptance of self-study packages. *Educational and Training Institute, 1*(2), 93-97.

Jones, B. F., Valdez, G., Nowakowski, J., & Rasmussen, C. (1995). *Plugging in: Choosing and using educational technology.* Washington, DC: Council for Educational Development and Research; Oak Brook, IL: North Central Regional Educational Laboratory.

Jordan, P. (2002). *Designing pleasure products.* Taylor & Francis.

Journal of Teaching in International Business. (n.d). Retrieved September 25, 2006 from, http://www.haworthpress.com/store/product.asp?sid=91D0P1HS6QEV

Jung, I. (2001). Building a theoretical framework of web-based instruction in the context of distance education. *British Journal of Educational Technology, 32,* 525-534.

Jurczyk, J., Benson, S. N. K., & Savery, J. R. (2004). Measuring student perception in web-based courses: a standard-based approach. *Online Journal of Distance Learning Administration, 7*(4). Retrieved March 22, 2006 from http://www.westga.edu/%7Edistance/ojdla/winter74/winter74.htm

Kaiden, R. (2002). A review of WebCT. *The Internet and Higher Education, 5*(4), 399–404. doi:10.1016/S1096-7516(02)00125-2

Kaiser, H. F. (1960). The application of electronic computers to factor analysis. *Educational and Psychological Measurement, 20,* 141-151.

Kali, Y. (2006). Collaborative knowledge-building using a design principles database. *International Journal of Computer-Supported Collaborative Learning, 1*(2), 187–201. doi:10.1007/s11412-006-8993-x

Kali, Y., Spitulnik, M., & Linn, M. (2004). Building community using the design principles database. In P. Gerjets, P. A. Kirschner, J. Elen & R. Joiner (Eds.), *Instructional design for effective and enjoyable computer-supported learning* (pp. 294-305). Tuebingen: Knowledge Media Research Center.

Kalifa, M., & Lam, R. (2002). Web-Based learning: Effects on learning process and outcomes. *IEEE Transactions on Education, 45*(4), 350–356. doi:10.1109/TE.2002.804395

Kanuka, H., & Anderson, T. (1998). Online interchange, discord, and knowledge construction. *Journal of Distance Education, 13*, 57–74.

Kanuka, H., Collett, D., & Caswell, C. (2002). University instructor perceptions of the use of asynchronous text-based discussion in distance courses. *The American Journal of Distance Education, 16*(3), 151-167.

Kanuka, H., Collett, D., & Caswell, C. (2002). University instructor perceptions of the use of asynchronous text-based discussion in distance courses. *The American Journal of Distance Education, 16*(3), 151-167.

Keegan, D. (1996). *Foundations of distance education* (3rd ed). London: Routledge.

Kelly, M. G., & McAnear, A. (Eds.). (2002) International Society for Technology in Education (ISTE): Eugene, OR.

Kennedy, D. M. (2002). Dimensions of distance: A comparison of classroom education and distance education. *Nurse Education Today, 22*(5), 409-416.

Kladko, B. (2005). *Lectures have to compete with Web.* Retrieved April 16, 2005 from, http://www.bergen.com/page.php?qstr=eXJpcnk3ZjczN2Y3dn FIZUVFeXkzJmZnYmVsN2Y3dnFIZUVFeXk2NjgwNjU1JnlyaXJ5N2Y3MTdmN3ZxZWVFRXI5NA

Kluev, V. (2004). Experiences teaching Japanese students. *The WSEAS Transactions on Advances in Engineering Education, 1*(1), 11–14.

Knade, S.L. (2001). Cyber school litigation expands. *Education Law Focus, 10*(5), 1-4.

Knight, W. E. (2006). Paper presented at the Annual Forum of the Association for Institutional Research (AIR). Chicago, IL.

Knowles, M. S. (1980). *The modern practice of adult education: From pedagogy to andragogy.* Englewood Cliffs, NJ: Cambridge Adult Education, Prentice Hall Regents.

Knowles, M. S., Holton, E. F., & Swanson, R. A. (1998). *The adult learner* (5th ed.). Woburn, MA: Butterworth-Heinman.

Knowlton, D. S. (2000). A theoretical framework for online classroom: A defense and delineation of a student-centered pedagogy. In R.E. Weiss, D.S. Knowlton, & B.W. Speck (Eds.). *Principles of effective teaching in the online classroom, 84.* San Francisco: Jossey-Bass, 5-14.

Knupfer, N. N. (1997). Participant analysis of a multi-class, multi-state, online, discussion list. In *Proceedings of Selected Research and Development Presentations at the 1997 National Convention of the Association for Educational Communications and Technology.* Albuquerque, USA, 133-137.

Kolodner, J. L. (2000). Theory and practice of case-based learning aids (pp. 215-242). In D. Jonassen (Ed.), *Theoretical foundations of learning environments.* Mahwah, New Jersey: Lawrence Erlbaum and Associates.

Koszalka, T. A., & Ganeson, R. (2004). Designing online courses: A taxonomy to guide strategic use of features available in course management systems (CMS) in distance education. *Distance Education, 25*(2), 245–256. doi:10.1080/0158791042000262111

Kozma, R. B. (1991). Learning with media. *Review of Educational Research, 61*(2), 179-211.

Krathwohl, D. (2002). A revision of Bloom's taxonomy: An overview. *Theory into Practice, 41*(4), 212-218.

Krathwohl, D., Bloom, B., & Bertram, B. (1973). *Taxonomy of educational objectives, the classification of educational goals. Handbook II: Affective domain.* New York: David McKay Company, Inc.

Krueger, A. & Sutton, J. (Eds.). (2001). *EDThoughts: What we know about science teaching and learning.* Aurora, CO: Mid-continent Research for Education and Learning.

Kuutti, K. (1996). Activity theory as potential framework for human-computer interaction research. In B. Nardi (Ed.), 1996. *Context and consciousness*. Cambridge, MA: The MIT Press, 17–44.

Lally, V., & Barrett, E. (1999). Building a learning community on-line: Towards socio-academic interaction. *Research Papers in Education*, *14*(2), 147-163.

Lam, W. (2005). Teaching e-business online: The Universitas 21 global approach. *Journal of Electronic Commerce in Organizations*, *3*(3), 18–41.

Landry, B., Griffeth, R., & Hartman, S. (2006) Measuring student perceptions of blackboard using the technology acceptance model. *Decision Sciences Journal of Innovative Education*, *4*(1), 87-99.

Laudato, N., & Nicoll, J. (1999). assessing the impact o n students of online materials in university courses. Retrieved October 3, 2006 from, http://www.educause.edu/ir/library/html/edu9952/edu9952.html

Lautenbach, G.V., & Van der Westhuizen, D. (2005a). *Investigating lecturers changing epistemologies and pedagogies during engagement with e-learning in an education faculty*. Proceedings of the ED-MEDIA Conference, Montreal, Canada, AACE, 2880–2887.

Lautenbach, G.V., & Van der Westhuizen, D. (2005b). Hybridising methods towards narratives in an educational ICT inquiry. *Education as Change. Special Issue: ICT in Education*, *9*(2), 46–73.

Lautenbach, G.V., Van der Westhuizen, D., & Luca, J. (2006). *Rags to riches and conflict on the playground: Contrasting narratives of e-learning in an education faculty*. Proceedings of the ED-MEDIA Conference, Orlando, Florida, AACE, 1700–1707.

Lave, J., & Wenger, E. (1991). Situated learning: Legitimate peripheral participation. *Cambridge University Press*.

Lawless, K. A., & Pellegrino, J. W. (2007). Professional Development in Integrating Technology into Teaching and Learning: Knowns, Unknowns, and Ways to Pursue better Questions and Answers. *Review of Educational Research*, *77*(4), 575–614. doi:10.3102/0034654307309921

Lawless, K. A., & Pellegrino, J. W. (2007). Professional development in integrating technology into teaching and learning: Knowns, Unknowns, and ways to pursue better questions and answers. *Review of Educational Research*, *77*(4), 575–614. doi:10.3102/0034654307309921

Lawless, K. A., & Pellegrino, J. W. (2007). Professional Development in Integrating Technology into Teaching and Learning: Knowns, Unknowns, and Ways to Pursue better Questions and Answers. *Review of Educational Research*, *77*(4), 575–614. doi:10.3102/0034654307309921

Lazar, J., Jones, A., & Hackley, M. (2006). Severity and impact of computer user frustration: A comparison of student and workplace users. *Interacting with Computers*, *18*, 187–207. doi:10.1016/j.intcom.2005.06.001

*learning mediated HRD baccalaureate program*. Paper presented at the Academy of Human Resource Development 2001 Conference, Fayetteville, AR. (AHRD Reference No. 021).

Learning Technologies, N. A. S. A. (2004). *Advanced technology applications for education benchmark study*. Retrieved January 9, 2009, from http://learn.arc.nasa.gov/benchmark/

Lee, C. S., Tan, D. T. H. T., & Goh, W. S. (2004). The next generation of e-learning: Strategies for media rich online teaching and engaged learning. *International Journal of Distance Education Technologies*, *2*(4), 1–16.

Lee, W. I., & Shih, B. Y. (2001). A case study of motivation theory on the web-based learning. In *Proceedings of the ASEE/IEEE Frontiers in Education Conference, October 10-13, 2001, Reno, NV, USA* (pp. 23-27).

Leemkuil, H. H., de Jong, T., de Hoog, R., & Christoph, N. (2003). KM quest: A collaborative internet-based simulation game. *Simulation & Gaming 34*, 89-111.

Legris, P., Ingham, J., & Collerette, P. (2003). Why do people use information technology? A critical review of the technology acceptance model. *Information & Management*, *40*(3), 191–204. doi:10.1016/S0378-7206(01)00143-4

Lei, X., Pahl, C., & Donnellan, D. (2003). An evaluation technique for content interaction in Web-based teaching and learning environments. In *Proceedings of the 3rd*

*IEEE International Conference on Advanced Learning Technologies (ICALT'03), July 9-11, 2003, Athens, Greece* (pp. 294- 295).

Lemke, C., & Coughin, E. (1998, July 10). *Technology in American schools: Seven dimensions for gauging progress*. Retrieved Sept. 25, 2006, from http://www.mff.org.publications

Lenhart, A., & Madden, M. (2005). *Teen content creators & consumers* [White paper, electronic version]. Retrieved October 8, 2007 from Pew Internet & American Life Project at http://www.pewinternet.org/PPF/r/166/report_display.asp

Lenhart, A., & Madden, M. (2007). *Social networking websites and teens: An overview* [White paper, electronic version]. Retrieved October 8, 2007 from Pew Internet & American Life Project at http://www.pewinternet.org/PPF/r/198/report_display.asp

Leont'ev, A. (1978). *Activity, consciousness and personality*. Englewood Cliffs, NJ: Prentice-Hall.

Lesser, E., & Prusak, L. (2000). Communities of practice, social capital and organizational knowledge. *Knowledge and communities* (pp. 123-131).

Levin, D. S., & Ben-Jacob, M. G. (1998). *Collaborative learning: A critical success factor in distance education.* Paper presented at the Annual Conference on Distance Teaching & Learning, August, Madison, Wi.

Liaw, S. (2002) Understanding user perceptions of World-wide Web environments, *Journal of Computer Assisted Learning*, *18*, 137-148.

Ligorio, B. & Veermans, M. (2005). Perspectives and patterns in developing and implementing international Web-based collaborative learning environments", *Computers and Education, 45*(3), 271-275.

Ligorio, B., Talamo, A. & Pontecorvo, C. (2005). Building intersubjectivity at a distance during the collaborative writing of fairytales, *Computers and Education, 45*(3), 357-374.

Ligorio, M.B., & Van Keen, K.( 2006). Constructing a successful cross-national Virtual Learning Environment in Primary and Secondary Education, *Association for the Advancement of Computing in Education Journal, 14*(2), 103-128.

Lim, D. H. (2006). *Online vs. blended learning: Differences in instructional outcomes and learner satisfactions.* Paper presented at the Academy of Human Resource Development International Conference (AHRD). Columbus, OH, USA, 809-816.

Lin, M. P., Ko, H. C., & Wu, J. Y. W. (2008). The role of postivie/negative outcome expectancy and refusal self-efficacy of internet use on internet addiction amond college students in Taiwan. *CyperPsychology and Behavior, 11*(4), 451–457. doi:10.1089/cpb.2007.0121

Lin, X. (2001). Designing metacognitive activities. *Educational Technology Research and Development, 49*(2), 23–40. doi:10.1007/BF02504926

Llamas-Nistal, M. *Fernández-Iglesias, Manuel J., Anido-Rifon, L.E. (Eds.) (2004).* Computers & education: Towards a lifelong learning society. *Kluwer Academic Publishers.*

Loch, K. & Conger, S. (1996). Evaluation ethical decision making and computer use. *Communications of the ACM, 39*(7), 74-83.

Logsdon, J., Thompson, J., & Reid, R. (1994). Software piracy: Is it related to level of moral judgment? *Journal of Business Ethics, 13*(11), 849-857.

Lowry, R. (1999). Electronic presentation of lectures—Effect upon student performance. *University Chemistry Education, 3*(1), 18-21.

Luarn, P., & Lin, H. H. (2005). Toward an understanding of the behavioral intention to use mobile banking. *Computers in Human Behavior, 21*(6), 873–891. doi:10.1016/j.chb.2004.03.003

Ludwig, B. (2000). *Web-based instruction: Theoretical differences in treatment of subject matter.* Paper presented at the Annual Meeting of the Psychological Association, August, Washington, DC. (ERIC Document Reproduction Service No. ED453708).

Lux, M., & Klieber, W., & Granitzer. (2006). On the complexity of high level metadata. *Journal of Universal Knowledge Management, 1*(1), 54–58.

MacDonald, J. (2001). Exploiting online interactivity to enhance assignment development and feedback in distance education. *Open Learning, 16*(2), 181-189.

Makitalo, K. (2002). Mechanisms of common ground in case-based web discussions in teacher education. *The Internet and Higher Education, 5*(3), 247–265. doi:10.1016/S1096-7516(02)00112-4

Makrakis, V. (1998). *Guidelines for the design and development of computer-mediated collaborative open distance learning courseware.* Paper presented at the 1998 World Conference on Educational Multimedia and Hypermedia & World Conference on Educational Telecommunications, June, Freiburg, Germany. (ERIC Document Reproduction Service No. ED428694).

Mangan, K. (2003). Accrediting board endorses stronger focus on ethics in business-school curriculum. *Chronicle of Higher Education* Retrieved from http://chronicle.com/prm/daily/2003/01/2003010805n.htm.

Marks, R. B., Sibley, S. D., & Arbaugh, J. B. (2005). A structural equation model of predictors for effective online learning. *Journal of Management Education, 29*(4), 531–563. doi:10.1177/1052562904271199

Martinko, M. J., Henry, J. W., & Zmud, R. W. (1996). An attributional explanation of individual resistance to the introduction of information technologies in the workplace. *Behaviour & Information Technology, 15*(5), 313–330. doi:10.1080/014492996120094

Martins, L, & Kellerman, F. (2004). A model of business school students' acceptance of a Web-based course management system. *Academy of Management Learning and Education, 3*(1) 7-26.

Martyn, M. A. (2004). *The effect of online threaded discussion on student perceptions and learning outcomes in both face-to-face and online courses.* Doctoral dissertation, University of Akron. (UMI No. 3123389)

Masiello, I., Ramberg, R., & Lonka, K. (2005). Attitudes to the application of a Web-based learning system in a microbiology course. *Computers & Education, 45,* 171–185. doi:10.1016/j.compedu.2004.07.001

Mathes, A. (2004). *Folksonomies – Cooperative classification and communication through shared metadata.* Retrieved January 1 2009 from Retrhttp://www.adammathes.com/academic/computer-mediated-communication/folksonomies.html

Mathieson, K. (1991). Predicting user intentions: comparing the technology acceptance model with the theory of planned behavior. *Information Systems Research, 2*(3), 173–191. doi:10.1287/isre.2.3.173

Matthews, M. R. (2002). Constructivism and science education: A further appraisal. *Journal of Science Education and Technology,* 11.

Mattos, F. L. (2005). *Concepção e desenvolvimento de uma abordagem pedagógica para processos colaborativos a distancia utilizando a internet.* PhD. UFC. McGuinness, D., Li Ding, Furtado, V., Glass, A., Zeng, H., & Chang, C. (2006). Explanation interfaces for the semantic web: Issues and models. In *Proceedings of SWUI,* V.1. Berlin: Springer Verlag.

Matuga, J. M. (2001). Electronic pedagogical practice: The art and science of teaching and learning on-line. *Educational Technology and Society, 4*(3), 77–84.

Mazza, R., & Dimitrova, V. (2004). Visualising student tracking data to support instructor in web-based distance education. In *Proceedings of the 13th international World Wide Web Conference on Alternate Track Papers & Posters. SESSION: Student tracking and personalization, May 17-22, 2004, New York, USA* (pp.154-161).

McAlpine, I., & Ashcroft, B. (2002). Turning points: Learning from online discussions in an off-campus course. In P. Kommers, & G. Richards (Eds.), *Proceedings of world conference on educational multimedia, hypermedia & telecommunications*pp. 1251-1257, Chesapeake, VA: AACE.

Means, B., & Haertel, G. (2003). Cross-cutting themes in research designs for technology. In G. D. Haertel & B. Means (Eds.), *Evaluating educational technology: Effective research designs for improving learning* (pp. 257-269). New York: Teachers College Press.

Means, B., Wagner, M., Haertel, G. D., & Javitz, H. S. (2003). Studying the cumulative impacts of educational technology. In G. D. Haertel & B. Means (Eds.), *Evaluating educational technology* (pp. 230-256). New York: Teachers College Press.

Medlin, B. D., Vannoy, S. A., & Dave, D. S. (2004). An Internet-based approach to the teaching of information technology: A study of student attitudes in the United States. *International Journal of Management, 21*(4), 427–434.

Mendenhall, R. W. (2003). *Technology: Creating new models in higher education*: National Governors Association. Retrieved Jan. 14, 2009, from http://www.nga.org/Files/pdf/HIGHEREDTECH.pdf

Mentzer, G. A., Cryan, J. R., & Teclehaimanot, B. (2007). Two peas in a pod? A comparison of face-to-face and Web-based classrooms. Journal of Technology and Teacher Education, 15(2), 233–246.

Merrill, M. D. (2000). Instructional transaction theory (ITT): Instructional design based on knowledge objects. In *Instructional design theories and models: A new paradigm of instructional theory*. Mahwah, NJ: Lawrence Erlbaum Associates.

Miles, M. (Ed.). (1983). *Qualitative data as an attractive nuisance*. Beverly Hills, CA: Sage Publications.

Miles, M. B., & Huberman, A. M. (1994). *Qualitative data analysis* (2nd ed.). Thousand Oaks, CA: Sage.

Milken Family Foundation. (1997). *A call for a national research agenda.* Retrieved January 9, 2009, from http://www.mff.org/edtech/article.taf?_function=detail&Content_uid1=100

Minton, T. M., & Willett, L. S. (2003). Student preferences for academic structure and content in a distance education setting. *Online Journal of Distance Learning Administration, 6*(1). Retrieved March 22, 2006 from http://www.westga.edu/%7Edistance/ojdla/spring61/spring61.htm

Molenda, M. (2002). *A new framework for teaching in the cognitive domain.* Syracuse, NY: ERIC Clearinghouse on Information and Technology (ERIC Document Reproduction Service No. ED470983)

Moodle. (2009). Retrieved January 2009 from, http://moodle.org/

Moore, G. C., & Benbasat, I. (1991). Development of instrument to measure the perceptions of adopting an information technology innovation. *Information Systems Research, 2*(3), 192–222. doi:10.1287/isre.2.3.192

Moore, M. G. (1973). Towards a theory of independent learning and teaching. *Journal of Higher Education, 44*, 661-679.

Moore, M. G. (1989). Three types of interaction [Editorial]. *The American Journal of Distance Education, 3*(2) 1-6.

Moore, M. G. (1991). Distance education theory [Editorial]. *The American Journal of Distance Education, 5*(3) 1-6.

Moore, M. G., & Kearsley, G. (1996). *Distance education: A systems view.* New York: Wadsworth.

Morgan, G. (2003). *Faculty use of course management systems.* 2. Retrieved March 16, 2006 from, http://www.educause.edu/ir/library/pdf/ers0302/rs/ers0302w.pdf

Moskal, P. (2006). Assessing online learning: What one university learned about student success, persistence, and satisfaction. *Peer Review, 8*(4), 26–30.

Motiwalla, L., & Tello, S. (2000). Distance learning on the Internet: An exploratory study. *The Internet and Higher Education, 2*(4), 253–264. doi:10.1016/S1096-7516(00)00026-9

Motolet, O., & Baloian, N. (2006). Hybrid systems for generating learning object metadata. In *Proceedings IEEE International Conference on Advanced Learning Technologies (ICALT)* (pp. 563-567).

Motteram, G., & Forrester, G. (2005). Becoming an online distance learner: What can be learned from students' experiences of induction to distance programmes? *Distance Education, 26*(3), 281–298. doi:10.1080/01587910500291330

Muse, H. E. (2003). The web-based community college student: An examination of factors that lead to success and risk. *The Internet and Higher Education, 6*, 241–261. doi:10.1016/S1096-7516(03)00044-7

Nageswaran, N., Dreher, H., & Chang, V. (2000). The use of information technology to support a cast of a thousand. In A. Herrmann & M.M. Kulski (Eds.), Flexible futures in tertiary teaching. In *Proceedings of the 9th Annual Teaching Learning Forum, February 2-4, 2000, Perth: Curtin University of Technology.*

Nardi, B. A., & O'Day, V. L. (1999). *Information ecologies: Using technology with heart.* Cambridge, MA: MIT Press.

Nardi, B.A., & O'Day, V.L. (1999). *Information ecologies. Using technology with heart.* Cambridge, MA: The MIT Press.

Nate Robins' OpenGL Demos in Ch. Retrieved from http://iel.ucdavis.edu/projects/chopengl/demos.html

National Governors Association. (2001, June 6). *The state of e-learning in the states.* Retrieved January 14, 2009, from http://www.nga.org/Files/pdf/060601ELEARNING.pdf

National Research Council. (1996). *National Science Education Standards.* Washington, DC: National Academy Press.

Navarro, P., & Shoemaker, J. (2000). Policy issues in the teaching of economics in cyberspace: Research design, course design, and research results. Contemporary Economic Policy, 18(3), 359–366. doi:10.1093/cep/18.3.359doi:10.1093/cep/18.3.359

NCES. (2003). *Computer and Internet use by children and adolescents in 2001 statistical analysis report*: National Centre for Education Statistics.

NCES. (National Center for Educational Statistics). (2003). *Distance education at degree-granting postsecondary institutions: 2000-2001.* Retrieved November 9, 2006 from, http://nces.ed.gov/pubs2003/2003017.pdf

NCH. (2005). *Putting U in the picture: Mobile bullying survey 2005.* NCH, the Children's Charity.

NCTE. (2002). *Be wise on the Web: Information and advice for schools*: Irish National Centre of Technology in Education.

NCTE. (2006). *Survey of children's use of the Internet: Investigating online risk behaviour*: Irish Department of Education & Science.

NetAlert. (2004). *A teacher's guide to Internet safety: How to teach Internet safety in our schools*: Australia's Internet safety advisory body.

Ng'ambi, D. (2002). Pre-empting user questions through anticipation—Data mining FAQ lists. *Proceedings of the ACM International Conference on SAICSIT Enablement through Technology,* (pp. 101–109) Port Elizabeth, South Africa.

Ng'ambi, D. (2003). Students helping students learn—The case for a Web-based dynamic FAQ environment. *Proceedings of the International Association of Science and Technology for Development (IASTED) Conference on Information and Knowledge Sharing* (pp. 293–298). Scottsdale, Arizona.

Ng'ambi, D. (2004). Towards an organizational intelligent framework based on Giddens' structuration theory: A case of using an organizational memory system—DFAQ. *Proceedings of the IFIP World Congress Symposium on Professional Practice in Artificial Intelligence,* Toulouse, France.

Ng'ambi, D., & Hardman, J. (2004). Towards a knowledge-sharing scaffolding environment based on learners' questions. *British Journal of Educational Technology, 35*(2), 187–196.

Nicenet. (2009). Retrieved January 2009 from, http://www.nicenet.org/

Nicol, D., Minty, I., & Sinclair, C. (2003). The social dimensions of online learning. *Innovations in Education and Teaching International, 40*(3), 270–280. doi:10.1080/1470329032000103807

Norman, D. (2004). *Emotional design.* Basic Books.

North Central Regional Technology in Education Consortium. (1997). *Learning with technology profile tool.* Retrieved January 14, 2009, from http://www.ncrtec.org/capacity/profile/profwww.htm

NSBF. (2000). *Safe and smart: Research and Guidelines for children's use of the Internet*: National School Boards Foundation. O' Reilly, D., & O' Neill, C. (2007). *An Analysis of Irish Primary School Children's Internet Usage and the Associated Safety Implications.* Paper presented at the e-Learning.

Nunnally, J. (1967). *Psychometric theory.* New York: McGraw-Hill.

Nunnally, J. (1978). *Psychometric Methods* (2nd ed.). New York: McGraw-Hill.

O'Reilly, T. (2005). *What is Web 2.0.* Retrieved January 1, 2009 from http://www.oreillynet.com/pub/a/oreilly/tim/news/2005/09/30/what-is-web-20.html

OCGn. Online computer graphics notes. Retrieved from http://graphics.idav.ucdavis.edu/education/Graphics-Notes/homepage.html

Ochoa, X., & Duval, E. (2006). Quality metrics for learning object metadata. In *Proceedings World Conference on Educational Multimedia, Hypermedia and Telecommunications* (pp. 1004-1011).

Olsen, F. (2001) Getting ready for a new generation of course management systems.*Chronicle of Higher Education*, Vol. 48 (17), A25-A27. Retrieved March 16, 2001 from, http://reddog.rmu.edu:2065/login.aspx?direct=true&db=eric&an=EJ639384

Olson, T., & Wisher, R. A. (2002). The effectiveness of Web-based instruction: An initial inquiry. *International Review of Research in Open and Distance Learning, 3*(2), 1–17.

Omar, N., et al. (2005). *Towards a model for knowledge certification in engineering.* CEE- Congress on Engineering Education. Lisbon, Portugal

OpenGL tutors. Retrieved from http://www.xmission.com/~nate/tutors.html

OpenGL. Retrieved from www.opengl.org

OpenSceneGraph. Retrieved from www.openscenegraph.org

Or-Bach, R. (2005). Educational Benefits of Metadata Creation by Students. *inroads –. SIGCSE Bulletin, 37*(4), 83–97. doi:10.1145/1113847.1113885

Owston, R. D. (2000). Evaluating web-based learning environments: Strategies and insights. *Cyberpsychology & Behavior, 3*(1), 79–87. doi:10.1089/109493100316256

Oz, E. (1990). The attitude of managers-to-be toward software piracy. *OR/MS Today, 17*(4), 24-26.

Pahl, C. (2003). Managing evolution and change in web-based teaching and learning environments. *Computers & Education, 40*, 99–114. doi:10.1016/S0360-1315(02)00100-8

Paivarinta, T. (2001). The concept of genre within the critical approach to information systems development. *Information and Organization, 11*, 207–234.

Paivio, A. (1979). *Imagery and verbal processes.* Hillsdale, NJ: Lawrence Erlbaum Associates.

Paivio, A. (1986). *Mental representations.* New York: Oxford University Press.

Palais, R., & Benard, L. (2006). Still life: Five glass surfaces on a tabletop. *Science.* Retrieved Aug. 3, 2006, from http://www.sciencemag.org/cgi/content/full/313/5794/1730

Palinscar, A. S., & Brown, A. (1984). Reciprocal teaching of comprehension-fostering and comprehension-monitoring activities. *Cognition and Instruction, 1*(2), 117–175. doi:10.1207/s1532690xci0102_1

Palloff, R. M., & Pratt, K. (2003). *The virtual student: A profile and guide to working with online learners.* Wiley: San Francisco.

Palloff, R., & Pratt, K. (1999). *Building learning communities in cyberspace: Effective strategies for the online classroom.* San Francisco: Jossey-Bass.

Panteli, N. (2003). *Discourse analysis in IS research: Constructing presence in virtual organizing.* Proceedings of the 11th European Conference on Information Systems, Naples.

Pappas, J. P., & Jerman, J. (2004). *Developing and delivering adult degree programs: New directions for adult and continuing education.* Jossey-Bass: San Francisco.

Parmar, N. (2005). Gifted students connect online with colleges. *Wall Street Journal* (Eastern Edition), B1.

Partnership for 21st Century Skills. (2004). Assessment of 21st century skills. Retrieved from http://www.21stcenturyskills.org

Partnership for 21st Century Skills. (Winter 2008). *News 2.1.* Retrieved January 14, 2009, from http://

www.21stcenturyskills.org/newsletters/P21-winter2008/

Peace, A., Galletta, D., & Thong, J. (2003). Software piracy in the workplace: A model and empirical Test. *Journal of Management Information Systems, 20*(1), 153-177.

Pepper, S. C. (1942). *World hypotheses: A study in evidence*: University of California Press.

Perez, R. S., Cherniavsky, J., & Hamilton, E. R. (2006). Summary and conclusions. In H. F. O'Neil & R. S. Perez (Eds.), *Web-based learning* (pp. 393-404). Mahwah, NJ: Erlbaum.

Perez,Y., Gamboa R., & Ibarra, O. (2004). Modeling affective responses in intelligent tutoring systems. In *Proceedings of IEEE ICALT.*

Perkins, M. & Pfaffman, J. (2006). Using a course management system to improve classroom communication. *The Science Teacher, 73*(7), 33-37.

Perrenoud, P. (2000). *Dez Novas Competências para Ensinar.* Porto Alegre: Artmed, Pimentel, E.P. (2006). *Um Modelo para Avaliação e Acompanhamento Contínuo do Nível de Aquisição de Conhecimentos do Aprendiz.* Doutorado (Tese). Instituto Tecnológico da Aeronática (ITA), Brasil.

Peternier, A.., Thalmann, D., et al. (2006). Mental vision: A computer graphics teaching plattform. Edutainment.

Phillips, L., & Jorgensen, W.M. (2002). *Discourse analysis as theory and method.* London: Sage Publications.

Piaget, J. (1976). *Le comportement, moteur de l'évolution.* Paris: Ed. Gallimard.

Pilkington, M R. (1999). *Analysis educational discourse: The DISCOUNT scheme* [technical report No.99/2].

Pillay, H., Irving, K., & Tones, M. (2007). Validation of the diagnostic tool for assessing tertiary students'readiness for online learning. Higher Education Research & Development, 26(2), 217–234. doi:10.1080/07294360701310821doi:10.1080/07294360701310821

Pimentel, E. P., Omar, N., & França, V. F. (2005). *Monitoring the knowledge acquisition level in distance learning education.* Paper presented at the International Conference on Methods and Technologies for Learning, Palermo, Italy.

President's Committee of Advisors on Science and Technology. (1997). *Report to the president on the use of technology to strengthen K-12 education in the United States, March 1997.* Retrieved Sept. 25, 2006, from http://www.ostp.gov/PCAST/k-12ed.html#2.1

President's Committee of Advisors on Science and Technology. (1997). *Report to the president on the use of technology to strengthen K-12 education in the United States, March 1997.* Retrieved January 9, 2009, from http://www.ostp.gov/cs/pcast/documents_reports/archive

Pressley, M., Wood, E., Woloshyn, V. E., Martin, V., King, A., & Menke, D. (1992). Encouraging mindful use of prior knowledge: Attempting to construct explanatory answers facilitates learning. *Educational Psychologist, 27*(1), 91–109. doi:10.1207/s15326985ep2701_7

Prestera, G. E., & Moller, L. A. (2001). *Facilitating asynchronous distance learning: Exploiting opportunities for knowledge building in asynchronous distance learning environments.* Paper presented at the Annual Mid-South Instructional Technology Conference, April, Murfreesboro, TN. (ERIC Document Reproduction Service No. ED463723.

Price, R. V. (2005). A model for designing web-based units of instruction. *Distance Learning, 2*(1), 5–8.

Proserpio, L. & Gioia, D. (2007). Teaching the virtual generation. *Academy of Management Learning & Education, 6*(1), 69-80.

Rada, R. (1999). Sharing standards: IT skills standards. *Communications of the ACM, 42*(4), 21–26. doi:10.1145/299157.299162

Raffaele, M. (2004, June 21). Some districts resist paying for cyberschools. The Pittsburgh Post-Gazette.

Rai, A., Lang, S. S., & Welker, R. B. (2002). Assessing the validity of IS Success models: An empirical test and theoretical analysis. *Information Systems Research, 13*, 50–69. doi:10.1287/isre.13.1.50.96

Rainer, R. K. Jr, & Harrison, A. W. (1993). Toward development of the end user computing construct in a university setting. *Decision Sciences Journal, 24*, 1187–1202. doi:10.1111/j.1540-5915.1993.tb00510.x

Ramakrishna, H., Kini, R., & Vijayraman, B. (2001). Shaping of moral intensity regarding software piracy in university students: Immediate community effect. *Journal of Computer Information Systems, 41*(4), 47-51.

Reading.V, (2002). *Knowledge is Power*, European Commissioner for Education and Culture. Extracts from the Public Service Review, European Union, Autumn 2002.

Reeves, M. (1990). An application of Bloom's taxonomy to the teaching of business ethics. *Journal of Business Ethics, 9*(7), 609-616.

*Rehberg, S. D., Ferguson, D. M., & McQuillan, J. M. (2004).* The ultimate WebCT handbook, a practical and pedagogical guide to 4.x WebCT.

Reiser, R. A. (2001). A history of instructional design and technology, Part I: A history of instructional media. *Educational Technology, Research and Development, 49*(1), 53-64.

Rheingold, H. (2002). *Smart mobs: The next social revolution*. Cambridge: Perseus Publishing.

Rhem, J. (2002). Of diagrams and models: Learning as a game of pinball. *The National Teaching and Learning Forum, 11*(4). Retrieved February 2, 2007, from http://lihini.sjp.ac.lk/careers/edreform/n_amer/carnegie/diagrams.htm

Richardson, W. (2007). Teaching in a web 2.0 world. *Kappa Delta Pi Record, 43*(4), 150-151.

Riddle, W. C., et al. (2007) *No Child Left Behind Act* (pp. 107-110). Washington, DC: Congressional Research Service.

Riessman, C.K. (2002). *Analysis of personal narratives*. In J..F. Gubrium, & J.A. Holstein (Eds.), 2002. *Handbook of interview research: Context and method*. Thousand Oaks, CA: Sage Publications, 695–709.

Rivard, S., & Huff, S. (1988). Factors of success for end-user computing. *Communications of the ACM, 31*, 552–561. doi:10.1145/42411.42418

Rivera, J. C., & Rice, M. L. (2002). A comparison of student outcomes and satisfaction between traditional & web based course offerings. Online Journal of Distance Learning Administration, 5(3).

Rivera, J. C., McAlister, M. K., & Rice, M. L. (2002). A comparison of student outcomes & satisfaction between traditional & web-based course offerings. *Online Journal of Distance Learning Administration, 5*(3). Retrieved March 22, 2006 from http://www.westga.edu/%7Edistance/ojdla/fall53/fall53.html

Roach, V., & Lemastus, L. (2006). Satisfaction with online learning: A comparative descriptive study. *Journal of Interactive Online Learning, 5*(3), 317–332.

Roblyer, M. D., & Wiencke, W. (2003). Design and use of a rubric to assess and encourage interactive qualities in distance courses. American Journal of Distance Education, 17(2), 77–98. doi:10.1207/S15389286AJDE1702_2doi:10.1207/S15389286AJDE1702_2

Rockman, S., et al. (1998). *Leader's guide to educational technology*. Alexandria, VA: EDvancenet.

Rogers, C. (1969). *Freedom to learn*. Columbus, OH: Merrill.

Rogers, M. (1999). Psychology of hackers: Steps toward a new taxonomy. Retrieved from http://www.infowar.com.

Roode, D., Speight, H., Pollock, M., & Webber, R. (2004). It's not the digital divide—It's the socio-techno divide! *Proceedings of 12th European Conference on Information Systems (ECIS2004)*, Turku, Finland.

Rosenberg, M.J. (2001). *e-Learning: Strategies for delivering knowledge in the digital age*. New York: McGraw Hill.

Rourke, L., Anderson, T., Garrison, D. R., & Archer, W. (2001). Assessing social presentation asynchronous text-based computer conferencing. *Journal of Distance Education, 14*(2), 50-71.

Rovai, A. P. (2001). Building a sense of community at a distance. *International Review of Research in Open and Distance Learning, 3*(1), 1–16.

Rovai, A. P., & Barnum, K. T. (2003). Online course effectiveness: An analysis of student interactions and perceptions of learning. *Journal of Distance Education, 18*(1), 57-73.

Ruberg, L. (April, 2009). Blending globalization and curriculum analysis to study technology-focused teacher training. Paper will be presented at the annual meeting of American Educational Research Association, San Diego.

Rudestam, K. E., & Schoenholtz-Read, J. (2002). The coming of age of adult online education. In K.E. Rudestam & J. Schoenholtz-Read (Eds.), *Handbook of online learning: Innovations in higher education and corporate training* (pp. 3-28). Thousand Oaks, California: Sage Publications.

Russell Educational Consultancy and Productions. (2007). *Linklogs and social bookmarking.* Retrieved from http://recap.ltd.uk/Web2/bookmarking.php

Russell, M. (2001). Framing technology program evaluations. In W. Heineke, L. Blasi, & J. Willis (Eds.), *Methods of evaluating educational technology* (pp. 149-162). Greenwich, CT: Information Age.

Russell, M. (2001). Framing technology program evaluations. In W. Heineke & L. Blasi (Eds.), *Methods of evaluating educational technology* (pp. 149-162). Greenwich, CT: Information Age Publishing.

Russell, M. (2001). Framing technology program evaluations. In W. Heineke, L. Blasi, & J. Willis (Eds.), *Methods of evaluating educational technology* (pp. 149-162). Greenwich, CT: Information Age.

Russell, T. (1999). *The no significant difference phenomenon.* Office of Instructional

Sacks, H., Schegloff, E. A., & Jefferson, G. (1974). A simplest systematics for the organization of turn-taking for conversation. *Language, 50*(4), 696–735. doi:10.2307/412243

Salmon, G. (2000). *E-moderating: The key to teaching and learning online*. London: Kogan Page.

Salmon, G. (2000). *E-moderating: The key to teaching and learning online*. London: Kogan.

Salter, G. (2003). Comparing online and traditional teaching- a different approach. *Campus-Wide Information Systems, 20*(4), 137–145. doi:10.1108/10650740310491306

Saunders, G. & Christopher, J. (2003). Teaching outside the box: A look at the use of some nontraditional teaching models in accounting principles courses. *Journal of American Academy of Business, 3*(1-2), 162.

Savery, J. R., & Duffy, T. M. (1996). Problem based learning: An instructional model and its constructivist framework. In B. Wilson (Ed.), Constructivist learning environments: Case studies in instructional design, educational technology publications. Englewood Cliffs, NJ: Prentice Hall.

Sayers, H. M., Nicell, M. A., & Hagan, S. J. (2004). Supporting and assessing first year programming: The use of WebCT. *Italics eJournal, 3*(1). Retrieved March 22, 2006 from http://www.ics.heacademy.ac.uk/italics/Vol3-1/sayers/Italics.htm

Schifter, C. C. (2008). *Infusing Technology into the classroom: Continuous practice improvement.* Hershey, PA: IGI Global.

Schulman, A. H., & Sims, R. L. (1999). Learning in an online format versus an in-class format: An experimental study. T.H.E. Journal, 26(11), 54–56.

Shank, G. (2007). Personal communication. 03/19/2007.

Shank, P., (2004). When to use instructional technology and when not to? *T+D, 58*(9), 30-37.

Shea, P., Chun, S. L., & Pickett, A. (2006). A Study of teaching presence and student sense of learning community in fully online and web-enhanced college courses. *The Internet and Higher Education, 9*, 175–190. doi:10.1016/j.iheduc.2006.06.005

Shelley, D. J., Swartz, L. B., & Cole, M. T. (2007). A comparative analysis of online and traditional undergraduate business law classes. International Journal of Information and Communication Technology Education, 3(1), 10–21.

Shelley, D. J., Swartz, L. B., & Cole, M. T. (2008). Learning business law online vs. onland: A mixed method analysis.

International Journal of Information and Communication Technology Education, 4(2), 54–66.

Shepard, L. A. (2001). The role of classroom assessment in teaching and learning. In V. Richardson (Ed.), *The handbook of research on teaching* (4th ed.). Washington, DC: American Educational Research Association.

Sherry, L. (2000). Good online conversation: Building on research to inform practice. *Journal of Interactive Learning Research, 11*(1), 85.

Shih, C., & Gamon, J. (2002). Relationships among Learning Strategies, Patterns, Styles, and Achievement in Web-based Courses. *Journal of Agricultural Education, 43*(4), 1–11.

Shih, H.-P. (2003). Extended technology acceptance model of Internet utilization behavior. *Information & Management, 41*, 719–729. doi:10.1016/j.im.2003.08.009

Shirky,C (2003). *A Group is its own Worst Enemy.* Paper given at Etech, April 2003. Available from http://www.shirky.com/writings/group_enemy.html

Shneiderman, B. (1998). *Designing the user interface - Strategies for effective human-computer interaction.* Addison-Wesley.

Shulman, L. (2002). Making differences: A table of learning. *Change, 34*(6), 36–44.

Sifry, D. (2007). *The state of the live Web, April 2007.* Retrieved October 8, 2007 from http://www.sifry.com/alerts/archives/000493.html

Silva, D. R., & Vieira, M. T. P. (2001). An ongoing assessment model in distance learning. In *Proceedings of Internet and Multimedia Systems and Applications.* Honolulu, USA.

Silveira, I. F., Mustaro, P. N., & Omar, N. (2004). Aprendizagem significativa baseada em uma arquitetura multicamadas de objetos de aprendizagem adaptativos. In *VII Congresso IberoAmericano de Informática Educativa, Monterrey. Anais do RIBIE.*

Silver, J.D. (2007, March 18). Cyber-school empire under attack: Beaver County educator fighting grand juries, suits and legislators. *The Pittsburgh Post-Gazette.*

Simonson, M., Smaldino, S., Albright, M., & Zvacek, S. (2000). *Teaching and learning at a distance: Foundation of distance education.* Upper Saddle River, NJ: Merrill.

Simpson, T. J. (1995). *Message into medium: An extension of the dual coding hypothesis.* Paper presented at the Annual Conference of the International Visual Literacy Association, October, Tempe, AZ. (ERIC Document Reproduction Service No. ED380084).

Sims, R. (2002). *Teaching business ethics for effective learning.* Westport: Quorum Books.

Sims, R., Cheng, H., & Teegen, H. (1996). Toward a profile of student software pirates. *Journal of Business Ethics, 15*(8), 839-849.

Sivo, S. A., Pan, C. C. S., & Hahs-Vaughn, D. L. (2007). Combined longitudinal effects of attitude and subjective norms on student outcomes in a web-enhanced course: A structural equation modelling approach. *British Journal of Educational Technology, 38*(5), 861–875. doi:10.1111/j.1467-8535.2006.00672.x

Smith, C. (2005, December 18). State-funded cyber school draws concern. *The Pittsburgh Tribune-Review.*

Smith, P. (2001). Understanding self-regulated learning and its implications for accounting educators and researchers. *Issues in Accounting Education, 16*(4), 663-701.

Smith, P. J., Murphy, K. L., & Mahoney, S. E. (2003). Towards identifying factors underlying readiness for online learning: An exploratory study. *Distance Education, 24*(1), 57–67. doi:10.1080/01587910303043

Smith, P. L., & Ragan, T. J. (1996). *Impact of R. M. Gagne's work on instructional theory.* Paper presented at the 1996 National Convention of the Association for Educational Communications and Technology, Indianapolis, IN. (ERIC Document Reproduction Service No. ED397841).

Sorenson, E. K., & Takle, E. S. (2002). Collaborative knowledge building in Web-based learning: Assessing the quality of dialogue. *International Journal on E-Learning, 1*(1).

Spring, J. (2008). Research on Globalization and Education. *Review of Educational Research*, 78.

Stacey, E. (2002). Social Presence Online: Networking Learners at a Distance. *Education and Information Technologies*, *7*(4), 287–294. doi:10.1023/A:1020901202588

Stacey, E. (2002). Social presence online: Networking learners at a distance. *Education and Information Technologies*, *7*(4), 287–294. doi:10.1023/A:1020901202588

Stokes, S. P. (2001). Satisfaction of college students with the digital learning environment: Do learners' temperaments make a difference? *The Internet and Higher Education*, *4*(1), 31–44. doi:10.1016/S1096-7516(01)00046-X

Stone, A. (in press). Learning object based instruction (LOBI). *Encyclopedia of Information Technology Curriculum Integration*. Hershey, PA: Idea Group Publishing.

Stone, R. W., & Bailey, J. J. (2007). Team conflict self-efficacy and outcome expectancy of business students. *Journal of Education for Business*, *82*(5), 258–266. doi:10.3200/JOEB.82.5.258-266

Stone, R. W., & Henry, J. W. (2003). The roles of computer self-efficacy and outcome expectancy in influencing the computer end-users' organizational commitment. *Journal of End User Computing*, *15*(1), 38–53.

Storey, M. A., Phillips, B., Maczewski, M., & Wang, M. (2002). Evaluating the usability of web-based learning tools. *Educational Technology & Society*, *5*(3), 91–100.

*Storey, M.-A., Phillips, B., MaczewskiM., & WangM. (2002). Evaluating the usability of Web-based learning tools.* Educational Technology & Society, 5*(3).*

Suanpang, P., Petocz, P., & Kalceff, W. (2004). Student attitudes to learning business statistics: comparison of online and traditional methods. Educational Technology & Society, 7(3), 9–20.

Summerville, J., & Reid-Griffin, A. (2008). Technology Integration and Instructional Design. *TechTrends*, *52*(5), 45–51. doi:10.1007/s11528-008-0196-z

Summerville, J., & Reid-Griffin, A. (2008). Technology Integration and Instructional Design. *TechTrends*, *52*(5), 45–51. doi:10.1007/s11528-008-0196-z

Susskind, J. (2005). PowerPoint's power in the classroom: Enhancing students' self-efficacy and attitudes. *Computers & Education, 50,* 2, 203-215. Retrieved September 25, 2006 from, ACM Portal database.

Swan, K. (2001). Virtual interaction: Design factors affecting student satisfaction and perceived learning in asynchronous online courses. *Distance Education, 22*(2), 306-331.

Swanson, D. & Frederick, W. (2003). Are business schools silent partners in corporate crime?. *Journal of Corporate Citizenship, 9,* 24-27.

Swanson, D. (2003). Business education puts corporate reputations at risk. *The Corporate Citizen, 3,* 34-35.

Swanson, D. (2004). The buck stops here: Why universities must reclaim business ethics education. *The Journal of Academic Ethics, 2*(1), 1-19.

Szabo, A., & Hastings, N. (2000). Using IT in the undergraduate classroom: Should we replace the blackboard with PowerPoint? *Computers & Education, 35,* 175-187.

Tallent-Runnels, M. K. (2005). How to teach online: What the research says. *Distance Learning*, *2*(1), 21–27.

Tastle, W. J., White, B. A., & Shackleton, P. (2005). E-learning in higher education: The challenge, effort, and return on investment. International Journal on E-Learning, 4(2), 241–251.

Taylor, S., & Todd, P. A. (1995). Understanding information technology usage: A test of competing models. *Information Systems Research*, *6*, 144–176. doi:10.1287/isre.6.2.144

Tebbutt, D. (2007). RSS is the new radar. *Information World Review*, Retrieved from the Library, Information Science & Technology Abstracts database.

Telecommunications, North Carolina State University Chapel Hill, N.C. Ryan, R.C. (2000). Student assessment comparison of lecture and online construction equipment and methods classes. *THE Journal, 6*(27).

Tennent, B., Windeknecht, K., & Kehoe, J. (2004). Teaching with technology: Value-added innovation or necessity? *Campus-Wide Information Systems, 21*(4), 144. doi:10.1108/10650740410555016

The Milken Family Foundation. (1997). *A call for a national research agenda.* Retrieved January 14, 2009, from http://www.mff.org/edtech/article.taf?_function=detail&Content_uid1=100

Thompson, M. (2002). ICT, power, and development discourse: A critical analysis. In E. Wynn, A.E. Whitley, D.M. Myers, & D.M. DeGross, *Global and organizational discourse about information technology.*

Thompson, M.P.A. (2004). ICT, power, and developmental discourse: A critical analysis. *The Electronic Journal on Information Systems in Developing Countries, 20*(4), 1–25. Retrieved August 11, 2005, from http://www.ejisdc.org

Thornburg, D. (2007). Connected learning: Changing education through technology use. *Interactive Educator, 3*(2), 20-21.

Tian, S. (2001). The World Wide Web: a vehicle to develop interactive learning and teaching applications. *Internet research: Electronic networking, applications and policy, 11*(1), 74-83.

Tobias, S., & Everson, H. T. (2002). *Knowing what you don't know: Further research on metacognitive knowledge monitoring.* College Board Research Report 2002-3, College Entrance Examination Board: New York.

Tomei, L. (2005). *Taxonomy for the technology domain.* Hershey, PA: Information Science Publishing.

Tomei, L. A. (2002). *The technology facade: Overcoming barriers to effective instructional technology.* Boston: Allyn and Bacon.

Tomei, L. A. (2005). *Taxonomy for the technology domain.* Hershey, PA: Information Science Reference.

Top Class E-learning Suite. (2009). Retrieved January 2009 from, http://www.wbtsystems.com/solutions/technology

Townsend, E., Campbell, C., Curran-Smith, J., McGinn, F., Persaud, D., Peters, P., Bower, I., & Sheffield, S. (2002) Accessibility and interactivity in distance education programs for health professionals. *Journal of Distance Education, 17*(2), 1-24.

Tractinsky, N. (1997). Aesthetics and apparent usability: Empirically assessing cultural and methodological issues. In *Proceedings of the CHI 97*, Atlanta, GA.

Trinidad, S. (2003). Working with technology-riched learning environments: Strategies for success. In M. S. Khine & D. Fisher (Eds.), *Technology-rich learning environments: A future perspective.* Hackensack, NJ: World Scientific.

Tripp, S., & Bichelmeyer, B. (1990). Rapid prototyping: An alternative instructional design strategy. *Educational Technology Research & Development, 38*(1), 31-44.

Tu, C. H. (1997). The measurement of social presence in an online learning environment. *International Journal of eLearning, corporate, Government, Healthcare, &. Higher Education, 2*(1), 34–45.

Tuckman, B. W. (2002). Evaluating ADAPT: A hybrid instructional model combining Web-based and classroom components. *Computers & Education, 39*, 261–269. doi:10.1016/S0360-1315(02)00045-3

U.S. Department of Education. (2006). *No Child Left Behind Act.* Retrieved Sept. 25, 2006, from http://www.ed.gov/nclb/landing.html

Udod, S. A., & Care, W. D. (2002). Lessons learned in developing and delivering Web-based graduate courses: A faculty perspective. *Journal of Continuing Education in Nursing, 33*(1), 19-23.

UK Department for Education andScience (2007). *Diversity and citizenship curriculum review.* Retrieved http://publications.teachernet.gov.uk/default.aspx?PageFunction=productdetails&PageMode=publications&ProductId=DFES-00045-2007

Ullrich, T., & Fellner, D. (2005). Computer graphics courseware. Eurographics Education Papers.

Umbach, J. (2006). Web 2.0: The new commons. *Feliciter, 52*(5), 192.*What is a blog?* Retrieved from http://www.blogger.com/tour_start.g

Underwood, J. S., Hoadley, C., Lee, H. S., Hollebrands, K., DiGiano, C., & Renninger, K. A. (2005). IDEA:

Identifying design principles in educational applets. *Educational Technology Research and Development, 53*(2), 99–112. doi:10.1007/BF02504868

University of Phoenix (2007). *Internal SEOCS report.* Retrieved on 2/11/07 from internal data source.

University of Phoenix. (2006). *Student rEsource Page.* Retrieved on 2/22/06 from https://mycampus.phoenix.edu.

US Department of Education (USDED). (2007). *Effectiveness of reading and mathematics software products: Findings from the first student cohort.* Online version. Retrieved May 29, 2007 from, http://ies.ed.gov/ncee/pdf/20074005.pdf

Using PowerPoint. (n.d.). *PowerPoint tips and techniques.* Retrieved 2002. http://www.pitt.edu/~ciddeweb/fds/tech_presentation_software.htm

Vacca, R. T., & Vacca, J. L. (1998). *Content area reading: Literacy and learning across the curriculum* (6th ed.). New York: Addison Wesley Longman, Inc.

Van Dijk, A.T. (1996). Discourse, power and access. In R.C. Caldas-Couthard, & M. Coulthard, *Texts and practices: Readings in critical discourse analysis* [pp. 85–104]. London: Routledge.

Van Gumster, J. (2003). *Blender as an educational tool.* Proceedings of the International Conference on Computer Graphics and Interactive Techniques.

Vandergrift, K. E. (2002). The anatomy of a distance education course: A case study analysis. *Journal of Asynchronous Learning Networks, 6*(1). Retrieved January 28, 2003 from, http://www.sloan-.org/publications/jaln/v6n1_vandergrift.asp

Vasconcelos, E., & Furtado V. (2006). Educational geosimulation. In Brasileiro, F., & Mendes, M. (Eds.), *Advances in computer-supported learning.* IGI Global: Hershey.

Venkatesh, V., & Davis, F. D. (1996). A model of the antecedents of perceived ease of use: Development and test. *Decision Sciences, 27*(3), 451–481. doi:10.1111/j.1540-5915.1996.tb01822.x

Verbert, K., & Duval, E. (2004). *Towards a global architecture for learning objects: A comparative analysis of learning object content models.* Retrieved May 24, 2007 from, http://www.cs.kuleuven.ac.be/~hmdb/publications/files/pdfversion/41315.pdf

Viljoen, J-M., Preez, C., & Cook, A. (2005). The case for using SMS technologies to support distance education students in South Africa. *Perspectives in Education, 23*(4), 115–122.

Virtanen, P., Niemi, H., Nevgi, A., Raehalme, O., & Launonem, A. (2003). Towards strategic learning skills throught self-assessment and tutoring in web-based environment. In *Proceedings of ECER 2003 - European Conference on Educational Research*, Hamburg, Germany.

Visser, J. A. (2000). Faculty work in developing and teaching web-based distance courses: A case study of time and effort. *American Journal of Distance Education, 14*(3), 21-32.

Volery, T., & Lord, D. (2000). Critical success factors in online education, the international. *Journal of Educational Management, 14*(5), 216–223.

Voos, R. (2003). Blended learning—What is it and where might it take us? *Sloan-C View, 2*(1), 3-5.

Vrasidas, C. (1999). *Meanings of online and face-to-face interactions in a graduate course.* Unpublished Ph.D. Dissertation, Arizona State University.

Vrasidas, C., & McIsaac, M. S. (1999). Factors influencing interaction in an online course. *The American Journal of Distance Education 13*(3), 22-36.

Vygotsky, L. (1978). *Mind in society. The development of higher psychological processes.* In M. Cole, V.J. Steiner, S. Scribner, & E. Souberman (Eds.). Cambridge, MA: Harvard University Press.

Wager, W. (1992). Educational technology: A broader vision. *Education and Urban Society, 24*(4), 454–465. doi:10.1177/0013124592024004003

Wang, W. (2004). How university students view online study: A PCP perspective. *Campus-Wide Information Systems, 21*(3), 108. doi:10.1108/10650740410544009

Weaver-Kaulis, A., & Crutsinger, C. (2006). Assessment of student learning outcomes in fcs programs. Journal of Family and Consumer Sciences, 98(1), 74–81.

Web, C. T. (2009). Retrieved January 2009 from, http://www.blackboard.com/us/index.bbb /

Webb, E., Jones, A., Barker, P., & van Schaik, P. (2004). Using e-learning dialogues in higher education. *Innovations in Education and Teaching International*, *41*(1). doi:10.1080/1470329032000172748

Webb, N. M. (1982). Student interaction and learning in small groups. *Review of Educational Research*, *52*(3), 421–445.

Wegerif, R. (1998). The social dimension of asynchronous learning networks. *Journal of Asynchronous Learning Networks*, *2*, 34–49.

Wegner, S. B., Holloway, K. C., & Gordon, E. M. (1999). The effects of Internet-based instruction on student learning. *Journal of Asynchronous Learning Networks*, *3*(2), 98–106.

Welsch, E. (2002). SCORM: Clarity or calamity? *Online Learning Magazine*, *07*(1), 4.

Wenglinsky, H. (2005). *Using technology wisely: The keys to success in schools.* New York: Teachers College Press.

White, B. Y., Shimoda, T. A., & Frederiksen, J. R. (1999). Enabling students to construct theories of collaborative inquiry and reflective learning: Computer support for metacognitive development. *International Journal of Artificial Intelligence in Education*, *10*, 151–182.

Wiesenberg, F. P., & Stacey, E. (2005). Reflections on teaching and learning online. Quality program design, delivery and support issues from cross-global perspective. *Distance Education*, *26*(3), 385–404. doi:10.1080/01587910500291496

Wiggins, G. (1998). *Educative assessment: Designing assessments to inform and improve student performance.* San Francisco: Jossey-Bass Publishers.

Wiley, D. (2003). *Getting axiomatic about learning objects*. Retrieved January 26, 2007 from, http://www.reusability.org/axiomatic.pdf

Wiley, D. A. (2000). Connecting learning objects to instructional design theory: A definition, a metaphor, and a taxonomy. In D. A. Wiley (Ed.), *The instructional use of learning objects.* Online version. Retrieved January 26, 2007 from, http://reusability.org/read/chapters/wiley.doc

Wiley, D. A. (2002). The instructional use of learning objects. *Agency for Instructional Technology.*[1]http://www-i5.informatik.rwth-aachen.de/i5new/staff/chatti/LOMEditor/index.html

Willard, N. (2006). Flame Retardant: Cyberbullies Torment Their Victims 24/7. Here's How To Stop The Abuse. *School Library Journal*, *52*(4), 55–56.

Williams, C. M., & Linda, M. S. (2006). The gendered nature of career related learning experiences: a social cognitive career theory perspective. *Journal of Vocational Behavior*, *69*, 262–275. doi:10.1016/j.jvb.2006.02.007

Williams, S. W. (2001). *Experiences of web-based instruction among African-American students enrolled in training and development graduate courses.* Paper presented at the Academy of Human Resource Development Conference, Raleigh, NC. (AHRD Reference No. 059).

Willig, C. (1999). Introduction: Making a difference. In C. Willig (Ed.), *Applied discourse analysis—Social and psychological interventions* (pp. 1–21), Buckingham: Open University Press.

Wilson, B. G. (1995). *Situated instructional design: Blurring the distinctions between theory and practice, design and implementation, curriculum and instruction.* Paper presented at the Association for Educational Communications and Technology, Washington DC.

Wilson, Mark. (2001). *Distance degrees.* Kearney, NE: Morris Publishing Company.

Winn, W. (1993) *Instructional message design.* In. Fleming & Levie (Eds.), Englewood Cliffs, NJ: Educational Technology Publications, pp. 55-119.

Wiske, M. S. (2006). Teaching for meaningful learning with new technologies. In E. A. Ashburn & R. E. Floden (Eds.), *Meaningful learning using technology* (pp. 26-44). New York: Teachers College Press.

Wlodkowski, R. J. (1999, 1985). *Enhancing adult motivation to learn*. San Francisco: Jossey-Bass Publishers.

Woods, R., Baker, J. D., & Hopper, D. (2004). Hybrid structures: Faculty use and perception of web-based courseware as a supplement to face-to-face instruction. *The Internet and Higher Education*, *7*(4), 281–297. doi:10.1016/j.iheduc.2004.09.002

Wright, T., & Thompson, L. (2002). Cost, access and quality in online nursing and allied health professions. *Journal of Asynchronous Networks 6*(2). Retrieved January 3, 2003 from, http://www.sloan-c.org/publications/jaln/v6n2_wright.asp

Wu, D., & Hiltz, S. (2004). Predicting learning from asynchronous online discussions. *Journal of Asynchronous Learning Networks*, 139–152.

Wu, F. (2002). Complexity and urban simulation: Towards a computational laboratory. *Geography Research Forum 22*, 22-40.

Yang, L., Mustafa Sanver, M. (2002). Web-based interactive 3D visualization for computer graphics education. *ICWL*, 423–432.

Ybarra, M. (2004). Sex Differences In Youth-Reported Depressive Symptomatology And Unwanted Internet Sexual Solicitation. *Journal of Medical Internet Research*, *6*(1). doi:10.2196/jmir.6.1.e5

Yen, C., & Wu-Jeng, L. (2003). Web-based learning and instruction support system for pneumatics. *Computers & Education*, *41*, 107–120. doi:10.1016/S0360-1315(03)00029-0

Zhang, D. (2004). Can e-learning replace classroom learning? *Communications of the ACM*, *47*, 75. doi:10.1145/986213.986216

Zhang, P., & Li, N. (2004). The importance of affective quality. *CACM'04*. 2004.

Zhang, P., Benbasat, I., Carey, J., Davis, F., Galletta, D., & Strong, D. (2002). Human-computer interaction research in the MIS discipline. *Communications of the AIS*, *9*(20), 334–355.

Zhao, Y., Byers, J., Pugh, K., & Sheldon, S. (2001). What is worth looking for? Issues in educational technology research. In W. Heineke & L. Blasi (Eds.), *Methods of evaluating educational technology* (pp. 163-179). Greenwich, CT: Information Age Publishing.

Zhao, Y., Byers, J., Pugh, K., & Sheldon, S. (2001). What's worth looking for? Issues in educational technology research. In W. F. Heineke, L. Blasi, & J. Willis (Eds.), *Methods of evaluating educational technology* (p. 260). Greenwich, CT: Information Age.

Zillman, D., & Bryant, J. (1988). Effects of prolonged consumption of pornography on family values. *Journal of Family Issues*, *9*, 518–544. doi:10.1177/019251388009004006

Zillman, D., & Bryant, J. (1988b). Pornography's impact on sexual satisfaction. *Journal of Applied Social Psychology*, *18*, 438–453. doi:10.1111/j.1559-1816.1988.tb00027.x

Zucker, A. A., & Light, D. (2009). Laptop programs for students. *Science*, *323*, 82–85. doi:10.1126/science.1167705

Zugner, L. C. (1987). A reappraisal of instructional television. *ERIC Digest*, *12*(00), 1-6.

# About the Contributors

**Lawrence A. Tomei** is the Associate Provost for Academic Affairs and Professor of Education at Robert Morris University, Pittsburgh, PA. Born in Akron, Ohio, he earned a BSBA from the University of Akron (1972). He entered the US Air Force in September 1972 and served on active duty for 22 years until his retirement as a Lieutenant Colonel in 1994. Dr Tomei completed his Master's of Public Administration and Master's of Education, both at the University of Oklahoma (1975, 1978). He was awarded his Doctorate of Education degree from the University of Southern California in 1983. As Associate Provost for Academic Affairs, Dr. Tomei directs the activities of academic services including the registrar's office, academic technology, the honors and cooperative education programs, international and study abroad programs. He holds a dual appointment as a professor of education in the School of Education and Social Sciences teaching graduate and doctoral courses. His articles and books on instructional technology include: ICT for Enhanced Education and Learning (2009), Online and Distance Learning (2008), Integrating ICT Into the Classroom (2007), Taxonomy for the Technology Domain (2005), Challenges of Teaching with Technology Across the Curriculum (2003); Technology Facade (2002); Teaching Digitally: Integrating Technology Into the Classroom (2001); Professional Portfolios for Teachers (1999); and, Technology Literacy Applications in Learning Environments (Chapter 1, Defining Instructional Technology Literacy) (2004).

* * *

**Lori Baker-Eveleth** (PhD Washington State University) is an assistant professor of information systems at the University of Idaho, Moscow, Idaho. Her current research interest is in third place development, technology acceptance, laptop use in the classroom, and computer absorption/addiction. She has published recent articles in the *Behaviour & Information Technology*, *EduCause Quarterly*, *Journal of Informatics Research Education*, and *Journal of Business and Management*.

**Manetta Calinger**, curriculum writer at the Center for Educational Technologies®, focuses on writing STEM-G curriculum for a variety of national educational programs after a 20-year career as a secondary science teacher. Her experience implementing technology-based science instruction in K-12 science programs has given her practical insights into both the benefits and difficulties involved in using technology to improve science instruction and student learning.

**Chuleeporn Changchit** is an Associate Professor of Management Information Systems at Texas A&M University – Corpus Christi. Drawing from experiences working for over five years as a senior consultant in industry, she provides her students with practical and comprehensive instructions, shares ideas and techniques enthusiastically, and enjoys substantial successes as an educator. Dr. Changchit is actively engaged in the scholarly activities. She has published articles in journals such as Decision Support Systems, Information Systems Journal, Expert Systems with Applications, the Journal of Computer Information Systems, and International Journal of Intelligent Systems in Accounting, Finance, and Management. She also serves as an Editor-in-Chief for the Journal of Information Privacy and Security (JIPS), serves as an Associate Editor for Journal of Electronic Commerce in Organization (JECO), and serves on editorial review board for the Information Resources Management Journal (IRMJ), the Journal of Global Information Technology Management (JGITM), and the International Journal of E-Business Research (IJEBR).

**Ching-Huei Chen** is an instructional designer at the Center for Educational Technologies. Her scholarly interests include instructional design and technology, professional development, and program evaluation. She received her Ph.D. in the Instructional Psychology and Technology at the University of Oklahoma where she also taught pre-service teachers on how technology could be incorporated into their teaching.

**Michele T. Cole**, J.D., PhD, Director of the Masters program in Nonprofit Management and Assistant Professor of Nonprofit Management at Robert Morris University, Moon Township, PA. Her research interests include effective online instruction, nonprofit sector curriculum development, legal issues in personnel management and application of business best practices and research to the nonprofit sector as well as the application of technology to learning strategies.

**Laura Curtis**, a curriculum writer at the Center for Educational Technologies, develops multimedia instructional materials for various NASA and other science-based programs. She also designs and facilitates workshops for K-12 educators.

**Daniel Dahl** obtained a diploma in Computer Science and a Master of Science in Media Informatics, both from the University of Applied Sciences, Cologne, Germany. In 2005 he joined Prof. Vossen's Database Group at the University of Muenster, Germany, where he received his Ph.D. in Computer Science in 2008. His thesis focussing on Web 2.0 and its implications on technology enhanced learning introduces a Web 2.0-coined approach of integrating learning objects into a heterogeneous environment based on their metadata. In 2008 Daniel Dahl joined the Bayer Group. At Bayer Business Services he is leading workplace collaboration and innovation projects, especially in the context of Web 2.0, resp. Enterprise 2.0.

**Daniel M. Eveleth** (PhD Washington State University) is an associate professor of management and human resources at the University of Idaho, Moscow, Idaho. His current research interests include expertise development, knowledge management, analogical reasoning, and using technology to enhance pedagogy. He has published in a variety of journals, including *EduCause Quarterly*, *Journal of Service Research*, *Journal of Information System Education*, *Journal of Education for Business*, *Journal of Ethics and Behavior*, and *Journal of Interactive Marketing*.

**Marcio Porto Feitosa** is a graduate student on Computer Engeneering, has a Specialist in Information Technology and a BS on Electronic Engineering from Mackenzie Presbyterian University. He has large experience in systems development and application on data bases to social information systems.

**Nathan Garrett** is a 6th year PhD student in the School of Information Systems and Technology at Claremont Graduate University, having finished an MSIS at CGU in fall 2005. His current research focuses on educational technology. He also maintains a wiki/ePortfolio plug-in for the Elgg social networking system. Nathan works full-time at Woodbury University as an Institutional Researcher, and lives with his wife Corrie, son Silas, and daughter Rose in Burbank, CA. Website: http://conversation.cgu.edu/garrettn/weblog.

**Lamis Hammoud** is a researcher student in the School of Information Systems, Computing and Mathematics at Brunel University, West London, UK. Her main research interests are the area of educational technology and web-based learning.

**Bruce C. Howard** is the assistant director at the Center for Educational Technologies® He is actively involved in developing innovative ways to integrate emerging technologies into formal and informal learning environments. In addition to his research on evaluating effective technology use, he manages curriculum development projects that combine simulations and videoconferencing for high-impact student learning.

**Tim Klaus** is an Assistant Professor of Management Information Systems at Texas A&M University – Corpus Christi. He earned his MBA (Finance) and MS (Computer Science) from Illinois State University and his PhD (Management Information Systems) from University of South Florida. His primary research interests are User Resistance, ERP implementations, IT personnel, and Web Usage. He has published papers in journals such as Communications of the ACM (CACM), Journal of International Technology and Information Management (JITIM), International Journal of Information and Communication Technology Education (IJICTE), and Human Systems Management (HSM). He also is a consultant in the area of IT-enabled change, helping organizations better understand the process of change as well as the impact of user attitudes and behaviors.

**Vitaly Klyuev** (vkluev@u-aizu.ac.jp) is an Associate Professor at the University of Aizu, Japan. He received his Ph.D. in Physics and Mathematics from St. Petersburg State University, Russia. Prof. Klyuev has published more than 50 papers in refereed journals, conference proceedings, and chapters of books. His research domain includes information retrieval, software engineering, analysis of computer algorithms, and programming languages.

**Steve Love** works in the School of Information Systems, Computing and Mathematics at Brunel University, West London, UK. His research interests primarily focus on looking at the social impact of mobile technology on people's behaviour and its affect on their perception of the usability of mobile services and applications. His other main area of research interest is looking at the effectiveness of web-enhanced technology from a student and teacher perspective.

**Gennadiy Nikishkov** (niki@u-aizu.ac.jp) is a Professor at the University of Aizu, Japan. He holds PhD and DSci from Moscow Engineering Physics Institute, Russia. Professor Nikishkov has more than one hundred publications in refereed journals and conference proceedings. His research interests include computational modeling, computer graphics and visualization, and e-learning.

**Michele O'Neill** (PhD Pennsylvania State University) is currently an associate professor of finance at the University of Idaho in Moscow, Idaho. Her teaching and research interests include entrepreneurship, market structures, and international finance. She has published recently in *Journal of Economics and Finance, The Journal of Financial Research, Journal of Business Management, Journal of Financial Services Research*, and *The Financial Review*.

**Nizam Omar** has a PhD in Informatics from PUC University in Rio de Janeiro, Brazil, and he is currently a Professor at Mackenzie Presbyterian University. He is a researcher in the field of Computer Science and Electrical Engineering and his main interest is Artificial Intelligence and its application in Education.

**Derek O' Reilly** is a Senior Lecturer in Dundalk Institute of Technology, Ireland. He has lectured in the areas of Multimedia and Computer Graphics since the early 1990s. He has an M.Sc. in Computer Science from Dublin City University, Ireland and an M.A. in Education from Durham University, UK. He has served on various computing and multimedia related boards and panels. He is active as an industrial software consultant. Derek is interested in two research areas: equality through internet accessibility and equality through educational accessibility. He has published various research papers that discuss how the internet can be made more accessible for various groups, such as children, old people and the visually impaired. He has published research discussing the effects that various educational policies, such as grading methodologies and marketisation, can have on accessibility to education. Derek has published research that incorporates both equality through internet accessibility and equality through educational accessibility. He has designed tools that can be used to provide children with safe access to the internet. He has shown how the internet, blended learning and active learning can lead to improved university student retention.

**Anna Oskorus** is an instructional design consultant for TiER 1 Performance Solutions in Covington, KY. She currently creates and deploys blended learning and knowledge management solutions for commercial, non-profit, and government organizations.

**Edson Pinheiro Pimentel** has a PhD in Computer Science from Aeronautical Technological Institute (ITA), Brazil, and he is currently a Professor at IMES University of Computer Science and System Information courses. He is a researcher in Computer Science and his main interests include Artificial Intelligence, Computers in Education and software development for people with special needs.

**Laurie F. Ruberg**, associate director of the Center for Educational Technologies, has led the design of award-winning, multimedia development and multisite, educational technology evaluation projects. She is currently involved in design research involving application of Web 2.0 technologies and examining science education affordances of virtual worlds such as Second Life. Ruberg teaches an online course in educational technologies to pre-service students at Wheeling Jesuit University.

**Terry Ryan** is Dean and Professor in the School of Information Systems and Technology at Claremont Graduate University. He does research on a variety of topics, including the design, development, and evaluation of IS applications and the determinants of IS effectiveness. He is particularly interested in these matters in the context of: teaching and learning in university and organizational settings, online discussion and dialogue, and preparing for and responding to emergencies. Website: http://administration.cgu.edu/fac/ryant/.

**Thomas M. Schmidt** is currently the campus college chairman for information systems and technology at the Jersey City Campus of the University of Phoenix. Prior to this, he worked in several roles in the information technology industry around New York City, including as President of his own corporation servicing the IT industry, and as an Officer of Harvard University as an instructional technology specialist. He has previously taught classes at Fordham University and Iona College in databases, Web design and overall information technologies. His work for different financial firms has included Equity and Fixed Income analytics, and data warehouses and databases in MS SQL Server, Sybase, and flat-file systems. He holds a BA degree from Carleton College and a Doctor of Science degree from Tulane University in Computer Science, with a concentration in databases. He is the author of several published articles and book chapters. He was a part of the initial council meeting where the University of Phoenix's Doctor of Management in Information Systems and Technology degree was designed, and has served as a Council member guiding the degree program since then, along with serving as course designer for all but one of the courses in the program. His research interests include databases and information theory. A trivia buff, he has appeared on the Jeopardy! program. His interests outside of academics include opera, biking the trails built on former rail beds, botany and horticulture, and passive-solar and other advanced energy technologies. He and his wife live 25 minutes north of Grand Central (terminal, not station, please) in the suburbs of New York City, Metropolis, where he was born.

**Michael Shaw** is currently the president of SilkWeb Consulting & Development and the founder of Value Based Training. SilkWeb is an education and business consultancy with a special focus on management consulting and corporate training services. Prior to his role at SilkWeb, he developed leadership curriculum and managed new program development for University of Phoenix. Michael remains with the University as an adjunct faculty member. His research interests include online faculty quality, employee turnover in call centers, knowledge management and values in the workplace. Michael recently developed an online faculty quality model that is quickly becoming an industry standard. Michael was born in Bethpage, Long Island, but has lived in Arizona for over 15 years. He currently lives in New River Arizona with his wife and two daughters. He is an avid outdoorsman, involved in hunting and fishing.

**Daniel J. Shelley**, Ph.D., Professor of Education at Robert Morris University, Moon Township, PA. Dr. Shelley earned his BS in Elementary Education from Penn State University in 1971. He completed a Masters Degree in Social Science with an emphasis in American History at Penn State in 1972. He earned his PhD in Education at the University of Pittsburgh in 1986. Dr. Shelley is also a certified Elementary Principal and a Curriculum Program Specialist. His research interests include enhancing pre-service teacher's skills and expertise in applying educational technology to their teaching, and the integration of technology into classroom teaching. He currently serves as the Director of the Southwestern region of the Pennsylvania Association for Educational Computing and Technology (PAECT).

His classroom interests include robotics and instructional software and authoring. He has also written numerous articles and given presentations at national and international conferences on the integration of technology into classroom teaching.

**Robert W. Stone** (PhD Purdue University) is currently a professor of information systems at the University of Idaho in Moscow, Idaho. His teaching interests are information systems and strategic management. His research interests include organizational impacts from information system use and user acceptance of information systems. Professor Stone has published numerous research works, some of which have appeared in the *International Journal of Technology Management*, *Review of Accounting Information Systems*, *Behavior, Research Methods, Instruments, & Computers*, *Journal of Business Research*, *Behaviour & Information Technology*, *Journal of Informatics Education Research*, and *Information Resources Management Journal*.

**Louis B. Swartz**, J.D., Associate Professor of Legal Studies at Robert Morris University, Moon Township, PA. Mr. Swartz teaches Legal Environment of Business and The Constitution and Current Legal Issues at the undergraduate level and Legal Issues of Executive Management in the M.B.A. program. He received his Bachelors degree from the University of Wisconsin in Madison, Wisconsin (1966) and his Juris Doctorate from Duquesne University in Pittsburgh, PA (1969). He is the Coordinator of the Robert Morris University Pre-Law Advisory Program and a member of the Northeast Association of Pre-Law Advisers (NAPLA) and the Academy of Legal Studies in Business (ALSB). His research interests include online education, legal studies and business law.

**Brian Thoms** is a 4th year PhD student in the School of Information Systems and Technology at Claremont Graduate University. His current research focuses on the impact of social software on teaching and learning in higher education. Brian works part-time developing software for small-to-midsize organizations in southern California, incorporating five years of software development experience from the private sector. Brian also teaches part time at San Diego State University. Website: http://brianthoms.com.

**Gottfried Vossen** is a Professor of Computer Science in the Department of Information Systems and a Director of the European Research Center for Information Systems (ERCIS) at the University of Muenster in Germany. He is the European Editor-in-Chief of Elsevier's Information Systems - An International Journal. His research interests include conceptual as well as application-oriented problems concerning databases, information systems, electronic learning, the Semantic Web as well as Web 2.0. Dr. Vossen has been a member in numerous program committees of international conferences and workshops. He is an author or co-author of more than 150 publications, and an author, co-author, or co-editor of 22 books on databases, business process modeling, the Web, e-commerce, and computer architecture.

**Charles Wood**, director of the NASA-sponsored Classroom of the Future, has had a varied career, always associated with NASA science and education. Currently he analyses data from the Cassini spacecraft orbiting Saturn, as well as directing an educational research center. He worked as a planetary geologist at Brown University, the Smithsonian Institution and at NASA's Goddard Spaceflight Center and Johnson Space Center. Since 1990, he has been primarily an educator, developing online programs, simulations and instrumentation. His life-long passion is the Moon and he regrets that will not get to teach there.

# Index

## D



## E

## F



## G



## H



## I

## N



## O



## P

## Q



## R



## S



## T